T0135308

Advances in Intelligent Systems and Computing

Volume 787

Series editor

Janusz Kacprzyk, Polish Academy of Sciences, Warsaw, Poland
e-mail: kacprzyk@ibspan.waw.pl

The series "Advances in Intelligent Systems and Computing" contains publications on theory, applications, and design methods of Intelligent Systems and Intelligent Computing. Virtually all disciplines such as engineering, natural sciences, computer and information science, ICT, economics, business, e-commerce, environment, healthcare, life science are covered. The list of topics spans all the areas of modern intelligent systems and computing such as: computational intelligence, soft computing including neural networks, fuzzy systems, evolutionary computing and the fusion of these paradigms, social intelligence, ambient intelligence, computational neuroscience, artificial life, virtual worlds and society, cognitive science and systems, Perception and Vision, DNA and immune based systems, self-organizing and adaptive systems, e-Learning and teaching, human-centered and human-centric computing, recommender systems, intelligent control, robotics and mechatronics including human-machine teaming, knowledge-based paradigms, learning paradigms, machine ethics, intelligent data analysis, knowledge management, intelligent agents, intelligent decision making and support, intelligent network security, trust management, interactive entertainment, Web intelligence and multimedia.

The publications within "Advances in Intelligent Systems and Computing" are primarily proceedings of important conferences, symposia and congresses. They cover significant recent developments in the field, both of a foundational and applicable character. An important characteristic feature of the series is the short publication time and world-wide distribution. This permits a rapid and broad dissemination of research results.

Advisory Board

Chairman

Nikhil R. Pal, Indian Statistical Institute, Kolkata, India
e-mail: nikhil@isical.ac.in

Members

Rafael Bello Perez, Universidad Central "Marta Abreu" de Las Villas, Santa Clara, Cuba
e-mail: rbellop@uclv.edu.cu

Emilio S. Corchado, University of Salamanca, Salamanca, Spain
e-mail: escorchado@usal.es

Hani Hagras, University of Essex, Colchester, UK
e-mail: hani@essex.ac.uk

László T. Kóczy, Széchenyi István University, Győr, Hungary
e-mail: koczy@sze.hu

Vladik Kreinovich, University of Texas at El Paso, El Paso, USA
e-mail: vladik@utep.edu

Chin-Teng Lin, National Chiao Tung University, Hsinchu, Taiwan
e-mail: ctlin@mail.nctu.edu.tw

Jie Lu, University of Technology, Sydney, Australia
e-mail: Jie.Lu@uts.edu.au

Patricia Melin, Tijuana Institute of Technology, Tijuana, Mexico
e-mail: epmelin@hafsamx.org

Nadia Nedjah, State University of Rio de Janeiro, Rio de Janeiro, Brazil
e-mail: nadia@eng.uerj.br

Ngoc Thanh Nguyen, Wroclaw University of Technology, Wroclaw, Poland
e-mail: Ngoc-Thanh.Nguyen@pwr.edu.pl

Jun Wang, The Chinese University of Hong Kong, Shatin, Hong Kong
e-mail: jwang@mae.cuhk.edu.hk

More information about this series at http://www.springer.com/series/11156

Tareq Z. Ahram

Editor

Advances in Artificial Intelligence, Software and Systems Engineering

Joint Proceedings of the AHFE 2018 International
Conference on Human Factors in Artificial
Intelligence and Social Computing, Software and Systems
Engineering, The Human Side of Service Engineering
and Human Factors in Energy, July 21–25, 2018,
Loews Sapphire Falls Resort at Universal Studios,
Orlando, Florida, USA

 Springer

Editor
Tareq Z. Ahram
University of Central Florida
Orlando, FL, USA

ISSN 2194-5357 ISSN 2194-5365 (electronic)
Advances in Intelligent Systems and Computing
ISBN 978-3-319-94228-5 ISBN 978-3-319-94229-2 (eBook)
https://doi.org/10.1007/978-3-319-94229-2

Library of Congress Control Number: 201894743

© Springer International Publishing AG, part of Springer Nature 2019
This work is subject to copyright. All rights are reserved by the Publisher, whether the whole or part
of the material is concerned, specifically the rights of translation, reprinting, reuse of illustrations,
recitation, broadcasting, reproduction on microfilms or in any other physical way, and transmission
or information storage and retrieval, electronic adaptation, computer software, or by similar or dissimilar
methodology now known or hereafter developed.
The use of general descriptive names, registered names, trademarks, service marks, etc. in this
publication does not imply, even in the absence of a specific statement, that such names are exempt from
the relevant protective laws and regulations and therefore free for general use.
The publisher, the authors, and the editors are safe to assume that the advice and information in this
book are believed to be true and accurate at the date of publication. Neither the publisher nor the
authors or the editors give a warranty, express or implied, with respect to the material contained herein or
for any errors or omissions that may have been made. The publisher remains neutral with regard to
jurisdictional claims in published maps and institutional affiliations.

Printed on acid-free paper

This Springer imprint is published by the registered company Springer International Publishing AG
part of Springer Nature
The registered company address is: Gewerbestrasse 11, 6330 Cham, Switzerland

Advances in Human Factors and Ergonomics 2018

AHFE 2018 Series Editors

Tareq Z. Ahram, Florida, USA
Waldemar Karwowski, Florida, USA

9th International Conference on Applied Human Factors and Ergonomics and the Affiliated Conferences

Proceedings of the AHFE 2018 International Conferences on Human Factors, Software, and Systems Engineering, Artificial Intelligence and Social Computing, Human Side of Service Engineering and Human Factors in Energy, held on July 21–25, 2018, in Loews Sapphire Falls Resort at Universal Studios, Orlando, Florida, USA

Advances in Affective and Pleasurable Design	Shuichi Fukuda
Advances in Neuroergonomics and Cognitive Engineering	Hasan Ayaz and Lukasz Mazur
Advances in Design for Inclusion	Giuseppe Di Bucchianico
Advances in Ergonomics in Design	Francisco Rebelo and Marcelo M. Soares
Advances in Human Error, Reliability, Resilience, and Performance	Ronald L. Boring
Advances in Human Factors and Ergonomics in Healthcare and Medical Devices	Nancy J. Lightner
Advances in Human Factors in Simulation and Modeling	Daniel N. Cassenti
Advances in Human Factors and Systems Interaction	Isabel L. Nunes
Advances in Human Factors in Cybersecurity	Tareq Z. Ahram and Denise Nicholson
Advances in Human Factors, Business Management and Society	Jussi Ilari Kantola, Salman Nazir and Tibor Barath
Advances in Human Factors in Robots and Unmanned Systems	Jessie Chen
Advances in Human Factors in Training, Education, and Learning Sciences	Salman Nazir, Anna-Maria Teperi and Aleksandra Polak-Sopińska
Advances in Human Aspects of Transportation	Neville Stanton

(continued)

(continued)

Advances in Artificial Intelligence, Software and Systems Engineering	*Tareq Z. Ahram*
Advances in Human Factors, Sustainable Urban Planning and Infrastructure	*Jerzy Charytonowicz and Christianne Falcão*
Advances in Physical Ergonomics & Human Factors	*Ravindra S. Goonetilleke and Waldemar Karwowski*
Advances in Interdisciplinary Practice in Industrial Design	*WonJoon Chung and Cliff Sungsoo Shin*
Advances in Safety Management and Human Factors	*Pedro Miguel Ferreira Martins Arezes*
Advances in Social and Occupational Ergonomics	*Richard H. M. Goossens*
Advances in Manufacturing, Production Management and Process Control	*Waldemar Karwowski, Stefan Trzcielinski, Beata Mrugalska, Massimo Di Nicolantonio and Emilio Rossi*
Advances in Usability, User Experience and Assistive Technology	*Tareq Z. Ahram and Christianne Falcão*
Advances in Human Factors in Wearable Technologies and Game Design	*Tareq Z. Ahram*
Advances in Human Factors in Communication of Design	*Amic G. Ho*

Preface

This book includes contributions from four AHFE affiliated conferences, Human Factors, Software, and Systems Engineering, Artificial Intelligence and Social Computing, Human Side of Service Engineering and Human Factors in Energy. The book is divided into seven main sections:

Section 1: Software and Systems Engineering Applications
Section 2: Advancing Smart Service Systems and The Contributions of AI and
         T-Shape Paradigm
Section 3: Innovations in Service Delivery and Assessment
Section 4: Artificial Intelligence and Social Computing
Section 5: Social Network Modeling
Section 6: Human Factors in Energy Systems: Nuclear Industry
Section 7: Applications in Energy Systems

The discipline of Human Factors, Software, and Systems Engineering provides a platform for addressing challenges in human factors, software, and systems engineering that both pushes the boundaries of current research and responds to new challenges, fostering new research ideas. The first section focuses on software and systems engineering applications. In this book, researchers, professional software and systems engineers, human factors and human systems integration experts from around the world addressed societal challenges and next-generation systems and applications for meeting them. This book focuses on the advances in the Human Factors, Software, and Systems Engineering, which are a critical aspect in the design of any human-centered technological system. The ideas and practical solutions described in the book are the outcome of dedicated research by academics and practitioners aiming to advance theory and practice in this dynamic and all-encompassing discipline. The books address topics from evolutionary and complex systems, human systems integration to smart grid and infrastructure, workforce training requirements, systems engineering education and even defense and aerospace. It is sure to be one of the most informative systems engineering events of the year.

If there is any one element to the engineering of service systems that is unique, it is the extent to which the suitability of the system for human use, human service, and for providing an excellent human experience has been and must always be considered. Section two addresses topics related to Advancing Smart Service Systems and The Contributions of AI and T-Shape Paradigm, and section three covers topics related to Innovations in Service Delivery and Assessment. Sections two and three focus on the Human Side of Service Engineering (HSSE 2018) which took place at the Loews Sapphire Falls Resort, Universal Studios Orlando, Florida, from July 21 to 25, 2018. The AHFE HSSE 2018 conference was co-chaired by Louis E. Freund and Wojciech Cellary.

As Artificial Intelligence (AI) and Social Computing (SC) become more prevalent in the workplace environment and daily lives, researchers and business leaders will need to address the challenges it brings. Roles that have traditionally required a high level of cognitive abilities, decision making and training (human intelligence) are now being automated. The AHFE International Conference on Human Factors in Artificial Intelligence and Social Computing (AISC) promotes the exchange of ideas and technology, which enables humans to communicate and interact with machines in almost every aspect. The recent increase in machine and systems intelligence leads to a shift from interaction to a much more complex cooperative human–system work environment requiring a multidisciplinary approach. Section four addresses topics related to Artificial Intelligence and Social Computing, and section five focuses on Social Network Modeling.

Human Factors in Energy focuses on the Oil, Gas, Nuclear and Electric Power Industries and aims to address the critical application of human factors knowledge to the design, construction, and operation of oil and gas assets, to ensure that systems are designed in a way that optimizes human performance and minimizes risks to health, personal or process safety, or environmental performance. The conference focuses on delivering significant value to the design and operation of both onshore and offshore facilities. Sections six and seven address topics related to Human Factors in Energy Systems: Nuclear Industry and Applications in Energy Systems.

Each section contains research papers that have been reviewed by members of the International Editorial Board. Our sincere thanks and appreciation to the board members as listed below:

Software, and Systems Engineering/Artificial Intelligence and Social Computing

A. Al-Rawas, Oman
T. Alexander, Germany
Sergey Belov, Russia
O. Bouhali, Qatar

Henry Broodney, Israel
Anthony Cauvin, France
S. Cetiner, USA
P. Fechtelkotter, USA
F. Fischer, Brazil
S. Fukuzumi, Japan
Ravi Goonetilleke, Hong Kong
C. Grecco, Brazil
N. Jochems, Germany
G. J. Lim, USA
D. Long, USA
M. Mochimaru, Japan
C. O'Connor, USA
C. Orłowski, Poland
Hamid Parsaei, Qatar
Stefan Pickl, Germany
S. Ramakrishnan, USA
Jose San Martin Lopez, Spain
K. Santarek, Poland
M. Shahir Liew, Malaysia
Duncan Speight, UK
Martin Stenkilde, Sweden
Teodor Winkler, Poland
Hazel Woodcock, UK

Human Side of Service Engineering

Alison Amos, USA
Clara Bassano, Italy
Freimut Bodendorf, Germany
Carolyn Brown, USA
Bo Edvardsson, Sweden
Walter Ganz, Germany
Dolly Goel, USA
Kazuyoshi Hidaka, Japan
Keisuke Honda, Japan
Kendra Johnson, USA
Kozo Kitamura, Japan
Eunji Lee, Norway
Christine Leitner, UK
Aura C. Matias, Philippines
Prithima Mosaly, USA
U. Narain, India

Shrikant Parikh, India
Paolo Piciocchi, Italy
Regiane Romano, Brazil
Debra Satterfield, USA
Yuriko Sawatani, Japan
Jim Spohrer, USA
Kinley Taylor, USA
Gregg Tracton, USA
Christian Zagel, Germany

Human Factors in Energy

Saif Al Rawahi, Oman
Ronald Boring, USA
Paulo Carvalho, Brazil
Sacit Cetiner, USA
David Desaulniers, USA
Gino Lim, USA
Peng Liu, China
Esau Perez, USA
Lauren Reinerman-Jones, USA
Kristiina Söderholm, Finland

Contents

**The Human Side of Service Engineering: Advancing Smart Service
Systems and the Contributions of AI and T-Shape Paradigm**

Software and Systems Engineering
Applications

Development of a Web Based Framework to Objectively Compare and Evaluate Software Solutions

Maximilian Barta$^{(\boxtimes)}$, Sigmund Schimanski, Julian Buchhorn,
and Adalbert Nawrot

Human Factors Engineering, Bergische Universität Wuppertal,
Rainer-Gruenter-Str. 21, 42119 Wuppertal, Germany
{barta,schimanski,j.buchhorn,nawrot}@uni-wuppertal.de

Abstract. We created a web based framework to objectively evaluate different software solutions using different approaches to the same problem. Solutions can either be single algorithms, software modules or entire programs. The evaluation strongly relies on the human component. In this paper we describe the conceptual structure as well as the evaluation of the first field test of the developed framework by utilizing the implementation of two distinct georouting algorithms as well as their validation with test users. No modification to the tested software needs to be done, as the state of each software component is saved and synchronized by the framework. The aim of the investigation of our implementation is to gather usage data, which then allows us to derive proposals for improvements upon the area of usability and user experience and to identify properties that increases the productivity of users as well as finding out about bottlenecks. The framework is validated by assessing the collected metrics and the user feedback produced by the test, as distinctions in usage, user experience and usability between the two test algorithms were identified.

Keywords: Algorithm comparison · Evaluation framework
Human factors engineering · Systems engineering · User centered
Web technologies

1 Introduction

In software development it is not uncommon to have more than one variation of an approach to solve a specific problem, albeit it an algorithm or a user interface. But as "Estimation of project development effort is most often performed by expert judgment rather than by using an empirically derived model" [1] and "Goals such as simplicity, aesthetics, expressiveness, and naturalness are often mentioned in the literature, but these are vaguely defined and highly subjective" [2], often subjective choices are made by the development team [3] in order to select one approach over another without being able to base the decisions made on falsifiable data. Even when user-centered design or user-driven development processes are incorporated – as often users participating in

© Springer International Publishing AG, part of Springer Nature 2019
T. Z. Ahram (Ed.): AHFE 2018, AISC 787, pp. 3–11, 2019.
https://doi.org/10.1007/978-3-319-94229-2_1

software testing declare what they *like* or *don't like* and these values are then used to iterate over the software, although user opinion still is a highly subjective value [2]. To solve this problem and harden decisions for objectivity in the development process, we propose a web based framework that tries to be a solution for this problem for web based software solutions or software components by comparing any two given pieces of software and producing quantified user data in the review process.

The proposed framework is aimed to objectively compare and evaluate software with a web endpoint. The overall goal is to measure, gather and quantify objective user data from two given pieces of (web based) software in order to identify which of the tested solutions quantitatively performs better (e.g., which solution allows users to complete certain defined tasks faster, or what solution takes up less bandwidth, etc.).

Additionally subjective user opinions are collected as well, so that data needed by more conventional approaches is not lost and can be used to determine user opinions. As the output of subjective user opinions also rely on the method of how it is acquired, we tried to follow established guidelines on how to formulate questions in our field test by following propositions made by Figl [4].

As a result the researched framework can also be used to harden user-centered or user-driven development against the highly subjective and suggestive nature of the human factor in such a way, that the grade of the fulfillment of requirements of the tested software can be measured objectively and used to improve the user experience and usability of a to be developed end product, as defined by ISO 9241-210 [5] and ISO/TR 16982:2002 [6] respectively, but also to identify possible missing or faulty functionality.

2   Methodology

Our framework is being actively developed and built around modern web technologies without relying on any 3rd party libraries or frameworks. This and the overall architecture of the framework enables virtually any software with a web interface to be tested with our framework. Also the software to be tested does not need to be hosted on the same machine as the framework itself, although some security restrictions apply (see Sect. 2.1 Framework).

2.1   Framework

While the main reason to develop this framework was to improve the usability and user experience of another piece of software – and thus making it a kind of *meta software* – we intentionally modeled the framework's end-user interface to be as minimal as possible, resulting in only a small toggle switch to switch between the two pieces of software that are about to be tested as shown in Fig. 1. As there is an application programming interface (API) endpoint for this feature, this functionality can be embedded within the test subject software if desired.

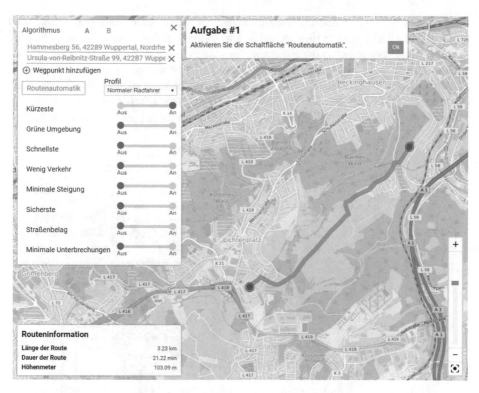

Fig. 1. Single user interface configuration with a toggle to switch between the two target algorithms. The switch is indicated by the arrow.

The small code footprint of our frameworks client-side code (the uncompressed code size is currently smaller than 50 KiB) and architecture (see Fig. 2) enables easy adding of software that should be tested – the only real restriction, aside from the software being web-based, is that the added test software needs to allow cross domain requests from and to the host the framework is located at; this can be done for example by setting the Access-Control-Allow-Origin header at the target software's host to an appropriate URL. The framework transparently wraps itself around the externally loaded software and immediately starts gathering telemetry, once a user opens it. Some of the data we gather consists of the load time of the software, several sub-modules, mouse position, clicked elements et cetera (see Table 1).

Table 1. Non exhaustive list of gathered telemetry in no particular order

Load time (server)	Execution time (client)	Clicks
Pointer position	Events triggered	Keyboard input
Form post data	Operating system/device	Screen resolution
Traffic produced	Browser used	Location of the user
AJAX calls (URL/parameters)	Latency (e.g., click to action)	Exposed test software data (e.g., REST-responses)

Network activity is intercepted and logged as well. All of this predefined, well formed, quantifiable data is then stored in a relational database. All other data, like for example output produced by the tested software, is stored in a NoSQL [7] database, which enables us to store virtually any information that can be produced by web based software.

Despite being able to gather all this data, the evaluation of unstructured data can become very complex. In order to ease processing the data further we provide an API endpoint for both the unstructured and the already structured and evaluated data.

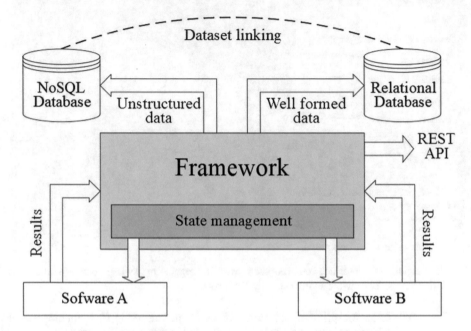

Fig. 2. Schematic representation of the framework's architecture.

In order to add and compare two distinct software programs, a simple edit in a configuration file needs to be done:

```
// config.json
{
    "software": {
        "first": {
            "url": "https://www.asoftware.com"
        },
        "second": {
            "url": "https://www.someothersoftware.net"
        }
    },
    /* ... */
}
```

Where *first.url* and *second.url* respectively define the entry point of the software
that should be added. The framework supports adding tasks that are presented to the
user in the form of a notification box (see Fig. 1) by default, as well as prompt the user
for feedback. The overall design and presentation can also be modified with standard
web techniques, i.e. Cascading Style Sheets (CSS) or JavaScript. User input can either
be gathered via a free form text input field, a range slider, a simple 1–5 star rating, a list
of choices or a combination of all of the former (see Fig. 3).

Tasks and feedback prompts can be triggered when a specific state in the software
to be tested is reached – this can be anything from either receiving a specified
web-request, a time based trigger, the user clicking on a specific element, performing a
specific series of inputs or the lack of inputs.

As the framework wraps itself around the test software every piece of the software
that is available to the client is also accessible by the framework.

Fig. 3. Live field test snapshot. Demographic data being collected with various input methods.

3 Field Test and Framework Evaluation

One of the advantages of the framework being web based is that testing does not rely
on a special laboratory setup but can be done by users at any given time anywhere a
web capable device and web access is available; for as long as the test is enabled.
Naturally this also greatly improves the diversity of devices participating in the evaluation compared to the amount of distinct devices that would be used for testing if the
test was conducted in a setup environment where the same devices are used across
various different test users.

3.1   Setup

In order to test the viability of the framework itself we applied two distinct georouting algorithms to the framework, focusing on routing users with pedal electric cycles (pedelecs) in and around the city of Wuppertal, Germany. The two algorithms were developed for the Electric Mobility centered research project EmoTal [8]. To keep the focus on the algorithms rather than the user interface the same and visually reduced user interface was used for both routing algorithms – resulting in very few and minor usability differences.

The field test was not conducted in a prepared environment (i.e., laboratory setup), giving all participants the degree of freedom to participate from no bound location (like home, office, on the road, etc.) and with the device each individual user prefers (for example desktop computer, smartphone or tablet device). The test itself consisted of a blinded-experiment, more specifically a double-blind test as proposed by Rivers and Webber [9]. This means that in our field test example the two algorithms that were examined were randomly mapped to the first or second choice on each new attempt, so that a bias towards *Algorithm A* or *Algorithm B* could be ruled out. Also tasks presented to the user were permuted (where applicable) in each run for the same reason.

In order to extend the reach of the field test, the city of Wuppertal advertised the test on their homepage. Overall a participation count of n = 351 was recorded – tests where the participant only opened the framework and did not enter anything were not saved and are not considered in the final results. Tests that were only completed partially were also discarded in the final results, but can still be used for analyzing (for example when and why the user jumped off).

In its current state (as of writing this paper) the framework does not provide an administrative back end. That means that all evaluation was computed manually by the researchers on the data stored in the database. In future versions of the framework we will provide tools to easily generate reports containing the most commonly used statistics as well as the ability to customize queries. Intriguingly the framework itself could be used to aid in the development process of such a back end (see Sect. 4 Outlook).

3.2   Field Test Results

The field test results presented in this paper are not exhaustive and should rather give a general overview. While evaluating the test results a bias towards one of the two tested algorithms could be identified. Users also tended to either omit rating the algorithms altogether (which was perfectly legit as we did not enforce a rating) or gave both algorithms the maximum score of 5 points (see Table 2).

Giving demographic and personal information was voluntary, for example this resulted in a ratio of approximately 75% of all users giving information about their age.

The average age of all participants was 45 years, with the youngest participant being 15 years old and the oldest being 77 years old (see Fig. 4). The top three occupations for the testers were student, official and employee. Users rated themselves on average a 3.5 out of 5 in smartphone proficiency.

Table 2. User evaluation of the algorithms based on ratings from 1–5.

Rating	Algorithm A	Algorithm B
1	8	3
2	8	14
3	13	18
4	19	14
5	122	156

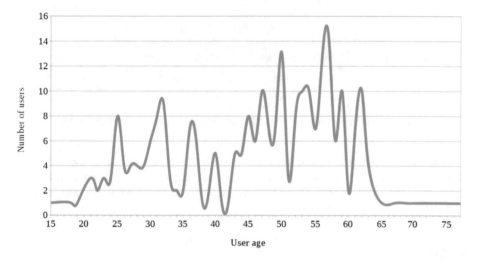

Fig. 4. Field test results – test users by age

Although users were encouraged to do so, only 9 out of the 351 participants gave unsolicited feedback. In contrast to that, 109 free form text notes were made. Some recurring remarks that were made, concerned the routing of either algorithms with some route section being *too steep* or allegedly taking *some detours*. Less common were remarks about the generated routes routing over *too many main streets*, or that *both algorithms generated the same route altogether*. Overall users took an average of about 80 min to complete the entire test.

3.3   Framework Validation

While the first field test proved the framework to be robust from a technical point of view – no major performance issues or system down times occurred during testing – some other issues were identified when the test results were being evaluated. As the structure produced by the tested software can change rapidly (as in general the tested software is software in active development) we utilized a non-relational database for this purposes. But even within our controlled field test the logged amount of unstructured data per user often exceeded 100 KiB, this could become a performance issue when importing the data set as a whole into statistics- or spreadsheet-software.

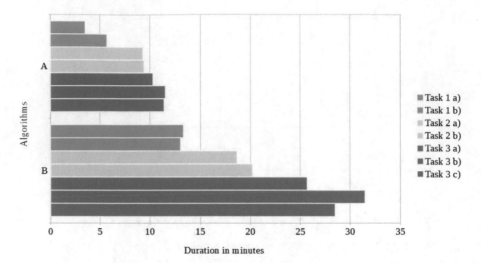

Fig. 5. Field test results – Average task completion duration by algorithm.

Validation of the framework itself was done using the evaluation of the field test. It could be shown that an algorithm with which users were the most productive (time-wise) could be identified (see Fig. 5). Also the objective data gathered correlates with the user subjective feedback.

4 Outlook

The framework could not only be used to aid in software development, other uses are also conceivable and practical. For example the framework could be used to aid in making commitment decisions for companies on what product should be chosen over another by comparing two very similar software solutions.

Future versions of the framework could also incorporate the ability to test multiple software solutions at once (up from the two that are currently mandatory and possible). Also only allowing one software solution to be tested at once could be a possibility. This would enable the framework be solely used as a telemetry data gathering tool. In order to give users a better impression of the tested algorithms/programs a split screen mode will be implemented, so that both test subjects can be displayed at once.

5 Conclusion

This paper proposed a new approach to objectively compare and evaluate the usability of software components which aids in the development process by using the generated results to improve the overall usability and user experience of said software. The first field test demonstrated not only how the framework can be used for evaluation and improvement purposes, but also showed the framework's limitations. Based on the results of the first field test, plausible improvements were made and implemented into the test subject software. This was done by taking higher rated algorithm and tuning it in such a way that the routes generated integrate the most often made user remarks like tuning the steepness of some routes and removing perceived detours (see Sect. 3.2 Field test results) and thus improving the user experience. Also properties of the overall lower scoring algorithm that users rated better, were implemented in the other algorithm. With further iterations we will add more improvements to the framework which will help in evaluating and comparing target software.

References

1. Gray, A.R., MacDonell, S.G., Shepperd, M.J.: Factors systematically associated with errors in subjective estimates of software development effort: the stability of expert judgment. In: Proceedings of Sixth International IEEE Software Metrics Symposium, pp. 216–227 (1999)
2. Moody, D.: The "physics" of notations: toward a scientific basis for constructing visual notations in software engineering. IEEE Trans. Softw. Eng. **35**(6), 756–779 (2009)
3. Albrecht, A.J., Gaffney, J.E.: Software function, source lines of code, and development effort prediction: a software science validation. IEEE Trans. Softw. Eng. **SE-9**(6), 639–648 (1983)
4. Figl, K.: Deutschsprachige Fragebögen zur Usability-Evaluation im Vergleich. Zeitschrift für Arbeitswissenschaft **4**, 321–337 (2010)
5. International Organization for Standardization: Ergonomics of human system interaction – Part 210: Human-centered design for interactive systems (formerly known as 13407). ISO F ±DIS 9241-210:2009 (2009)
6. International Organization for Standardization: Ergonomics of human system interaction – Usability methods supporting human-centered design. ISO/TR 16982:2002 (2005)
7. Cattell, R.: Scalable SQL and NoSQL data stores. ACM SIGMOD Rec. **39**(4), 12–27 (2011)
8. Schimanski, S., et al.: EmoTal – Nutzerzentrierte Elektromobilität Wuppertal. In: Dienstleistungen als Erfolgsfaktor für Elektromobilität, pp. 120–127. Fraunhofer Verlag, Stuttgart (2017)
9. Rivers, W.H.R., Webber, H.N.: The action of caffeine on the capacity of muscular work. J. Physiol. **36**, 33–47 (1907)

A Case Study of User Adherence and Software Project Performance Barriers from a Sociotechnical Viewpoint

Nicole A. Costa[⊠], Florian Vesting, Joakim Dahlman,
and Scott N. MacKinnon

Chalmers University of Technology, Hörselgången 4,
412 96 Gothenburg, Sweden
{nicole.costa, florian.vesting,
joakim.dahlman, scottm}@chalmers.se

Abstract. A marine propeller company and a technical university collaborated to optimize the company's existing propeller design software. This paper reviews the project based on a sociotechnical perspective to organizational change on (a) how the university-company project and user involvement were organized, and (b) what the main management barriers were and why they may have occurred. Fieldwork included interviews and observations with university and company stakeholders over thirteen months. The data was analyzed and sorted into themes describing the barriers, such as lack of a planned strategy for deliverables or resource use in the project; the users exhibited low adherence towards the optimized software, as well as there was limited time and training allocated for them to test it. Lessons learned suggest clarifying stakeholder roles and contributions, and engaging the users earlier and beyond testing the software for malfunctions to enhance knowledge mobilization, involve them in the change and increase acceptance.

Keywords: Software development · User participation
Sociotechnical systems · Organizational change · Knowledge transfer

1 Introduction

This paper describes and examines the case study of a software optimization project that took place at a well-established company that had developed in-house marine propeller design software and used it for many years. The software required its users (the propeller designers) to manually insert and adjust propeller blade parameters based on professional experience and skill and according to customer requirements, reiterating until a best-fit was found [1]. The number of design alternatives tested was thus limited and time-consuming (commonly around seven baseline designs each iterated 3 to 8 times, which might take about a week). Alternatively, the automation of this process via algorithms has made optimization more feasible. Hence, the company in collaboration with a technical university initiated a project for the development of algorithms for marine propeller design optimization. The optimization would allow the software to automatically run more than hundreds of iterations of each baseline design

© Springer International Publishing AG, part of Springer Nature 2019
T. Z. Ahram (Ed.): AHFE 2018, AISC 787, pp. 12–23, 2019.
https://doi.org/10.1007/978-3-319-94229-2_2

and run an element of self-assessment regarding the most suitable design solutions (and therefore provide the users with decision support) within about two days, while the users were free to engage in other tasks. The optimization was intended to render the use of the software more efficient and objective, as well as increase the overall performance of propeller design in terms of production costs, efficiency and passenger comfort, ultimately amplifying the company's competitive advantage and the possibility of attending to a larger number of projects in a shorter period.

At the university, a marine technology researcher was responsible for the development, testing and implementation of the software optimization (whose role is hereby defined as *developer*). Several visits to the company took place during the project, as well as regular meetings on both premises. The developer was in contact with the company's project coordinator, the propeller design team manager, and a propeller designer as a user representative. The project coordinator ensured that the software would fulfil the company's needs and monitored the progress of the developer; the design team manager and the propeller designer aided the developer in terms of the software modules needed and bugs. The software optimization was to be developed and tested with users iteratively to spot software bugs and correct them.

Retrospectively, barriers to the software project management and user engagement practices were identified and thus the intent of this paper is to describe and examine (a) how the university-company collaboration project and user involvement were organized, and (b) what the main management barriers were and why they may have occurred. The discussion of the barriers draws from references – and contributes to the research and body of knowledge – in software development risk management, sociotechnical theories and participatory approaches.

2   Theoretical Background

Software development risks occur when technology, tasks, structures and actors are in conflicting states and harm the stability of the sociotechnical[1] system [4] (see also Leavitt [5]). This instability can happen due to limited information, skills or resources necessary for the developers to make alternate and useful design decisions, as well as to appropriately implement them. Inadequate software performance, delays, implementation processes exceeding the budget, being discontinued or requiring significant and expensive programming adjustments after implementation so that the users can accept it are examples of potential circumstances that render software development projects unsuccessful [4, 6]. In addition, collaborative research and development (R&D) projects between universities and companies are a common setting where lack of communication and of mutual understanding of the collective objectives can generate inefficiencies in the project outcomes [7]. To address similar threats, literature has suggested a sociotechnical risk-based approach to organizational change by technology

---

[1] A sociotechnical approach to systems is one where the technological and the social are mutually dependent [2, 3]. Linking the technology and the personnel are routines, organizational structures and processes, and beyond the boundaries of this system is an external environment in which the system exists, upon which it depends and to which it must adapt [3].

by collecting more information about the three environments in which the software development is taking place, to improve knowledge mobilization and diminish uncertainty [4, 5]: the system/use environment where the software is to be operated; the development environment where the software is developed; and the management environment that determines how the software development management occurs. Software development risks within the system/use environment can range from users not knowing how to use the software to not accepting it. Risks within the development environment might be associated with the lack of capabilities of the developers to analyze the system/use environment concerning how the software can be used and implemented. Lastly, risks within the management environment might be related to the managers' lack of knowledge or attention to certain available information necessary for the success of the development process [4].

Definition of the important stakeholders to project success, appropriate stakeholder management and direction through the project is required to respond to the ambitions of the project [4, 6, 8, 9]. There must be a feedback loop throughout the process between environments to improve experiences and methods; to accentuate teamwork and user participation [4]. Joint problem-solving must take place to ensure that goals are mutually understood among the stakeholders who are involved in the process and will be affected by the change [10]. Frequent interactions and strong partnership ties should stimulate more knowledge transfer [7, 11] and enhanced technology and innovation [11], and help clarify the goals of all parties [7, 10, 11].

The success of a technology might be determined by whether its impact on the users and the ways they work were captured [12], ensuring that the technology is congruent with the users performing the tasks and not triggering negative effects in the efficiency of the overall system [3]. This suggests a triangular relationship: the developer (actor) will use research and development methods (means) to develop a software tool (end), and the software is in itself a goal. On the other hand, the users of this software will simply see it as a means to achieve their own end-goal. In other words, one person's work outcome becomes another person's work tool necessary to complete work tasks that will in turn help maximize company success. The development environment should inquire the use environment and understand the impact of the technology from a sociotechnical viewpoint to help predict risks, prepare the organization for the change and reduce resistance [4, 13]. This can be accomplished by optimizing the work system that comprises the technology and the personnel [3] by involving the end-users to facilitate the transmission of relevant information and knowledge within and between organizations.

User involvement has indeed been suggested to help mitigate the risk of misspending resources on the progress of a product that is based on incomplete or misunderstood requirements, thus reducing the product's risk of failing to meet stakeholder requirements or being rejected by its end-users. It has been suggested to help boost the users' acceptance, commitment and trust in the product as they understand the characteristics of the system and its value and that the design is being suited to them rather than imposed [14]. It has been suggested to help elevate the reputation and competitive advantage of the organization [14]. User participation, promoted at the early stages of the process, has also been found to help reduce uncertainty by filling in information gaps in collaborative projects [10], as well as determine the usefulness of the project outcomes [9], since the end-users possess expert knowledge of the operations. Tait and Vessey [6] advised that, to predict the success of the user involvement and of the introduction of a computer system into an organization, accounting for environmental factors in the development process (e.g., project resource constraints), the technical subsystem (e.g., system complexity) as well as the user subsystem (e.g., system impact and user attitudes) is key.

3 Methods

Following the initiation of the project, a fellow researcher at the university – from the field of Maritime Human Factors (main author of this paper, whose role is hereby defined as *researcher*) – was given the opportunity to observe the project. Fieldwork and data collection included fourteen meetings at university and company premises over thirteen months, particularly twelve semi-structured interviews with project stakeholders (nine of which with the developer, one with the project coordinator, one with the propeller design team manager, and one with a group of three available propeller designers), and attendance of two general project review meetings [15–17]. The interview with the project coordinator was aimed at capturing the company's objectives and requirements of the project; with the propeller design team manager and the propeller designers at understanding propeller design procedures pre- and post-software upgrade, and exploring the propeller designers' expectations of how their design practice could be affected and optimal design solutions found. The interviews with the developer happened regularly to capture his work and the organizational structure of the optimization and deployment process. Events missed by the researcher were reported back by the developer. Each meeting had the approximate duration of one hour. The interviews were audio-recorded and all data collection events annotated [16]. Field notes were incrementally and chronologically documented, highlighted where relevant, and memos and codes helped to reduce the data and identify the main themes iteratively [16, 18–20] based on a thematic analysis [20, 21]. Literature was then studied to help understand the observed barriers in the project and potential mitigation.

4 Results

Before the official launch of the project (during a definition phase), the optimization developer investigated the original software and the tasks of its users (propeller designers), with the help of the company's appointed user representative. Once the project was launched [year 3], the developer also kept contact with the consultant developer in charge of the Graphical User Interface (GUI) and with the project coordinator and the design team manager at the company. Together they refined the technical requirements to achieve the project's specific goals, affecting the development of the optimization routine.

The computational power available to the developer required a certain time for running test trials and this hindered spotting errors and debugging. Considering this and the company's intention that all users should begin to integrate the upgrade into their daily work activities, the beta upgrade was deployed [year 4]. This was an explanatory session from the developer, passive from the users' side. The users were then requested by the company to explore and test the beta version of the optimization upgrade on real propeller design cases, and to report bugs and suggestions to the design team manager if wanted. For this to be done, the developer and company needed user buy-in, but this did not occur as expected.

For another year, the developer performed optimization code modifications and testing, and then provided the users at the company with a more thorough introduction and a tutorial session of the optimized software as a way to push for the use of the new software, to transmit how it was intended to be used, and for the propeller designers to get familiarized with the product, find bugs and propose remedies [year 5]. The tutorial session included an explanatory section and typical scenarios for user familiarization (only half of the ten users were present). During the tutorial session, information was provided in terms of technical functionalities and of interface characteristics compared to the original software. The developer also prepared an instructions manual for the usage of the software. Months later [year 6], the developer presented to the company the results of his latest developments based on previous meetings. He showed the group of users (in total four were present) that the optimization upgrade was working successfully on typical propeller design tasks. The automated optimization yielded solutions comparable to the manual benchmark designs. The developer received encouraging feedback from the management team, who believed that the recent modifications would solve the long-time reluctance to usage by the users by expanding functionality automation. The users also seemed positive, but emphasized their lack of time for exploring and using the upgrade.

Less than a half year later, the design team manager moved to university premises and changed his role to work more closely together with the developer. The company began allocating specific time for a couple of users to test the software in real-life propeller projects and compare the results against manual designs. The GUI was also further developed before the developer's university affiliation voided and he handed the optimized software over for further development [year 8]. The following themes describe the set of organizational barriers hypothesized to have factored into the limited adherence of the users to the new software and into the project delays.

Time Allocation and Technical Readiness: The users repeatedly emphasized the high workload with propeller orders and lack of time for trying out the new software upgrade (*"It takes time to learn, and we have very many projects to handle"*). The motivation to do so also seemed negatively affected by the fact that the software was not finalized; it was deployed at an ongoing development phase with the intent of testing it for bugs with the users, receive feedback, correct, and deploy again. Considering this, learning an *alpha* or *beta* version that was likely to suffer alterations and have to be relearned was not prioritized among dominating tasks. At this stage, the users were pushed by the company management to begin integrating the new software upgrade into their daily design activities, which the users also deemed problematic from an automation reliability perspective. The integration into daily work relied on the users' own initiative, as it was not made mandatory nor was a deadline established by the management. In relation to this, the developer encountered problems getting all of the users together in the same room for project workshops dedicated to providing software information and training, testing and troubleshooting.

Perceived Relevance and Trust in Automation: Applying the software upgrade, especially an unfinished version, to real-life propeller projects in progress represented a high-risk activity in the sense that it could cause a negative impact on their work performance if the output was not reliable. The users did not know to what level the new software solutions could be trusted, and this implicated more time and resources for double-checking until they could gain trust in the system: *"We should run the regular process and see if we can trust the new software. At least in the beginning to confirm if it runs properly"*.

The interviewed users were eager and optimistic concerning improved efficiency and time-saving aspects of the optimization (e.g., *"Less time and more iterations. Good for doing ten times as many iterations and fine-tune the results"*; *"it goes into areas that you wouldn't get into if you do it manually"*). Yet, they expressed reservations towards automation that the project failed to account for. The possibility of automating a set of decisions that were originally made by the designers based on their experience and knowledge was questioned. This was also perceived by the users as a replacement of their work, therefore becoming a risk for future designers who would not be trained to judge the suitability of the software results. This called into question what role they would be fulfilling and what skills they would be expected to have (*"Everyone can press the button and run it"*; *"(...) now you don't know what the tool is doing and in order to decide if the result is ok, you have to know what it does"*; *"The risk is that people doing the design don't have the knowledge, so they can't judge when there's something wrong. If a module doesn't work, you should know why. You can't run everything automatically because you lose control of it"*). The users expressed, therefore, an increased need for training with higher levels of automation, to be able to understand what the computer is running and how.

Training Needs: Users reported that they wanted more instances with the developer for instructing them on how the software was intended to be used. That way, the users felt they would not have required so much time for learning it by themselves, and integration in their daily work might have become an easier and faster process.

Project Management Strategy: The project did not use any particular management strategy or specific deadlines for delivery. There was no clear separation of the design stages or a clarification of when and how the users were expected to intervene. Only during the last year of the project did the management team begin to implement certain milestones and deliverables and to document, which had not been prioritized in previous years. The developer reported that defining milestones was helpful in the project but that he understood the unpredictability and difficulty of setting deadlines in this R&D endeavour. At this stage, the developer chose to look up the human-centred design (HCD) cycle on Google – knowing that it was one of the human factors researcher's topics – and claimed he perceived his own optimization development process to mirror the HCD concept in terms of having the user in mind. However, he recognized that this approach could have potentially represented a solution for certain issues encountered in the project had it been considered from the beginning. For example, the developer realized by the end of the project that the GUI had not been appropriately tailored for the users and that this might have been an impact factor on the users' hesitation to using the optimization upgrade in their daily work (*"usability of the GUI is not optimal (...) it could be better, more flexible and self-explanatory"*). At a separate instance, the project coordinator was convinced of the benefits and main constraints of the optimization upgrade, but also believed that the process of designing this upgrade was critical to its *"user-friendliness"*. Such realizations and modifications to the structure of the project reflected the need for and interest in a user-oriented project management strategy to help attain the project's objectives successfully.

Stakeholder Roles and Task Centralization: During his involvement, the developer led tasks of a technical and logistic nature, not having a software development background. Although this project was collaborative, the principal tasks converged to the developer. The developer maintained close ties with the rest of the team, although the strength of these ties varied. For example, the ties with the group of end-users were not continuous, as there was a physical separation between premises and the end-users were only involved in a software testing capacity. In addition, test reports from the group of users, for instance, were to be passed on to the design team manager who would only subsequently transmit them to the developer.

Those developing the software optimization and the GUI did not have an internal affiliation with the firm where the software was to be used. So as the developer was not within the company, core information or more practice-based knowledge were not as accessible.

User Requirements and Expectations: Having appointed a user as the developer's contact person and requested the remaining users to test and report system errors reflected the developer's and company's will to incentivize direct user participation. The contact person's role was informative regarding the functions of the original software to discuss what could be altered towards the automation objectives. There were emerging requirements through the process, and technical availability was discussed among the team. The participation of the remaining users was requested on a testing basis, and their reluctance to contribute was a continuous barrier in the timely progression of the project. The input of the user representative could not fully cover the other users' design experience or their perceptions, expectations or reservations towards the software upgrade.

5   Discussion

The outcome of the developer's work (which occurred within the development environment) would become the propeller designers' new work tool to fulfil their tasks at the firm, which would in turn play an essential part in meeting overall company goals (within the system/use environment), as illustrated in Fig. 1[2]. The precondition for the

---

[2] The management environment encompasses both the stakeholders at the university and the managers at the company in a continuous back-and-forth of information and joint decision-making.

company to attain their company strategy is thus by having their designers successfully fulfil their work goals. This points at the designers as users of the optimized software as key stakeholders in the success of the project [9], and the optimized software would come to impact the propeller designers' required skills and ways of performing their tasks, and in turn the activities of the company as a whole, in a chain of changes [22, 23]. Hence, lack of vertical alignment or mismatches can represent risks in the sociotechnical system. The definition and alignment of objectives between the university and the firm occurs with communication and collaboration [10], with information sharing across environments [4] and with maintaining strong ties through the process [7, 11].

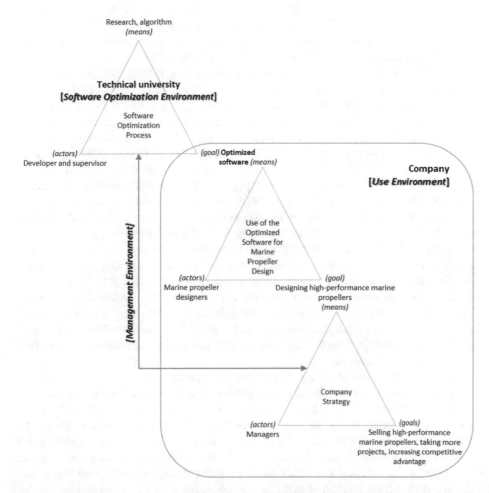

Fig. 1. The risk-based software development model combined with the activity theory model demonstrating the sociotechnical nature of the project with the triangulation between actors, tasks and tools in the project environments (adapted from Andersson et al. [24]; Karlsson [25]; Lyytinen et al. [4]).

The software would increase productivity at the company by automating much of the workload in propeller design, and decreasing diversity and subjectivity [13] of propeller design dependent on the experience and choices of the individual marine propeller designer. Ultimately, the software upgrade brought an extra level of automation to the work of the users, modifying the way they were expected to accomplish their tasks and the skills necessary. Since the original software had been consistently used for several years, it was well known to its users. Its lower level of automation allowed them to have higher level of control and more transparency to the process. The optimization upgrade, on the other hand, partially automated the task of the users through algorithms, turning it into a cryptic process for them [3]. Consequently, the users were now presented with a process that they could not control sufficiently to interpret whether the output was reliable, at least at the initial stages of the testing. This caused resistance to usage in real propeller design cases [26, 27], and lowered perceived system usefulness [28] linked to some reservations as to what this automation would implicate for them as expert designers.

The development of such algorithms entails capturing the knowledge of the users, which was attempted through user representativeness. Yet, as the objective was to standardize/automate the users' subjective knowledge into the software code, the representativeness of one user is questioned. The remaining users were asked to provide input mostly at a testing and deployment stage, but had technical difficulties with the system, did not know how to use it or have time to try it or learn it; did not prioritize it or trust it since the algorithm integration and GUI were not fully matured yet. The users wanted more training, but the developer had a hard time getting them together in one room. As is portrayed in Fig. 1, the developer was external to the use environment, and hence needed support from the management at the firm to establish and help him maintain ties with the users, allocate more time for joint activities and problem-solving within the project. The users themselves needed more support from the management to orientate themselves with the optimized software rather than just getting a push to use it with limited time and training, and no particular deadlines. In retrospect, the team could see the benefit in more structured planning for clearly defined stakeholder contributions, milestones and deadlines for delivering the optimization upgrade [8, 9], and in requesting more user input, for example, on the new GUI, and how these adjustments might have incentivized user adherence to the specified project events where their participation was requested.

The user involvement, being that of the user representative or of the remaining users, was focused on the technological variable (what functions the algorithms could cover, the performance of design outcomes, system malfunctions). What could not be captured or predicted to the same extent was the human element/actor variable [4, 5]. This in the sense of (a) what concerns and attitudes the users had towards the automation, (b) what it would mean for the users' skill development and expert knowledge to have a software tool make propeller design decisions for them, and (c) how their design work and skills would be required to adapt. Heuristics can be helpful for the project team to initiate reflection about the tasks to be performed within the use environment [4]: *What is the impact on user tasks? Will actors be critically dependent on the system? Is there much unarticulated, tacit knowledge involved? How much variation and flexibility are involved in the tasks? Are the actors' expectations realistic? Do actors personally benefit from the change?* To capture these aspects at the human element/personnel level and assure acceptance [14, 28, 29], an early analysis of the impact of the technology is recommended to occur to help establish design alternatives [3]. Moreover, involving users not just in a testing capacity might have, from an earlier stage, amplified the capabilities necessary for appropriate and effective knowledge transfer in the system [30], and better prepared the users for the change [6, 13].

Introducing a new technology in favour of company goals not in alignment with the tasks and needs of the personnel subsystem may have a negative impact on the output of the entire work system. Optimizing the work system as a unit, then, means reimagining the technology and the personnel as a unit [3]. For this, the project could have benefited from the software optimization environment engaging the personnel subsystem earlier to foster communication among them and obtain more user expert input to inform both the conceptualization and the development processes of the software [3, 4, 31]. This should help to ensure that the workers are not left with the residual functions of the technological automations [3], which was a concern expressed by the users. User participation before implementation at the firm, and training after implementation are recommended interventions [28, 32], as per the training needs expressed by the users. A sociotechnical and participatory approach can not only be used to improve upon the usability of the end-product, but also as a way to refine and lead the management strategy of the project [33, 34].

6 Conclusions

This case study contributes to research and to practicing managers with empirical evidence of barriers to software maintenance management and user engagement practices in joint R&D projects, emphasizing the relevance of a sociotechnical and participatory perspective in organizational change. Lessons learned emphasize the importance of a management structure that can (a) balance the objectives of the project stakeholders, including those of the users; (b) stipulate responsibilities, lines of communication, and allocate time and support to maximize the capacities necessary for appropriate and effective knowledge transfer; (c) prepare the users for the change; and (d) increase software adoption and use. Recommendations include facilitating user input at an earlier stage and beyond testing the software for malfunctions, to capture user needs and attitudes towards the software and its future impact on their work. Future research is needed in similar in-house settings to investigate whether all company users should be involved in the organizational change or whether a subset of those users would suffice, and in that case what the ideal number [28] of users involved would be representative of user perceptions and attitudes. This would perhaps shed more light on the challenges of user representation concerning the facilitation of risk management and knowledge transfer.

References

1. Vesting, F.: Marine propeller optimisation - strategy and algorithm development. Doctor of Philosophy. Chalmers University of Technology, Gothenburg, Sweden (2015)
2. Fox, W.M.: Sociotechnical system principles and guidelines: past and present. J. Appl. Behav. Sci. **31**(1), 91–105 (1995)
3. Hendrick, H.W., Kleiner, B.M.: Macroergonomics: An Introduction to Work System Design. Human Factors and Ergonomics Society, Philadelphia (2001)
4. Lyytinen, K., Mathiassen, L., Ropponen, J.: A framework for software risk management. Scand. J. Inf. Syst. **8**(1), 53–68 (1996)
5. Leavitt, H.J.: Applied organizational change in industry: structural, technological and humanistic approaches. In: March, J.G. (ed.) Handbook of Organizations, pp. 1144–1170. Rand McNally & Company, Chicago (1965)
6. Tait, P., Vessey, I.: The effect of user involvement on system success: a contingency approach. MIS Q. **12**(1), 91–108 (1988). Management Information Systems Research Center, University of Minnesota
7. Arroyabe, M.F., Arranz, N., de Arroyabe, J.C.F.: R&D partnerships: an exploratory approach to the role of structural variables in joint project performance. Technol. Forecast. Soc. Change **90**, 623–634 (2015)
8. Curedale, R.A.: Design Thinking: Process and Methods Manual. Design Community College, Woodland Hills (2013)
9. Karlsen, J.T.: Project stakeholder management. Eng. Manag. J. **14**(4), 19–24 (2002)
10. Sjödin, D.R., Frishammar, J., Eriksson, P.E.: Managing uncertainty and equivocality in joint process development projects. J. Eng. Technol. Manag. **39**, 13–25 (2016)
11. Lee, Y.S.: The sustainability of university-industry research collaboration: an empirical assessment. J. Technol. Transf. **25**, 111–133 (2000)
12. Kirwan, B.: Soft systems, hard lessons. Appl. Ergon. **31**, 663–678 (2000)
13. Schein, E.H.: Organizational Culture and Leadership. Wiley, Hoboken (2010)
14. Maguire, M.: Methods to support human-centred design. Int. J. Hum.-Comput. Stud. **55**(4), 587–634 (2001)
15. Czarniawska, B.: Social Science Research: From Field to Desk. SAGE, London (2014)
16. Creswell, J.W.: Research Design: Qualitative, Quantitative, and Mixed Methods Approaches. SAGE Publications Inc., Thousand Oaks (2014)
17. Patton, M.Q.: Qualitative Research and Evaluation Methods. Sage Publications Inc., Thousand Oaks (2002)
18. Silverman, D.: Interpreting Qualitative Data. SAGE, Thousand Oaks (2014)
19. Corbin, J.M., Strauss, A.L.: Basics of Qualitative Research: Techniques and Procedures for Developing Grounded Theory. Sage Publications Inc., Thousand Oaks (2008)
20. Charmaz, K.: Constructing Grounded Theory. SAGE, Thousand Oaks (2014)
21. Joffe, H., Yardley, L.: Content and thematic analysis. In: Marks, D.F., Yardley, L. (ed.) Research Methods for Clinical and Health Psychology. SAGE Publications Ltd., London (2004)
22. Rasmussen, J.: Risk management in a dynamic society: a modelling problem. Saf. Sci. **27**(2/3), 183–213 (1997)
23. Vicente, K.J.: The Human Factor: Revolutionizing the Way People Live with Technology. Routledge, Taylor & Francis Group, LLC, New York (2006)
24. Andersson, J., Bligård, L.-O., Osvalder, A.-L., Rissanen, M.J., Tripathi, S.: To develop viable human factors engineering methods for improved industrial use. Springer (2011)

25. Karlsson, M.: User requirements elicitation: a framework for the study of the relation between user and artefact. Chalmers University of Technology, Gothenburg (1996)
26. Norman, D.A.: The 'problem' of automation: inappropriate feedback and interaction, not 'over-automation'. Philos. Trans. R. Soc. London. Ser. B Biol. Sci. **327**(1241), 585–593 (1990)
27. Bainbridge, L.: Ironies of automation. In: Rasmussen, J., Duncan, K., Leplat, J. (eds.) New Technology and Human Error. Wiley, New York (1987)
28. Venkatesh, V., Bala, H.: Technology acceptance model 3 and a research agenda on interventions. Decis. Sci. **39**(2), 273–315 (2008)
29. Norman, D.A.: The Design of Everyday Things. Basic Books, New York City (2013)
30. Parent, R., Roy, M., St-Jacques, D.: A systems-based dynamic knowledge transfer capacity model. J. Knowl. Manag. **11**(6), 81–93 (2007)
31. Bligård, L.-O., Österman, C., Berlin, C.: Using 2D and 3D models as tools during a workplace design process - a question of how and when. In: 46th Nordic Ergonomics Society Annual Conference, Copenhagen, Denmark (2014)
32. Long, S., Spurlock, D.G.: Motivation and stakeholder acceptance in technology-driven change management: implications for the engineering manager. Eng. Manag. J. **20**(2), 30–36 (2008)
33. Giacomin, J.: What is human centred design? Des. J.: Int. J. All Asp. Des. **17**(4), 606–623 (2014)
34. ISO: ISO 9241-210 Ergonomics of human-system interaction—part 210: human-centred design for interactive systems. In: ISO 9241-210. International Organization for Standardization, Geneva (2010)

Objectification of Assembly Planning for the Implementation of Human-Robot Cooperation

Rainer Müller, Richard Peifer, and Ortwin Mailahn[✉]

Research Group: Assembly Planning,
ZeMA – Zentrum für Mechatronik und Automatisierungstechnik gemeinnützige
GmbH, Eschberger Weg 46, 66121 Saarbrücken, Germany
`{rainer.mueller,r.peifer,o.mailahn}@zema.de`

Abstract. This article will show how the planning of assembly processes by translating the planning problem into computer-readable domains contributes to the objectification of planning. Especially in the planning of human-robot cooperation, this is necessary due to the complexity of the relationships of the planning contents. For this purpose, the planning problem is transformed into an ontology domain of which an appropriate reasoner can draw logical conclusions from.

Keywords: Task assignment · Human-robot cooperation · Work description
Assembly planning assistance · Tool chain · Optimization · Objectification
Decision making

1 Introduction

Maintaining competitiveness of manufacturing companies in high-wage countries is an ongoing challenge in the context of globalized markets. Highest quality requirements, fast reaction to market changes, mastery of variant diversity as well as the reduction of manufacturing costs are the essential strategies to succeed in the market [1].

The assembly still has a high proportion of manual activities compared to other areas of production [2]. Particularly in the field of large component assembly, there are great rationalization potentials [3]. However, these do not consist in automating the largest possible number of processes, as was the case in the 1990s [4]. Current assembly concepts pursue the goal of "adapted automation" [5].

Due to the increasingly available sensitive robots, which fulfill the criteria and requirements for a secure cooperation with humans, previously not automatable tasks can be handed over to the robot in the future [6]. The drivers can have a variety of objectives. For example, an increasing average age of the working people requires ergonomically-designed workplaces, and robots can become assistants by taking on tasks. In addition, for example, the assurance of quality, a shortage of workers or rationalization efforts can lead to increased use of robots in production.

Accordingly, the decision for or against an automated process no longer has to be done station by station, but can be broken down to the partial tasks, according to the

© Springer International Publishing AG, part of Springer Nature 2019
T. Z. Ahram (Ed.): AHFE 2018, AISC 787, pp. 24–34, 2019.
https://doi.org/10.1007/978-3-319-94229-2_3

specific capabilities of human or robot [7]. In this way, tasks can be assigned as needed, depending on the availability of resources, and the system can react flexibly to changes in the production situation [8].

To give planners who are faced with the challenge of the most efficient use of resources in the context of re-planning or by changing boundary conditions, a skill-oriented assembly process planning is required [9]. Against the background of Industrie 4.0, the consistent use of planning data is recommended [10].

This article presents a method for determining the assignment of assembly tasks to resources. It is based on a holistic approach that includes the product, the assembly processes and the performing resources. The main focus of the method is the further development of the skill-based approach for the planning of adapted automation, the description of planning problems with the help of ontologies and the multi-criteria optimization of the precedence of operations and processes. This computer-aided processing of the planning problems enables the automatic minimization of set-up times for tools and facilitates the implementation of adaptation planning on the basis of already existing knowledge representations.

For the resolution of the specified planning problems, that take weighted target criteria into account, so-called reasoners and solvers are used in the context of this method. They are able to capture, solve and optimize planning problems that are present in a specific planning language. For the integration of the planning description and planning resolution, a separate assembly planning language (TOOL) was developed, which can be accessed via a web interface.

At the Center for Mechatronics and Automation Technology (ZeMA) in Saarbrücken, the method described here is currently being tested using the example of assembly processes from the aerospace and the automotive industry.

2 State of the Art

The planning of assembly systems differs in parts significantly from the planning of production systems. Due to its complexity, assembly planning has developed into an independent field of research. The description methods that are classed as classical systematics today have been developed by REFA, Bullinger, Konold & Reger and Lotter. They structure the planning process' entirety into six to eight phases [11–14]. Roughly speaking, planning begins by collecting the data required for planning. Subsequently, the planning task is specified and the assembly system basically planned. In a further step, the planning is detailed and worked out. Subsequently the system can be realized and introduced. The planning concludes with the system operation and the evaluation of the real planning results.

These systematics form the basis for the research-specific approaches that have emerged in recent years with a deeper focus on certain aspects of assembly planning. From the variety of existing approaches, some papers with particular relevance for the method presented in this paper will be described.

As part of his work, Weidner developed a concept for the knowledge-based planning and evaluation of assembly systems [15]. His concept "PEAS" (Planning and Evaluating Assembly Systems) is based on modular resources in order to calculate the required resource combinations against the background of the planning problem. In the course of a computer-assisted analysis, stochastic factors of influence are taken into account and a multi-criteria assessment procedure for the assessment of the planning result is presented.

Rudolf also researched the topic of "knowledge-based assembly planning" with another focus [16]. The description of the relationship knowledge between processes and resources depending on product parameters is of particular importance to him. He developed a concept by which these interdependencies between objects can be described. Dependencies described once can be used in future new or adaptation planning.

With the help of the main classes "product", "process" and "resource", Jonas described a method for the integrated computer-aided planning of assembly tasks [17]. These main classes are extended in his concept by connection information and object attributes in order to generate a data model, which contains all planning-relevant data. The streamlining of the planning process and the avoidance of redundant data generation is his concept's aim.

Ross develops a method for determining the degree of economic automation in the early phase of assembly planning [18]. First, Ross sets criteria that affect the automation decision. Through a utility analysis of the considered criteria, effort values are determined. By comparing the cost values with the threshold values of already implemented automation solutions, a decision can be made with regard to the automation capacity.

Beumelburg developed a system for the skill-based assembly planning of hybrid assembly systems [7]. She limits the focus to human-robot cooperation (HRC) scenarios. Beumelburg uses a catalog of criteria to determine the suitability of the actors human and robot for a specific operation. To evaluate the assignment alternatives, a genetic algorithm is used. In addition to suitability levels, the process time is also taken into account.

Kluge dealt with the development of a system for the skill-based planning of modular assembly systems [19]. He described a generic system of descriptions for resources and processes. On the basis of attributes, alternative combinations between processes and resources are determined, taking into account the experience of the planners. Subsequently, the scenario-based assessment of the alternatives follows.

A skill-based analysis of the required assembly modules, as described in the approach of Kluge, represents an efficient approach to the assignment of assembly resources [19]. However, Kluge's concept does not look at modules in which humans and robots stand in some sort of cooperative relation. Therefore, it is of relevant interest to include in the analysis the skills of people and those of robotic systems that are allowed to cooperate with humans.

Beumelburg started here and realized this missing link [7]. In her analysis, however, she confines herself to the statement whether a human worker or a robot better fulfills the requirements of an operation. In practice, a variety of different resources are available today, which make a review for the planner very complex. Furthermore, the valuation always takes place on the basis of a single operation to be considered. More operations, which follow each other directly, are grouped into group of operations and evaluated together. In practice, it also requires consideration of the dependency relationships of operations among each other that result from the choice of a particular resource. Required setup operations may, for example, influence that, in addition to a determined first action for one resource, it may also make sense to carry out another action, even though the second action, taken individually, would have been assigned to another resource.

The influences on assembly planning have been demonstrated by Ross [18]. Weidner and Rudolf developed possibilities to formalize planning knowledge and Jonas showed how planning problems can be made computer-processible [15–17]. From this an interest for the design of an own assembly planning specific domain language can be derived. This could describe computer-readable relations in the sense of an ontology and in this way represent the planning knowledge. To solve the planning problems described by the planner, moreover, efficient solution algorithms are needed that can solve multi-criteria problems as described by Beumelburg and Weidner [7, 15].

Building on the work described here, the method developed by ZeMA addresses the aforementioned challenges and is thus intended to contribute to the objectification of the planning of assembly systems.

3   Method and Software Implementation

In terms of classic planning systems, the method presented here is a further development of the rough planning and assessment phase. In particular, this section of the planning process was focused on the software implementation. The concept of the method also includes the provision of interfaces for the detailed planning phase and indicates further necessary planning steps for the generation of a final assessment basis. The method presented here is based on the following development goals:

- Planning knowledge should be formalized as semantic contexts in order to be processed computer-aided
- Adaptation planning based on existing knowledge representations and external ontologies should be supported
- Assignment options from resources to operations should be based on skill matching
- The representation of the hardware system components should be based on a modular design principle, which allows skills to be aggregated over several levels
- Criteria-based planning results that were determined by a computer should always be supplemented by the individual experience of the planner
- For the purposes of detailing or adaptation, a return of planning results should be possible
- Dependencies between operations should be considered when calculating the results
- The alternative combinations should be calculated against the background of the four target criteria (time, costs, process capability, quality of work) by means of a multi-criteria evaluation and optimization procedure and should provide for both weighting and extension of the target criteria
- The aim is to develop an assembly planning specific language that can be used to describe knowledge representation as well as evaluation and optimization

- The operation should be as simple and intuitive as possible for the planner with the help of a web browser.

The structure of the method presented here comprises eleven steps. It begins with the development of assembly scenarios for the assembly-product along the product life cycle. In this way, the output quantity of the assembly system can be determined. In the second step, all other planning-relevant data, such as the weighting of the target criteria as well as the boundary conditions for the assembly system are gathered. Then, the description of the planning problem is made on the basis of the product structure and the assembly order derived from this structure. With the help of this information, it is possible to gather the required operations, which can be aggregated to form processes. An assembly precedence graph is formed that structures the processes and operations in a predecessor-successor relationship.

All processes and operations must now be described with skills that are required for their execution. Likewise, the resources must be described with the skills and abilities offered as part of a skill matching to serve the needs of the processes and operations. Figure 1 shows that resources always consist of an actor and optionally of a task-specific tool.

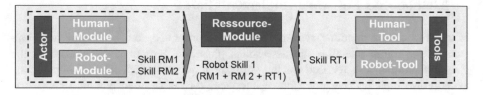

Fig. 1. Aggregation of skills of actors and tools to skills of resources.

The skills of the tool are attributed to the actor when used. This is followed by the skill matching, which provides an indication of which resource combinations are potentially eligible for the performance of an assembly activity. The result of the skill matching is included in a suitability assessment, in which the planner's practical knowledge is included in the planning.

For the assessment of suitability, the interest groups responsible for assembly determine a set of criteria in the context of a workshop, on the basis of which the planner then assesses the appropriateness of the assignment option. The suitability assessment criteria formulated in this context require the planner to assess the impact on the target criteria defined at the beginning of the planning. A basic set of suitable criteria has already been defined by Beumelburg [7].

In order to consider the cooperation of humans and robots in this method, evaluation criteria for safe cooperation were added. Depending on the type of cooperation (self-sufficient/co-existent, synchronized, cooperative, collaborative), the planner must evaluate appropriate criteria. In the later scheduling it is taken care, that only initially assessed forms of cooperation are actually generated. The assessment of suitability leads to a supplement of the parameters time, costs, process capability and quality of work. With regard to these target criteria, the planning result is also optimized.

In order to make the evaluation effort in the rough planning phase justifiable, a compromise between level of detail and feasibility had to be found. For this reason, only the parameters time (by comparing, estimating or assembling) and cost (based on the actual cost) are evaluated on cardinal scales. Process capability (through comparison, estimation) and ergonomics (through comparison, estimation or composition) are assessed ordinal scaled.

After this step, the planning problem (product, process, resource) is described in sufficient detail to be solved in the following step. So-called alternative combinations have to be generated taking into account the precedence graphs, which are subsequently evaluated and can be further detailed. The schedule is the solution of a flexible job shop problem (FJSSP) with multi-criteria evaluation. Since the complexity of these problems quickly exceeds the capacity of the human brain to solve problems, a computer-aided solution is an aspired option. For this purpose, possibilities for computer-readable description of planning problems were first identified.

AutomationML is a neutral, XML-based data format, based on which the CAEX module can be used to map topologies and structures [20]. Because CAEX does not support formal semantics for reasoning and querying of engineering knowledge, the W3C standardized Web Ontology Language (OWL) was chosen. This language allows the precise description of assembly planning domain relationships in ontologies that can be validated with appropriate OWL reasoners [21]. HermiT was selected for its scope of services, good documentation and well-maintained codebase. HermiT was selected from seven reasoners (including CEL, ELepHant, Fact++, Pellet), two of which (Konclude and MORe) support the OWL language scope. In contrast to Konclude, HermiT has the advantage of supporting the OWL API, which, among other things, makes it possible to use it in the comfortable "Protégé" editor [22]. For the time being, the use of the more complex MORe has been avoided to reduce the error potential, as it is based on both HermiT and ELK. The OWL API was used in version 3.4, as this version was supported by most reasoners.

For the scheduling of processes and operations taking into account the precedence graph as well as the optimization, methods of AI Planning were evaluated. Here are mainly STRIPS and PDDL to call. PDDL subsumes STRIPS, is well documented, complements numerous features such as numerical sizes and scheduling and was therefore favored. As a PDDL solver, LPG-TD was chosen because it provided the most advantages in terms of time and quality of solution finding compared to solvers such as MIPS-XXL or Optic, which were also tested [23]. The performance was only tested against TOOL domains. It was not searched for PDDL all-rounders.

For the LPG-TD solver to be able to solve the planning problem, a PDDL domain of the planning problem must first be available. With the TOOL API developed at ZeMA, a bridge was built between OWL and PDDL so that planning problems can be formulated in TOOL and the planning knowledge gets stored in OWL.

HermiT is used for skill matching because using a PDDL solver at this point would require a full specification of skills in PDDL. This is hardly possible in practice, or the possibility of continuously detailing the planning problem represents a significant advantage of the method. Based on the results of the HermiT Reasoner, a PDDL domain is then generated by TOOL and with the help of LPG-TD with regard to the target criteria solved.

TOOL is a Domain Specific Language and stands for "Tool Ontology and Optimization Language"; TOOL is implemented as a Kotlin library and interoperable with Java. A TOOL domain includes the following key aspects:

- **Module**, which consists of some humans and robots that may have tools
- **Action**, the super-type of assembly group and component
- **Skill**, skills can be required by actions and can be offered by modules
- **Assembly Item**, the common super-type of components and assembly groups
- **Action Condition**, models especially the order in which actions can be performed
- **Assessment**, assigns cost, duration, error-probability, process capability and quality of work to actions for a given module.

Since a large number of planning data are already digitally available in the digital factory through PLM systems such as Siemens Teamcenter or DS DELMIA, a dispatcher as middleware was developed above the TOOL backend, which supports the grpc protocol. To simplify the work with TOOL and the creation of planning domains, a python django-based web frontend, called "MoPlaTo" was developed to assist the planner. Here the planner can execute the steps of the method browser-supported and the planning result data is graphically given.

Since the planning result, especially with regard to possible HRC forms, does not include a geometric assessment, an task-based evaluation, for example with the aid of simulation tools such as IMK ema, is a good option. This step can also be used to further specify the ergonomic rating. It should also be noted that a risk assessment for the selected alternative combination must take place before implementation. The picture below shows the tool chain described here (Fig. 2).

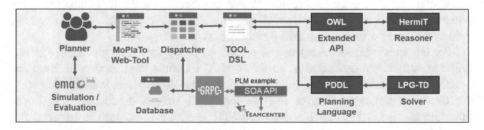

Fig. 2. The blue arrows represent the tool chain which is part of the method. Gray arrows are optional interfaces.

4 Results

For validation purposes, the assembly of an aircraft shell was planned. This takes place in two stages. First, an assembly of skinner, stringer and clips is assembled and is secondly completed with the frames. In Fig. 3, the user interface of MoPlaTo for the planner is shown. Section 1 (left half of the picture) shows the modeling of the product structure and section 2 (right half of the picture) shows the assembly processes derived from this product structure. These were supplemented in further steps by their operations and their skill requirements.

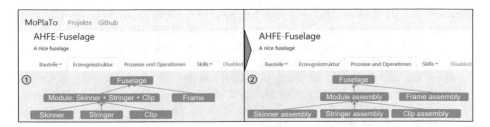

Fig. 3. Designing the precedence graph of a fuselage assembly with MoPlaTo

Each product is assembled by an adhesive bonding process. This consists of single processes (P), which can be subdivided into further operations (O):

Plasma activation of the surface (P) → Application of glue (P) → Joining of the component (P) → Crabbing (P) → Thermal inspection (P).

On the resource side, there are up to three UR10 robots that are able to work in the presence of humans. In addition, up to three people can be selected who have different physical profiles and skills. Both robots and humans have access to appropriate tools to handle the identified processes and operations. This information was entered into the ontology via MoPlaTo using TOOL. The classes defined for this are briefly presented below.

To represent the product structure, the classes *Component* and *AssemblyGroup* exist. These classes contain the attributes and methods of the objects they designate. In this way, the skinner is given a dimension, gripping points, surface properties and a weight, for example. In order to ensure uniform data processing, the class definition is based on that of common PLM systems. From these also the concept of blueprints was adopted. The *ComponentBlueprint* and *AssemblyGroupBlueprint* classes provide a definition of abstract objects from which individuals can be derived. This allows the same object concept (such as stringer) to be generated multiple times. Variants of objects are created as separate individual objects.

Skills are defined in the *Skill* class. The attributes and methods of *Skill* can also be inherited. In this way, for example, the ability to carry out a plasma activation can be linked with the requirement that a corresponding ozone extraction must then be provided. The description language specifies the specification of attributes as intervals. This allows the planner to first make the description roughly and later detail it.

There are also classes for processes and operations: *Process* and *Operation*. Objects of these classes can also be generated from blueprints or directly created individually. Both have their name, their skills, and their relationship to the components or assembly groups they assemble as properties.

The structure of the processes and operations is defined in the sense of the precedence graph with the local *Order* method. The predecessor and successor relationships can be derived from a tree structure with edges and nodes and are stored in the action conditions of the domain.

In addition to the actors and tools, resources also consist of basic, transport, supply and assistance modules. These resources were implemented by the class *Modules*.

Humans, robots and their tools are represented in their own sub-classes (e.g., *HumanWorker*, *RobotWorker*, *ToolType*) and aggregated into the *Module* class. Again, the concept of blueprints has been used so that e.g. the same type of tool, which differs only in its parameters, does not have to be recreated every time.

With this set of classes, attributes and methods the domain is basically described. Additional aspects can be added or changed at any time. The created ontology can already be displayed with the protégé editor. With the function *planOnto* the axioms of the domain are summarized in an ontology.

Now the skill matching can be processed. This is called with the method *SkillMatchingClassification*. This method invokes the required, offered and unassignable skills. If required and offered skills match, these relations are issued. Unknown skills are counted as non-resolvable skills and must be checked individually. The resulting operation resource combinations are now checked for suitability as described in the method.

The filtered and assessed result can then be fed to the *Assess* method. It builds objects that can be evaluated from the alternative assignment variants of the processes, operations, and resources, which also include the costs, the time, the process capability, and the work quality of the assignment variant.

The scheduling of resources against the background of the precedence graph is now taken over by the PDDL solver. A so-called optimization task is created with the function *optJob* in which, for example, it is defined under which boundary conditions (such as cycle times or target costs) the planning problem is to be solved and which target criterion is included in the solution finding with which weighting. The optimization task is translated by TOOL into a PDDL domain and then solved by the solver. As a result, an output file is generated and can be displayed by MoPlaTo as a Gantt chart. It is a good idea to check the geometric relationships on the basis of the result and to include any insights into a further assessment of suitability in order to detail the planning.

5   Discussion

By describing planning problems with descriptive logic and integrating the correlations in a computer-readable ontology, it was shown that planning problems can be solved automatically with suitable reasoners or solvers. If the planner enters the logical relationships correctly, planning decisions are made exactly on the basis of these logical relationships. The subjective influence of the planner, which grows when the planner can no longer objectively evaluate planning problems due to their complexity, is thereby substantially reduced. At the points where a subjective assessment still has to take place, this was made possible by the suitability assessment.

Since the description of planning problems in TOOL requires basic Kotlin programming knowledge and familiarization with the TOOL API, MoPlaTo was used to create a user interface that appeals to planners in the range of their daily methods and does not require a substantial training period.

The separation of knowledge representation in OWL and scheduling or optimization in PDDL has proven to be efficient. Especially due to the possibility of being able to modify the semantic description of the planning problem in the ontology at any time and to call the reasoner for further conclusions, there is a significant advantage over today's conventional planning practice.

The improvement of the evaluation method with regard to planning optimization is currently the focus of our research interest. Currently, only the variable costs of resource usage are minimized. However, it is important to place the resources in the context of the planned operating life of the assembly system, therefore the evaluation method is currently being extended appropriately.

Since this paper focuses on the method, a number of aspects of TOOL have not been described, or only briefly. Nevertheless, it has been shown that the use of TOOL helps to avoid design errors when transferring planning knowledge into an ontology. The possibilities of error compared to a direct design via Protégé Editor could be significantly reduced. The same is true for the design of the PDDL problem. If other methods, such as the consideration of geometric information, are taken into account, the planner must only be able to operate TOOL, which in turn has been translated into the web interface MoPlaTo with the aim of easy operation.

Acknowledgements. This paper was written in the framework of the research project IProGro 2, which is funded by the European Union (EU). The project is part of initiative "Investment in growth and jobs", which takes place within the structural fund for development (EFRE). It is supervised by the state of Saarland.

References

1. Müller, R., Vette, M., Quinders, S., M'Barek, T., Schneider, T., Loser, R.: Wandlungsfähiges Montagesystem für Großbauteile am Beispiel der Flugzeugstrukturmontage. In: Spath, D., Müller, R., Reinhart, G. (eds.) Zukunftsfähige Montagesysteme. Wirtschaftlich, wandlungsfähig und rekonfigurierbar, pp. 251–259. Fraunhofer-Verlag, Stuttgart (2013)
2. Westkämper, E.: Modulare Produkte – Modulare Montage. wt Werkstattstechnik Online **91** (8), 479–482 (2001)
3. Müller, R., Vette, M., Quinders, S.: Handhabung großer Bauteile zur Flugzeugmontage mittels eines Verbunds kinematischer Einheiten unterschiedlicher Struktur. In: VDI e.V. (eds.) Bewegungstechnik 2012: Koppelgetriebe, Kurvengetriebe und gesteuerte Antriebe im Maschinen-, Fahrzeug-, und Gerätebau. VDI Produkt- und Prozessgestaltung, VDI-Berichte, vol. 2175, pp. 3–18. VDI Verlag, Düsseldorf (2012)
4. Lay, G., Schirrmeister, E.: Sackgasse Hochautomatisierung? Praxis des Abbaus von Overengineering in der Produktion. Mitteilungen aus der Produktionsinnovationserhebung, no. 22, pp. 1–12 (2001)
5. Lotter, B.: Der Wirtschaftlichkeit angepasster Automatisierungsgrad. In: Deutscher Montagekongress, vol. 17, München, 10 October 2002
6. Andelfinger, V.P., Hänisch, T.: Industrie 4.0. Wie cyber-physische Systeme die Arbeitswelt verändern. Springer Gabler, Wiesbaden (2017)
7. Beumelburg, K.: Fähigkeitsorientierte Montageablaufplanung in der direkten Mensch-Roboter-Kooperation. In: Zugl.: Univ. Diss., Stuttgart. IPA-IAO-Forschung und Praxis, vol. 413. Jost-Jetter Verlag, Heimsheim (2005)

8. Thiemermann, S.: Direkte Mensch-Roboter-Kooperation in der Kleinteilemontage mit einem SCARA-Roboter. In: IPA-IAO Forschung und Praxis, vol. 411. Jost-Jetter, Heimsheim (2005)

9. Beumelburg, K., Spingler, J.C.: Automatisierungspotential-Analyse. Eine Methode zur technischen und wirtschaftlichen Klassifizierung von Automatisierungspotentialen. wt Werkstattstechnik online 92(3), 62–64 (2002)

10. Bracht, U., Geckler, D., Wenzel, S.: Digitale Fabrik. In: Methoden und praxisbeispiele. Morgan Kaufmann, Berlin (2017)

11. Lotter, B., Wiendahl, H.-P.: Montage in der industriellen Produktion. Ein Handbuch für die Praxis, 2nd edn. In: VDI-Buch. Springer, Heidelberg (2012)

12. Konold, P., Reger, H.: Praxis der Montagetechnik. Produktdesign, Planung, Systemgestaltung, 2nd edn. In: Vieweg Praxiswissen. Vieweg, Wiesbaden (2003)

13. Bullinger, H.-J., Ammer, D. (eds.): Systematische Montageplanung. In: Handbuch für die Praxis. Hanser, München (1986)

14. REFA - Verband für Arbeitsstudien und Betriebsorganisation: Planung und Gestaltung komplexer Produktionssysteme. In: Methodenlehre der Betriebsorganisation, REFA, Verband für Arbeitsstudien und Betriebsorganisation. Hanser, München (1987)

15. Weidner, R.S.: Wissensbasierte Planung und Beurteilung von Montagesystemen in der Luftfahrtindustrie. In: Zugl.: Univ. der Bundeswehr, Diss., Hamburg. Berichte aus dem Institut für Konstruktions- und Fertigungstechnik, vol. 32. Shaker, Aachen (2014)

16. Rudolf, H.: Wissensbasierte Montageplanung in der digitalen Fabrik am Beispiel der Automobilindustrie. In: Forschungsberichte IWB, vol. 204. Utz, München (2007)

17. Jonas, C.: Konzept einer durchgängigen, rechnergestützten Planung von Montageanlagen. In: Forschungsberichte/IWB, Bd. 145. Utz, München (2000)

18. Ross, P.: Bestimmung des wirtschaftlichen Automatisierungsgrades von Montageprozessen in der frühen Phase der Montageplanung. In: Zugl.: Techn. Univ., Diss., München. Forschungsberichte/IWB, vol. 170. Utz, München (2002)

19. Kluge, S.: Methodik zur fähigkeitsbasierten Planung modularer Montagesysteme. In: Zugl.: Univ., Diss., Stuttgart. IPA-IAO Forschungs und Praxis, vol. 510. Jost-Jetter, Heimsheim (2011)

20. Persson, J., Gallois, A., Björkelund, A., Hafdell, L., Haage, M., Malec, J., Nilsson, K., Nugues, P.: A knowledge integration framework for robotics. In: ISR/ROBOTIK 2010 (eds.) Proceedings for the joint conference of ISR 2010, (41st International Symposium on Robotics) und ROBOTIK 2010 (6th German Conference on Robotics). VDE Verlag, Berlin (2010)

21. Abele, L., Legat, C., Grimm, S., Müller, Andreas, W.: Cognitive and computational intelligence. In: IEEE (eds.) Proceeding of the 11th IEEE International Conference on Industrial Informatics, INDIN, Bochum, pp. 236–241. IEEE, Piscataway (2013)

22. Glimm, B., Horrocks, I., Motik, B., Stoilos, G., Wang, Z.: HermiT: an OWL 2 reasoner. J. Autom. Reason. 53, 245–269 (2014)

23. Gerevini, A., Saetti, A., Serina, I.: Planning through stochastic local search and temporal action graphs in LPG. J. Artif. Intell. Res. 20, 239–290 (2003)

DevOps for Containerized Applications

Adam S. Biener[✉] and Andrea C. Crawford[✉]

IBM, New York, USA
{biener,acm}@us.ibm.com

Abstract. The term DevOps is a combination of the words "development" and "operations", referring to a model of software development and delivery that is cross-functional, spanning from development to running it in production. Containerized applications require unique toolchain and pipeline processes for deployment that are different from conventional applications. This paper provides an overview of the tools, processes and anti-patterns for DevOps with containerized applications.

Keywords: DevOps · Docker · Kubernetes · Containers · Containerization Microservices · Software development lifecycle · Continuous integration Continuous delivery · Pattern

1 Introduction

Containerization of application components, using technology such as Docker, for instance, provides an alternative to running them on separate physical computers or virtual machines. Multiple containers may run on a single host sharing the operating system, or in a clustered environment, with each component consisting of application code and middleware.

Containers should not be seen as a complete replacement for virtual machines, but they can be a convenient way of packaging application components in a way that allows them to be easily deployed to multiple environments. Container deployments are congruent with more modern leading practices in application development, such as Twelve-Factor Apps[1], which rightfully advise externalizing deployment configurations, deploying stateless components, binding services to ports, easing startup and shutdown, and establishing parity between deployment environments.

As an example, consider a complex web application, depicted by Fig. 1, with the following elements:

1. A user interface, written as a collection of single-page applications (SPA).
2. A collection of web service APIs (e.g. microservices) called by the SPAs that get data from one or more back-end systems to display to the users.

Putting each SPA and each web service API in its own container makes each component a deployable unit which has the following benefits from a DevOps point of view:

---

[1] Twelve-Factor Apps. https://12factor.net/.

© Springer International Publishing AG, part of Springer Nature 2019
T. Z. Ahram (Ed.): AHFE 2018, AISC 787, pp. 35–44, 2019.
https://doi.org/10.1007/978-3-319-94229-2_4

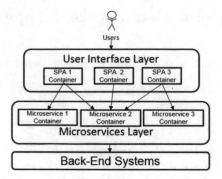

Fig. 1. A web application with a containerized architecture.

1. With a one-component-per-container scheme, component development can be spread across the development team.
2. Containerized components can be tested in isolation of other components.
3. The container image can be built just once and deployed in any environment, such as development, test and production. Environment-specific attributes could be specified at run-time by deployment automation tools when the container is started.
4. Modern container management systems can scale the application automatically by launching new instances or replicas of a container.
5. Automated testing, static code analysis, and security scanning can be part of the image build process. This shortens the testing cycle and improves security as more issues are found before code gets to the test environment.

2   Software Development Lifecycle (SDLC) Process Flows

This section presents a software development lifecycle (SDLC) process for containerized applications. While every project will have different requirements, the process presented here contains elements that address many common characteristics of containerized applications. In general, the SDLC follows a high level flow depicted in Fig. 2.

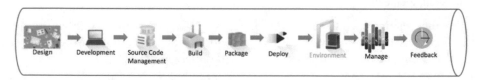

Fig. 2. The software development lifecycle (SDLC)

Once the application has gone through an initial design phase, development begins with source code updates stored in a source code management system. It is good practice to design APIs early in the design phase, and a well-defined API should be mocked and validated for service virtualization. Once the code is ready for deployment, a build process prepares it for packaging and deployment into a target environment. The lifecycle of APIs should also be considered in the SDLC, including creating, registering, and versioning in an API gateway. As the running application is used, it is continually evaluated by stakeholders. User and operational feedback is input back into the design process so that new features and fixes can be deployed iteratively over time.

1. **Feature Intake to Code Commit in a Development Environment**

The process begins with a stakeholder submitting a new requirement, such as a new feature request or problem found with the application, via an Issue/Requirements Tracking System. This system assigns the item to the developer based on what was submitted and sends a notification to the developer. The developer would review the item, perform development and testing tasks, and then commit the changes into the source code repository when complete. Figure 3 details this workflow.

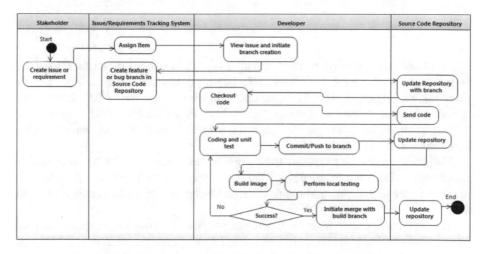

Fig. 3. Feature intake to code commit in a development environment

Once the code is committed to the repository, automation takes over with the build process depicted in the next Section.

2. **Build Process and Deployment to a Development Environment**

Figure 4 depicts the automated process for the build and deployment to the development environment. While the process is primarily triggered by the commit of code into the build branch of the source control system, it can also be triggered manually by a developer when necessary.

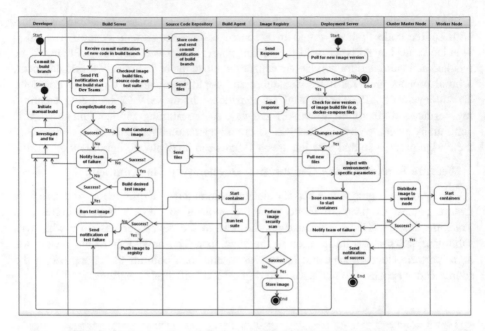

Fig. 4. Build process and deployment to a development environment.

The build process refers to the packaging of code and preparation for its deployment into the development environment. The details of the build process depend on the programming languages used. For instance, code compilation would be a key step for compiled languages such as Java or Go. A NodeJS application build may involve, among other tasks, running an `npm install` command to build module dependencies.

The build process for containerized applications includes building the container image (e.g. a Docker image) that includes the prepared and unit-tested code. This image, with the application and required dependencies, would comprise the deployable unit.

The build agent provides an environment with the container engine installed, separate from the build server, to run automated tests against the built image. Such tests are separate from the unit tests done after code compilation but before the image is built. For instance, with a Java-based microservice, JUnit tests would be run after code compilation to perform unit tests. Once the image is built, the build agent would start the newly built microservice container and test web service invocations would be done against that as a functional or performance test.

This task could theoretically be done on the build server itself, but we recommend use of a separate build agent server to ensure the build server's capacity is not strained due to many tests which may run simultaneously in a complex environment.

Once the image passes all automated test cases, it would be stored in an image registry, staging it for deployment to various target environments.

3. Deployment to Higher Environments

The term *higher environment* refers to any target server deployment environment beyond the development environment depicted in the previous section. While every organization defines environments differently, test and production environments are typically the minimum set that would comprise the higher environments. The "highest" environment would be the production environment.

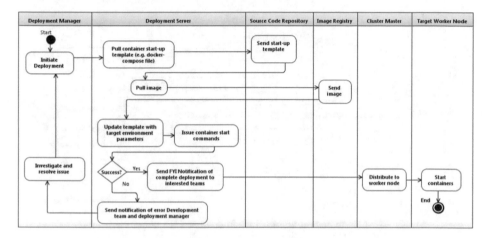

Fig. 5. Deployment process into higher environments.

The process depicted in Fig. 5 shows the deployments as a fairly lightweight process that merely involves starting already-tested containers in the target environment[2]. Any differences between the environments would be accounted for by specifying parameters (e.g. host names, gateway information) at run time and deploying between environments should not require the images to be rebuilt.

The deployment is often triggered manually by someone with appropriate authorization, depicted as the Deployment Manager in Fig. 5. While it is possible to automate deployments through higher environments to production, we recommend caution as automated deployments add an element of risk. Depending on the tools used, it is possible to selectively automate deployments for certain types of low-risk changes.

3 Toolchain Architecture

Figure 6 depicts the SDLC workflow and the flow of deployable artifacts in the context of the installed tools that support the process described in the *SDLC Process Flows* section. Artifacts include code and other assets that would be stored in a source code management system, and would become built into a container image that would be deployed into the target environments.

---

[2] This follows the 10th tenant of Twelve-Factor Apps "Dev/prod parity". https://12factor.net/.

The Developers and Stakeholders are the primary end users of the DevOps tool-chain, with developers handling the requests submitted by the stakeholders, and invoking and defining the build automation. The Tools Support Team would be responsible for the DevOps toolchain installation and upgrades. The Site Reliability Engineer is responsible for the overall operational health of the application in the target environments and would work with developers to ensure reliability is built into the application architecture and infrastructure.

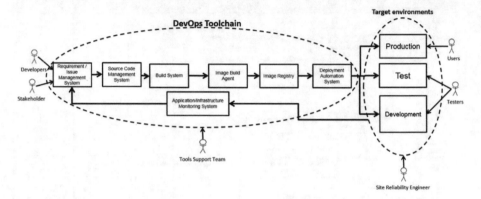

Fig. 6. DevOps toolchain architecture in context with target environments.

Table 1 lists some example tools that would fill the toolchain roles depicted above. We list some common tools at the time of writing.

Table 1. DevOps tool roles and example products.

Tool role	Example products
Requirement/issue management system	Atlassian JIRA IBM rational team concert Github issue tracking Gitlab issue tracking
Build system	Jenkins Atlassian bamboo Gitlab continuous integration IBM Rational BuildForge
Source code management system	Git Subversion CVS IBM jazz source control
Image build agent	A virtual machine running the docker engine
Image registry	Docker trusted registry JFrog artifactory
Deployment automation system	IBM UrbanCode deploy Gitlab continuous delivery Atlassian bamboo Jenkins
Application/infrastructure monitoring	ELK Splunk

4 Do this, Not that: Common Pitfalls

1. "Reinventing the Wheel (Monolith)"

The definition of "application" is becoming looser in its interpretation. Applications are typically collections of functions that serve a common business purpose. Many applications can be broken down into services that can be reused for other applications. Some examples might be an authentication or authorization set of functions. Common functions can be grouped into components that have common service characteristics. If designed and implemented thoughtfully, components lend themselves to being mapped to containers. Taking an entire application and implementing it as one container could very well be undermining the advantages of component-based design.

Since componentization is a key benefit of using containers, putting everything into one monolithic container defeats the purpose. Loose coupling among components can be achieved with a proper containerization strategy[3].

Consider: Spending thoughtful effort in examining an application and decomposing it into domains, components, services and APIs.

Stateless components[4], like service APIs and web user interfaces with loose coupling to other services or data stores, are the best candidates for containerization since those are the most portable between environments. Consider refactoring a monolithic application into stateless components to gain the benefits of containerization.

Benefits:

- Time well spent in this stage may yield benefits by identifying components that can be reused for other applications.
- Agility in performing, testing and deploying updates to components can be achieved greater than a monolithic application (multi-speed IT).
- Services mapped to containers can be managed individually, allowing granular control of container characteristics.

2. Trying to Containerize Everything

Not every component of an application is appropriate for containerization.

For instance, a web application may include an application server as part of an N-tier architecture, serving the front-end. This front-end may not be conducive to containerization due to the middleware being used with a tight coupling to data stores, where a dedicated virtual machine may be more appropriate. If a monolithic API is being used to provide data to this front-end instead, consider containerizing just the API or refactoring to a microservices architecture that can be containerized.

Consider: Applications may have a heterogeneous mixture of container and non-containerized components and services. There may be different DevOps tools and processes depending on uniqueness of the application's component diversity. The good

---

[3] This follows the 7th tenant of Twelve-Factor Apps "Port binding". https://12factor.net/.

[4] This follows the 6th tenant of Twelve-Factor Apps for stateless "Processes". https://12factor.net/.

news is many of the same automation tools and slight modifications to the SDLC process can automate delivery for non-container components of an application.
Benefits:

- Flexibility to use containerization, where it makes sense.
- Leverage some of the same DevOps tools for non-container deployments of components and services.

3. Turning a Blind-Eye to Container Management

Developing a containerization strategy not only involves deciding which application components should be containerized, but also how containers will be managed at scale and across the organization's infrastructure. Not using build-in container engine functions for deployment orchestration, communication and service discovery can lead to re-inventing these functions outside of what already exists. Orchestration refers to how multiple containers are created across the infrastructure, communicate with each other, updated and made highly available.

Consider: An orchestration solution designed for containerized applications should be selected for this role, such as Kubernetes or use of the orchestration features of Docker Enterprise. Early adopters, may find that using API service discovery products that pre-date container orchestration solutions is a common pitfall, as these products are not meant for a containerized environment. They may operate outside the containerized environment's network, making service discovery difficult and unreliable.
Benefits:

- Container management with Docker Enterprise provides built-in service discovery, clustering (with swarm mode), traffic routing and load balancing.
- Cluster and container management with Kubernetes provides many of the same benefits[5], such as: persistent volume management, application health check framework, auto-scaling, naming and service discovery and load balancing.
- Many PaaS platforms provide wrappers of container and cluster management implementations that provide command line or GUI interfaces to make orchestration and management easy.

4. Not Scanning Images for Security Vulnerabilities

With news of high-profile information security breaches becoming more common, it is imperative that organizations adopt robust, multi-faceted security strategies. Deployment of containers should include a security scanning step that should fail the image build and deployment if the scan fails. Introducing Docker images into a solution

---

[5] Full list of Kubernetes features listed at "Why do I need Kubernetes and what can it do?". https://kubernetes.io/docs/concepts/overview/what-is-kubernetes/#why-do-i-need-kubernetes-and-what-can-it-do.

stack is another opportunity for introducing vulnerability and should be another opportunity to mitigate risk through image scanning.

Consider: Modern image registry products would include security scanning functions, either built-in, or as an add-on, that check for vulnerabilities in all layers of the image. Docker-based examples include JFrog's XRay (which integrates with the Docker Registry function of JFrog Artifactory) product and the security scanning functions of Docker Trusted Registry. These products scan for vulnerabilities against the Common Vulnerabilities and Exposures (CVE) database.

Benefits:

- Mitigate the risk of introducing image vulnerabilities using security tools.
- Higher degree of confidence that the right image is being used.

5. Always Using the Latest Image from a Public Registry

Simply using the latest version of an image as a base image can introduce unexpected or undesirable effects, break functionality or introduce security vulnerabilities.

Consider a `Dockerfile` with this line, using the latest NodeJS image from the public Docker registry:

```
FROM node:latest
```

It is not possible to tell which version of the image is really being used because it depends on which version is the latest at the time the image is pulled from the registry. Developers within an organization will likely be using different versions of the base image depending on timing, causing inconsistencies across the applications that use it.

Consider: Using an explicit version of the base image that has been tested and qualified will ensure consistency across the environment.

Our Dockerfile could be changed to specify the version of the NodeJS base image:

```
FROM node:8.7.0-alpine
```

This line specifies the 8.7.0 version of NodeJS, running on Alpine Linux, a minimalistic distribution of Linux. Minimalistic images are not only more resource-efficient, but improve security with less attack surface area.

Benefits:

- Opportunity to limit Docker images to a useful subset for the enterprise.
- Consistency of base image versions across enterprise.
- Controlled base image upgrades across the enterprise.

6. Rebuilding the Image for Each Deployment to Higher Environments

A common pitfall is rebuilding images with environment-specific characteristics prior to deployment.

Since containers are meant to be easily transferrable between environments, rebuilding the image defeats the purpose of containerization. Use an image registry to store built images, and simply pull the built images from the registry to the target environment and run them from there. Environment-specific variables should be specified at run time.

Consider: Stage built and tested images in a private registry for deployment, and use continuous deployment tools to inject environment-specific parameters at run time during deployment.

In a Docker environment, the sequence of events may follow this pattern:

1. Continuous Integration tools invoke the image build and run the automated tests against them.
2. Successfully built and tested images are pushed into the private registry.
3. Continuous Deployment tools set environment variables that are passed to the docker-compose.yml file used to start the containers using images pulled from the private registry.

Benefits:

- The same image that is successfully tested in the build process is the same one that gets deployed in all environments.
- Deployment is simplified because the activity mainly involves starting containers in different environments. A simplified process means fewer deployment issues.

References

1. Docker. https://www.docker.com/
2. Kubernetes. https://kubernetes.io/
3. Common Vulnerabilities and Exposures. https://cve.mitre.org/
4. JFrog Xray. https://www.jfrog.com/xray/
5. The Twelve-Factor App. https://12factor.net/

Modelling of Polymorphic User Interfaces at the Appropriate Level of Abstraction

Daniel Ziegler$^{(\boxtimes)}$ and Matthias Peissner

Fraunhofer IAO,
Nobelstr. 12, 70569 Stuttgart, Germany
{daniel.ziegler,matthias.peissner}@iao.fraunhofer.de

Abstract. Polymorphic user interfaces (UIs) can offer different modes of display and interaction for different devices, situations and user needs. This increased variety adds complexity to UI development, which is often addressed by model-based UI development approaches. However, existing approaches do not offer an attractive balance of required abstraction and a graphical and vivid representation for developers. In this paper, we present the Model-with-Example approach that combines abstract interaction modelling with a wireflow-like concrete visualization. The results of a user study with industrial front-end developers show that this concrete visualization can improve development efficiency.

Keywords: Model-based user interface development · Development tool
Adaptive user interface · Context-aware user interface · Personalization
Abstract interaction model · Market adoption

1 Introduction

For obvious reasons, a one-size-fits-all approach to user interface (UI) design cannot provide an optimum usability and user experience for a broad variety of users and usage situations. Thus, the concept of polymorphic UIs allows to provide multiple context-specific UI alternatives [1]. Taking into account diverse aspects of the context of use including characteristics of users, tasks to be performed, used equipment as well as the physical and social environment [2] results in very complex design and development processes for polymorphic UIs.

Model-based UI development (MBUID) is a widely known approach to reduce development efforts while maintaining the flexibility to produce diverse UIs [3]. Instead of directly implementing all UI variants by themselves, developers specify models of the required interaction on more abstract levels. Based on those models the MBUID system is able to generate multiple UIs for specific contexts of use. Over years, many MBUID systems have been created for different application domains and focusing on different aspects of the context of use.

While MBUID decreases development efforts when being used, it also introduces a new hurdle for developers. Because of the abstract nature of the required models, "it is difficult to understand and control how the specifications are connected with the final UI. Therefore, the results may be unpredictable" [4]. To overcome this issue,

© Springer International Publishing AG, part of Springer Nature 2019
T. Z. Ahram (Ed.): AHFE 2018, AISC 787, pp. 45–56, 2019.
https://doi.org/10.1007/978-3-319-94229-2_5

Akiki et al. "consider tool support to be crucial for the adoption of adaptive model-driven UI development by the software industry" [5]. The tools of existing MBUID systems do not seem to be sufficient to eliminate this barrier to entry. Until today, there is only limited adoption of MBUID systems for polymorphic UIs in the market of professional software development [6].

In this paper, we first analyze existing MBUID systems based on the level of abstraction that developers predominantly work on. We then describe the concepts, meta model and default graphical syntax of the Abstract Application Interaction Model (AAIM), which our new modelling approach is based on. Next, we present the Model-with-Example approach and describe how the related modelling environment advances over the current state of the art. Finally, we present the results of a user study with industrial front-end developers.

2 Existing Modelling Approaches

The CAMELEON reference framework [7] has been established to set a common structure and language for the description of MBUID systems. It defines four levels of abstraction: the UI users finally interact with is called Final User Interface (FUI), the Concrete User Interface (CUI) abstracts from the specific implementation platform, the Abstract User Interface (AUI) correspondingly is independent of specific interaction modalities, and task & domain models describe processes and concepts relevant for an application completely abstracted from human-computer interaction.

While the direct implementation of an FUI describes the traditional UI development approach, existing MBUID approaches can be categorized by the predominant level of abstraction from which developers start to create models.

2.1 Starting from Concrete User Interfaces

As CUIs abstract from technical implementation platforms, the creation of UIs based on those models is a common approach for multi-platform UI design and development. Model editors for CUIs often use a wireframe-like visualization and work like graphical UI (GUI) builders incorporated in many modern integrated development environments (IDE). Developers use common UI elements like text boxes and buttons and place them into different types of containers.

Damask [8], for example, is a prototyping tool for cross-device UIs based on sketches. It addresses the design of different UIs for diverse screen sizes and interaction techniques by providing a separate layer for each of their relevant combinations. It provides an extensible set of cross-device design patterns for devices like desktop computers and smartphones to support designers. Adding one of these patterns to a specific device layer will add the corresponding patterns to all other layers. However, designers are allowed to edit individual elements, potentially changing the original purpose of the pattern instance and thus breaking the consistency between the layers.

Another concept is used by Gummy [9], a multi-platform GUI builder that is able to generate FUIs using pluggable renderers. Unlike Damask, Gummy adapts its whole GUI-Builder-like workspace according to a specifically selected target platform. When switching to another platform, Gummy uses a rule-based mechanism to transform the existing CUI into an initial design for this new platform.

In both cases, the GUI-Builder-like approach results in a close connection between the created CUI models and the FUIs (potentially) being presented to users at runtime. On the other side, developers are still responsible to ensure consistency between all UI variants since pattern instances might have been changed or transformation rules might be incomplete.

2.2   Starting from Task Structures

Systems for the creation of context-aware UIs strive to enable a wide variety of differing UIs aligned to diverse contexts of use. To allow the adaptation of all aspects of UIs these systems typically incorporate all CAMELEON levels of abstraction. Commonly the notation of ConcurTaskTrees [10] is used to describe task structures as the most abstract modelling level.

While UsiComp [11] generally is designed to allow transformations between all levels of abstraction only forward transformations into more concrete levels have been realized for now. Thus, to use the adaptation capabilities of all transformation steps, developers have to start on the most abstract task & domain level. Additionally, it does not include visual editors for CUIs itself but allows to assign wireframes created with an external tool to certain nodes of the task model.

Quill [12] addresses this limitations by requiring all transformations to be applicable in both directions. This generally allows for middle-out modelling approaches starting at any level of abstraction. Nevertheless, it still requires developers to work with models on all levels, thus resulting in a high complexity of the development process.

2.3   Starting from Abstract Interactions

The previously described categories represent the two predominant modelling approaches in current development systems for polymorphic UIs [13]. Some of those systems claim to enable developers to start the development process from a model on any level of abstraction but in turn require them to check and potentially edit models on all levels of abstraction [11, 12].

In contrast, the MyUI system for accessible adaptive UIs [14] has explicitly been designed to require only one single application specific AUI model to be manually created by developers. This Abstract Application Interaction Model (AAIM) fulfils four major roles. It is the interface between developers and the adaptive UI system defining all possible interactions between the user and the application. It serves as basis of the UI generation and adaptation since it contains the aspects common to all UI variants that may be generated. Additionally, it wires together UI events and implemented application specific functions that may set variables, manipulate the applications data or call external services like sending an e-mail. Finally, it enables UI controls to access application data stored in databases or other storage systems.

3   The Abstract Application Interaction Model

3.1   Abstract Syntax: Concepts and Relations in the AAIM

The abstract syntax of the AAIM is based on behavioral state machines as defined in the Unified Modeling Language (UML) Superstructure [15] which themselves are based on statecharts described by Harel [16]. Statecharts provide the opportunity to model an UI without any references to interaction modality or concrete interaction elements. Because of that statecharts have been broadly used for specification of UI behavior [17, 18]. As a part of the UML, their general concepts and notations are widely known and used in software engineering practice.

The abstract syntax specifies the concepts used in the AAIM and their relations using UML class diagrams in three conceptual groups as follows:

Interaction Situations (see Fig. 1). A key concept of the AAIM is the usage of interaction situations (IS). They represent interaction possibilities the system offers to the user at a certain point during interaction to fulfil a certain purpose. Interaction possibilities include all information presented to or requested from the user as well as all actions the user may perform. At runtime the MyUI system selects the concrete realization of the specified IS (called interaction pattern) to be presented to the user that fits best to the user profile and device capabilities. The definition of concrete interaction situation instances is not part of the AAIM modelling itself but preconfigured in the MyUI development toolkit.

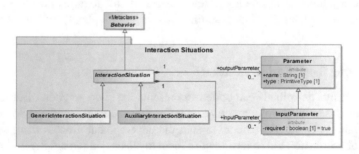

Fig. 1. Meta model of interaction situations in the AAIM based on the UML metaclass *Behavior*.

In UML state machines, the metaclass *Behavior* is used to represent what happens while a certain state is active. Accordingly, in the AAIM the abstract class *InteractionSituation* represents what will be presented to the user while the application is in a certain state. Therefore it inherits from *Behavior* which itself is a subclass of *NamedElement*. The attribute *inputParameter* defines the parameters that can be provided to the interaction situation. Conversely, the attribute *outputParameter* defines the parameters that constitute the result of the interaction situation.

Two concrete subclasses of *InteractionSituation* represent two different types of interaction situations. The class *GenericInteractionSituation* represents interaction situations that can form the main interaction purpose of a certain application state. Complementary the class *AuxiliaryInteractionSituation* is used for additional interaction options that can optionally be used together with any *GenericInteractionSituation*.

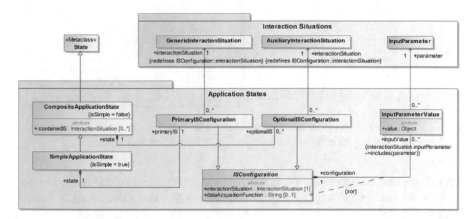

Fig. 2. Meta model of application states and their configuration in the AAIM based on the UML metaclass *State* including its relations to the concepts of interaction situations.

Application States (Fig. 2). Application states build up the structure of an interactive application defined by an ΛΛIM. Each state represents a possible point of the user's interaction sequence with the application. The AAIM defines the active interaction situations for each state and therefore the interaction possibilities of that state.

Classes representing application states are modelled as subclasses of the UML metaclass *State*. Therefore, the main structural aspects like being a *NamedElement*, the relation through transitions and the possibility to build composite structures by nesting states into other states are inherited from UML.

There are two classes for application states in the AAIM meta model. The class *SimpleApplicationState* represents distinct elementary states that do not contain any states or submachine. The main purpose of *SimpleApplicationStates* is to define the primary interaction situation of the state via a *PrimaryISConfiguration*. This class holds a reference to a *GenericInteractionSituation* together with the definition of the input parameter values handed over to the interaction situation when executed or a data acquisition function providing the input parameters as return value.

Complementary to *SimpleApplicationState* the class *CompositeApplicationState* does contain nested states or a submachine. Composite states do not contain primary interaction situations themselves. However, they refer to *AuxiliaryInteractionSituations* via *OptionalISConfiguration* defining their parameter values the same way *PrimaryISConfiguration* does for the primary interaction situation. Optional interaction situations defined in a composite state are interpreted as if defined in every contained simple state.

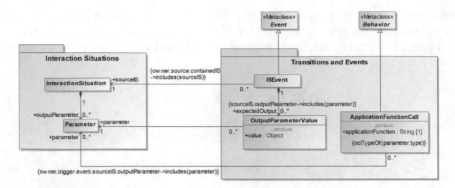

Fig. 3. Meta model of transition and events in the AAIM based on the UML metaclasses *Event* and *Behavior* including its relations to the concepts of interaction situations.

Transitions and Events (Fig. 3). Just like in statecharts in an AAIM transitions from one state to another are triggered by events. In AAIMs these events result from the interaction situations, e.g. when the user performs an action or provides an input. Therefore, the class *ISEvent* is a subclass of the UML metaclass *Event*. Its *sourceIS* has to be a primary or optional interaction situation of the state the transition originates from. The expected output parameter values are defined by instances of the class *OutputParameterValue* and may only refer to parameters defined as output parameters of respective *sourceIS*. The transition will only be triggered if the actual values match the specified ones. This way it is possible to trigger different transitions depending on user input, e.g. a command typed into a command line or different options of a payment form.

To perform additional application functions when a transition is executed the AAIM meta model specifies the class *ApplicationFunctionCall*. It is a subclass of the UML metaclass *Behavior* and refers to a function by its name. Output parameters of the *sourceIS* can be passed to the function for further processing.

3.2 Graphical Editor in the MyUI Development Toolkit

Despite the definition of the concepts and relations contained in the AAIM as described above, MyUI also provides a development toolkit. It incorporates a graphical model editor for AAIMs as well as preview function.

The model editor uses a visualization similar to usual UML state machine diagrams (see Fig. 4). Boxes are used to represent application states, while arrows between these boxes stand for available transitions. The interaction situations of each state as well as the trigger events are defined by textual expressions inside the boxes respectivly near to the arrows.

While the states can be placed on the modelling canvas using drag-and-drop, the MyUI model editor makes heavy use of context menus to add and manipulate AAIM elements and popup dialogs for data entry. Interaction situations can be assigned to application states by drag & drop from a library. Transitions, however, have to be added via the context menu of the source state.

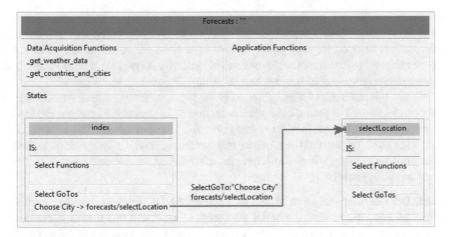

Fig. 4. Example of a simple AAIM of a weather forecast application as presented in the AAIM editor of the MyUI development toolkit [19]. The model is shown in an intermediate state during editing and is therefore incomplete.

To give the developer an impression of what the resulting user interfaces look like the MyUI development toolkit provides a preview function following the "What You See Is What Others Get" (WYSIWOG) principle [14]. The developer can run the AAIM and directly manipulate the user profile variables to see the individual resulting UI.

To be able to use that function, the AAIM has to be complete including data acquisition functions supplying data. In addition it requires a running installation of the MyUI runtime infrastructure.

4 The Model-with-Example Approach

In the analysis of existing modelling approaches described before has identified advantages and limitations of related systems for the development of polymorphic UIs. Based on these insights, in the following we define three design goals for our new Model-with-Example (MwX) approach that avoid the limitations and emphasize the advantages. Thereafter, we present the concept of a model editor following this approach.

4.1 Design Goals

Prevent Interpretation in Transformations. The CAMELEON reference framework generally allows for transformations between its levels of abstraction in both directions [7]. The forward transformation from a high-level abstraction to more concrete models is called "reification". Vice versa, the reverse transformation from concrete models to more abstract ones is referred to as "abstraction".

While the generation of FUIs from abstract models through reification has been often demonstrated, the automatic abstraction still seems to be an open issue [11, 12].

One reason for this problem is that one specific configuration in a concrete model might be the result of different abstract models. For example, in voice menus entering a number might be used for navigation options as well as for confirmation of an action. In consequence, reverse transformations might need to interpret what purpose the initial designer might have had in mind when creating the concrete input model.

Besides this technical issue, potentially ambiguous interpretations also limit the ability of developers to gain a clear understanding of the relations between the models. In consequence, such systems do not support the transparency of the UI generation mechanism [20]. To prevent the need for interpretation, MyUI's generation mechanism by design is only using reification. Thus, its AAIM is an appropriate base to start from for our MwX approach.

Reduce the Distance Between Model and Resulting UI. GUI-Builder-like modelling systems address the preference of UI designers to start and work with concrete representations [8, 9]. Developers create UI models in a "What You See Is What You Get" fashion with representations that directly provide an impression of the final results and thus match developers' mental models of UIs. In contrast, task-based approaches force developers to work with models that do not only abstract from concrete UI elements but also from a certain interaction modality.

To shorten the learning curve towards MBUID for polymorphic UIs [20], the distance between the abstraction level the developer is working on and the FUI presented to users at runtime should be as small as possible. Therefore, our MwX approach sets the focus on a CUI visualization while modelling.

Provide Instant Feedback. The WYSIWOG feature of the MyUI development toolkit allows running a preview of their current AAIM to give developers an idea of the FUIs potentially being presented to users. This results in the need for frequent context switches for developers between modelling and preview. Hence, the feedback regarding the resulting UI is delayed during the development process. Additionally, when using the preview developers have to navigate from the initial state to the point in the interaction flow they are currently working on.

In order to improve the efficiency of the development process [20], MwX provides an impression of the resulting UIs immediately while modelling. Receiving feedback for all states at once instead of just single states, developers will gain an overview over the whole model more easily. This will further support the systems transparency and shorten the learning curve for developers.

4.2   Model Editor

MwX is a new approach on how to work with AUI models in the model-based development of polymorphic UIs. Therefore, in the first place it does neither define a new AUI model language nor extend the concepts of an existing one. Instead, the model editor is built upon the AAIM meta model as presented before.

Instead of using the abstract boxes-and-arrows visualization known from the MyUI development toolkit, the MwX editor renders a specific wireframe. This wireframe represents a preview of one of the possible CUIs for the interaction situation modeled in the respective state. This results in a wireflow-like visualization of the model as depicted in Fig. 5. In addition to the modelling canvas the editor provides a palette containing the tools for creating states and transitions as well as the available inter-action situations. Interaction situations can be assigned to application states by dragging them from the palette.

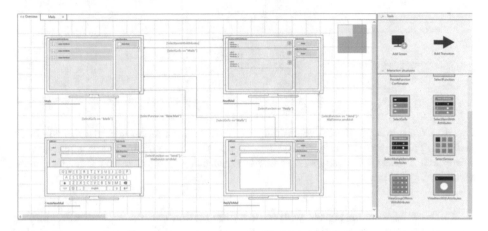

Fig. 5. Overview of the MwX editor showing the visualization for the case of smart TV as the selected device profile. Application states are represented by wireframes of the defined interaction situation according to the selected profiles [21].

In fact, the concept of interaction situations together with pattern-based nature of the MyUI system provide a consistent base for rendering specific wireframes. When the developer changes the user preferences by choosing another predefined persona or fine-tuning individual preferences in the model editor, all wireframes presented on the modelling canvas as well as the palette icons are adapted accordingly. The mechanism also applies when the developer selects a different device profile like a smart TV or a specific smartphone.

5   Evaluation Study with Developers

To evaluate the suitability of the MwX approach and the concept of the related model editor, we carried out a user study with front-end developers. In this study, we especially addressed the questions if the MwX's combination of abstract model concepts and concrete presentation is perceived to be the appropriate level of abstraction for the model-based development of polymorphic UIs.

In total, eight front-end developers from companies located in three European countries (Germany, Poland, and Denmark) volunteered to participate in one of three group sessions. In their daily work, building UIs has a share of 10% to 80% (Mean $M = 41.25$, Standard Deviation $SD = 24.16$) with only one of them having a dedicated User Experience design role. On a five point scale their self-rating of knowledge regarding the UML covered the complete range ($M = 2.63$, $SD = 1.41$).

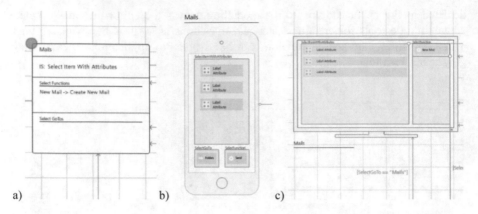

Fig. 6. Comparison of different application state visualizations: (a) abstract visualization in the first prototype similar to the original MyUI notation; (b) concrete visualization as wireframe for a smartphone in the MwX prototype; (c) concrete visualization as wireframe for a smart TV in the MwX prototype [21].

Each of the sessions started with a short introduction into the topic of polymorphic UIs and MBUID. Then, ordered by random, two animated prototypes demonstrating the creation of the AAIM for a simple e-mail application. One prototype was based on the original abstract boxes-and-arrows notation used in the MyUI development toolkit (see Fig. 6a). The other one demonstrated the same modelling task based on the MwX approach using device specific wireframes for smartphones and smart TVs (see Fig. 6b and c).

After each presentation of a prototype, the participants were asked to indicate their level of agreement regarding the following four items on a five-point Likert scale:

- *Impression of resulting UI:* "The tool gives an impression of how the user interfaces generated at runtime might look."
- *Clarity of abstract character:* "It remains clear at all time that at runtime different user interfaces can be generated out of the initially created AAIM."
- *Mental effort:* "Creating an AAIM with this tool requires a high mental effort."
- *Appropriate representation:* "The representation of the AAIM is appropriate for doing the modelling."

When comparing the quantitative results for both prototypes the mean value is higher and therefore better[1] rated for the MwX prototype as depicted in Fig. 7. The highest difference can be observed for the impression of the resulting UI ($M = 1.75$, $SD = 1.49$), followed by the clarity of the abstract character ($M = 0.88, SD = 1.55$) and the mental effort ($M = 0.75$, $SD = 1.28$). Finally, in case of the appropriate representation, there only is a small difference visible ($M = 2.63$, $SD = 1.06$).

---

[1] The scale for the item "Mental effort" has been reversed before further calculations to match the principle of higher values representing a positive tendency of the other items.

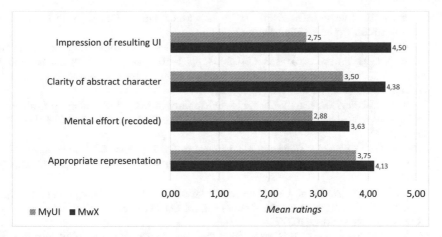

Fig. 7. Quantitative comparison of the mean ratings for four items related to the level of abstraction of both presented modelling approaches. Note that the values for mental effort have been reversed to match the principle of higher values representing a positive tendency.

In total over all four items, only one participant preferred the abstract notation of the MyUI-oriented notation to the concrete visualization of the MwX prototype. Interestingly this participant also had wide knowledge of UML.

6   Conclusion and Outlook

The results of the developer study show that the Model-with-Example approach is able to provide developers a modelling experience at the appropriate level of abstraction. It has a major impact on the transparency of the UI generation mechanisms by making the variety of generated FUIs more predictable.

However, the concept of the MwX model editor has been limited to simple application states and GUI representations. Further research will be required on how to deal with composite states and representations of non-graphical interaction modalities like voice interfaces. Additionally, it might be interesting to transfer the Model-with-Example approach to other AUI models than the AAIM.

Acknowledgments. The research leading to these results has received funding from the European Union's Seventh Framework Program under grant agreements no. 248606, "MyUI" and no. 610510, "Prosperity4All". The opinions herein are those of the authors and not necessarily those of the funding agency.

References

1. Savidis, A., Stephanidis, C.: Unified user interface design: designing universally accessible interactions. Interact. Comput. **16**, 243–270 (2004)
2. International Organization for Standardization: ISO 9241: Ergonomic requirements for office work with visual display terminals (VDTs) – Part 11: Guidance on usability (1998). https://www.iso.org/standard/16883.html

3. Hußmann, H., Meixner, G., Zühlke, D. (eds.): Model-Driven Development of Advanced User Interfaces. Springer, Heidelberg (2011)
4. Vanderdonckt, J.: Model-driven engineering of user interfaces: promises, successes, failures, and challenges. In: Proceedings RoCHI 2008, pp. 1–10. Matrix ROM, Bucharest (2008)
5. Akiki, P.A., Bandara, A.K., Yu, Y.: Adaptive model-driven user interface development systems. ACM Comput. Surv. **47**, 1–33 (2014)
6. Céret, E., Calvary, G., Dupuy-Chessa, S.: Flexibility in MDE for scaling up from simple applications to real case studies. In: Proceedings IHM 2013, pp. 33–42. ACM Press, New York (2013)
7. Calvary, G., Coutaz, J., Thevenin, D., Limbourg, Q., Bouillon, L., Vanderdonckt, J.: A unifying reference framework for multi-target user interfaces. Interact. Comput. **15**, 289–308 (2003)
8. Lin, J., Landay, J.A.: Employing patterns and layers for early-stage design and prototyping of cross-device user interfaces. In: Czerwinski, M. (ed.) Proceedings CHI 2008, pp. 1313–1322. ACM, New York (2008)
9. Meskens, J., Vermeulen, J., Luyten, K., Coninx, K.: Gummy for multi-platform user interface designs. In: Levialdi, S. (ed.) Proceedings AVI 2008, p. 233. ACM, New York (2008)
10. Paterno, F., Mancini, C., Meniconi, S.: ConcurTaskTrees: a diagrammatic notation for specifying task models. In: Howard, S., Hammond, J., Lindgaard, G. (eds.) Human-Computer Interaction INTERACT 1997, pp. 362–369. Springer, Boston (1997)
11. García Frey, A., Céret, E., Dupuy-Chessa, S., Calvary, G., Gabillon, Y.: UsiComp: an extensible model-driven composer. In: Sukaviriya, N., Vanderdonckt, J., Harrison, M. (eds.) Proceedings EICS 2010, pp. 263–268. ACM Press, New York (2010)
12. Genaro Motti, V., Raggett, D., van Cauwelaert, S., Vanderdonckt, J.: Simplifying the development of cross-platform web user interfaces by collaborative model-based design. In: Albers, M.J., Gossett, K. (eds.) Proceedings SIGDOC 2013, p. 55 (2013)
13. Nguyen, T.-D., Vanderdonckt, J., Seffah, A.: Generative patterns for designing multiple user interfaces. In: Proceedings MOBILESoft 2016, pp. 151–159. ACM Press, New York (2016)
14. Peissner, M., Häbe, D., Janssen, D., Sellner, T.: MyUI: generating accessible user interfaces from multimodal design patterns. In: Barbosa, S.D.J., Campos, J.C., Kazman, R., Palanque, P., Harrison, M., Reeves, S., Barbosa, S.D.J. (eds.) Proceedings EICS 2012, p. 81. ACM Press, New York (2012)
15. Object Management Group (OMG): Unified Modeling Language Superstructure. Version 2.4.1. http://www.omg.org/spec/UML/2.4.1/Superstructure/PDF
16. Harel, D.: Statecharts: a visual formalism for complex systems. Sci. Comput. Program. **8**, 231–274 (1987)
17. Horrocks, I.: Constructing the User Interface with Statecharts. Addison-Wesley, Harlow (1999)
18. World Wide Web Consortium (W3C): State Chart XML (SCXML). State Machine Notation for Control Abstraction. https://www.w3.org/TR/scxml/
19. MyUI Project: Development Toolkit Guide. http://myui.eu/index.php?content=dev_toolkit.html
20. Peissner, M., Schuller, A., Ziegler, D., Knecht, C., Zimmermann, G.: Requirements for the successful market adoption of adaptive user interfaces for accessibility. In: Hutchison, D., et al. (eds.) Universal Access in Human-Computer Interaction. Design for All and Accessibility Practice, vol. 8516, pp. 431–442. Springer, Cham (2014)
21. Ziegler, D., Peissner, M.: Enabling accessibility through model-based user interface development. In: Cudd, P., de Witte, L. (eds.) Harnessing the Power of Technology to Improve Lives, 242, pp. 1067–1074. IOS Press Incorporated, Amsterdam (2017)

Guided Terrain Synthesis Through Distance Transforms

Caleb Holloway and Ebru Celikel Cankaya[(⊠)]

Department of Computer Science, University of Texas at Dallas,
Richardson, TX, USA
{cbhl30030, exc067000}@utdallas.edu

Abstract. We present a novel approach for terrain synthesis where the terrain is created only through procedural techniques and an initial input sketch. We employ a heightmap function that creates a distance map of the source image. The sketch is semantically annotated so that sections of the image are seen as a terrain type. Once a distance map of the source image is created, the algorithm can begin defining the heights inside the heightmap based on the distance values of each underlying pixel, the terrain type of that underlying pixel, and which terrain type that is different from its own it is nearest to. The results we obtain are promising as the terrain creation is fast, and the input system is non-complex.

Keywords: Terrain synthesis · Distance transformation

1   Introduction

Procedural terrain synthesis has become a major component of all fields that use terrain. The creation of terrain can take considerable time and effort, and by allowing the computer to create the terrain for the designer, a large amount of time can be saved. One of the major issues of this is that the terrain can be much more random than desired. A procedural system may use noise or mathematical systems too complex to predict the final look of the terrain. One way in which this can be combated is by having an input or source influence the procedural output. A common form of this is to combine an input sketch, alongside a heightmap, and combine them through machine learning to synthesize a new terrain heightmap that resembles the sketch.

This paper presents a new approach for terrain synthesis where the terrain is created only through procedural techniques and an initial input sketch. The main heightmap function that is used to accomplish this task is through a distance map of this source image. The sketch is semantically annotated so that sections of the image are seen as a terrain type. Once a distance map of the source image is created, the algorithm can begin defining the heights inside the heightmap based on the distance values of each underlying pixel, the terrain type of that underlying pixel, and which terrain type that is different from its own it is nearest to.

© Springer International Publishing AG, part of Springer Nature 2019
T. Z. Ahram (Ed.): AHFE 2018, AISC 787, pp. 57–65, 2019.
https://doi.org/10.1007/978-3-319-94229-2_6

2 Background and Related Work

Zhou et al. [1] focused on the idea of combining Digital Elevation Models with sketches provided by the user. This system uses patch placement and patch matching to accomplish realistic terrain based on the simple input image. Cruz et al. [2] creates terrain through patch-based synthesis like Zhou et al., but uses a guide input and exemplars to create the final terrain. The exemplars are then compared to a "categorization" map, where colored areas are set to become similar to the exemplars. Tasse et al. [3] is focused on taking an input sketch that resembles a side-view of mountain tops, and creating this view inside 3d terrain. Gain et al.'s work [4] is another system that focused on sketching into terrain, but also allows vertical control of sketches. Rusnell et al. [5] uses a distance function against a node graph, with the inputs of starting nodes and simple side profile sketches, to create a 3D terrain. James Gain et al. [6] uses a widget system to apply constraints directly onto parts of the 3D terrain. The system also uses an input function where sections can be labelled as specific colors, which represent exemplars to use in the terrain generation.

These systems are all procedural systems that are focused on the creation of realistic and detail-heavy terrain creation. Unlike those systems, an important aspect of this program is the level of control of it. Other than a small amount of noise to make mountains seem rocky, the output is deterministic given an input. The distance transform system provides a much more parameterized system, so that many aspects can be influenced by the user. The strong parameterization helps create a tool with a strong leaning towards artist use. This program is designed around simplistic use where the user can know what the output may look like before it is created.

In comparison to Zhou et al.'s system [1], the distance map system used in this paper uses a sketch where different terrain types are placed throughout the image. This provides more control than Zhou et al.'s input of simple lines and curves.

Cruz et al. and Gain et al. [2, 6] both provide a system for categorization, where sections can be defined as representing a specific terrain type. This does not provide the same amount of control as the distance map system, because the terrain types are defined by example input heightmaps, and therefore there is a much larger amount of randomness to both of these systems. Rusnell et al. [5] uses a distance system similar to the distance transform synthesis, but provides the ability to define only a set of nodes in the graph alongside a few simple profile sketches, instead of each node or pixel.

3 Experimental Work

3.1 Test Bed

We test our terrain synthesis program in Unity version 4.7 and 5.3. The hardware used to run the program includes an Intel i5-4590 CPU at 3.30 GHz, containing 4 CPUs, an NVidia GeForce GTX 660 Ti, and 16 gigabytes of DDR3 RAM. The operating system used to test the program is Windows 7. We use GIMP image editing software to create the input sketches.

3.2   Algorithm

The algorithm begins by creating a color map integer matrix. This is the matrix that shows what ground-type corresponds to each pixel in the input sketch. The algorithm makes sure to expand the input sketch out through a nearest-neighbor method to fill the full terrain, which must be a power of 2 in pixel width. After this, the system needs to define 2 more pieces of data per pixel: the distance to a pixel of a different ground-type, and what that ground-type is. The algorithm does this through a local propagation algorithm. The program loops through the color map matrix twice, once from top left to bottom right, and then from bottom right to top left. During the first loop, the pixel is compared to the pixel to the left and the pixel above it. The second loop has the pixel compared to the pixel below and the pixel to the right of it. This comparison is how Manhattan distance is accomplished. Chessboard/Chebyshev distance can be calculated by including diagonally located matrix cells in the local propagation loop.

Once both the distance matrix and the edge type matrix are calculated, the algorithm begins using those matrices to create the terrain. The algorithms start by creating a height map matrix of floats. Each cell of the new matrix is compared to the ground type matrix, the distance matrix, and the matrix that contains the nearest different ground type for that cell. For water ground type and mountain ground type, the number placed is just the distance value multiplied by 0.02. On the other hand, the "field" or grassy areas will slope down towards the water or up towards the mountain areas. This is done by looking at the nearest different field type/color matrix, and then using a smoothstep function to give it the sloped look. If the distance values along the edges between grass area nearest to water and nearest to mountains, this will create cliffs and edges, as shown in Fig. 1.

Fig. 1. A sample terrain sketch of a colored grassy area terrain created by distance transformation guided terrain synthesis algorithm.

After the step of creating an initial heightmap values matrix is complete, the next step done is to find out what the minimum and maximum value is of each ground type. Using these values, the program will define what each cell's heightmap value is, between the numbers of 0 and 1. The values are placed based on their field type. Water values will be less than other values, continuously decreasing as they move away from their edges. Mountain values will be above ground/field values, sloping up the higher the distance value gets. Then, the matrix is smoothed out, averaging with the pixels next to it, giving cliffs and slopes a more natural look. The final step taken with the heightmap matrix is to add a bit of random noise to the cells, making the surface more uneven and realistic.

Finally, the heightmap is sent into Unity's terrain system to create the 3D model of the terrain. The terrain is then looped through, adding textures and grass based on what the heights of each point on the map are. The 3D terrain is then sent into the Unity world to render onto the screen.

3.3   User Interface

As shown in Fig. 2, the system comes with many settings to help define the final world output. The "Chess" checkmark box is to switch between Manhattan distance and chess distance. The terrain width/length boxes define the size of the terrain in both horizontal axes in meters. The "Tex" box is where the input sketch is defined. The Texture List is the list of textures to use to place on the world, based on what field type it is. Element 0 is for the ground for the flat parts of grasslands. Element 1 is for mountains, under-water, and steep slopes. Element 2 defines the highest mountain snow peaks.

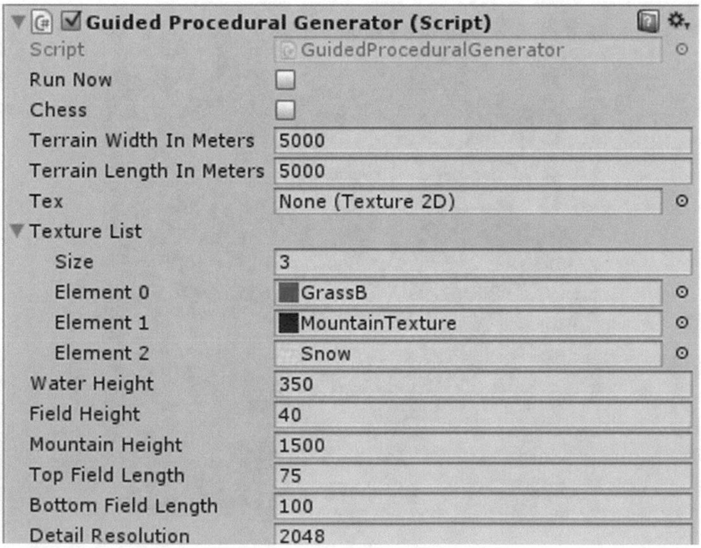

▼ G ☑ **Guided Procedural Generator (Script)**		🔲 ⚙
Script	GuidedProceduralGenerator	⊙
Run Now	☐	
Chess	☐	
Terrain Width In Meters	5000	
Terrain Length In Meters	5000	
Tex	None (Texture 2D)	⊙
▼ Texture List		
Size	3	
Element 0	◼ GrassB	⊙
Element 1	◼ MountainTexture	⊙
Element 2	Snow	⊙
Water Height	350	
Field Height	40	
Mountain Height	1500	
Top Field Length	75	
Bottom Field Length	100	
Detail Resolution	2048	

Fig. 2. Sample settings of the distance transformation guided terrain synthesis algorithm.

After this, the 3 "height" inputs define how much vertical space they take up in the world. The "Top Field Length" and "Bottom Field Length" define how much horizontal space the slopes take up within the green field areas. The detail resolution defines how many final heightmap points there are, representing the number of pixels per axis. Finally, the first checkmark box of "Run Now" is pressed when the user wants to start running the program.

3.4   Input

The input for the algorithm is a simple image, as shown in Figs. 3 and 5. The user can create these images using any image editing or digital painting tool, like GIMP or MS Paint, and export the images to PNG. When the user draws the sketch, they must remember that there are 3 colors as input to the system: red, green and blue. When the user places Red down, this will be treated as mountain, and rise away from the center of the vertical space of the terrain. Blue is treated as water, and will sink into the vertical space of the terrain. Green will be treated as grasslands or fields, and will occupy the space in-between mountain and water, having small vertical differences.

Fig. 3. Example input sketch and output (Manhattan distance) used in timing calculations

4   Results

After running the program using a test input, as shown in Fig. 3, it can be shown that the speed of the program is dependent on the size of the input sketch. As seen in Fig. 4, the distance transform used against the input will dramatically slow down when given the 4096 pixel-wide input sketch. On the other hand, as can be seen in Fig. 5, there is a floor to the amount of time it takes for the program to run, no matter what the sketch size. Even one of the smallest inputs of 512×512 still takes over 8 s to complete, and increasing that to 1024 only increases the average time by 0.5 s.

One issue in the generation algorithm is the existence of uniform, rough lines and edges throughout the map, as is visible in Figs. 1, 3, 4, 5, and 6. These are created by the currently used Manhattan and chess distance type transforms. This can be mitigated by implementing a Euclidean distance transform. Also visible in Figs. 5 and 6, are the fact that the mountain that takes up the most land mass, also dwarfs the other mountainous areas. A small mountain height input can make the smaller land mass mountains be

Fig. 4. Example output with grass

Fig. 5. Input and output as viewed from above, with water turned off

much tinier than expected. Changing the mountain height algorithm in order to make more realistic mountain heights might be ideal.

To measure distance transform speed accurately, we repeated five rounds of test run for our algorithm on various square matrix pixel levels ranging between 512 square pixels to 4098 square pixels and recorded each, then averaging the speed. Table 1 presents these values, where we observe that the difference per round per matrix level is statistically insignificant.

We further investigated how distance transform speed changes as we increase the matrix square pixels. The plot in Fig. 7 demonstrates the behavior of this change: The distance transform speed increases linearly in between 512 and 1024 pixel square

Fig. 6. Output close up with grass, pond, and hills visible

Table 1. The distance transform speed (in milliseconds)

Width of matrix in pixels being processed	1st round	2nd round	3rd round	4th round	5th round	Average
512	85	84	84	83	83	83.8
1024	447	444	445	445	454	447
2048	1383	1386	1389	1391	1387	1387.2
4098	8046	8024	7995	7987	8012	8012.8

matrices. The speed preserves its linear property of change from 1024 pixel square matrix to 2048 pixel square matrix, but this time with a steeper linearity. And for the last interval, i.e. from 2048 pixel square matrix to 4096 pixel square matrix, the rate of change is even steeper though still remains linear.

We performed a similar experiment on the same square pixel levels, i.e. 512 square pixels, 1024 square pixels, 2048 square pixels, and 4096 square pixels, to measure time to create whole terrain in Unity (including textures and grass) this time. The results we obtained are listed in Table 2. Comparing with the pure distance transform speeds from Table 1, we see that there is a hundred-fold increase in each round for the lowest pixel level 512, and a twenty-fold increase in each round for pixel level 1024. For the pixel level 2048, the time to create whole terrain in Unity outperforms pure distance transform speeds by around 33%. And for the last pixel level tested (4096), the speed to measure time to create whole terrain in Unity is double the pure distance transform time.

When we plot the graph for the time it takes to generate the entire terrain in Fig. 8, we see that the rate of change between pixel levels are less dramatic as opposed to the time it takes to complete the integer-matrix distance transform (Fig. 7).

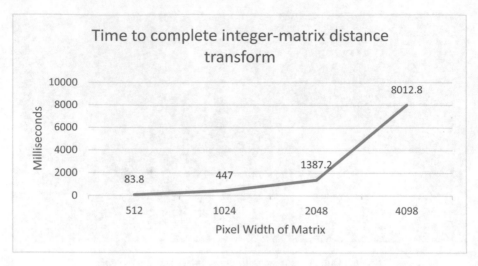

Fig. 7. Table on distance transform speed (in milliseconds)

Table 2. Time to create whole terrain in Unity (including textures and grass) in milliseconds

Width of matrix in pixels being processed	1st round	2nd round	3rd round	4th round	5th round	Average
512	8313	8323	8270	8294	8274	8294.8
1024	8709	8787	8689	8693	8713	8718.2
2048	10521	10489	10469	10464	10489	10486.4
4098	15090	14946	14925	14918	14917	14959.2

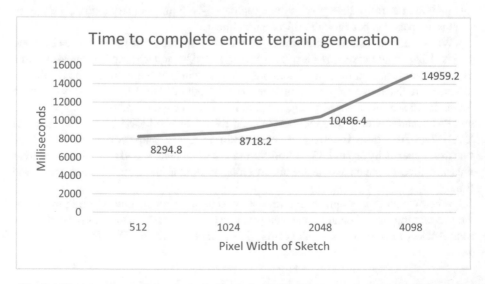

Fig. 8. Chart to create whole terrain in Unity (including textures and grass) in milliseconds

5 Conclusion and Future Work

Distance transforms give a promising result when used in procedural generation. The terrain creation is quick, and the input system is simple to use. While the program does not provide realistic terrain styles, the artistic control allowed by the system can create diverse and interesting opportunities for those in need of guided terrain synthesis.

There are many features that could be added to this system. The ability to control the heights of each half of the grassland areas (parts closer to water and parts closer to mountain) individually would be useful. The ability to calculate Euclidean distances inside of the distance transform is an important step towards creating pleasant landscapes. One large feature would be adding more colors. Alongside green, blue, and red, it would be possible to add black, yellow, cyan, magenta, and white as colors for field types.

It is even possible to add a system that allows the user to define colors and their meanings in the system. A simple scripting language could be created that allows the user to manipulate the distance and different field type matrices in order to define what a color means for an area. For example, a user could define a dark reddish color to mean a plateau, and then in the scripting language would define the distance of the dark reddish pixels to matter less, and instead script height reactions to what pixel color is closest that is not dark red.

For distant work, it could be possible to make a system where rivers rise in elevation as they get farther away from the ocean. Alongside this, it could be doable to have bodies of water at different levels of elevation. It could be made so that settings are changeable from the UI without needing to recalculate the entire distance transform. There are also more procedural synthesis systems that are possible. A section to add boulders and small rocks to the landscape could dramatically change the look of the maps. Caves that can be done using a sketch input are a possible feature. Eventually, a road and town/building system could be added to the program using procedural methods.

References

1. Zhou, H., Sun, J., Turk, G., Rehg, J.M.: Terrain synthesis from digital elevation models. IEEE Trans. Vis. Comput. Graph. **13**(4), 834–848 (2007)
2. Cruz, L., Velho, L., Galin, E., Peytavie, A., Guerin, E.: Patch-based terrain synthesis. In: International Conference on Computer Graphics Theory and Applications, pp. 189–194, March 2015
3. Tasse, F.P., Emilien, A., Cani, M.-P., Hahmann, S., Dodgson, N.: Feature-based terrain editing from complex sketches. Comput. Graph. **45**(1), 101–115 (2014)
4. Gain, J., Marais, P., Straßer, W.: Terrain sketching. In: Symposium on Interactive 3D Graphics and Games, Boston, MA, 27 February–01 March 2009
5. Rusnell, B., Mould, D., Eramian, M.: Feature-rich distance-based terrain synthesis. Vis. Comput. Int. J. Comput. Graph. **25**(5-7), 573–579 (2009)
6. Gain, J., Merry, B., Marais, P.: Parallel, realistic and controllable terrain synthesis. Comput. Graph. Forum **34**(2), 105–116 (2015)

Interactive Mining for Learning Analytics by Automated Generation of Pivot Table

Konomu Dobashi[✉]

Faculty of Modern Chinese Studies, Aichi University, 4-60-6 Hiraike-cho
Nakamura-ku, Nagoya-shi, Aichi-ken 453-8777, Japan
dobashi@vega.aichi-u.ac.jp

Abstract. This paper describes a method to reproduce and visualize student course material page views chronologically as a basis for improving lessons and supporting learning analysis. Interactive mining was conducted on Moodle course logs downloaded in an Excel format. The method uses a time-series cross-section (TSCS) analysis framework; in the resulting TSCS table, the page view status of students can be represented numerically across multiple time intervals. The TSCS table, generated by an Excel macro that the author calls TSCS Monitor, makes it possible to switch from an overall, class-wide viewpoint to more narrowly-focused partial viewpoints. Using numerical values and graph, the approach enables a teacher to capture the course material page view status of students and observe student responses to the teacher's instructions to open various teaching materials. It allows the teacher to identify students who fail to open particular materials during the lesson or who are late opening them.

Keywords: Interactive mining · Time-series · Cross-section · Visualization
Educational data mining · Learning analytics · Pivot table

1 Introduction

Recently, various educational institutions have adopted e-books and Course Management Systems (CMS) or Learning Management Systems (LMS) in an effort to enhance student learning. In order to support analysis of the large amounts of data that can be accumulated from student learning logs, research has been conducted to develop an effective learning dashboard. A number of studies have focused on how student characteristics obtained from an analysis of learning logs, including such factors as the number and duration of student views of various course materials and quiz performance, can be used to improve classroom teaching and learning [1]. In CMS/LMS and e-book systems, learning support functions and learning analysis functions are necessary for both teachers and students in order to enhance the educational effect [2].

The author uses face-to-face blended lessons in a PC classroom, with course materials uploaded on Moodle, to test a method to collect and analyze student learning logs [3]. In the proposed method, an Excel macro, which the author calls TSCS Monitor, generates a time-series cross-section (TSCS) table in an Excel format from the Moodle learning log. Analysis of the Moodle course log from a number of different

© Springer International Publishing AG, part of Springer Nature 2019
T. Z. Ahram (Ed.): AHFE 2018, AISC 787, pp. 66–77, 2019.
https://doi.org/10.1007/978-3-319-94229-2_7

perspectives is then conducted through various pivot table operations, and results are visualized with tables and graphs.

In the character string processing performed on the time data recorded in the Moodle course log, a method to create discrete time categories, such as year, month, day, hour, and minute, is developed [4]. By discretizing the time data, multiple timelines can be included and manipulated in the resulting pivot tables. The method makes it possible to perform automatic extraction and visualization and produce the in-class page view status of students chronologically. Moreover, by utilizing the functions of the pivot table, it is possible for the user to switch between an overall, whole-class viewpoint and a more narrowly-focused partial viewpoint, and to conduct multi-faceted analyses through various trial-and-error repetitions.

2 Related Research

Research is being conducted on the efficacy of student support systems that integrate CMS/LMS data with student management and grading management systems. Course Signals at Purdue University is an early-intervention system developed to provide real-time student feedback based in part on student records accumulated in Blackboard and past learning logs. The system evaluates the learning behavior of students and provides ongoing feedback in the form of personalized emails from the teacher; it also uses a colored signal light to indicate how the student is doing [5]. Krumm and colleagues are developing and applying systems to support learning advice to under-graduates by utilizing data accumulated in CMS/LMS [6]. A system called E2Coach at the University of Michigan also sends messages to students based on their course score data. These messages motivate students to take the actions necessary for success, reminding them, for example, to ensure sufficient time to prepare for the next exam [7].

Dawson et al. advocated the use of data collected by CMS/LMS to visualize student on-line activity and to use the information for student instruction and guidance [8]. May et al. developed a system with a user interface that enables real-time tracking of group discussions in language learning and visualizes each student's interaction level with a radar chart [9]. Hardy and colleagues developed a tool to track student browsing situations by using supplementary online teaching materials for students who registered for face-to-face blended classes in introductory physics and measured the results of final exams and re-sits [10]. Konstantinidis [12] and Dobashi have also developed Excel macros to process Moodle logs in order to analyze page views and overall usage [11]. Moodog, developed by Zhang and Almeroth, incorporates an analysis function for logs in Moodle. Moodog is a plugin application to analyze Moodle logs and make a multidimensional analysis of their content page views. The system is able to analyze student course material browsing rates, page views and time spent. Analytical results are displayed on the Moodle screens, representing the interaction of students and Moodle with graphs and tables.

A system that uses the access logs for online teaching materials and displays the results in an easily read graph has been developed. Mazza and Dimitrova have devised a system called CourseVis that tracks student behavior in an online class [13]. Student behavior can be visualized graphically, along with the status of student access to content pages following the course schedule. GISMO, a similar tool, was developed as a plug-in system for Moodle and today is used by many Moodle users [14]. By installing GISMO into the Moodle reporting tool, Moodle course administrators can analyze student access activity by specific materials and resources, the number of times a student accesses a forum, and quiz results. Analysis results are presented in tables and graphs, allowing users to grasp the state of the class from an overall viewpoint, individual viewpoints, or for each of the course materials.

Currently, Google Analytics provides a wide range of access analysis services for websites, which makes it possible to analyze access logs from various perspectives [15]. Verbert and Govaerts, who investigated various learning dashboard systems and developed a system of their own, reported that these systems are being used to support class improvement and monitor the learning state of students [16]. Duval has developed a system to support the selection of metadata necessary for learning by incorporating the information visualization method into a dashboard system and highlights the need to find a support system that identifies and promotes the most effective learning methods for teachers and students [17].

3 Course Overview and Material Setup

In this paper, we demonstrate our approach in a blended learning environment in which face-to-face classes using course materials uploaded to Moodle were conducted in a computer classroom. The approach can be adapted to most classes that allow students to log in to Moodle from a personal computer in order to browse the teaching materials and access quizzes (Fig. 1). In order to collect and accumulate Moodle course logs, course materials need to be uploaded to Moodle beforehand. The Moodle topic format is used to create a table of contents for the course; students are then able to click on a table of contents link to browse specific course materials, which is a common procedure in Moodle.

"China Data Analysis" is one of the subjects in which the author uses Moodle. For the lessons described in this paper, course materials were prepared to explain the basic theory of statistics and provide exercises related to geographical information systems in text and figures. The course materials were prepared as 15 lessons, with 14 chapters, 77 commentary files (including figures), 13 exercise files, eight quizzes and one final exam. In addition, there were nine statistical data files and outside links. In total, the PDF files consisted of 158 pages, pre-divided into 14 chapters and 77 sections. Using Moodle's topic mode, headlines corresponding to chapters, sections, and items of course materials were entered, and the course material files were uploaded.

Various course materials were made available on Moodle so that students could browse them both during and outside class hours, on or off campus. When students open any of the course materials on Moodle from their PC or tablet, a learning log is automatically collected and accumulated. While listening to the teacher's in-class explanations, students are expected to open related course materials according to the teacher's instruction. At the beginning of class, a brief quiz of approximately five minutes duration is given to confirm the contents of the previous lesson; in the second half of the class, students are given exercises applying the current day's learning.

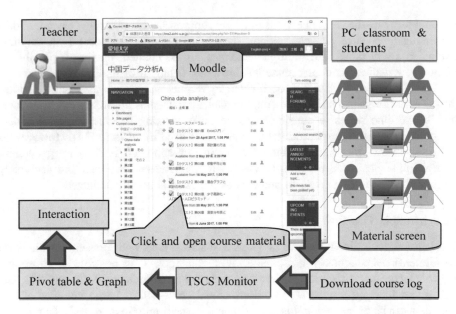

Fig. 1. Moodle entry page and course overview (China Data Analysis, 2017 spring)

In the classroom lessons addressed in this paper, we set up one computer for each student and conducted class using one material screen for every two students (Fig. 1). The material screen were used to display teacher demonstrations and relevant course materials. The teacher opens the materials on Moodle and explains their contents to the students. The students are then able to open the materials on their own computer, but could also see the materials on the teacher's screen. In order to collect student page views, the teacher instructed the students to open the Moodle materials on their own computers.

4   Utilization of Multiple Timelines by the Pivot Table

The Moodle learning log records when the teacher or student opened specific course materials. The log items and data are summarized in a time series in table form. The Moodle course log consists of the following nine items: *Time, User full name, Affected user, Event context, Component, Event name, Description, Origin, and IP address.*

4.1   TSCS Analysis

Although there are various forms of classes in which collecting learning log may be possible, the research reported here assumes a face-to-face blended lesson where a large number of students are being taught in a classroom equipped with PCs. In such a situation, many students select multiple course materials and browse them during class hours, so that a large volume of time-series data is generated at the same time. To compile and display these time-series data efficiently, two-dimensional frequency cross-tables can be extremely useful. By appropriately utilizing category data expressing time, such as year, month, day, hour, minute, and second, in the time display, a TSCS table with multiple timelines can be created, making it possible to efficiently visualize the course materials browsing behavior of multiple students.

The duration of a class can be divided in various ways, such as 90-min periods, half-term or full year. During a lesson, the data are treated as not infinitely continuous; that is, the time series is considered to be finite, making it easier to process than continuous data. A similar cross-table approach is used in GISMO [14]. However, as described below, the distinguishing feature in our approach is that, by generating multiple timelines, it is possible to easily switch between timelines and accurately monitor the viewing behavior of students as the class is being conducted.

4.2   Multiple Timelines

Excel Pivot table is a tool to aggregate discrete data and present results in a two-dimensional frequency cross-table. With this tool, it is possible to determine multiple frequency distributions simultaneously for a multiplicity of discrete data sets and to simultaneously generate two-dimensional cross-tables. Large volumes of discrete data can be processed in this way. When there are multiple items to be counted, as in this paper, a pivot table can be set up by selecting the items and cross-tabulating. Once a table is produced, various pivot table functions are available: for example, columns and rows can be switched and filters can be applied to narrow the data. Since these functions are available after the table has been created, analysis from various viewpoints—overall and individual—is easily performed.

The time data in Moodle as recorded in the Moodle course log is continuous in minutes and seconds; moreover, it is represented in a character string format rather than a numerical format. Because time in a Moodle log is recorded in minutes and seconds, when the time data are collected to produce a cross-table, a huge, unwieldy table is likely to result. In order to display the data in a manageable pivot table, we preprocess the data using time categories such as month, day, hour, or minute, thus discretizing the data at a fixed time interval. By having a number of processable time categories, it is possible to display multiple timelines in the same pivot table and to use one or more as a filter function. For example, it is possible to create a table to show activity in 15-min intervals and to specify that one particular 15-min interval show minute-by-minute activity (Fig. 3). Furthermore, by using a pivot table filter, selecting multiple time categories enables aggregation and analysis in various time zones, allowing more detailed and multifaceted analysis. The process of discretizing the time data is detailed below.

4.3   Discretization of Time Data

In the case of extracting year, month and day data, specific time categories can be isolated by extracting parts of the time data in the Moodle learning log. For example, from the Moodle time 29/07/17, 23:58:07, we can extract (1) the year, month, and day as 29/07/17; (2) the month and year as 07/17, or (3) the month and day as 29/07, and have the appropriate pipeline processing performed.

In the case of extracting hour, minute, second and time, the time categories can be used to specify particular time zones. We may, for instance, want to analyze activity only during lessons, perhaps specifying one-minute intervals during the time that a quiz is being given and intervals of 30 s, two minutes, three minutes, or 15 min, elsewhere in the lesson; pipeline processing is performed here as well. We simply need to extract the appropriate portion of the Moodle Time data. For example, if the Time is given as 29/07/17, 23:58:07, we could use (1) 13:00, 14:00, ..., for hourly activity (2) 13:00, 13:30, ..., for activity in 30-min intervals, (3) 13:01, 13:02, 13:03, ..., for one-minute intervals, etc. Using 23:58:07 would be appropriate when want to see more detail than is shown in the one-minute display: when, for example, we might want to see how far students are behind the instructions of the teacher.

By discretizing the time data as described above, TSCS analysis using a pivot table becomes manageable. The more frequent the items of data in the crossover scale, the more complex and detailed become the aggregation possibilities, increasing the multidimensional analysis possibilities for the Moodle learning log. Figure 2 shows a part of the Moodle course log after discretizing the time data: Date is shown in column C, Month in column D, Time in column E, Hour in column F, Minute in column G; all data are generated using the developed macros. The label at the top of each column is a heading in the pivot table and must be created with an appropriate name. In the original Moodle course log, these columns are followed by column K and Description, Origin, and IP address, which have been omitted here due to the limited page width.

	A	B	C	D	E	F	G	H	I	J	K
1	Date/Time	Year	Date	Month	Time	Hour	Minute	30Seconds	30Minutes	15Minutes	User full name
2	29/07/17, 23:58:07	2017	17/07/29	2017/07	23:58:07	23:00	23:58	23:58:00	23:30	23:45	Stud01
3	29/07/17, 23:56:37	2017	17/07/29	2017/07	23:56:37	23:00	23:56	23:56:30	23:30	23:45	Stud01
4	29/07/17, 23:56:37	2017	17/07/29	2017/07	23:56:37	23:00	23:56	23:56:30	23:30	23:45	Stud01
5	29/07/17, 23:56:30	2017	17/07/29	2017/07	23:56:30	23:00	23:56	23:56:30	23:30	23:45	Stud01
6	29/07/17, 23:52:26	2017	17/07/29	2017/07	23:52:26	23:00	23:52	23:52:00	23:30	23:45	Stud01
7			More data cut here						More data cut here		

Fig. 2. After preprocessing data and adding new labels year, date, month, time, hour, minute, 30 s, 30 min, 15 min. Hard copy of Excel work sheet.

Adding time data for every two minutes or every three minutes to Fig. 2 makes it easier to find students who open the teaching materials late when extracting and analyzing the course material page views during the lesson, but there are a wide range of applications. The time interval for data extraction should be determined according to the purpose of the analysis before creating the pivot table. By using the "field" selection function provided in the pivot table, it is possible to select the time zone to be analyzed and to perform the filter function.

5   Interactive Mining by Pivot Table and the TSCS Table

With Moodle, a student's learning log record begins at the time of registration. Once a log is created, the course administrator can freely access, browse and download the log at any time during or after classes. Figure 3 shows student page views of course materials for the first lesson of the spring semester. As indicated, multiple timelines were used in the table. The process of producing Fig. 3 proceeded as follows: First, the course log was downloaded from Moodle, and a pivot table was automatically generated by the developed Excel macro. The timeline was then manually manipulated to create the table as it appears here.

	A	B	D	E	F	G	H	I	J	K	L	M	N	Q	R	S	T
1	Date	17/04/11															
2																	
3	Sum of User full name / User full na	Column label															
4		13:00												14:00	15:00	23:00	Total
5		13:00	13:15	13:30	13:45												
6	Row label	13:14			13:45	13:46	13:49	13:50	13:52	13:53	13:54	13:55	13:57				
7	Course: China data analysis	148	34	2			2							1	1	3	2	193
8	File: 1.01 Starting and closing Excel		67	2			1								1		71
9	File: 1.02 Loading and saving files		1	42	2				2	1			3		1		52
10	File: 1.03 Screen and function		1	2		1	36	1	1	1	1			2	1		47
11	File: 1.04 Data input		1	1							23	9	1	7	1		43
12	File: 1.05 Worksheet		1											39	1		41
13	File: 1.06 Correcting and erasing data		1											33	1		35
14	File: 1.07 Copy and Paste			1										31	1		33
15	File: 1.08 Discontinuous data			1										1	2		4
16	File: 1.09 Auto fill and continuous data			1											2		3
17	File: 1.10 Create graph															1	1
18	File: Exercise 1													1	1		2
19	File: Exercise13		1														1
20	URL: China Statistical Yearbook		1														1
21	Total	148	108	52	2	1	39	1	3	2	24	9	5	115	15	3	527

Fig. 3. An example of a multiple-timeline TSCS table showing student openings of the various teaching materials (China Data Analysis, 04/11/2017). Columns C, O, and P are not displayed because it is subtotal. Screen copy of Excel pivot table.

Cell A1 (Date) indicates that the date item was selected by using the pivot table filter function. The entry in cell B1 indicates that the table data are from April 11, 2017. We could display a different date by pulling down the B1 cell. The date in cell B1 is displayed as 17/04/11 since it corresponds to the original sort function of the pivot table. The time line for cells B4 to S4 is one hour, broken down into 15 min intervals in cells B5 to F5, and one-minute intervals in cells B6 to N6. The lesson started at 13:00 and was held until 14:30. Subtotal columns C, O, and P were set to non-display. Cell B7 shows the number of accesses to the entry page of the Moodle course. At the start of class, all students accessed this page since it is from this page that students initiate their browsing of class materials.

During the class, the teacher was logged in to Moodle at the teacher's desk. While the teacher explained the lesson and demonstrated various features of Excel, students were asked to open the appropriate course materials. Cells D8, E9, H10, and L11 cells relate to points in the lesson where the teacher requested that particular course materials be opened, which explains why the values in these cells are relatively larger than those in surrounding cells. Cells L11 and M11 cells relate to the same course material opening. However, the Moodle course log shows that the teacher opened "File: 1.04 Data input" at 13:54:04. The entry in cell M11 thus indicates that nine students took one minute before opening the file. In Fig. 3, the course items used during the lesson are identified in cells A7 to A14; the numerical values in Column Q and T are relatively large. Cells A15 to A20 show course items that the teacher did not require students to open during the class; however, it can be seen that there were a few student views of some of these items.

Figure 4 shows the number of page views that the students and the teacher (identified as au172002 in cell A61) opened course materials. The table was created using the same procedure as was used to produce Fig. 3; the same time periods are displayed for comparison. Given the total number of student page views (cell T62: 527) and the number of attending students indicated in cell A60 (Stud53), the average number of page views per student was 9.717; the standard deviation (using the entries in cells T7 to T60) was 3.302 views. In Column J, M, and N, it can be seen that there were students who delayed opening materials that the teacher had opened.

	A	B	D	E	F	G	H	I	J	K	L	M	N	Q	R	S	T
1	Date	17/04/11															
2																	
3	Sum of Ever	Column label															
4		13:00												14:00	15:00	23:00	Total
5		13:00	13:15	13:30	13:45												
6	Row label	13:14			13:45	13:46	13:49	13:50	13:52	13:53	13:54	13:55	13:57				
7	Stud01	3	8	3													14
8	Stud02	3	1	3													7
9	Stud03	3	1														4
10	Stud04	3	1	1			1				1				3		10
11	Stud05		5	1			1				1				3		11
12	Stud06	3	2				1				1						7
13	Stud07	3	1	1													5
14	Stud08	3	2	2			1				1						9
15	Stud09	3	1	1		1									1		7
16	Stud10	3	1	1			1				1				3		10
17	Stud11	3	1				1								1		6
18	Stud12		4	2			1						1		3		11
19	Stud13	3	2	1											1		7
20	Stud14	3	3	1							1				6		14
21	Stud15	3	1	1									4		2		11
22	Stud16	3	1	2	1		1						1		3		12
23	Stud17	3	1												1		5
24	Stud18		4				2								2		8
25	Stud19	3	1	1			1				1				5		12
26	Stud20	3	4	2						3	1				3		16
27	Stud21	4	1	1			1								3		10
59		Part of the data is omitted								Part of the data is omitted							
60	Stud53	3	2	1			1				1				3		11
61	au172002	3	3	1			1				1				3		12
62	Total	148	108	52	2	1	39	1	3	2	24	9	5	115	15	3	527

Fig. 4. Example of a TSCS table featuring each user (China Data Analysis, 04/11/2017). Subtotal columns C, O, and P are not displayed. The times at which the teacher opened files were 13:34:12 (E61), 13:49:16 (H61), and 13:54:04 (L61).

Figure 5 shows a graph of activity during the final lesson of the spring semester (July 18, 2017). It was created by extracting page views of the course items used in the lesson as well the final exam that was given at the end of the lesson. The line graphs in various dotted lines show the number of course item page views; the scale for these graphs is on the left. The bar graph in red at the far right of the figure shows the number of page views of the final exam; the scale is to the right. The final exam started at 14:05 and ended at 14:25. The small number of page views before and after the test period are likely associated with exam preparation prior to the exam and confirmation of results afterwards. The exam consisted of 25 questions, with five choices offered for each question. Page views during the final exam are shown here to be much more concentrated than is the case for the other course items. In Figs. 3 and 4, it is possible to observe the time until most of the students have opened the materials that the teacher had instructed them to open. Of course, it is possible that some students open course items after the teacher's explanation or demo is finished. If the analysis is conducted using a pivot table, we would be able to see precisely who has opened the course items and when they opened them.

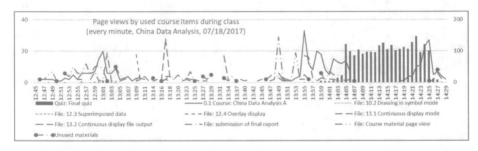

Fig. 5. Example of a time series graph showing the number of page views by course item and end-of-term final exam (the number of attending students are 54, China Data Analysis, 07/18/2017).

6 Discussion

In this paper, we analyzed Moodle learning logs covering 15 lessons in a single course over one semester (half-year) at a Japanese university. Using Excel to conduct the analysis, we found that the rather extensive data that was gathered could be processed and analyzed without difficulty. However, if the time intervals used in the analysis are shortened to seconds or minutes, or if the number of students increases significantly, the generated TSCS table will eventually become too large to fit on a PC screen. It may be inconvenient under these conditions to generate a TSCS table that requires time to manipulate and is easily viewable. In such cases, a function to automatically create a graph is needed to display a graphic that is intuitive and understandable. If the volume of data is large, the generated graph can be displayed in a reduced format; details would need to be checked in the table.

In this paper, an improved pre-processing procedure is applied, making it easy to display results on the screen of a PC using multiple timelines. The filter function of the pivot table allows the manipulation of multiple timelines so that a specific day or time can be displayed in a limited manner and a TSCS table that easily fits on a PC screen can be generated. To generate these multiple timelines, character string processing was performed applying Excel functions to the time data of the Moodle course log. Although, in some approaches, the character string processing used for English cannot be applied to other languages, the procedure developed in this paper is applicable to languages other than English, including Japanese.

In our one-semester experiment, we observed that many of the students tended to access the course materials only during the lessons. When we examined the number of students who were logged into Moodle on a daily basis, we also discovered a tendency for students not to log in on days other than class days. Moreover, in the lessons analyzed, students often logged in to Moodle a few minutes before the day's scheduled quiz to browse the course materials in preparation for the quiz and the lesson. This suggests that assigning homework that would support and encourage such preparation is needed. Throughout the semester, the teacher showed the course materials under discussion on the teacher's screen. The teacher delivered the day's lesson while demonstrating the required operations on the teacher's computer. From the TSCS tables, it was apparent that some students did not open the related course materials on their individual computers, or delayed their viewing. Such behavior is indicated in Fig. 4.

Because students in the class were expected to use Excel during virtually all of the lesson, there was a tendency to open Excel on the full screen of their individual computers, which effectively prevent the display of any of the related course materials. This meant that the students were likely viewing the teaching materials on the material screen rather than on their own computer screens. Another reason for the failure of some students to open some of the course materials may be that the content of the course was fairly basic and opening materials on topics already understood was seen by students as unnecessary. It should be noted, too, that a large number of students took at least a minute or two to open the teaching materials when instructed to do so by the teacher. All of these issues should be considered before the teacher proceeds to the next part of the lesson.

The method proposed in this paper is based on the assumption that the teaching materials are prepared on Moodle and that the students have a computer or tablet on which to view the materials. It is also assumed that the teacher and students will browse the course materials and engage during the lessons. Consequently, the approach described here cannot be used in all classes. Furthermore, student page view behavior is likely influenced by the quality and relevance of the materials, factors not addressed in this paper.

Prior to the development of the macro system proposed by the author, a TSCS table similar to those described in this paper would have to be created by hand, taking thirty minutes or more from the downloading of the log to the generation of the table. By contrast, using the macro described here requires only a few seconds. Thus, producing a TSCS table during class time can be rather easily managed. Nevertheless, depending on the manner in which a teacher conducts the lesson, in certain cases there may not be enough time to use the macro or act on the results.

7 Conclusion

Analysis of student in-class page views of teaching materials is extremely important as it directly affects class evaluations. In the generated pivot and TSCS tables, student page view behavior following the instruction of the teacher can be visualized and subsequently used to clearly identify students who did not open the course material during class or who opened the course material later than instructed by the teacher. The results of the analysis can be used by teachers to review the progress of their lessons and provide suggestions for improvement. Additionally, by seeing variations in the browsing situation for each of the course materials, teachers are in a position to develop more effective teaching materials and improve lesson management.

Acknowledgments. This work was supported by JSPS KAKENHI Grant Number 15K00498.

References

1. Romero, C., Ventura, S.: Educational data mining: a survey from 1995 to 2005. Expert Syst. Appl. **33**, 135–146 (2007)
2. Mostow, J., Beck, J., Cen, H., Cuneo, A., Gouvea, E., Heiner, C.: An educational data mining tool to browse tutor-student interactions: time will tell!. Educational data mining. In: 2005 AAAI Workshop. Technical report WS-05-02, pp. 15–22 (2005)
3. Dobashi, K.: Development and trial of excel macros for time series cross section monitoring of student engagement: analyzing students' page views of course materials. Proc. Comput. Sci. **96**, 1086–1095 (2016)
4. Dobashi, K.: Automatic data integration from Moodle course logs to pivot tables for time series cross section analysis. Proc. Comput. Sci. **112**, 1835–1844 (2017)
5. Kimberley, E.A., Matthew, D.P.: Course signals at purdue: using learning analytics to increase student success. In: LAK 2012 Proceedings of the 2nd International Conference on Learning Analytics and Knowledge, pp. 267–270. Vancouver, Canada (2012)
6. Krumm, A.E., Waddington, R.J., Teasley, S.D., Lonn, S.: A learning management system-based early warning system for academic advising in undergraduate engineering. In: Larusson, J.A., White, B. (eds.) Learning Analytics: From Research to Practice, pp. 103–119. Springer, New York (2014)
7. McKay, T., Miller, K., Tritz, J.: What to do with actionable intelligence: E2Coach as an intervention engine. In: LAK 2012 Proceedings of the 2nd International Conference on Learning Analytics and Knowledge, pp. 88–91, Vancouver, Canada (2012)
8. Dawson, S.P., McWilliam, E., Tan, J.: Teaching smarter: how mining ICT data can inform and improve learning and teaching practice. In: Annual Conference of the Australasian Society for Computers in Learning in Tertiary Education, pp. 221—230, Melbourne, Australia (2008)
9. May, M., Sebastien, G., Patrick, P.: TrAVis to enhance online tutoring and learning activities: real-time visualization of students tracking data. Technol. Smart Educ. **8**(1), 52–69 (2011)
10. Hardy, J., Bates, S., Hill, J., Antonioletti, M.: Tracking and visualization of student use of online learning materials in a large undergraduate course. In: ICWL. LNCS, vol. 4823, pp. 464–474. Springer (2008)

11. Konstantinidis, A., Grafton, C.: Using excel macros to analyses Moodle logs, In: Proceedings of the 2nd Moodle Research Conference (MRC2013), pp. 33–39, Sousse, Tunisia (2013)
12. Zhang, H., Almeroth, K.: Moodog: tracking student activity in online course management systems. J. Interact. Learn. Res. **21**(3), 407–429 (2010)
13. Mazza, R., Dimitrova, V.: CourseVis: externalising student information to facilitate instructors in distance learning. In: Proceedings of the International Conference in Artificial Intelligence in Education, pp. 279–286, Sydney, Australia (2003)
14. Mazza, R., Milani, C.: GISMO: a graphical interactive student monitoring tool for course management systems. In: Technology Enhanced Learning 04 International Conference (T.E. L. 04), pp. 1–8, Milan, Italia (2004)
15. Google Analytics. http://www.google.com/analytics/
16. Govaerts, S., Verbert, K., Duval, E., Pardo, A.: The student activity meter for awareness and self-reflection. In: CHI EA 2012 CHI 2012 Extended Abstracts on Human Factors in Computing Systems, pp. 869–884, Austin, Texas, USA (2012)
17. Duval, E.: Attention please! Learning analytics for visualization and recommendation. In: Proceedings of the 1st International Conference on Learning Analytics and Knowledge (LAK 2011), pp. 9–17, Banff, Alberta, Canada (2011)

Slot-Ultra-Wideband Patch Antenna
for Wireless Body Area Networks Applications

Javier Procel-Feijóo, Edgar Chuva-Gómez,
and Paúl Chasi-Pesántez$^{(\boxtimes)}$

Grupo De Investigación En Telecomunicaciones Y Telemática,
Universidad Politécnica Salesiana sede Cuenca, Cuenca, Ecuador
`{cprocelf,echuva}@est.ups.edu.ec, pchasi@ups.edu.ec`

Abstract. The patch antennas in wireless networks of the body are widely used in the health sector and for the monitoring of a person, for the diagnosis and control of diseases. Its applicability is due to its small size and its low power, but designing this type of antennas has some disadvantages, since by contacting the human body it is very difficult to transmit the data correctly. Therefore, a slot-ultra-wideband (S-UWB) patch antenna will be analyzed and designed for applications in biomedicine that were used in the UWB wireless body area network (WBAN) and where its reliability is tested. Its frequency of operation is 4.9 and 7.1 GHz. The dimensions of the S-UWB patch antenna are 27 mm 27 mm × 1.1 mm. This antenna was designed and simulated to verify the results of the measurements. The S-UWB has been simulated with a mathematical software to obtain the possible results.

Keywords: WBAN · UWB · Patch antenna

1 Introduction

The patch antennas are widely used for small applications, they are known as microstrips since they are based on this technology [1]. The current tendency to make each time smaller devices makes these antennas are in a big boom [2]. But designing a very small antenna has some disadvantages such as its low power, low efficiency, considerable losses and a narrow bandwidth, which makes it difficult to transmit data [3].

Patch antennas in wireless body area networks (WBAN) is one of the applications in which we will focus this research [4]. The WBANs have been very useful in the health sector, and they have been applied for the monitoring of a person, the diagnosis of diseases and the control of treatments [5]. With the trend of these networks have emerged communication standards such as Bluetooth, Zigbee, Utra Wide Band (UWB), Wi-Fi, among others that aim to solve the needs that are generated by incorporating the WBAN in everyday life [6]. Wireless body area networks carry data transmissions around, inside or on the human body with very small communication systems so it is necessary that the antennas to be used have special characteristics [7].

Therefore, the design of a slot-ultra-wideband patch antenna (S-UWB) is proposed, where the ultra-wide band (UWB) technology allows a higher rate and a high data transmission speed, and thus can achieve better innovations in the WBANs [8].

© Springer International Publishing AG, part of Springer Nature 2019
T. Z. Ahram (Ed.): AHFE 2018, AISC 787, pp. 78–84, 2019.
https://doi.org/10.1007/978-3-319-94229-2_8

By having a range of high precision and robustness of fading [9], UWB technology is ideal for the implementation of applications in the field of medicine and monitoring of people [10].

A slot-ultra-wideband antenna (S-UWB) will be analyzed for applications in bio-medicine that will be used in the UWB wireless body area network (WBAN) and its operating frequency is 4.9 and 7.1 GHz [11]. The frequency sweep is in the range of 2 Ghz to 6 Ghz. This article pretend to show how our S-UWB patch antenna works, where a simulation software will be used to see some important factors such as S11 parameter, radiation pattern, directivity and gain [12]. The dimensions used are shown in Fig. 1:

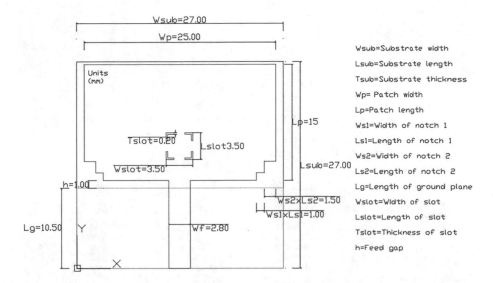

Fig. 1. S-UWB patch antenna dimensions.

The dimensions of the SUWB patch antenna are 27 mm × 27 mm × 1.1 mm. This antenna was designed and simulated to verify the results of measurements and will also be tested in free space so that your measurements are realistic. The S-UWB has been simulated with a mathematical software to obtain the possible results. The results of the antenna such as S11, the radiation pattern, the efficiency of the antenna and the gain were measured, obtained and verified with the simulated results in the software and the results obtained in the free space.

2   S-UWB Patch Antenna Design Parameters

The design of our antenna in the software is shown in the Fig. 2, it can see that the patch antenna has a very special feature and is the groove in the center of the area where the antenna radiates, this slot allows us to have a better conductance between the rectangular patch and the ground plane so that a small resistor may be able to distribute

the surface current along the symmetric groove [13, 14]. This simulation was performed in the HFSS software [15].

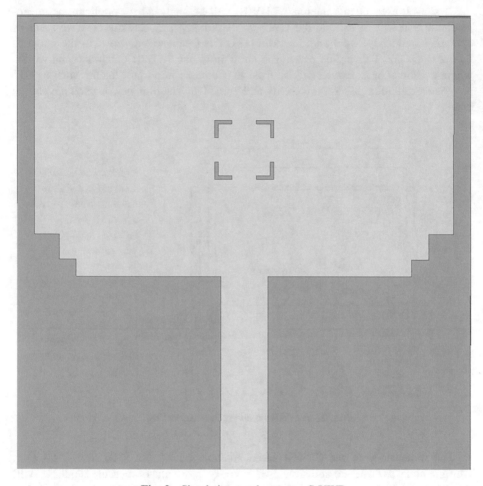

Fig. 2. Simulation patch antenna S-UWB.

The partial ground plane printed on the back of the substrate serves as an impedance matching element. In Fig. 3 we can see that the best frequency of coupling is of
4.9 GHz. The measures taken are the most appropriate for the patch antenna operation
[16, 17]. It is shown that the antenna slot of the width of 0.2 mm has greatly improved
the return loss, compared to other slot width values. Most of the variations were
visualized at lower frequencies [18, 19].

We can see in the Fig. 4 in which direction radiate the patch antenna S-UWB. The
directivity of the antenna is the association between the power density radiated in one
direction, at a distance. So we can see that the antenna has a big transmission power.

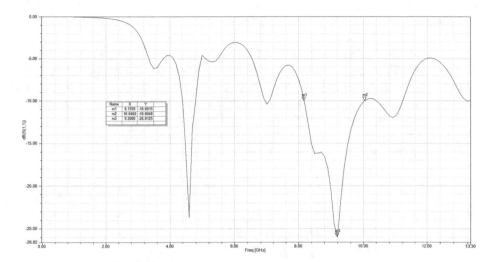

Fig. 3. S11-parameter.

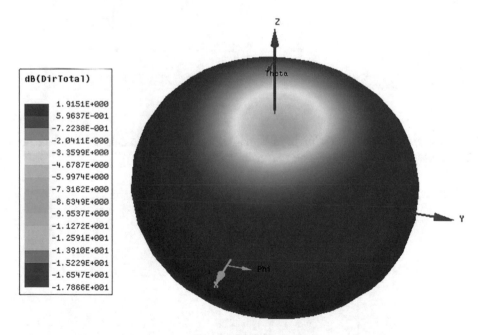

Fig. 4. Directivity.

This antenna design will be very useful for the monitoring of vital signs in the health area, where it will allow us to measure the heart rate, body temperature and also to monitor athletes. Corporal area networks do not allow the implementation of small transistors, which allows us to interconnect several devices, such as sensors with a PC or a telephone to process all the information received.

3 Results

It is observed that the radiated power of the antenna 1 with respect to the antenna 2 have a slight difference in both the power and the frequency selectivity, although the antennas were designed in the same way and at the moment of the impression undergoes a slight change from one device to another; With respect to the frequency selectivity the design was made to work at 4.9 GHz and can be saw that at the time of physical testing they undergo a slight displacement. In the Fig. 5 we can see the difference in radiated power of the two antennas in dBs.

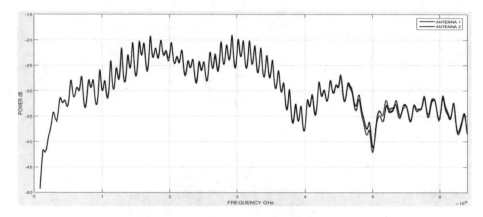

Fig. 5. Power radiated in dB.

To have a clearer perception of this small difference between antenna 1 and antenna 2 is shown in Fig. 6 the radiated power in watts.

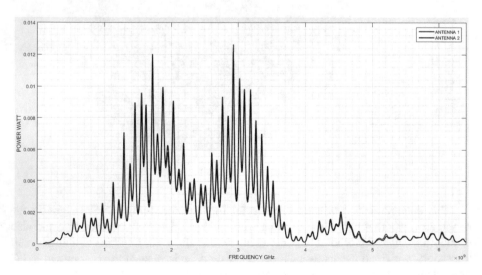

Fig. 6. Power radiated in Watts.

4   Conclusions

The development of the patch antenna S-UWB with a slot in the central part where it radiates, doesn't allow to see its operation to the dimensions of 27 mm × 27 mm × 1 mm. This research gives us a broader view of the behavior of this antenna in free space, what is intended to determine in future research is how it reacts to human contact and see if it is viable the use of these antennas. The antenna has an operating frequency of around 9.2 GHz and a bandwidth of 1.8 Ghz, in the simulation was made a frequency sweep of 1 to 15 GHz. What remains to be said is that in wireless body area networks (WBANs) a study of the antennas is essential, in order to achieve a more efficient monitoring of the sensors. The design of a small, cheaper and high-power patch antenna is a little complicated, it needs to taking into account many factors such as the coupling, its impedance and determinate if it has a good gain. The proposed antenna could be used efficiently for medical application in WBAN and personal communication in WPAN. With the design and analysis made of this antenna, we can corroborate its utility for biomedicine and body area networks (WBAN), since it is of a very small size, its design was low cost and has a very good power for the data transmission.

Acknowledgments. This research was supported by the Telecommunications and telematics Investigation Group (GITEL for its acronym in Spanish) and the teachers of Electronic Engineering of the Salesian Polytechnic University who provided insight and expertise that greatly assisted the research, although they may not agree with all the interpretations of this paper. We would also like to show our gratitude to Diego Cuji for sharing his pearls of wisdom with us during the course of this research, and we thank to all our classmates for their ideas that helped us developing this paper in a proper way. We are also immensely grateful to Juan Pablo Bermeo and Paúl Chazi for their comments on an earlier version of the manuscript, although any errors are our own and should not tarnish the reputations of these esteemed persons. Engineering gives us the opportunity to learn many things and to engineer others, so engineering students or engineers have the obligation to look for solutions and help our society to get ahead.

References

1. Kumar, V., Gupta, B.: On-body measurements of SS-UWB patch antenna for WBAN applications. AEU-Int. J. Electron. Commun. **70**(5), 668–675 (2016)
2. Bradai, N., Fourati, L.C., Kamoun, L.: Investigation and performance analysis of MAC protocols for WBAN networks. J. Netw. Comput. Appl. **46**, 362–373 (2004)
3. Hu, F., Liu, X., Shao, M., Sui, D., Wang, L.: Wireless energy and information transfer in WBAN: an overview. IEEE Network **31**(3), 90–96 (2017)
4. Jin, Z., Han, Y., Cho, J., Lee, B.: A prediction algorithm for coexistence problem in multiple-WBAN environment, Int. J. Distrib. Sens. Netw. **11** (2015)
5. Jovanov, E., Milenkovic, A., Otto, C., De Groen, P.C.: A wireless body area network of intelligent motion sensors for computer assisted physical rehabilitation. J. NeuroEngineering Rehabil. **2**(1), 6 (2005)
6. Yang, J., Geller, B., Arbi, T.: Proposal of a multi-standard transceiver for the WBAN internet of things. In: International Symposium on Signal, Image, Video and Communications (ISIVC), pp. 369–373. IEEE (2016)

7. Kang, D.-G., Tak, J., Choi, J.: Planar MIMO antenna with slits for WBAN applications. Int. J. Antennas Propag. **2014**, 7 (2014)
8. Lakshmanan, R., Sukumaran, S.K.: Flexible ultra wide band antenna for WBAN applications. Procedia Technol. **24**, 880–887 (2016)
9. Kang, D.-G., Tak, J., Choi, J.: MIMO antenna with high isolation for WBAN applications, Int. J. Antennas Propag. **2015** (2015)
10. Khan, F.A., Haldar, N., Ali, A., Iftikhar, M., Zia, T., Zomaya, A.: A continuous change detection mechanism to identify anomalies in ECG signals for WBAN-based healthcare environments. IEEE Access **5**, 13531–13544 (2017)
11. Kim, Y., Lee, S., Lee, S.: Coexistence of Zigbee-based WBAN and WiFi for health telemonitoring systems. IEEE J Biomed. Health Inform. **20**(1), 222–230 (2016)
12. Kong, H.Y., Van Khuong, H., Jeong, H.J. Lee, D.-U.: Design of a novel SS (spread spectrum) system based on cooperative communications. In: 2006 IEEE Region 10 Conference on TENCON 2006, pp. 1–4. IEEE (2006)
13. Naganawa, J.-I., Wangchuk, K., Kim, M., Aoyagi, T., Takada, J.-I.: Simulation-based scenario-specific channel modeling for WBAN cooperative transmission schemes. IEEE J. Biomed. Health Inform. **19**(2), 559–570 (2015)
14. Naganawa, J.-I., Takada, J.-I., Aoyagi, T., Kim, M.: Antenna deembedding in WBAN channel modeling using spherical wave functions. IEEE Tran. Antennas Propag. **65**(3), 1289–1300 (2017)
15. ANSYS: High frequency structure simulator (2017). http://www.ansys.com
16. Pandey, B., Jain, A.: Self-sustaining WBAN implants for biomedical applications. In: 2016 2nd International Conference on Applied and Theoretical Computing and Communication Technology (iCATccT), pp. 494–503. IEEE (2016)
17. Park, D.-S., Chao, H.-C., Jeong, Y.-S., Park, J.J.J.H.: Advances in computer science and ubiquitous computing. In: CSA & CUTE, vol. 373. Springer (2015)
18. Pathak, S., Kumar, M., Mohan, A., Kumar, B.: Energy optimization of Zigbee based WBAN for patient monitoring. Procedia Comput. Sci. **70**, 414–420 (2015)
19. Rezvani, S., Ali Ghorashi, S.: A novel WBAN MAC protocol with improved energy consumption and data rate. KSII Trans. Internet Inf. Syst. **6**(9) (2012)

Multi-view Model Contour Matching Based Food Volume Estimation

Xin Zheng[✉], Yifei Gong, Qinyi Lei, Run Yao, and Qian Yin[✉]

Image Processing and Pattern Recognition Laboratory,
Beijing Normal University, Beijing, China
{zhengxin,yinqian}@bnu.edu.cn, {201511210126,
201511210220,201511210201}@mail.bnu.edu.cn

Abstract. In this paper, an automatic food volume estimation method based on outer contour matching is proposed, which avoids the complicated calculation. We pre-defined a simple 3D model library and stored the projections and the user's hand. Users took three images containing their hands from three views. The contour of segmented image was compared with the projections to find the best match. Meanwhile, we took the user's hand as the scale and calculated the volume. As the method is easy to operate, less space-consuming, it is quite suitable for integrated application in the mobile app.

Keywords: Food volume estimation · Image segmentation · Contour matching
Image registration · Mobile application

1 Introduction

As early as 1997, the World Health Organization (WHO) already classified obesity as a disease. In 2016, over 1.9 billion adults aged 18 and over were overweight, of whom more than 650 million were obese, 39% were overweight and 13% were obese adults aged 18 and over [1]. Therefore, the problem of obesity needs to be solved urgently. WHO research shows that at for the individual level, one can limit energy intake from total fat and sugar and increase consumption of fruit and vegetables, as well as legumes, whole grains and nuts. This shows that controlling diet has a great effect on managing the health of obese patients [2].

At present, with the improvement of people's living standard, more and more food types and more perfect nutrition, and fast-paced living makes people unable to attend to and monitor daily intake of nutrients, which often leads people to eat foods that are too high in fat and sugar unconsciously, and leads to obesity. Therefore, a convenient and accurate mobile phone application is needed, and the popularity of mobile phones creates this possibility. Applying the idea of food volume estimation to mobile phone app enables users to understand the amount of calories they will consume before a meal so that users can manage their diet conveniently and quickly and thus this method plays an important role of automatic diet control.

Globally, imbalances between high-calorie, high-calorie food intake and low calorie consumption are the underlying causes of obesity and overweight. Therefore, this paper starts from the aspect of food and calculates the components of food through the methods

© Springer International Publishing AG, part of Springer Nature 2019
T. Z. Ahram (Ed.): AHFE 2018, AISC 787, pp. 85–93, 2019.
https://doi.org/10.1007/978-3-319-94229-2_9

of image segmentation, contour matching and volume estimation, so as to let people know the content of each component of food they eat and to monitor their own diet status, to automatically control the function of the diet, better manage their own health.

2 Related Work

At present, there are many ways to estimate the volume of food. In 2003, Puri et al. Proposed to restore the food model by using 3D point cloud and RANSAC to extract depth information [3]. Jia et al. also proposed the use of outer contour extract, combined with simple geometry matching [4]. Taken together, these two major methods are more widely used.

In order to simplify the calculation and make the project results better applied to the mobile app, in this paper, we use three views of the food to estimate the food volume. Each of the three views of the food is nested in a simple model to find the most contoured item. These simple models are our pre-defined libraries for multi-angle projection of simple three-dimensional graphics (including rectangles, spheres, cylinders, etc.), and we also need information about the multi-angle projection of the user's hand as a scale, and finalize the food volume according to the formula.

Compared with the former two schemes, the modeling process of point cloud is more complex, and often does not need to be very accurate when matching food, so it is not very suitable for integration in the mobile app. The general idea in this article is nested using a simple model, but we've made some optimizations for this approach, and we've repeatedly nested from three different angles to improve accuracy. At the same time, most studies now often require some extra specific items when choosing a reference, which is often where the most inconvenient area for users to know the ingredients of food at any time, at anywhere. Therefore, we record the user's hand data, using the user's hand as a scale. The user only needs to include his hand information in the shooting process.

3 Method

3.1 Overview

The proposed method involves three stages: preparation, food segmentation and volume estimation based on simple 3D model. The flowchart of our method is shown in Fig. 1. Firstly, we pre-define a simple 3D model library. These 3D models are projected at every particular angle and the projected outer contour images are stored. Testers need to enter their hands information. Next, preprocess the three photos taken by testers from top view, and convert them into L*a*b color space, and segment food and palm based on K-means method. As for the irregularly shaped food, divide it into simple geometric shapes. After extracting the outer contour of each spilt part, match them with projected outer contour images of 3D models at every particular angle. Find the two images fit best to make sure the 3D model corresponding the food. Using tester's palm as a scale to determine the model parameters, and calculate the food volume. The following sections describe the details of these computational steps.

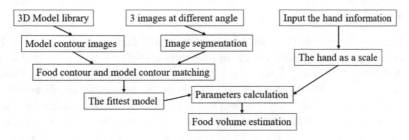

Fig. 1. Algorithm flowchart

3.2 Preparation

Before the first time calculating the food volume, some preparations need to be made to ensure that food volume can be estimated more accurately and conveniently in the future.

3.2.1 3D Model Library

According to our research, cooked food can be modeled by a set of shapes, especially healthy food. Therefore, we select a certain number of 3D models through statistical analysis, and build a 3D food-model library.

The detailed information of this 3D food-model library is shown in Table 1. The first column shows the model name, model parameters, formula to calculate the volume and examples. Meanwhile, a camera is added to each model and it can project the model at every particular angle at any time in order to get the outer contour images.

Table 1. 3D model library

Model	Parameters	Formula	Example
Cuboid	Length l, width w, height h	$V = l * w * h$	Sliced bread
Cylinder	Radius r, height h	$V = \pi r^2 * \cdot h$	Hamburger
Wedge	Radius r, height h, angle a	$V = \frac{1}{2} r^2 * w * h$	Cake, Pie
Sphere	Radius r	$V = \frac{4}{3} \pi * r^2$	Cherry, Tomato
Half sphere	Radius r	$V = \frac{2}{3} \pi * r^2$	Rice, Jelly
Ellipsoid	Principal semi-axes a, b, c	$V = \frac{4}{3} \pi * a * b$	Egg, Potato
Half ellipsoid	Principal semi-axes a, b, c	$V = \frac{2}{3} \pi * a * b$	Half egg

3.2.2 Input Hand Information

In our method, a hand is the scale to calculate every parameters' actual value in the model. Testers need to input the length and width of their hands. Meanwhile, it is necessary to take photos of their hands at several particular angle, which will be used to calculate the camera angle of food images.

3.3   Food Segmentation

After finishing the preparation, we can start our work. Testers need to put their hands and food in the same horizontal plane to take pictures. After that, the original image is converted into L*a*b color space, the brightness l is defined as a constant which we selected after a number of experiments, to ensure the segmentation effect meets the need. Next, the image in L*a*b color space is segmented based on K-means method. After the image we get is filtered and filled color, extract its outer contour. According to the idea of the connected domain of binary image, each food and hand are cut into separate images. The specific process can be seen in Fig. 2:

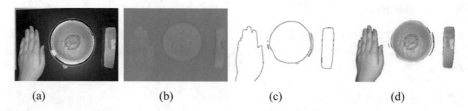

(a)   (b)   (c)   (d)

Fig. 2. Image segmentation process: (a) Original image, (b) L*a*b color space, (c) outer contour, (d) segmentation

3.4   Volume Estimation

The subjects were required in the task of estimating food volume based on three different angle images of both hand and food in the same horizontal plane.

3.4.1   Outer Contour Matching

In our paper, the main method of outer contour matching is to calculate the Hu moments of two contour images. This method was first proposed by Ming-Kuei Hu in 1962. He used the normalized 2 and 3 center moments, and introduced 7 local transformations, rotation and scale independent moments (Hu invariant moments) [5].

First, the segmented food contour image is converted into a binary image to calculate the central moment of the boundary and target region. Normalize the two groups of central moments, and calculate 7 invariant moments M1–M7 on the basis of normalization. Finally, these 7 moments compose the eigenvector of the target of food image and 3D model contour image.

Through our research and experiment, it can be found that in the recognition process of objects in the image, only M1 and M2 invariance remain relatively good, and the errors of other invariant moments are relatively large. Therefore, the Hu moments are only suitable for identifying images with little texture information, and our contour image just meets the requirements.

3.4.2   Nest Simple 3D Model

The contour food image is matched with the contour image of each model projected by the camera in a particular angle in a specified order.

According to a function we use in the program, it define I as the similarity of two images.

$$I(A, B) = \sum_{i=1}^{7} \left| \frac{1}{m_i^A} - \frac{1}{m_i^B} \right| \tag{1}$$

$$m_i^A = \text{sign}\left(h_i^A\right) * \log\left|h_i^A\right| \tag{2}$$

$$m_i^B = \text{sign}\left(h_i^B\right) * \log\left|h_i^B\right|. \tag{3}$$

h_i^A and h_i^B represent the Hu moments of the two images.

The smaller the value, the more similar they are. Calculate the similarity of food contour images and all model projections at the same angle, and find the minimum similarity I_{min}. Through our research and many experiments, we identify a threshold L that can maintain high accuracy. If $I_{min} < L$, the corresponding model of the projection image matches the food. If $I_{min} > L$, need to adjust the camera angle and do the previous operation again until $I_{min} < L$.

3.4.3   Using Hands as the Scale

After the matching model is determined, calculate the size of the palm in each image, and get the scale α which equals to the length of the palm in the image dividing the actual length of the palm of the tester. At the same time, calculate the length of the food projection contour in each image, and calculate model parameters using α.

3.4.4   Volume Estimation

After determine the corresponding model(s), put the parameters we get into the volume calculation formula of each model, which is shown in Table 1, and we can get the food volume. For complex foods composed of multiple simple models, add the simple model together. After getting the food volume, the nutritional value and calorie index of different foods found on the official website can be calculated to help people eat healthily.

4   Results and Discussion

4.1   Matching Results

In our experiments, we took small photographs of hands and food on the same level and got three pictures. Processed three pictures in turn. Take Fig. 3 as an example, K-means algorithm is applied to image segmentation. The result is shown in Fig. 3(b). Three kinds of things are numbered from left to right, where b1 is the hand, b2 is a cake and b3 is a cheesecake. The picture of the hand is matched with the pre stored pictures of the hand to find the stored picture which has the highest matching degree. Result shows that the angle between the food and the camera is 90°.

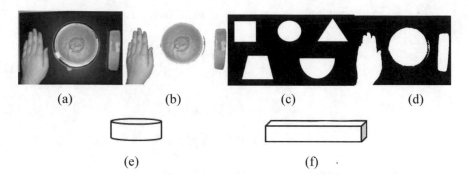

Fig. 3. Matching results. (a) Original picture, (b) segmentation, (c) projections of geometry, (d) outer contour of original picture, (e) matching results, (f) matching results

Next, the projection of the geometry in the 90° angle in the 3D graphics library is obtained and is shown in Fig. 3(c). The outer contour of food in Fig. 3(d) is matched with the outer contours of projections in Fig. 3(c). Result showed that b2 has the greatest matching degree with the circular on outer contour, the matching coefficient r is less than threshold L. The projection of both the cylinder and the ball in the direction of 90° is round, therefore, whether the matching geometry of b2 is the sphere or the cylinder is determined by the remaining two pictures. Do the same operation on b3. After processing three pictures, we can find that the cylinder shown in Fig. 3(e) has the biggest matching degree with the cake, and the largest match with the cheese is the cube, shown in Fig. 3(f).

4.2   Accuracy Evaluation of Food Volume Estimation

In the experiment, we measured the real volume of irregular food by a drainage method. The average value of the volume calculated by the three nested simple model is used as the estimated volume. Based on the relative error, the error rate under the volume estimation is calculated. The calculation formula is as follows:

$$\text{Error} = \frac{\overline{\text{VE}} - \text{VT}}{\text{VT}} * 100\% \qquad (4)$$

Among them, $\overline{\text{VE}}$ and VT represent the average of the estimated volume and real volume, respectively. Table 2 is the evaluation of 9 kinds of food.From the table, we can see that the estimated volume of three kinds of foods, such as bananas, fries, and ice cream, has a greater error than the actual volume. These errors are caused by the geometric shape of the food. These three kinds of foods represent the three kinds of common sources of error in estimating volume. The first is that the shape of the simple geometry is quite different from that of the food. Take banana as an example, the gap between the banana and the cylinder leads to a larger error rate in the volume calculation. Second, there is a gap in the simple geometry of food. French fries often appear on the table in the form of "heap". When estimating its volume, nest it with a semicircle or cube. The volume of geometry contains the volume of french fries and the space between it. It is the space that leads to the large error in volume estimation. In the third case, the surface of the food is uneven. Although ice cream can fit well with the hemispherical body, the volume of the concave surface in its surface can not be calculated, resulting in greater volume estimation than the true value. Of the 9 kinds of food measured, the average error rate is 7.54%, far below the average error of visual estimation 20%.

Table 2. Results of estimating volume

sample	true V	simple geometry	average V	error(%)
peach	151		147.8	2.12
cake	340		335	-1.47
banbana	99		117.41	18.6
cheesecake	106		108	1.89
ice cream	80		93	16.25
French fries	70		83.44	19.2
hamburger	307		319	3.91
Boxed milk	250		245	-2
potato	113		116.5	3.1

4.3   Analysis of Influencing Factors

4.3.1   The Influence of Shooting Distance on Volume Estimation

In this method, defaulting that the angle between camera and the hand is the same as the angle between camera and the food is the key to determine the projection angle and then determine the matching result. However, as Fig. 4 shown, the distance between camera and the hand is different from which between camera and food. In one picture, with the condition of the same position of the camera, the angle from camera to the hand and that from camera to the food is different. Consequently, the angle of projection is different. In our experiment, after matching the picture of the hand in Fig. 3(b) and stored pictures of the hand, the angle from camera to hand was calculated to be 90°. Assuming that the distance between the hand and the food in the picture is d, the difference between real angle and the estimated angle is α, $(0 < \alpha < 90)$, the distance between food and camera is x. Then the relationship is as follows:

$$\tan\alpha = \frac{d}{x} \qquad (5)$$

Amusing that the value of d is determined and remains unchanged, the relationship between the angle α and the distance x is shown in Fig. 5.

As can be seen from the figure, with the distance x increasing, α gradually decreases. When tanα is the maximum, match the projection with the food for the outer contour. The results show that the maximum matching degree of the cake is round, and the matching coefficient is less than the threshold L, also the matching coefficient is less than that under the condition of 90°. After operating all three pictures, we can come to conclusion that the simple geometry which is most matched with the cake is the cylinder. It is the same as the result under the condition of 90° projection in the actual operation. This shows that the shooting distance does affect the matching of single graphics, but it has little effect on the final matching results. So it is effective to use the hand angle as the food angle in the experiment.

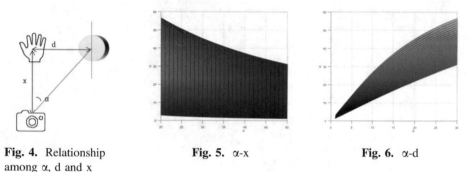

Fig. 4. Relationship among α, d and x   **Fig. 5.** α-x   **Fig. 6.** α-d

4.3.2 The Effect of Distance Between the Hand and Food on Volume Estimation

Figure 6 shows the relationship between α and d. When α takes its maximum, match the projections with the picture of food. The minimum of r is greater than L. In this situation, we chose the tube with maximum similarity, and then made projections of it in angles every 2° vary from α − 10 to α + 10°. Next, matching the outer contour of projections with the picture of food. Finally, we can find the actual angle from the camera to food. Because of the angle of shooting is different, the proportion of hands and food became inaccurate. So there is an error in the real size and estimated size of the food, and then affects the volume estimation. Consequently, on the premise that the hand and the food are not overlapped and the camera position is fixed, the nearer the hand to the food, the smaller the error of the estimate of the food volume.

5   Conclusion and Future Work

5.1   Conclusion

In this article, we present a method for calculating the volume and nutrient content of foods that are easily applied to mobile apps. By looking at the three views of a simple, three-dimensional figure in life, we can see that the three views of these figures are composed of simple two-dimensional figures, such as rectangles, circles and semicircles. Some more complicated food images such as chicken legs, can be divided into these simple two-dimensional graphics. So, in our research, we pre-defined a library for storing 2D simple graphs. When photographing a food picture, the user needs to take three views of the food, that is, the front, the side and the top view of the food. At the same time, the user's picture needs to include the user's hand. Based on the user's three pictures, we use algorithm for contour matching, in order to find the best fit for volume calculation. When calculating the volume, we use the user's hand as a scale to calculate, which greatly facilitates the user's use and enhances the user's experience. Compared with the traditional method, there are several advantages. First, the traditional methods of food volume estimation often have high demands on the pictures taken and require more complicated calculations of parameters. Our three-view and predefined graphics library simplify computation and are suitable for integration into mobile phones. Second, most of the commonly used food volume estimation methods require users to select additional reference objects, which are inconvenient for users to use anytime, anywhere. The method used in this project only requires users to include their own hand while taking pictures, which greatly simplifies the user's use.

5.2 Future Work

At present, there are still many limitations in the research. In the future, we will make improvements in the following aspects:

- As the user can not guarantee accurate three-shot when shooting, it will have some impact on the actual calculation. We plan to further improve our concept by expanding our simple graphics library into a library of variability and performing image correction based on the user's hand angles.
- The current limitations of food volume-based research are large and this project is no exception. A lot of research is only applicable to western-style catering, and there are certain requirements for food display. In the future, we will further explore this aspect and conduct more research on Chinese food.

Acknowledgment. The research work described in this paper was fully supported by the grants from the National Natural Science Foundation of China (Project No. 61472043), National Key Research and Development Program Project: "The key technologies research and integrated demonstration of mountain torrent disaster monitoring and early warning (2017YFC1502505). Prof. Xin Zheng and Qian Yin are the authors to whom all correspondence should be addressed.

References

1. World Health Organization, February 2018. Obesity and overweight. http://www.who.int/mediacentre/factsheets/fs311/en/
2. World Health Organisation, February 2018. Obesity and overweight. http://www.who.int/mediacentre/factsheets/fs311/zh/
3. Puri, M., Zhu, Z., Yu, Q., Divakaran, A., Sawhney, H.: Recognition and volume estimation of food intake using a mobile device. Sarnoff Corporation (2009)
4. Jia, W., Yue, Y., Fernstrom, J.D., Zhang, Z., Yang, Y., Sun, M.: 3D localization of circular feature in 2D image and application to food volume estimation. In: 34th Annual International Conference of the IEEE EMBS
5. Image moment – Wikipedia. https://en.wikipedia.org/wiki/Image_moment

Visual Analysis on Macro Quality Data

Gang Wu[1,2], Chao Zhao[1,2], Wenxing Ding[1,2], Fan Zhang[1,2],
Jing Zhao[1,2], and Haitao Wang[1,2(✉)]

[1] China National Institution of Standardization, No. 4 of Zhichun Road,
Haidian District, Beijing, China
{wugang, zhaochao, dingwx, zhangfan, zhaoj,
wanght}@cnis.gov.com
[2] AQSIQ Key Laboratory of Human Factors and Ergonomics (CNIS),
Beijing, China

Abstract. Quality inspection and quarantine processing is an important process with extensive business impacts. Data visualization methods failed to fully combine standards and certifications or accreditation. While inspection, quarantine and other quality related measures are hard to analyze due to the multi-factor dimension and therefore it is difficult to reflect the macro quality status. Thus, in order to integrate quality data from quality supervisory inspection and quarantine departments, this paper propose three visualization methods to analyze quality data, such as tree models and histogram visualization, map and histogram visualization, the models implement quality visualization system based on the methods above to realize the comprehensive analysis of macro quality data.

Keywords: Data visualization · Macro quality data · Data analysis
Time series visualization

1 Introduction

With the current development and progress of China economy the country is transitioning into a modern society and economy focusing on quality and efficiency. Quality plays a basic role in the new economy, and quality problems are getting more attention. In China, quality inspection organizations accumulated a mass of quality data. How to represent quality data and identify the trends and analysis represents a challenge and one of the most important problems to be solved urgently.

Data visualization [1] is the science about data representation and form, it is representing data as graphs and the process of using data analysis technologies to find hidden information. The aim of data visualization mainly by the aid of graphical methods, clearly and efficiently delivery and communicate information. Efficient visualization methods convenient observe and analysis data for us, to find the hidden data features and law, helps people understand data and make efficient decision. In the big data era, the role of data visualization is becoming more and more important. By analysis macro quality data, extensity, timeliness and hierarchy are the features of data, many researchers have much works focus on these features visualization, but limited on research the visualization forms of combined these features.

© Springer International Publishing AG, part of Springer Nature 2019
T. Z. Ahram (Ed.): AHFE 2018, AISC 787, pp. 94–99, 2019.
https://doi.org/10.1007/978-3-319-94229-2_10

By studying the hierarchy visualization, timeline visualization, and extensity visualization technologies, this paper proposed two combined visualization methods, include tree and histogram visualization, map and histogram visualization, integrated multiple visualization methods [2] to represent quality data according to data characteristics. These visualization methods completely satisfy the macro quality data visualization requirements, like compares qualification rate of product quality inspection between 31 provinces of china, compares qualification rate of a certain class product during many years, and find the trends of product quality development and so on. These methods represent hierarchy structure and time series data, and in terms of user interaction, users observe the data by multiple interactive technologies, to discover the characteristics and law of data, such as the dependency, periodic and exception of data.

On this basis, the researchers designed and implemented a quality statistics data visualization system. For the usability and scalability of the system, we realized the authority control model based on RBAC (Role-Based Access Control), which linked role and user's authorities. We use a variety of traditional methods to represent data, and use the three methods above mentioned to implement data visualization, achieved good visual effect. This system strong the visual analysis of quality statistics, further mining the relationship between quality management and society economic development, will provide valuable reference base for government reinforce macro quality management.

The data source cited in this paper was retrieved from website published by General Administration of Quality Supervision, Inspection and Quarantine of the People's Republic of China, and the visualization graphs implemented by ECharts [3]. This visualization topic is quality development. Around this topic, analysis related data and make a conclusion by data.

2   Visualization Methods of Macro Quality Data

Based on visualization design principles and targets, this paper designed and implemented macro quality data visualization methods, which includes time sequences visualization method, map visualization method and integrated method.

2.1   Tree and Histogram Visualization Method

For the hierarchy structure visualization, nodes linked technology are used usually. treemap and radial tree diagram are common layout methods for data, as the Fig. 1(a). Generally, quality data includes dimensions like provinces, times, and values, it is hard to represent data use a certainty visualization method, like histogram, or tree graphic. In the Fig. 1(a), shows the percent of pass data for provinces, the data sorted by value descend, the effects of presentation is less than histogram. Users prefer the Fig. 1(b) to show the progressive increasing or decrease progressively data. This visualization form only present one year or one time interval data.

In the macro quality dataset, which cross many years or months, how to represent data as time sequence is the significant issue for quality data statistics. Design and implement time sequences visualization method, main to analysis quality state in the

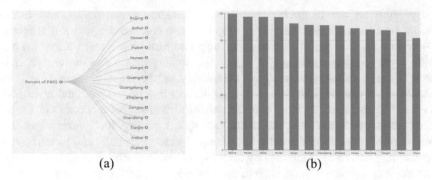

Fig. 1. Percent of pass data diagram of 31 provinces, (a) is tree graph, (b) is histogram of percent of pass data

special time and analysis one or more quality features development situation. This method solve these problems like: the eligible rate of products and consumer goods, the increase rate of goods value or pots of inspection and quarantine of imported and exported, the counts of standards draw up or revise, etc. We designed a way to present many years or many time interval data by histogram integrate tree graph that display each provinces' product supervision sampling rate from 2010 to 2017. The tree node present province, and in the right of province name, display a histogram to show the information of product supervision sampling rate value, create a function to implement show histogram when click tree node symbol. Then we can display all the product supervision sampling rate in one graph, users no need to click redundant.

Except integrate tree and histogram, other charts like heat map represent time series relationship also apply to this scene. As Fig. 2 showed data for eight years of each province, we regard these data as a matrix, provinces express x-axis, years express y-axis, the tables separated by axis show the value, and the dark color shows the bigger value.

2.2   Map and Histogram Visualization Method

Macro quality data not only provides value, it supports related regional, compared features. Map visualization method fit regional data comparison, generally, compared between macro quality data of 31 provinces or cites in china. This paper mapping value of provinces by color, the different color indicate interval value defined by legend setting. In Fig. 3, showed the eligible rates of product quality supervision of each provinces and municipality directly under the Central Government, red represent eligible rate between 8% and 84%, orange red represent 84% to 88%, pink represent from 88% to 92%, green represent 98% to 100%. Follow the color rule, green indicate excellent state, red indicate warning. In this chart, users conveniently know which province is the low eligible rate of product quality supervision, and which is the best, clearly distinguish between provinces. From the chart, the provinces Heilongjiang, Ningxia, Jilin is the top 3 of eligible rate, Qinghai and Jiangxi is the lowest rate.

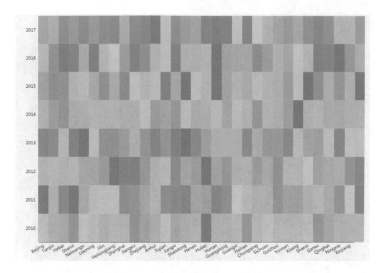

Fig. 2. Heatmap

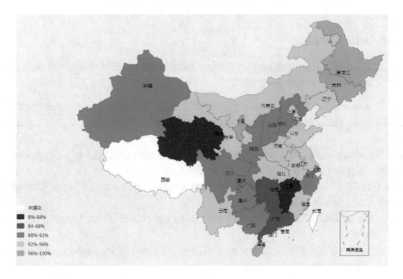

Fig. 3. Map visualization of eligible rate of product quality supervision

The second method focuses on applying map visualization with more complexity, in order to display more information, display histogram on the map for each province, as the Fig. 4 showed. On the map, the histogram represents four statistics value of metering standards for every provinces. This representation type show the multiple dimension data, compared more than one features of province.

Fig. 4. Integrate map and histogram

2.3 Interaction Design

We designed abundant interaction operation, user interact with charts get detail, to hold more information the chart delivered.

- Selection of date interval, users get special data via select date, and then the data filtered from date begin to end. The time granularity includes year, quarter, and month.
- Filter data based on interest. When the page includes multiple type data, the system function support users query and filter data.
- Chart transfer. Users switch multiple charts type among line charts, histogram and scatter chart, when they need.
- Zoom. The system allows users to zoom in or out of the chart, for display more data from database.
- Tooltips. Move on the data area, displays detailed information and custom the format of data string.

3   Visualization System Overview

Quality Statistics Visualization System (MQDVS) focus on macro quality data from government and internet published by authority organizations, use data visualization and visual analysis methods, analysis macro quality states. Data process and data visualization are main function models of this system, as Fig. 5.

Data process module responses data pump, process and analysis, extract quality related information. Transfer data from Excel to MySQL.

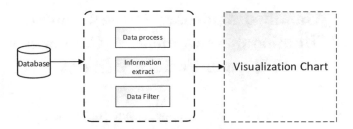

Fig. 5. MQDVS overview

Visualization chart module responses data representation includes three topics visualization graphics, and the two visualization methods mentioned in Sect. 2. The visualization topic includes quality infrastructure, quality development and quality security. Represent many kinds information graphics based on data type and data topic.

4 Conclusion

This paper proposes two visualization methods to represent quality data. To display quality data and data analysis results, the proposed methods were framed in dataset that includes regionals, times, and given values. The visual methods were designed and implemented in a quality statistics visualization system, which implement quality data representation, and can manage users' authorities or access based on RBAC. By this system, users can observe quality data and discover quality change trends, even find quality risk, provide a platform to users for data exploring.

Acknowledgements. This paper is supported by grants from National Key R&D Program of China (2016YFF0204205) and China National Institute of Standardization (712016Y-4941-2016, 522016Y-4681-2016).

References

1. Friendly, M.: Milestones in the history of thematic cartography, statistical graphics, and data visualization (2008)
2. Wu, Y., Cao, N., Gotz, D., Tan, Y.P., Keim, D.: A survey on visual analytics of social media data. IEEE Trans. Multimed. **18**, 2135–2148 (2016)
3. http://echarts.baidu.com
4. Smith, A., Hawes, T., Myers, M.: Hiearchie: visualization for hierarchical topic models. In: The Workshop on Interactive Language Learning, pp. 71–78 (2014)
5. Draper, G.M., Livnat, Y., Riesenfeld, R.F.: A survey of radial methods for information visualization. IEEE Trans. Vis. Comput. Graph. **15**(5), 759–776 (2009)
6. Coppola, A., Stewart, B.: A tool for structural topic model visualizations (2016)
7. Macro Quality Data. http://www.aqsiq.gov.cn

Gamified Approach in the Context of Situational Assessment: A Comparison of Human Factors Methods

Francesca de Rosa[1]([⊠]), Anne-Laure Jousselme[1],
and Alessandro De Gloria[2]

[1] NATO STO Centre for Maritime Research and Experimentation,
Viale San Bartolomeo 400, 19126 La Spezia, SP, Italy
{Francesca.deRosa, Anne-Laure.Jousselme}@cmre.nato.int
[2] University of Genoa, Via Opera Pia 11A, 16145 Genoa, Italy
adg@elios.unige.it

Abstract. Decision support tools are increasingly common in daily tasks, with a core component of enhancing user Situational Awareness required for an informed decision. With the goal of informing the design of automated reasoners or data fusion algorithms to be included in those tools, the authors developed the Reliability Game. This paper compares the Reliability Game to other Human Factor methods available in the context of Situational Awareness assessment. Although the Reliability Game shares many common elements with HF methods it also presents some unique features. Differently than those methods, the former focuses on the Situational Assessment process and on how source factors might influence human beliefs, which are considered as basic constructs building up Situational Awareness. Moreover, the gamified approach introduces an engaging component in the setup and the specific design of the method allows the collection of data expressing second-order uncertainty.

Keywords: Human factors methods · Situational awareness
Situational assessment · Reliability game

1 Introduction

Decision-making is a typical human task. While we move towards higher degrees of automation, many of the cognitive tasks on which decision-making is grounded are gradually delegated to machines to facilitate operators of different working environments (*e.g.* safety, security, crises management, health, first aid). For instance, the competent national maritime authorities established the Vessel Traffic Service (VTS), a system of systems for maritime traffic monitoring. Operational environments such as the VTS can entail a high degree of complexity in terms of *information quantity*, *information quality*, *information variety*, *communication means* and *communication formats*, rendering the information processing tasks (*e.g.* perception, correlation, filtering, sense making) sometimes beyond human ability. Support systems aim thus at assisting problem solving and decision-making tasks, acting as "enabler[s], facilitator [s], accelerator[s] and magnifier[s] of human capability, [but] not [as] its replacement"

© Springer International Publishing AG, part of Springer Nature 2019
T. Z. Ahram (Ed.): AHFE 2018, AISC 787, pp. 100–110, 2019.
https://doi.org/10.1007/978-3-319-94229-2_11

[1]. Therefore, one of their primary objectives is to enhance users' Situational Awareness and possibly propose candidate courses of action. Situational Awareness (SAW) is a state of knowledge defined by Endsley as "the perception of the elements in the environment within a volume of time and space, comprehension of their meaning and the projection of their status in the near future" [3]. SAW can be obtained through a cognitive process known as Situational Assessment (SA). Information processing tasks are progressively automated, and information fusion techniques come into play to reduce the cognitive burden placed on the operator, while maintaining an appropriate level of SAW. Typically, SA corresponds to high-level fusion, a process farther to sensors and closer to humans, with increased semantics.

For these reasons, many authors (e.g. [4, 5]) have underlined the importance of adopting a human-centered design approach for decision support systems as systems do not only include technological elements, but extend beyond hardware and software to include procedural and human elements (e.g. Christensen's system model [6]). In operational environment operators might play several roles, possibly concurrently, such as "decision maker, monitor, information processor, information encoder and storer, discriminator, pattern recognizer [,] . . . ingenious problem solver" [7] or disseminator. With respect to information processing humans can be assimilated to a processor with a single channel and limited-capacity [4]. Therefore, the reasoning and communication schemes implemented in the systems should be intelligible and possibly intuitive, in order to ensure algorithms *transparency* [8], accountability, user acceptance and optimal *human-machine synergy*. In order to bring humans and machines closer, Human Factors (HF) methods [2] have been developed with the intent of assessing operators SAW, concentrating on physiological and performance aspects, imbedded tasks, subjective ratings and questionnaires.

To support the design of transparent automated reasoners and information fusion algorithms, mimicking the human information combination process in a multisource context, the authors developed the Reliability Game [9]. This method aims at eliciting from the participants the impact of source factors (*type* and *quality*) on the Situational Assessment process and final Situational Awareness. As the remainder of this paper will develop, the Reliability Game is an innovative approach compared to other HF methods in the context of Situational Awareness, both in terms of its set-up (gamified approach) and focus.

After a brief survey of some Human Factors methods adopted in the context of Situational Awareness assessment in Sect. 2 and the presentation of relevant results of psychology and social science research on source factors in Sect. 3, the Reliability Game method is outlined in Sect. 4. Section 5 summarises a qualitative comparison with other Human Factors methods and highlights novelties and similarities. Finally, the conclusions and way ahead are presented in Sect. 6.

2   Situational Awareness Assessment Methods

A literary review by Stanton *et al.* [2] highlighted the existence of several human factors methods dedicated to the assessment of SAW. Most techniques concentrate on the assessment of individual SAW through measurement approaches for example by looking at physiological aspects, performance aspects, imbedded tasks, subjective ratings and questionnaires [2, 10]. Less emphasis has been put on distributed or team SAW techniques [2]. Following [2] the individual SAW assessment techniques can be categorised as:

1. SAW requirements analysis techniques;
2. Freeze probe technique;
3. Real-time probe technique;
4. Self-rating techniques;
5. Observer-rating techniques.

SAW requirement analysis techniques, which might be based on interviews with Subject Matter Experts (SMEs), questionnaires and goal-directed task analysis [11], aim at understanding which are the elements that contribute to SAW with respect to a specific task or environment. In freeze probe techniques (*e.g.* SACRI [12], SAGAT [10] and SALSA [13]) a task and/or scenario is simulated and participants have to respond to SAW related queries administered during a freeze of the simulation. Real-time probe techniques (*e.g.* SASHA [14] and SPAM [15]), differently from the previous ones, administer the SAW queries without freezing the simulation, while in the self-rating techniques (*e.g.* CARS [16], MARS [17], SARS [18], SART [19] and C-SAS [20]) the participants are requested, generally post-trial, to self-rate dimensions related to SAW. For example, in SART the dimensions include *familiarity*, *complexity of situation*, *information quality*, *information quantity* and *concentration of attention*. In observer-rating techniques (*e.g.* SABARS [21]), contrary to the ones previously mentioned, it is an appropriate SME who rates the participants SAW, while observing them performing a specific task.

The aim of the above-mentioned methods (with the exception of SAW requirements techniques) is to measure the level of SAW, often with the primary scope of assessing specific operational systems and/or innovative technologies and designs. Besides serving as key performance indicators of the effectiveness of novel technologies, those techniques and measurements allow also to investigate [22]:

1. the nature of SAW;
2. factors affecting SAW;
3. the strategies and processes adopted to acquire SAW.

The last two points are the objectives of the Reliability Game method, which is going to be detailed in the following sections. In fact, the Reliability Game method aims at characterising the impact of factors related to sources of information (*e.g.* source type and source quality) on the Situational Assessment process and final SAW.

3   Source Factors Impact on Human Assessment

Aspects related to the impact of source factors on human assessments have been the subject of social science and the experimental psychology for decades. Although the results of those studies cannot be directly incorporated within the modelling paradigms used in the context of information fusion, they served as basis for the interpretation and analysis of the data gathered through the Reliability Game method.

Persuasion literature reports on the mechanisms that determine the effectiveness of sources of information perceived as *credible, attractive, similar* or *powerful* [23–26]. Research has shown the complexity and dynamic nature of the processes taking place with respect to source factor impact on *attitude* change: there is not a linear mapping between a message provided by a more attractive or expert source and a higher degree of persuasion (or attitude change in the expected direction). Therefore, studies have been focusing on complementarity aspects, such as how persuasive sources might affect both primary levels of cognition (*e.g.* source serving as peripheral cue, source influencing the direction of thoughts or source influencing the amount of thoughts) and secondary or metacognition levels (*e.g.* thought confidence) [23]. It has to be underlined that most of the conducted research explores attitude change, as it is assumed to serve as key mediation construct with respect to other targets of change, such as *emotions, behaviors* and *beliefs* [23].

In the contemporary theories on attitude formation and update, such as Dual-Processing theories (*e.g.* [27, 28]) and Dual-System theories (*e.g.* [29]), researchers postulate that several factors, including *source factors*, can affect attitudes through processes such as:

1. acting as peripheral cue or heuristics [30, 31];
2. acting as issue-relevant argument [32];
3. impacting the amount of processing taking place [33–35];
4. biasing the nature of thoughts [36];
5. impacting structural properties of thoughts (*e.g.* thought confidence) [37].

With regard to the factors that might be used as cues a relevant role is played by *attractiveness of the source* [38] and *credibility of the source* [37], often referred to as source reliability [39].

4 The Reliability Game

4.1 The Reliability Game Method

The aim of the Reliability Game is to characterise the impact of source factors, such as source quality and source type, on human situational assessment process. The game, which has been inspired by the Risk Game [40], allows capturing belief changes using cards and a specific game board. The participant is requested to estimate which of the three possible hypotheses corresponds to a specific situation on the basis of available information. Information together with possible additional meta-information (*i.e.* *source type* and *source quality*) is provided to the participant through messages that are contained in cards.

Each game session is divided in four rounds, in which a set of eleven cards is sequentially provided to the player, displaying information about the situation. The rounds differ in the meta-information displayed on the cards. As can be seen in Table 1 during the first round, information is only provided without its origin source and quality. During Round 3, only the *source type* is provided and the participant is asked to rate the *source quality*, based on his/her background knowledge and experience of the *source type*. A six-point source quality rating scale (*1* = *bad* to *5* = *good* and *Unknown*) has been selected to align with most of the existing standards of source reliability rating in the intelligence domain [39]. The same rating scale has been adopted for the *confidence* level (expressing the final belief) in the three different hypotheses that the participants have to rate at the end of each round.

Table 1. Rounds and associated meta-information.

Round	Source type	Source quality
Round 1	Not provided	Not provided
Round 2	Not provided	Provided
Round 3	Provided	Rated by participant
Round 4	Provided	Provided

During each round the cards need to be positioned on a game board and the selected position reflects the weight of belief that the information in a card provides toward some subsets of the mutually exclusive and collectively exhaustive hypotheses (see Fig. 1). The game board has been specifically designed with the intent of providing to the participant the freedom to express their belief without being constrained by weights of belief that sum up to *1*. Moreover, the board enables a relatively straightforward modelling of the data collected within mathematical frameworks for second-order uncertainty handling (*e.g.* Evidential Theory [41]). The card positions, together with the participant's *source quality* rating (if any) and the participant's *confidence* in the hypotheses, are recorded at the end of each round.

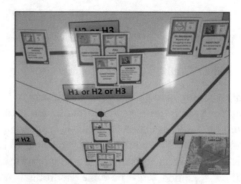

Fig. 1. Example of card positioning on the Reliability Game board

4.2   Training and Time

The game session has been designed to last around thirty to forty minutes, which is quite fast. The facilitator first introduces the participant to the game scope, rules and scenario. The game does not foresee a real pilot run. On the other hand the facilitator guides the player when positioning the first card, making the initial brief easy and short.

4.3   Domain of Application and Example

The method has been developed in the context of Maritime Situational Awareness, thus with the specific objective of assessing the impact of source factors on human Maritime Situational Assessment and resulting SAW. It has been played with twenty-one players, subject matter experts of Maritime Situational Awareness.

Although it has been developed with respect to the maritime domain, it is important to underline that the game could be easily tailored to other domains (*e.g.* air traffic control, medicine, emergency and disaster recovery). Moreover, the method shows its potential to assess the impact by other factors related to uncertainty and information quality, such as trueness and precision.

5   The Reliability Game and the Other Methods in the Context of Situational Awareness

Differently from the other Human Factor methods available in the context of Situational Awareness assessment (see Sect. 2), the Reliability Game main focus is not on SAW, but rather on the Situational Assessment process. Therefore, it does not provide a measure of SAW (contrary to SAGAT, SART, SALSA) or a set of SAW requirements (*e.g.* SAW requirement Analysis [11]). The Reliability Game objective is to evaluate *which* and *how much* information and source factor impact human beliefs, which are assumed as basic constructs that build up SAW. Given that the game is not measuring SAW, the correct estimation of the *true* hypotheses by the participant is of secondary importance. In fact, the analysis is concentrating mainly on the extent and direction of belief changes induced by the above-mentioned factors.

The Reliability Game is a simulation technique (each round corresponds to a simulation), however the simulation is not performed in a high-fidelity simulator. In fact, the participant is just presented with a scenario map, briefed on a scenario story and presented with incoming information reported on cards. The design choice was driven by the attempt to reduce the impact of information visualisation and system familiarity, while focusing on the information processing. Although the *complexity of the element design* can be regarded as medium, as it might require the support of SMEs to the scenario and cards design, the effort and required resources are less than for other common HF methods, such as SAGAT or SALSA. This has the advantage of making the Reliability Game a quick, easy to apply and low-cost approach (such as SART and C-SAS). Moreover, the method is characterized by a low training and facilitation complexity.

The Reliability Game presents both elements of freeze probe techniques (*e.g.* SAGAT, SACRI and SALSA) and self-rating techniques (*e.g.* CARS, MARS, SARS, SART and C-SAS). In fact, like in freeze probing techniques the simulation is frozen during query administration to the participant. The method has only a three item pre-defined set of queries that require a self-rating from the participant, that is:

Table 2. Comparison of the Reliability Game method and other HF methods in the context of situational awareness assessment

Criteria	Reliability game	SAW req. anal.	SACRI	SAGAT	SALSA	SASHA	SPAM	CARS	MARS	SARS	SART	C-SAS	SABARS	SA-SWORD
Technique	Freeze probe/self-rating	•	✓	✓	✓	•	•	✓	✓	✓	✓	✓	•	•
Main focus	Situational assessment	•	•	•	•	•	•	•	•	•	•	•	•	•
Dimensions analysed	Source factors impact	•	•	•	•	•	•	•	•	•	•	•	•	•
Set-up	Game	•	•	•	•	•	•	•	•	•	•	•	•	•
Query design	Simple	•	•	•	•	•	•	✓	✓	•	✓	✓	•	✓
Element design (e.g. scenario, cards)	Medium	•	•	•	•	•	•	✓	✓	•	•	•	✓	•
Execution time	Low	•	•	•	•	•	✓	•	•	✓	✓	✓	•	✓
Cost	Low	•	•	•	•	•	•	✓	✓	•	✓	✓	✓	✓
Training time	Low	•	✓	✓	✓	•	•	✓	✓	✓	✓	✓	•	✓
Facilitation complexity	Simple	•	✓	•	•	•	•	✓	✓	✓	✓	•	•	
Query administration complexity	Simple	•	✓	✓	✓	•	•	✓	✓	✓	✓	•	•	✓
Data Collection complexity	Simple	•	✓	✓	✓	•	•	✓	✓	✓	✓	•	•	✓
Possible Mathematical Modelling	Direct	•	•	•	•	✓	•	•	•	•	•	•	•	•
Domain of application	Multiple	✓	•	✓	•		✓	✓	✓	•	✓	✓	•	•

1. Request to position the card on the board at each freeze;
2. Request to rate the source quality at each freeze (only in specific simulations);
3. Request to rate the confidence in the different hypotheses at the end of each simulation.

Table 2 reports the main elements of a qualitative comparison between the Reliability Game and the other HF methods described in Sect. 2. The first column lists the comparison criteria, which are instantiated for the Reliability Game in the second column. The comparison criteria have all a binary outcome ($\vee$ = equal, • = not equal), with the exception of the *element design complexity* ($\vee$ = equal, • = not equal – Reliability Game lower, $•^*$ = not equal – Reliability Game higher). This table highlights how the Reliability Game shares with other HF methods important elements such as the *type of technique* (number of HF methods with equal value $n_\vee$ = 8), *simplicity of query design* ($n_\vee$ = 5), *simplicity of facilitation* ($n_\vee$ = 5) and the *simplicity of query administration and data collection* ($n_\vee$ = 8). Moreover, the method presents a low *cost* ($n_\vee$ = 6), low *execution time* ($n_\vee$ = 5) and low *training time* ($n_\vee$ = 9).

The comparative analysis showed that the Reliability Game presents many common elements with CARS, MARS and SART. On the other hand, it highlights some of the innovative aspects of this method, namely its main focus and the dimensions analysed. Those aspects are intrinsically linked to the innovative scope of the method that is to guide the design of reasoners and algorithms to be used in support systems. An additional innovative feature of the Reliability Game is the data collection technique. Although many techniques as previously mentioned present a low level of complexity with respect to the data gathering, to the best of our knowledge, positioning the card on the game board is an original way of answering a SAW related query and directly record the participant belief, while minimising the intrusiveness of the procedure. In fact, positioning the card is actually supporting the assessment process, instead of interrupting it to answer to the query. Moreover, the gamified approach creates an engaging context, as showed by the participant feedbacks. This has a direct impact on participants' message processing mechanism. In fact, it has been demonstrated that engagement enhances the information elaboration motivation [42, 43], leading to a more in-depth consideration of the message content and to minor reliance on cues.

As a new-born method, the Reliability Game has not been applied extensively and validation studies have not been performed. However, the feedbacks from the SMEs, which have been exposed to the Reliability Game [9], show that the game is not only perceived as *engaging*, but also as *realistic*, *relevant* with respect to operational needs and *effective* in the elicitation component.

6   Conclusions

This paper presents a comparison of the Reliability Game and other Human Factors methods available in the context of Situational Awareness assessment. The Reliability Game is an innovative method to gather data regarding the impact of factors related to sources of information (*e.g. source type* and *source quality*) on the Situational Assessment process and final Situational Awareness. The comparative analysis between the Reliability Game and other thirteen HF methods available in the context of Situational Awareness assessment, shows that although the former shares many common elements with some of the latter (*e.g.* CARS, MARS and SART) it also presents some unique features. In fact, the Reliability Game does not provide a measure of Situational Awareness, but rather an evaluation on how source factor might influence human beliefs. The gamified approach introduces an engaging component in the setup and the specific design of the method allows the collection of data expressing second-order uncertainty. The data collected in a first experiment is currently under analysis and the results will be included in the design of a Bayesian Network for maritime behaviour analysis.

Acknowledgments. This research was supported by NATO Allied Command Transformation (NATO-ACT). The authors wish to thank Cdr. Andrea Iacono for advice and feedback during the development of the game. Moreover, the authors would like to thank Dr. David Mandel for the fruitful technical discussions on the concept of source reliability.

References

1. Stikeleather, J.: Big data's human component. Harvard Bus. Rev. (2012). https://hbr.org/2012/09/big-datas-human-component
2. Stanton, N.A., Salmon, P.M., Walker, G.H., Baber, C., Jenkins, D.P.: Human Factors Methods: A Practical Guide for Engineering And Design. Ashgate Publishing Company, Brookfield (2006)
3. Endsley, R.M.: The application of human factors to the development of expert systems for advanced cockpits. In: Human Factors Society 31st Annual Meeting, pp. 1388—1392. Human Factor Society, Santa Monica (1987)
4. Nemeth, C.P.: Human Factors Methods for Design: Making Systems Human-Centered. CRC Press, Boca Raton (2004)
5. Hall, D.L., Jordan, J.M.: Human-centered Information Fusion. Artech House, Boston (2010)
6. Christensen, J.: The nature of systems development. In: Human Factors Engineering: Engineering Summer Conferences. University of Michigan, Ann Arbor (1985)
7. Pew, R.: Human skills and their utilization. In: Human Factors Engineering: Engineering Summer Conferences. University of Michigan, Ann Arbor (1985)
8. Explainable AI systems: understanding the decisions of the machines. https://www.bbvaopenmind.com/en/explainable-ai-systems-understanding-the-decisions-of-the-machines/#.WkEZgAJbI-A.twitter
9. de Rosa, F., Jousselme, A.-L.: A reliability game for source factors impact assessment. In: 2017 Conference on Decision Support and Risk Assessment for Operational Effectiveness (DeSRA). NATO STO CMRE – Centre for Maritime Research and Experimentation, La Spezia (2017)
10. Endsley, R.M.: Measurements of situation awareness in dynamic systems. Hum. Factors **37**(1), 65–84 (1995)
11. Endsley, M.R.: A survey of situation awareness requirements in air-to-air combat fighters. Int. J. Aviat. Psychol. **3**, 157–168 (1993)
12. Hogg, D.N., Folleso, K., Strand-Volden, F., Torralba, B.: Development of a situation awareness measure to evaluate advanced alarm systems in nuclear power plant control room. Ergonomics **38**(11), 2394–2413 (1995)

13. Hauss, Y., Gauss, B., Eyferth, K.: SALSA - a new approach to measure situational awareness in air traffic control. Focusing attention on aviation safety. In: 11th International Symposium on Aviation Psychology, Columbus (2001)
14. Jeannott, E., Kelly, C., Thompson, D.: The development of situation awareness measures in ATM systems. Technical report, EATMP (2003
15. Durso, F.T., Hackworth, C.A., Truitt, T., Crutchfield, J., Manning, C.A.: Situation awareness as a predictor of performance in en route air traffic controllers. Air Traffic Q. **6**, 1–20 (1998)
16. McGuinness, B., Foy, L.: A subjective measure of SA: the crew awareness rating scale (CARS). In: Human Performance, Situational Awareness and Automation Conference, Savannah (2000)
17. Matthews, M.D., Beal, S.A.: Assessing situation awareness in field training exercises. Research Report, U.S. Army Research Institute for the Behavioural and Social Sciences (2002)
18. Waag, W.L., Houck, M.R.: Tools for assessing situational awareness in an operational fighter environment. Aviat. Space Environ. Med. **65**(5), A13–A19 (1994)
19. Taylor, R.M.: Situational awareness rating technique (SART): the development of a tool for aircrew systems design. In: Situational Awareness in Aerospace Operations (AGARD-CP-478), pp. 3/1–3/17, Neuilly Sur Seine (1990)
20. Dennehy, K.: Cranfield - situation awareness scale user manual. Technical report, College of Aeronautics, Cranfield University, Bedford (1997)
21. Matthews, M.D., Pleban, R.J., Endsley, M.R., Strater, L.D.: Measures of infantry situation awareness for a virtual MOUT environment. In: Human Performance, Situation Awareness and Automation Conference (HPSAA II), Daytona, LEA (2000)
22. Endsley, R.M.: Toward a theory of situation awareness in dynamic systems. Hum. Factors **37**(1), 32–64 (1995)
23. Briñol, P., Petty, R.E.: Source factors in persuasion: a self-validation approach. Eur. Rev. Soc. Psychol. **20**, 49–96 (2009)
24. Kelman, H.C., Hovland, C.I.: Reinstatement of the communicator in delayed measurement of opinion change. J. Abnorm. Soc. Psychol. **48**, 327–335 (1953)
25. Rhine, R.J., Severance, L.J.: Ego-involvement, discrepancy, source credibility, and attitude change. J. Pers. Soc. Psychol. **16**(2), 175–190 (1970)
26. Sternthal, B., Dholakia, R., Leavitt, C.: The persuasive effect of source credibility: tests of cognitive response. J. Consum. Res. **4**(4), 252–260 (1978)
27. Petty, R.E., Cacioppo, J.T.: Source factors and the elaboration likelihood model of persuasion. Adv. Consum. Res. **11**, 668–672 (1984)
28. Chaiken, S., Liberman, A., Eagly, A.H.: Heuristic and systematic processing within and beyond the persuasion context. In: Uleman, J.S., Bargh, J.A. (eds.) Unintended thought, pp. 212–252. Guilford Press, New York (1989)
29. Epstein, S., Pacini, R., Denes-Raj, V., Heier, H.: Individual differences in intuitive-experiential and analytical-rational thinking styles. J. Pers. Soc. Psychol. **71**, 390–405 (1996)
30. Hovland, C.I., Weiss, W.: The influence of source credibility on communication effectiveness. Public Opin. Q. **15**, 635–650 (1951)
31. Petty, R.E., Cacioppo, J.T., Goldman, R.: Personal involvement as a determinant of argument-based persuasion. J. Pers. Soc. Psychol. **41**, 847–855 (1981)
32. Kruglanski, A.W., Thompson, E.P.: Persuasion by a single route: a view from the unimodel. Psychol. Inq. **10**, 83–110 (1999)
33. DeBono, K.G., Harnish, R.J.: Source expertise, source attractiveness, and the processing of persuasive information: a functional approach. J. Pers. Soc. Psychol. **55**, 541–546 (1988)

34. Heesacker, M.H., Petty, R.E., Cacioppo, J.T.: Field dependence and attitude change: source credibility can alter persuasion by affecting message-relevant thinking. J. Pers. **51**, 653–666 (1983)

35. Priester, J.R., Petty, R.E.: Source attributions and persuasion: perceived honesty as a determinant of message scrutiny. Pers. Soc. Psychol. Bull. **21**(6), 637–654 (1995)

36. Chaiken, S., Maheswaran, D.: Heuristic processing can bias systematic processing: effects of source credibility, argument ambiguity, and task importance on attitude judgment. J. Pers. Soc. Psychol. **66**(3), 460–473 (1994)

37. Briñol, P., Petty, R.E., Tormala, Z.L.: The self-validation of cognitive responses to advertisements. J. Consum. Res. **30**, 559–573 (2004)

38. Haugtvedt, C.P., Petty, R.E., Cacioppo, J.T.: Need for cognition and advertising: understanding the role of personality variables in consumer behavior. J. Consum. Psychol. **1**, 239–260 (1992)

39. de Rosa, F., Jousselme, A.-L.: Critical review of uncertainty communication standards in support to maritime situational awareness. Technical report, NATO STO Centre for Maritime Research and Experimentation, La Spezia (2018)

40. Jousselme, A.-L., Pallotta, G., Locke, J.: A risk game to study the impact of information quality on human threat assessment and decision making. Technical report CMRE-FR-2015-009, NATO STO Centre for Maritime Research and Experimentation, La Spezia (2015)

41. Shafer, G.: A Mathematical Theory of Evidence. Princeton University Press, Princeton (1976)

42. Bakker, A.B.: Persuasive communication about AIDS prevention: need for cognition determines the impact of message format. AIDS Educ. Prev. **11**(2), 150–162 (1999)

43. Stephan, J., Brockner, J.: Spaced out in cyberspace?: evaluations of computer-based information. J. Appl. Soc. Psychol. **37**, 210–226 (2007)

Estimation of Risks in Scrum Using Agile Software Development

Muhammad Ahmed$^{(\boxtimes)}$, Babur Hayat Malik, Rana M. Tahir,
Sidra Perveen, Rabia Imtiaz Alvi, Azra Rehmat, Qura Tul Ain,
and Mehrina Asghar

University of Lahore, Punjab, Pakistan
m.ahmed.3705@gmail.com, engr.ranatahir01@gmail.com,
sidra.perveen6@gmail.com, rabi.alvi421@gmail.com,
azrarehmat82@gmail.com, q.ainie052@gmail.com,
mehru46@gmail.com, baber.hayat@cs.uol.edu.pk

Abstract. In this era a number of latest trivial software development process methods have been developed. Agile software development is one of them. Agile is a time-dependent approach to software delivery. Scrum is the agile software development methodology process that is extensively utilized today in many of the software companies. It is an agile technique to handle a project, usually in agile software development. Agile software development by way of Scrum is frequently perceived as a methodology; however than showing Scrum as methodology, consider of it as a framework for handling a process. This paper will initiate with the background, it will cover the characteristics and definition of agile software development and highlights the major different agile software techniques. Various agile techniques will also elaborate in this paper. The core aim of this paper is to identify risk on agile development and improve the quality of the software by using agile methodologies.

Keywords: Scrum · Agile software development
Traditional software development · Risk management
Software development techniques

1 Introduction

Programming began through prepared languages like "*FORTRAN in 1954*" [1], and after developed towards the object-oriented languages in the 1960s [2]. Correspondingly, developmental approaches in software developments have developed over time. The main developed approach appeared later than the software crisis in the seventies. Software engineers introduce methodologies; methodologies split the software development process into five stages [3]. Agile be a phrase first established in "*2001*" with referring to gather of lightweight software development procedure developed in the "*mid-1990s including Scrum (1995), Crystal Clear, Extreme Programming (1996), Adaptive Software Development, Feature Driven Development, and Dynamic Systems Development Method (DSDM) (1995)*" [4]. A lot of information handling prompted the introduction of

© Springer International Publishing AG, part of Springer Nature 2019
T. Z. Ahram (Ed.): AHFE 2018, AISC 787, pp. 111–121, 2019.
https://doi.org/10.1007/978-3-319-94229-2_12

original database administer framework. The product framework turned out to be increasingly intricate and extensive; software reliability issue was very noticeable. The person unique own design, personal processes never again assemble the needs; software needs to modify method of development *"Software Crises broke out"* [5] (Fig. 1).

Software Crisis

Traditional Software Development

Agile Software Development

Fig. 1. Process of software development

Agile techniques are utilized to accomplish advanced quality software in a minimum timeframe, self-arranging groups, client coordinated effort few documentation and limited time to market [6]. Agile methods spilt errands in little steps with least scheduling known as Iteration. Iterations are brief time span that keeps running one to a month. Every iteration compromises groups functioning throughout complete software development sets, such as "planning, requirement analysis, design, coding, unit testing and acceptance testing". Majority of the agile implementation usage utilize a formal every day eye-to-eye correspondence among colleagues [7]. Whenever client or domain expert works specifically with the creation group stakeholder discovers some new information about the issue [8]. Scrum is an iterative and incremental software development way to deal with depicts a proficient and adaptable product development method. It was sophisticated enhancement, brought into software business by "Ken Schawaber and Jeff Sutherland", and developed into form [9]. Scrum's main features deceit in its iterative development procedure, as similar to other agile processes, the execution of Scrum depends to a vast amount on capacity of included colleagues. In Scrum, much is being done to draw the exhibition in the sprint review gathering, in which the administration becomes more familiar with the status of product [10].

This paper organizes as Sect. 2 will cover related work of the agile and SCRUM methodologies, Sect. 3 includes the overview of traditional development method, Sect. 4 will elaborates the Scrum methodology, Sect. 5 will cover the risks that may happened to Scrum methodology.

2   Literature Review

Different examination and reviews has been made that demonstrate the fame of agile methodology that supports on feature requirements little or large companies and occurrences of stakeholders. Agile methods have demonstrated viability and changing the product business. A few of the existing literature review is discussed below:

Begel and Nagappan [11] utilized a review base technique; agile technique is beneficial because of enhanced correspondence between colleagues, quick release and adaptability of plans. Scrum methodology has been very prominent; and test-driven development and some programming are the minimum exploited practices.

Zuo et al. [12] concerned proper strategies into agile software development. They connected "rCOS" a question-situated way to deal with agile methodology to enhance precision of the framework and encouraging framework development with object oriented ideas.

Livermore [13] demonstrated that XP is utilized vigorously in the companies. Scrum is utilized as function of "diminishing use, function oriented development, dynamic development methodology, adaptive software development and then other flexibility methodologies" which are altered according to organization utilization.

In additional explanation by Ahmed et al. [6] is commonly utilized, 50% of products are conducted by the energetic contribution of stakeholder, 66.7% of participants agreed that productivity improved would improve the quality by 50%.

Nakki et al. [14] showed that agile strategies were beneficial for some program foundation. Large stakeholder software, efficient requirements, clients have big security and huge codebase, traditional plan oriented product performing professionally. Thus, agile methods give finest outcomes when the group is not huge, the requirements are not efficient the code for the product is small and the client demonstrates huge interested for huge development advancement.

Solo and Abrahamsson [15] demonstrated that appropriation increment of agile procedures is associated with organization dimension. Additionally, the approval of agile procedures in big and distributed condition is being attended to all the more every now and again. 54% of the group members utilized XP and the five factors in XP utilized are "open office space, coding standards, 40 h week, consistent incorporation and aggregate code possession"; 27% utilized Scrum procedures and other used different agile practices.

Dyba and Dingsoyr [8] reported that XP it appeared to be hard to set up in big, complex companies; however, it is simpler in other association sorts. Pair programming is wasteful and XP is workings preeminent with experienced advance development team. There is absence of consideration regarding architectural plan and design problems.

Far [16] showed software charted reliability engineering in robust development process. According to research, test driven development looks miss matched with the reliability model.

Misra et al. [17] demonstrated structure for recognizing essential modification mandatory for accepting agile software development methods in traditional development associations.

Kohlbacher et al. [18] illustrated the negative effect of modification in client approval. The principle commitment of their research relates to collaboration impacts of alter in requirements and agile procedure on client's satisfaction. They originated that work environment, the adaptableness of the ending product and eagerness to adapt to change has positive directing impact on the affiliation.

Imreh and Raisinghani [19] assumed that agile software development significantly affects the quality of a software product. They recognized important quality aspects of agile software development and different methodologies of remediation suggested inside methodologies, hierarchical, cultural framework and company best practice.

Research accomplished by Syed-Abdullah et al. [20] demonstrated that agile technology is greater paying attention by people than on process orientation in a more unpredictable condition. But for the satisfaction of the designers, this is useful just when the requirements are indeterminate or unpredictable.

3   Traditional Development Method

Software crises have pretty much precede the development of software engineering. In nineties, product advancement of software development started to utilize with the documentation, in view of hypothesis of software technology. Waterfall show was delegated of traditional software project administration and had a possessed a critical spot. As indicated by the waterfall appeared into "*Fig. 2*" is a model, which has been created for software development that is to develop software. It is called as such because the model increases efficiently one stage to other in a descending way, similar to a waterfall [21]. Waterfall highlighted software development cycle as appeared in Fig. 2.

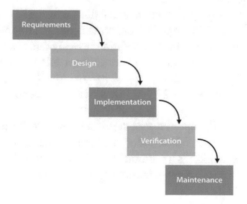

Fig. 2. Waterfall model

"*Structured Analysis and Structured Design (SASD)*" [6] is a software development technique, introduced in seventies by "*Yourdon, Constantine*". It highlighted complete venture or activity is separated into sub activities. This spares time and incredibly enhanced by proficiency. Aside from this waterfall, strategy has a few advantages and drawbacks, which are given in Fig. 3 [22].

Because children are only concerned about games that are easy to play, rather than focusing on innovation, developers consider their end-users.

Advantages	Disadvantage
The result is predictable	A large number of documents
Strict control	The product will appear in the end of project
High quality	Difficult to modify

Fig. 3. Advantages and disadvantages of waterfall model

3.1 Concept of Agile Software Development

The perception of agile advancement was anticipated by the agile team in 2001, afterwards numerous software advancement groups and organization perceived, acknowledged it, and bit by bit been generally utilized as a part of many tasks. Agile Software Development [23] distributed that Agile Manifesto appeared in Fig. 4; sequentially software development has entered in a new era.

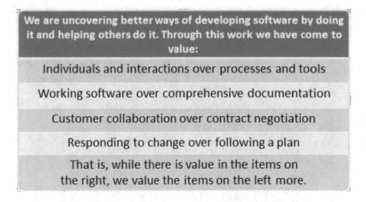

Fig. 4. Manifesto for agile software development

3.2 Rules and Laws of Agile Manifesto

Agile evaluation is an iterative strategy utilizing an augmentation and conveyance of important software. Indeed, agile development is additionally eager to roll out correct change with respect to the requirements. Software is profitable software, when it totally fulfills the customer needs. A few standards of agile manifesto [21] are given below:

1. Our most elevated need is to fulfill the client needs through right on time and incessant conveyance of profitable software.

2. Welcoming evolving requirements even late are being developed.
3. Always convey working software every now and again, from half a month to two or three months with an inclination to shorter time scale.
4. Business and designers have to cooperate every day all through the product.
5. Developing products around motivated people. Give them situation and shore up they need and believe them to take care of business.
6. The most proficient and viable strategy of transferring data to and inside a designing team is a face-to-face conversation.
7. Software working is essential measure of advancement.
8. Agile processes endorse maintainable improvement. The backers, designer and clients ought to have the capacity to keep up a consistent pace indefinitely.
9. Constant concentration regarding industrial brilliance and finest design for improvement of maneuverability.
10. Simplicity, craft of capitalizing measure of work not done- is basic.
11. Most excellent models, requirement and designs rise up out of self arranging groups.
12. At standard intermission, the stakeholder thinks about how it can be more efficient, and then it concludes and alters its actions consequently.

3.3   Agile Development Versus Traditional Development

We evaluated the contrasts between Agile Manifesto and Traditional Software Development. They are two aspects [24] which agile development clearly focuses. The one is adaptability and the other is teamwork (Fig. 5).

Model	Adaptability	Teamwork
Traditional	Too much reliance on documentation Design and impelment are separete	Lack of communication Manager take charge of process
Agile	Documentation, design and demand can be changed during the implement	Divide into small group Flexibel and efficient

Fig. 5. Differences between the agile development and traditional development

3.3.1   Adaptability

Distinction to traditional model, agile development places accentuation on adaptableness than on inevitability of a traditional model. A person who picks traditional always enjoys highly exhaustive and entire documentation when starting a project. They will examine whole development process and points of interest of each sub process for instance, e.g. how frequently or what number of individuals will be put resources into, lastly the results are documented in the document. Most importantly, the document cannot be changes once it has been recognized. All project developers are mandatory to firmly chase the document. When somebody needs to modify the document or schedule, he is not permitted for that. However, there are many advantages of this model but people still faces many issues in this model.

Agile keen to acknowledge modification, even in the last procedure of software development. Its own particular strategies for framework plan and framework developers can rapidly react to modify in client request. It guarantees that results of latest iteration are the clients truly wants, it also assemble modifications of business. Contrasts from waterfall model, agile development completely conforms to design. Agile development would toward start of project to build up coarse plan, giving more space to change of projects.

3.3.2 Teamwork

The purpose of traditional process management is to guarantee procedure inside association is performed as expected and that define process is entirely stayed. A document-oriented model is likely to characterize individual's parts as negotiable. Agile software processes are people centered instead of process driven. They trust that people and their communications essential than procedures and methods. Practice is the life of approaches. Key purpose of agile development is let individuals to acknowledge a procedure instead of enforce a process. Developers should have the full freedom to formulate all decision about the technical features. To make a team with optimistic staff is key to agile development.

4 Scrum

In 1995, Ken Swaber introduced the Scrum methodology. Before implementation, it was practice. Afterward it was incorporated into agile methods since it has similar basic idea principles of agile development. Scrum has been utilized by the goal of product management during basic procedures, simple to modify documentation and higher team iteration over comprehensive documentation [25]. Scrum imparts fundamental idea and practices to other agile methods; however, it includes project management as a major aspect of its practices. These practices direct development team to discover duties at every development cycle [26]. Moreover, practices characterized for agility, one primary methods prescribed by Scrum is to develop a backlog. "*A backlog is where one can see all requirements pending for a project for monitoring, estimated by complexity, days or some other unit of measure that the team chooses*". Inside a product backlog there is fundamental sentence for every requirement, somewhat that will be utilized by team to begin discussion and group for that requirement [25] should have actualized putting points of interest of what it.

For group of Scrum, three fundamental responsibilities are characterized as appeared in Fig. 6. The primary task is of product owner, who chiefly is influence of business. The second task is of SCRUM team, which includes "*developers, testers and other roles*". This team would reach client and distinguish the requirement for another product. Scrum Master, the third position, is in charge of keeping team concentrated on particular objectives and assists the colleagues to solve the issue when they faced [27].

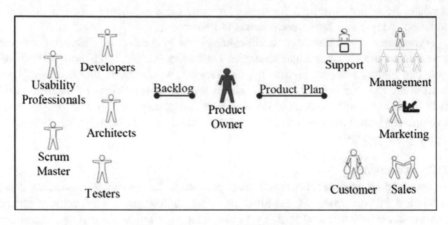

Fig. 6. Key roles in SCRUM [26]

The development procedure utilizing Scrum split project into stages. In every stage, 1 element is completely developed, tested and turned out to be prepared to go to construction. Team does not move to another stage until existing stage is completed. Regardless of whether what is being done increase worth of procedure or not, is fundamental anxiety of every stage. Recent investigation on traditional Scrum development have demonstrated that despite its point of interest, it is not most appropriate for products where the attention is on ease of use [27]. It does not meet the user's usability needs, as product owners keep their emphasis chiefly on business issues and disregard ease of the use. Because product owners typically originate from business foundation, they do not have the experience, abilities and inspiration to design for user experience. Besides traditional agile techniques are not worried about vision of user experience, which drives design and is most important for guaranteeing an intelligible set of user experience [25]. SCRUM has the highest usage percent as compared to other agile methodologies [28], Table 1 below shows some parameters of SCRUM and Traditional software development.

Table 1. Parameters of SCRUM and traditional software development.

Parameters	SCRUM	Traditional
Documentation	Low	High
Adaptability	High	Low
User involvement	High	Low
Cost	Low	High
Risk	Unknown risk	Well understood risk

5 Risk on Scrum

Scrum is an iterative and incremental project management method that gives a straightforward "*Inspect and Adapt*" method [29]. When utilizing Scrum, the software is conveyed in steps known as "*Sprints*". A sprint begins with plans and finishes in reviews. When using Scrum there are many risks that may happen or not, the risk are as follow:

5.1 Lack of Team Understanding

Software project managers utilize various practices to build team to encourage better team cooperation in different Scrum meeting session. We sort those practices are following:

5.1.1 Team Meeting

Project stakeholders of Scrum groups are accumulated in a solitary area and carry out introductory runs as a gathered group before the groups are disseminated [30]. Distributed project interests of Scrum groups are likewise assembled quarterly or every year for few days [25]. Here Scrum groups have meeting sessions including Scrum arranging, audit gathering, reviews [31].

5.1.2 Visit

Visits of product team among scattered sites are common practice to increase team spirit. Product owners frequently trip seaward locales to enable increase the learning area of project domain [32]. Arranged revolutions among disseminated colleagues facilitate "*cultural exchange, improve shared understanding, reduce miscommunication and improved distributed meeting sessions*" [30]. Practices as product owners sorting out quarterly products guide gathering are likewise successful for groups of completely comprehend project vision and strengthen the estimation of Scrum [33].

5.1.3 Training

Practice includes "Initial Scrum training" or even a "Specialized Scrum" to clarify new innovations issues additionally strengthens the estimation of Scrum and enhances software stakeholder to joint effort [34].

5.2 Poor Communication

To help rich correspondence and coordinated effort condition requirement required for Scrum, software project managers needs to give huge correspondence data transfer capacity and solid system sustain throughout development life cycle [35].

5.3 Number of Project Personnel

A Scrum group is normally consists of five to ten people, in spite of the fact that teams as huge as fifteen and as little as three have likewise revealed benefits. Therefore utilizing Scrum for a group of countless work forces is thought to be a risk. It is considerably more risky to utilize Scrum in a huge group dispersed over numerous destinations [33].

6 Conclusion

Agile software development has gives us numerous great effects in software development. The most natural is enhanced quality of products, enhanced effectiveness of developers and less mistakes makes an exceptional software product. Utilizing agile software development methodologies positively affects both the productivity and the quality. Scrum gives intense tools to specialized estimation, for adjusting to changing requirement and for inspiring stakeholders. They are compelling in large part since they urge the executing group to acknowledge responsibility for traditional managerial tasks for example estimation and team motivation. Risk management is a crucial part for traditional and agile methodology. Agile methodologies are generally embraced in enterprises as software development approach, furthermore, among every agile process. Scrum has broad request because of its various advantages.

Acknowledgments. We authors acknowledge with thanks assistance rendered by Prof. Dr. Javed Anjum Sheikh, University of Lahore, (Gujrat Campus) for providing crucial insight during the course of the research work, which greatly improved the manuscript.

References

1. Computer programming (2009). http://en.wikipedia.org/wiki/Progarming. Accessed 5 Apr 2009
2. Object oriented programming (2009). http://en.wikipedia.org/wiki/Object_oriented. Accessed 20 Mar 2009
3. Klimes, C., Prochazka, J.: New approaches in software development in acta elctrotechnica et informatics (2006)
4. Szalvay, V.: An introduction to agile software development. Technical report, Danube Technology (2004)
5. Royce, W.: Software Project Management: A Unified Framework (1998)
6. Ahmed, A., Ahmed, S., Ehsan, N., Mirza, E., Sarwar, S.Z.: Agile software development: impact on productivity and quality, pp. 287–290. IEEE (2010)
7. Kumar, G., Bhatia, P.K.: Impact of agile methodology on software development process. IJCTEE, **2** (2012)
8. Dyba, T., Dingsoyr, T.: What do we know about agile software development. IEEE Softw. **26**, 6–9 (2009)
9. Schawber, K., Beedle, M.: Agile Software Development with Scrum. Prentice Hall, Upper Saddle River (2001)
10. Rong, G., Shao, D.: SCRUM-PSP: embracing process agility and discipline. In: Asia Pacific Software Engineering Conference (2010)
11. Begel, A., Nagappan, N.: Usage and perceptions of agile software development in an industrial context: an exploratory study. In: 1st International Symposium on Empirical Software Engineering and Measurement, pp. 255–264 (2007)
12. Zuo, A., Yang, J., Chen, X.: Research of agile software development based on formal methods. In: International Conference on Multimedia Information Networking and Security, pp. 762–766 (2010)
13. Livermore, J.A.: Factors that impact implementing an agile software development mehtodolgy, pp. 82–85. IEEE (2007)

14. Nakki, P., Koskela, K., Pikkarainen, M.: Practical model for user driven innovation in agile software development. In: Proceedings of the of 17th International Conference on Concurrent Enterprising, pp. 1–8 (2011)
15. Solo, O., Abrahamsson, P.: Agile methods in european embedded software development organization: a survey on the actual use and usefulness of extreme programming and scrum. IET Softw. 2(1), 58–64 (2008)
16. Far, B.: Software reliability engineering for agile software development, pp. 694–607. IEEE (2007)
17. Misra, S.C., Kumar, U., Kumar, V., Grant, G.: The Organizational changes required and the challenges involved in adopting agile methodologies in traditional software development organization, pp. 25–28. IEEE (2006)
18. Kohlbacher, M., Stelzmann, E., Maierhofer, S.: Do agile software development practices increase customer satisfaction in system engineering porject? In: IEEE International System Conference (SysCon), pp. 168–172. IEEE (2011)
19. Imreh, R., Raisinghani, M.S.: Impact of agile software development on quality within information technology organization. J. Emerg. Trends Comput. Inf. Sci. 2(10), 460–475 (2011)
20. Syed-Abdullah, S., Holcombe, M., Gheorge, M.: The impact of an agile methodology on the well being of development teams. Empir. Softw. Eng. 11, 143–167 (2006)
21. Li, J.: Agile Software Development. Technische University, Berlin
22. Kruchten, P.: Introduction to the rational unified process. In: Proceedings of the 24th International Conference on Software Engineering, p. 703 (2002)
23. Flower, K.B.M., Highsmith, J.A.: Manifesto for agile software development (2011)
24. Highsmith, J.: Agile Software Development Ecosystem. Addison Wesley, Boston (2002)
25. Cristal, M., Wildt, D., Prikladnicki, R.: Usage of SCRUM practices within a global company. In: IEEE International Conference on 2008 Global Software Engineering, ICGSE, pp. 222–226 (2008)
26. Hneif, M., Ow, S.H.: Review of agile methodologies in software development. Int. J. Res. Rev. Appl. Sci. 1, 1–8 (2009)
27. Singh, M.: U-SCRUM: an agile methodology for promoting usability. In: Ag. Agile 2008 Conference, Toronto 2008, pp. 555–560 (2008)
28. West, D., Grant, T.: Agile Development Mainstream Adoption has Changed Agility. Forrester Research Inc., Cambridge (2010)
29. Sutherland, J., Schwaber, K.: The SCRUM papers: nuts, bolts and origin of an agile process (2009)
30. Therrien, E. Overcoming the challenges of building a distributed agile organiation. In: In: Proceedings of the Conference on Agile 2008, pp. 368–372 (2008)
31. Cottmeyer, M.: The good and bad of agile offshore development. In: Proceedings of the Conference on Agile 2008, pp. 362–367 (2008)
32. Paasivaara, M., Lassenius, C.: Cloud global software development benefit from agile method. In: Proceedings of the Conference on ICGSE 2006, pp. 109–113 (2006)
33. Berczuk, S.: Back to basics: the role of agile principles in success with a distributed SCRUM team. In: Proceeding of the Conference on Agile 2007, pp. 382–388 (2007)
34. Smits, H.: Implementing SCRUM in a distributed software development organization. In: Proceeding of the Conference on Agile 2007, pp. 371–375 (2007)
35. Drummond, B., Unson, J. F.: Yahoo! Distributed agile: notes from the world over. In: Proceedings of the Conference on Agile 2008, pp. 315–321 (2008)

Software Development Practices in Costa Rica: A Survey

Brenda Aymerich, Ignacio Díaz-Oreiro, Julio C. Guzmán,
Gustavo López(✉), and Diana Garbanzo

Research Center for Communication and Information Technologies,
University of Costa Rica, San Pedro, San José 11501, Costa Rica
{brenda.aymerich,ignacio.diazoreiro,julio.guzman,
gustavo.lopez_h,diana.garbanzo}@ucr.ac.cr

Abstract. In recent years, many studies have focused on software development practices around the world. The HELENA study is an international effort to gather quantitative data on software development practices and frameworks. In this paper, we present the Costa Rican results of the HELENA survey. We provide evidence of the practices and frameworks used in 51 different projects in Costa Rica. Participants in this survey represent companies ranging from 50 or fewer employees to companies with more than 2500 employees. Furthermore, the industries represented in the survey include software development, system development, IT consulting, research and development of IT services and software development for financial institutions. Results show that Scrum, Iterative Development, Kanban and Waterfall are the most used software development frameworks in Costa Rica. However, Scrum doubles the use of Waterfall and other methods.

Keywords: Software development approach · HELENA project
Scrum · Waterfall · Agile

1 Introduction

Many studies focused on software development practices in the past years. This research is part of an international study on the use of hybrid development approaches in software system development around the world called HELENA Study [1]. This study aims to investigate the use of hybrid development approaches in software system development in a variety of context (e.g., emerging companies, innovative sectors, and regulated domains).

Through an online survey, we studied the current state of software and system development. We collected data related to the development approaches (e.g., traditional, agile, mainstream, or homegrown), practices and methods. Furthermore, we characterized each response based on both the participant and the company they represented. We also focused on the goal pursued by the companies in the definition of their development practices.

© Springer International Publishing AG, part of Springer Nature 2019
T. Z. Ahram (Ed.): AHFE 2018, AISC 787, pp. 122–132, 2019.
https://doi.org/10.1007/978-3-319-94229-2_13

In this paper, we present the results of the HELENA survey in Costa Rica. We describe the results of an online survey with 51 responses from different Costa Rican software development companies.

Costa Rica's software market is vast. Companies such as Intel, IBM, Boston Scientific, HP, and Procter & Gamble host services and product development from this country. Some essential characteristics of the country that favors software development are the proximity to the US and the time difference with other service providers (e.g., India, China, and Brazil). Moreover, culture and high education levels allow people to work efficiently in real-time under a tight deadline [2].

Nine percent of participants of the survey represent small companies (50 or fewer employees), twenty-one percent are medium companies (from 50 to 250), thirty-seven percent are large companies (from 251 to 2500), and the remaining thirty-three very large companies (2500+). Furthermore, the industries represented in the survey include software development, system development, IT consulting, research and development of IT services and software development for financial institutions.

Results show that Scrum, Iterative Development, Kanban, and Waterfall are the most used software development frameworks in Costa Rica. However, Scrum doubles the use of Waterfall and other methods. The survey not only focuses on the application of software development frameworks but also delves into the goals and results of using hybrid approaches for software development.

The survey also provides information regarding the size and criticality of the projects developed. Responses evidence that almost 15% of the software developed is of high criticality (i.e., threatens human health or could have legal consequences for the development company). Twenty percent of the developments directly affect services (i.e., direct impact on clients) and the rest is of medium criticality (i.e., some kind of financial loss or impact on the company reputation).

In general, this research provides an overall understanding of the software development industry in Costa Rica and allows other researchers to compare and quantitatively assess the differences in software practices around the world.

2   Related Work

In the past, some research focus on the combination of software development approaches trends. West, Gilpin, Grant, and Anderson [3] described that organizations are adopting agile through a combination approaches. Moreover, authors state that companies develop software using a combination of Waterfall and Scrum due to a series of governance and cultural restrictions do not allow the use of a single software development approach.

Theocharis, Kuhrmann, Munch and Diebold [4], provided evidence that development approaches are used in combination (i.e., agile and traditional). Furthermore, authors mention that there is a lack of quantitative evidence about processes used nowadays.

To address the lack of quantitative data on the use of software development practices around the world a group of researchers conceived the HELENA project [1]. Some publications show results of the HELENA project around the world.

Paez, Fontdevila, and Oliveros published the initial observations of software development practices in Argentina [5]. Argentina achieved 53 data point in the HELENA Survey. Most of these 53 data points (40%) correspond to medium size companies (i.e., 51–250 employees) that focus on custom software development. In Argentina, Scrum, Iterative Development, and Kanban are the most used development methods.

Scott et al. [6], published the results of HELENA comparing Estonia and Sweden, two highly digitalized societies but different regarding the type of software industry. As the authors expected, many of the Estonian respondents work in small companies (11–50 employees) while most of the Swedish respondents work in large companies (251–2499 employees). Regarding the usage of development frameworks and methods, Estonian responses state a clear preference for a small set of agile frameworks, with Scrum "always used" by 58% of the respondents. Responses from Sweden also show frequent use of Scrum, but only 8% use it always. Iterative Development, Kanban and the Classic Waterfall Process are used as often as Scrum in Sweden.

Nakatumba-Nabende et al. [7], presented the results of the HELENA survey comparing Sweden and Uganda focused on demographics, processes usage, and standards. In Uganda, most of the respondents were developers. In Sweden, the most common roles were architect and project/team managers. The main finding of this research is that neither country adheres to one particular development model but rather employ hybrid approaches.

Tell, Pfeiffer and Pagh, published the results of HELENA in Denmark [8] showing that Danish respondents are mostly product managers/owners, developers, and architects. Authors also present the company size distribution, pointing out that not only small and medium-sized enterprises are represented, but also a third of respondents (approximately) work in large companies. Authors also indicate that traditional and agile methods and practices are combined regardless of the company size and industry sector. However, presented results seem to indicate that Danish enterprises might favor a more agile development environment.

Felderer, Winkler, and Biffl [9] published the results of HELENA in Austria. Results showed a balanced distribution of small/medium and large/very large organizations. Moreover, a majority of respondents' business areas are custom/standard and software/system development. The main industrial sectors are financial, public sector, medical, energy, web application, and automotive. Most participants in this survey have more than ten years of working experience. Authors, indicate that most respondents are familiar with iterative development and Scrum, while other approaches are used if required by the customer.

3   Methods

This paper presents the results of the HELENA survey in Costa Rica. To spread the survey in the country, a group of six researchers from the University of Costa Rica contacted some software development groups and invited them. Moreover, two prominent organizations that represent the country's software development industry invited their members. Table 1 shows the structure of the survey.

Table 1. HELENA survey structure.

Section	Question
Metadata	Organization's size
	Organization's business area
	Development dynamics
	Application domain
	Participant's role
	Risks associated with software development
Process use	Process and standards coverage
	Development approaches and practices
	Development approach selection motivation
	History of the development process definition
Process use and standards	Degree of use of worldwide standard activities
	Process compliance assessment
Process used in the lifecycle	Company's continuous improvement culture
	Motivation for the improvement programs
	Goals of the improvement programs
Other	Final comments and considerations

The survey is grounded in several sources [1]. The design of the survey targets practitioners and includes both their experiences and the companies they represent.

4   Results

This section presents the results of the HELENA survey for Costa Rica. In total, the survey had 51 participants. First, we will describe the demographics of participants; later we will describe the primary results addressing the applied software development approaches and practices in Costa Rica.

4.1   Demographics

Costa Rica is a small country (51,100 km^2) and around 4.8 million people. However, in the HELENA survey we gathered 51 data points representing different people involved in software development.

We conducted a characterization based on the company size (using as an indicator the number of employees). Figure 1 shows the distribution of companies by size according to the number of full-time employees and by business area. In addition, Fig. 1 shows a combination of business area and company size.

Small companies do not participate in the System Development area or the Finance area. However, they do have significant participation in Research & Development. It is interesting that only large or very large companies conduct System Development. A high percentage of the Costa Rican software industry dedicates to software development (36%) and Consulting (31%).

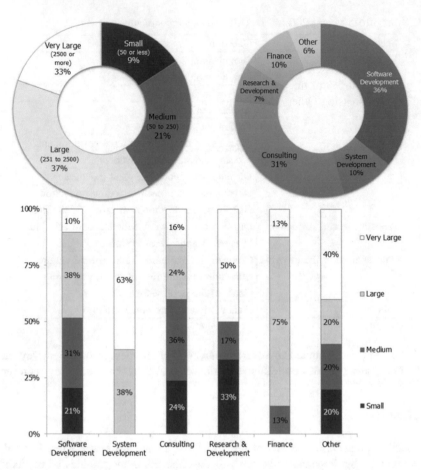

Fig. 1. Top-left: Distribution of companies by size based on the number of full-time employees. Top-right: Distribution of companies by business area. Bottom: Combination of company business area and size.

Regarding the size of the products or projects that are developed in Costa Rica, 51% are considered very large (i.e., more than one person-years to develop), 20% are large projects (i.e., between one person-year and six person-months) and the remaining 29% are small projects (i.e., less than six person-months). Medium, Large and Very Large companies only conduct very large projects. Small companies, on the other hand, participate in smaller projects. More than half the projects conducted by small companies are either small or medium size projects. Even though we do not intend to compare the results of Costa Rica with the rest of the world, this particular result differs with HELENA results from other countries.

Respondents of the research have two main roles: 31% are managerial roles (e.g., Product manager, CEO or architect) and 69% are operative (e.g., developer, tester or analyst). Additionally, 12% of participants have less than 2 years in their role, 35%

have from 3 to 5 years, 20% have 6 to 10 years and 33% have more than 10 years in their role.

Figure 2 shows the relation between the Target Domain and Company Size. We aligned the results based on the number of responses in each category: left 26% of the data points, right 5% of the data point. An interesting result here is that for the 9% of companies addressing Other Information Systems (e.g., ERP or SAP) 51% are small.

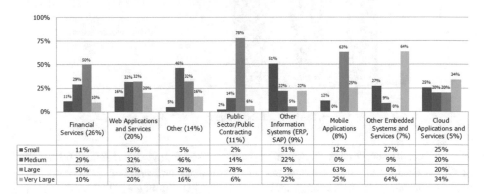

	Financial Services (26%)	Web Applications and Services (20%)	Other (14%)	Public Sector/Public Contracting (11%)	Other Information Systems (ERP, SAP) (9%)	Mobile Applications (8%)	Other Embedded Systems and Services (7%)	Cloud Applications and Services (5%)
Small	11%	16%	5%	2%	51%	12%	27%	25%
Medium	29%	32%	46%	14%	22%	0%	9%	20%
Large	50%	32%	32%	78%	5%	63%	0%	20%
Very Large	10%	20%	16%	6%	22%	25%	64%	34%

Fig. 2. Target domain by company size.

The category named other (46% in the hands of Medium size companies), covers the following domains: Games, Home Automation and Smart Buildings, Logistics and Transportation, Media and Entertainment, Medical Devices and Health Care, Space Systems, Telecommunication, Defense Systems and Energy.

The last characterization of the companies was by the criticality of their projects. We asked what would be the consequence of a failure in the system. The available responses were System Degradation, Impact Company Business and Reputation, Financial Loss, Legal Consequences. An interesting finding is that legal consequences are present in all sizes of companies. However, large companies significantly differ from others in System Degradation. Financial Loss and a possible impact on the company business is a risk in all company sizes, but it is especially important for very large companies. An expected finding is that the bigger the company, the lower the preoccupation on of a failure in the company reputation. Figure 3 shows a combined analysis of company size and possible consequences of an error in the development process.

4.2 Applied Methods and Practices in Costa Rica

This section describes the primary software development methods used in Costa Rica. Firstly, we depict which frameworks and methods are most frequently used; secondly, we show if the stages in the software lifecycle and different managerial tasks associated with the software development process are traditional or agile; finally, we provide an overview of the goals of each company to use the methods they use.

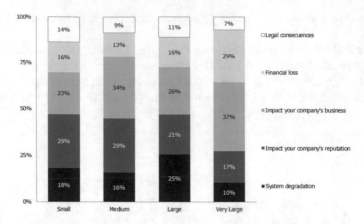

Fig. 3. Criticality of the project by company size.

As it can be seen in Fig. 4, Scrum is the most used Framework/Method by far: 42% of Costa Rican companies always use the framework and another 40% often use it. Iterative Development, Kanban, and DevOps appear in places 2, 3 and 4 of this classification. The Classical Waterfall Process ranks fifth, with a combined total of 44% of the companies using it sometimes, often and always.

Figure 5 shows the type of development method used in different stages of the development process and managerial tasks. Participants were asked if their process was traditional or agile (and anything in between). Results show that Integration and Testing, as well as Implementation, are mostly agile. An interesting result is that Management and Operations are the mainly traditional (Configuration Management, Risk Management, Project Management and Quality Management). The most balanced stage is Transition and Operation.

Besides the used practices, participants were asked for their primary goals to decide which practices to use or which ones to combine. The top goals mentioned were: Improved Planning and Estimation (51%), Improved Productivity (39%), Improved Frequency of Delivery to Customers (39%), Improved Client Involvement (39%) and Improved Adaptability and Flexibility of the Process to React to Change (37%).

For the five mentioned goals, Costa Rican companies interpret that the usage or combination of development approaches has generated the degrees of achievement shown in Fig. 6. Being ten the maximum value possible (goal fully achieved) and one the minimum value possible (not achieved at all), it can be seen that the median in all cases is between the values seven and eight. Moreover, the third quartile for Improved Planning and Estimation or Improved Frequency of Delivery to Customers is considerably large. These findings can be interpreted as a high degree of achievement.

Other goals also mentioned by the respondents such as Improved external product quality (35%), Improved internal artifact quality (33%), Improved project monitoring and controlling (27%) and Improved knowledge transfer and learning (25%) also presented high median values, but with bigger first quartiles. Therefore, the goals are achieved but the satisfaction degree with the results is lower.

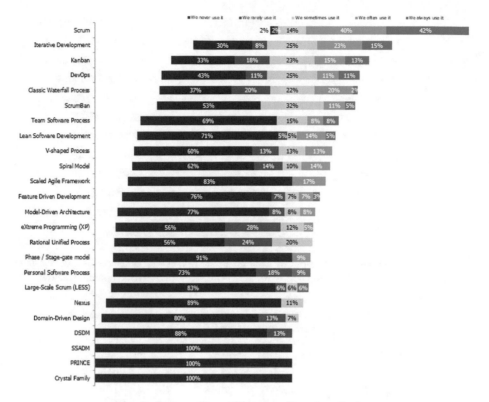

Fig. 4. Degree of use of frameworks and methods.

5 Discussion

The main finding of this survey is that the Costa Rican software development industry is mainly agile in the development process and traditional in the operation management. However, hybrid or balanced practices (a combination of agile and traditional) are commonly used. Scrum is, by far, the most common development approach. However, practices such as iterative development, Kanban, and DevOps are also applied.

Another interesting finding is that, even though Scrum is commonly used, extreme Programming is almost never used. Therefore, the predilection for agile is usually implemented using Scrum.

Small companies in Costa Rica focus on research and development. Many small software development companies are dedicated to innovation and sell their inventions worldwide. On the other hand, large companies dominate the finance market. Consulting is a common occupation for software companies in Costa Rica; in this area, the size of the company does not matter.

Financial services and web development are the most common areas of work for Costa Rican software development companies. Cloud, embedded and mobile have

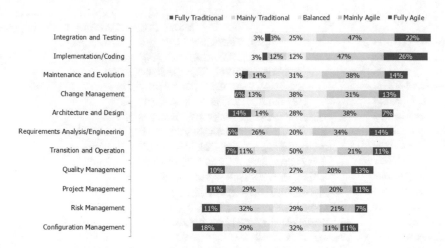

Fig. 5. Type of software development method used in different stages of the development process and different managerial tasks associated with the software development process.

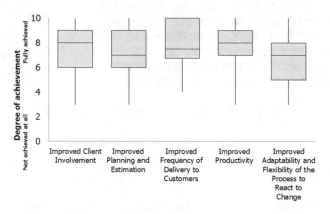

Fig. 6. Achievement degree of top goals pursued by using or combining development approaches.

some interest; however, none of them achieves 10% of the market (at least represented in this survey). This result was expected because most of the software development industry in Costa Rica provides services for companies around the world.

Even though most companies use Scrum as their development process, the main sessions of this agile framework are not always conducted. Table 2 shows the main sessions of Scrum and the use distribution of respondents.

Table 2 shows that almost none of the Scrum sessions are always conducted. Moreover, some companies never do reviews. Sprint reviews are usually not conducted because event tough the development is iterative there is no contact with the clients. Therefore, developers do not deliver to users. Surprisingly the daily standup is sometimes, often or always used in almost 85% of the cases.

Table 2. Scrum sessions degree of use.

	Never	Rarely	Sometimes	Often	Always
Daily standup	2%	11%	34%	32%	21%
Iteration planning	27%	20%	23%	23%	7%
Iteration/Sprint reviews	60%	30%	0%	10%	0%
Retrospectives	4%	33%	25%	21%	17%

6   Conclusions and Future Work

In this paper, we presented the results of the HELENA survey in Costa Rica. We focused on the usage of software development frameworks, methods, and practices. However, this paper also contains a characterization of the Costa Rican software development industry. Each one of the 51 responses to the survey were categorized based on the size of the company, the business area, and the target technologies.

Agile practices seem to be widely accepted in Costa Rica. Even though traditional and agile methods are sometimes combined, several participants stated that they usually use Scrum and almost never use the traditional Waterfall development process.

The main limitation of this research is the threat of its external validity due to the number of participants and the self-reporting structure of the survey for Costa Rica. However, future work includes the analysis of HELENA results in other countries compared with Costa Rica. These studies will provide further information to assess if the Costa Rican results are similar to other countries around the world and allow further analysis of the data described in this paper.

Acknowledgments. This work was partially supported by CITIC at the University of Costa Rica. Grant No. 834-B4-412. We would also like to thank all 51 participants of the survey and the rest of researchers promoting this international initiative (HELENA Project) Acabo.

References

1. Kuhrmann, M., Hanser, E., Prause, C.R., Diebold, P., Münch, J., Tell, P., Garousi, V., Felderer, M., Trektere, K., McCaffery, F., Linssen, O.: Hybrid software and system development in practice: waterfall, scrum, and beyond. In: Proceedings of the 2017 International Conference on Software and System Process - ICSSP 2017, pp. 30–39. ACM Press, New York (2017)
2. CAMTIC: Camara de Tecnologias de Informacion y Comunicacion. https://www.camtic.org
3. West, D., Gilpin, M., Grant, T., Anderson, A.: Water-Scrum-Fall is the reality of agile for most organizations today (2011)
4. Theocharis, G., Kuhrmann, M., Münch, J., Diebold, P.: Is water-scrum-fall reality? on the use of agile and traditional development practices. In: Abrahamsson, P., Corral, L., Oivo, M., Russo, B. (eds.) Product-Focused Software Process Improvement, pp. 149–166. Springer, Cham (2015)
5. Paez, N., Fontdevila, D., Oliveros, A.: HELENA study: initial observations of software development practices in Argentina. In: Felderer, M., Méndez-Fernández, D., Turhan, B., Kalinowski, M., Sarro, F., Winkler, D. (eds.) Product-Focused Software Process Improvement, pp. 443–449. Springer, Cham (2017)

6. Scott, E., Pfahl, D., Hebig, R., Heldal, R., Knauss, E.: Initial results of the HELENA survey conducted in Estonia with comparison to results from Sweden and worldwide. In: Felderer, M., Méndez-Fernández, D., Turhan, B., Kalinowski, M., Sarro, F., Winkler, D. (eds.) Product-Focused Software Process Improvement, pp. 404–412. Springer, Cham (2017)
7. Nakatumba-Nabende, J., Kanagwa, B., Hebig, R., Heldal, R., Knauss, E.: Hybrid software and systems development in practice: perspectives from Sweden and Uganda. In: Felderer, M., Méndez-Fernández, D., Turhan, B., Kalinowski, M., Sarro, F., Winkler, D. (eds.) Product-Focused Software Process Improvement, pp. 413–419. Springer, Cham (2017)
8. Tell, P., Pfeiffer, R.-H., Schultz, U.P.: HELENA stage 2—Danish overview. In: Felderer, M., Méndez-Fernández, D., Turhan, B., Kalinowski, M., Sarro, F., Winkler, D. (eds.) Product-Focused Software Process Improvement, pp. 420–427. Springer, Cham (2017)
9. Felderer, M., Winkler, D., Biffl, S.: Hybrid software and system development in practice: initial results from Austria. In: Felderer, M., Méndez-Fernández, D., Turhan, B., Kalinowski, M., Sarro, F., Winkler, D. (eds.) Product Focused Software Process Improvement, pp. 435–442. Springer, Cham (2017)

Convincing Systems Engineers to Use Human Factors During Process Design

Judi E. See[(⊠)]

Sandia National Laboratories, P.O. Box 5800, Albuquerque,
NM 87185-MS0151, USA
jesee@sandia.gov

Abstract. A controlled between-groups experiment was conducted to demonstrate the value of human factors for process design. Twenty-four Sandia National Laboratories employees completed a simple visual inspection task simulating receipt inspection. The experimental group process was designed to conform to human factors and visual inspection principles, whereas the control group process was designed without consideration of such principles. Results indicated the experimental group exhibited superior performance accuracy, lower workload, and more favorable usability ratings as compared to the control group. The study provides evidence to help human factors experts revitalize the critical message regarding the benefits of human factors involvement for a new generation of systems engineers.

Keywords: Human factors · Human-systems integration · Systems engineering
User-centered design · Mental workload · Usability

1 Introduction

It is well established that the discipline of human factors provides numerous benefits throughout the product lifecycle [1–9]. Such benefits have been demonstrated through positive examples, which highlight the value of including human factors experts early and often throughout the lifecycle; and negative examples, which underscore the adverse consequences of neglecting human factors [3]. Demonstrated benefits encompass the design process itself and the subsequent operations and maintenance phase. Design is impacted through reduced product development time and costs, by as much as 50% [7]. As just one example, Bailey demonstrated that interfaces developed by human factors experts had fewer design errors after a single iteration as compared to the same interfaces developed by programmers after three to five iterations [1]. In sum, the extra time, labor, and costs for programmers to develop an effective and usable

Sandia National Laboratories is a multimission laboratory managed and operated by National Technology and Engineering Solutions of Sandia, LLC, a wholly owned subsidiary of Honeywell International, Inc., for the U.S. Department of Energy's National Nuclear Security Administration under contract DE-NA-0003525.

© Springer International Publishing AG, part of Springer Nature (outside the USA) 2019
T. Z. Ahram (Ed.): AHFE 2018, AISC 787, pp. 133–145, 2019.
https://doi.org/10.1007/978-3-319-94229-2_14

product could have been saved by incorporating human factors experts from the beginning.

The operations and maintenance phase is impacted in terms of increases in advantageous states and reductions in detrimental states. Advantages include improved safety, effectiveness, efficiency, productivity, and operator satisfaction [3, 7]. Positive reductions include decreased training time and costs, accidents, error rates, maintenance costs, and equipment damage [8]. One particularly noteworthy positive reduction involves the number of errors requiring resolution during operations and maintenance, attributable explicitly to investing in human factors early in design. First, errors can be 30 to 1500 times costlier to correct in operations and maintenance as compared to early design phases [10]. Second, up to 67% of operations and maintenance costs stem from modifications to resolve operator dissatisfaction with the original system [11]. Ultimately, a large portion of detrimental impacts associated with error cost escalation could be avoided with proper attention to human factors during design.

Some benefits such as productivity can be expressed in quantitative cost savings; other less tangible benefits such as improved operator attitudes may be difficult to measure and quantify, but nevertheless have a positive influence. In a review of 24 human factors projects, Hendrick concluded that human factors has a direct cost benefit of 1:10+, with a typical payback period of 6 to 24 months [5]. He further inferred that earlier incorporation of human factors translates into even lower costs and greater benefits.

Despite a proven record of success, human factors experts continue to face challenges convincing personnel outside their field. Namely, "inadequate consideration of human factors engineering issues is a familiar problem…issues are not expressly considered, they are considered but their importance is underestimated, or they are considered too late in the design process" (p. 1) [7]. Hence, there is an ongoing need for evidence documenting the benefits of human factors.

1.1   Previous Research

Much of this ongoing evidence is derived from reactive case studies of fielded systems exhibiting signs of trouble, due to failure to properly include human factors during development. As one example of this approach, Sen and Yeow analyzed an existing electronic motherboard that suffered from 70% rejects, leading to low productivity and an overall loss at the factory [12]. Human factors interventions reduced fabrication time from six to two shifts, eliminated rejects, reduced repairs, improved lost business, and saved the factory approximately $582k per year. In a similar reactive study, Yeow and Sen investigated a failing inspection process at a printed circuit assembly factory that caused an annual rejection cost of nearly $300k, poor quality, customer dissatisfaction, and operator occupational safety and health issues [13]. Human factors interventions resolved three primary issues encompassing operator eye problems, insufficient time for inspection, and ineffective visual inspection processes; reduced customer site defects by 2.5%; improved customer satisfaction; and saved the factory over $250k per year.

While such reactive case studies are informative, they lack the rigor of controlled experiments to draw definitive conclusions about causality. To be sure, some researchers have conducted controlled experiments comparing "old designs" (without human factors involvement) and "new designs" (with human factors involvement) to provide additional confidence in the results from reactive studies. In one such study, a computer-telephony product developed without human factors engineers was compared to a version redesigned by human factors experts [14]. Participants using the new design rated overall ease of use more highly and successfully completed more tasks than participants using the old design. In a similar experiment, novice users had faster performance, fewer errors, and lower workload when programming doctors' orders in a patient-controlled analgesia machine interface redesigned to conform to human factors principles [15]. Another study demonstrated that applying human factors principles for the redesign of a medication alert interface improved performance time and usability, while reducing prescriber workload and prescribing errors [16].

1.2 Objectives of the Present Study

While some studies employed controlled experiments, their basis was still firmly rooted in reactive investigation of existing flawed systems designed with little or no human factors involvement. Further, existing research focused primarily on product or inter-face design, not process design. Finally, even experimental comparisons of old and new designs have been scarce since the 1990s. Most current human factors research implicitly attests to the value of human factors, but it is not typically an explicit goal. The present study was designed to fill these gaps and revitalize the message regarding the benefits of human factors for a new generation of systems engineers.

Toward that end, a controlled experiment was conducted to demonstrate the value of human factors for process design, using a visual inspection task designed with adherence to common human factors principles (experimental group) and without (control group). It was hypothesized the experimental group would perform more accurately and quickly, with lower workload, and higher usability ratings.

2   Methodology

2.1   Participants

Participants were 24 employees (14 males) at Sandia National Laboratories who responded to an advertisement in the Sandia Daily News. They ranged in age from their twenties to their sixties, with 42% of participants in their thirties.

2.2   Design

A between-groups design was used. The experimental group task was designed to conform to human factors principles, whereas the control group task was designed without consideration of human factors principles. The task simulated a receipt inspection process wherein a lot of vendor parts is visually inspected to accept quality parts and remove flawed items, based on pre-defined defects that might impact func-tionality during subsequent operations. Visual inspection is a difficult task susceptible to human error if not designed in accordance with human factors principles and research lessons learned [17, 18].

To eliminate requirements for prior experience and minimize participant training, parts for inspection in the present study consisted of 350 tiles from the Hasbro, Inc. Scrabble game. Acceptable parts contained any one of six different Roman characters, and rejectable parts contained any one of four different Cyrillic characters. The inspector's task was to sort the tiles by acceptability and letter type, count the quantities of each letter type, and calculate vendor fees. Fees were calculated using the number on each tile to represent its dollar value. The product of value and quantity equaled the total dollar amount for each letter type, constituting either an amount to pay (acceptable parts) or charge (rejectable parts) the vendor. To simulate the low defect rates typically seen in visual inspection, only 15 tiles (4%) were rejectable [18].

Tasks were designed with and without adherence to general human factors and specific visual inspection principles and confirmed through independent heuristic evaluations. Accordingly, the experimental group task was structured to accommodate the range of participant physical dimensions and preferences and to facilitate all components of the process (sort, count, and derive fees) to maximize accuracy, speed, and usability and minimize workload. The control group task provided the minimum tools necessary to complete the task, without consideration of usability or user preferences. Every effort was made to avoid intentionally exaggerating task difficulty in the control group (i.e., to prevent artificially biasing outcomes in favor of the experimental group). Toward that end, control group process design was grounded in issues commonly reported in the research literature.

2.3 Procedure

The experiment occurred in a private enclosed office with two sit-stand tables that provided a large adjustable workspace for task completion. Each participant individually completed a session lasting approximately 1.3 h. Participants were randomly assigned to the experimental or control group, with 12 participants per group. The experimenter described the purpose of receipt inspection and reviewed the work instruction, informing participants that both task speed and accuracy were important. Participants practiced the inspection task by first verbally providing inspection decisions for five tiles and identifying their values, uniformly achieving 100% accuracy. Next, participants completed the entire inspection process with a set of 20 tiles. The structure of the practice session was congruent with the experimental and control group process designs in terms of incorporation of human factors and visual inspection principles. Ninety-two percent of participants achieved 100% accuracy. The experimenter reviewed any errors before the main task began.

Before starting the main task, experimental group participants had an opportunity to customize their workspace. Customization included the ability to change the heights of the sit-stand tables to support a preference for sitting or standing during the task and arrange the sorting and counting trays in the workspace. Control group participants were informed they could use any of the available workspace to compete the task, but were not offered customization options.

During the main task, the experimenter remained in the room to document observations for post-session interviews. All participants received a single bin containing the lot of tiles for inspection in a random arrangement. At that time, they were informed the entire task takes approximately 20 min, a benchmark based on pilot testing and designed to amplify task demand. Experimental group participants used labeled trays with numbered and color-coded slots to sort and categorize tiles. Control group participants received only the bin containing the lot of tiles for inspection and had to develop their own methods to sort and categorize tiles within the available workspace (Fig. 1). Control group techniques included sorting tiles into rows, columns, piles, or vertical stacks according to letter type.

Fig. 1. Experimental and control group sorting techniques.

Following sorting and categorization, participants calculated vendor fees. Experimental group participants used electronic forms pre-populated with letter type and tile values as well as built-in formulas to automatically calculate totals, based on tile quantity entries. Control group participants manually recorded letter types, tile values, and quantities on paper forms and computed totals with a handheld calculator. Three primary elements on the forms supported subsequent analyses of task accuracy: tile values, quantities, and dollar amounts (Fig. 2).

Experimental Group

Tile	Value	Quantity	Total Amount to Be Paid
A	1	100	$100
H	1	70	$70
M	2	42	$84
P	1	60	$60
T	1	49	$49
X	5	14	$70
	Totals	335	$433

Control Group

Tile	Value	Quantity	Total Amount to Be Paid
A	1	100	100
H	1	70	70
P	1	60	60
M	2	42	84
T	1	49	49
X	5	14	70

Total $: 433

Fig. 2. Experimental and control group data entry forms for acceptable tiles.

At the end of the task, participants provided NASA-TLX workload ratings [19] and rated inspection task ease of completion, amount of time, and task work instructions on a usability scale ranging from *Strongly Disagree* (1) to *Strongly Agree* (7) [20]. Before concluding the session, the experimenter interviewed participants to gain insight into their thought processes throughout the experiment, collect subjective descriptions of any errors that occurred, and discuss experimenter observations.

3   Results

Incorporating human factors in process design led to superior performance accuracy, lower workload, and more favorable usability ratings in the experimental group as compared to the control group. The experimental group process design promoted more uniform task approaches among participants, effectively reducing process variation and mitigating or eliminating errors observed in the control group.

3.1   Task Accuracy and Speed

Task Accuracy. Accuracy was addressed by examining errors recording tile values, quantities, and dollar amounts. For acceptable tiles, differences emerged between the two groups only for quantities and dollar amounts (Table 1). Control group participants either miscounted acceptable tiles or mistakenly categorized a rejectable letter type as acceptable. These quantity errors led to under payments, ranging from $44 to $98, and over payments of up to $24. The single experimental group error resulted from a simple miscount, which led to a $5 over payment. None of the differences in accuracy was statistically significant.

Table 1.   Acceptable tile accuracy results.

Dependent variable	Group	Incorrect responses	Statistical significance
Tile values	Experimental	0	Dependent variable is a constant; no statistics computed
	Control	0	
Quantities[a]	Experimental	1	$p = .295$, Fisher's exact test, one-tailed
	Control	3	
Dollar amounts	Experimental	1	$p = .077$, Fisher's exact test, one-tailed
	Control	5	

[a]Signal detection theory analysis was not possible due to zero false alarms in the experimental group.

For rejectable tiles, errors in tile values, quantities, and dollar amounts were confined to the control group (Table 2). Value errors occurred when value and quantity entries for a single tile were transposed. Rejectable quantities were all under-recorded by 1 to 3 tiles due to the transposition error, misclassifying one rejectable tile type as acceptable, and miscounting tiles. All four dollar amount errors consisted of under-charging, ranging from $2 to $24. Differences in accuracy for quantities and dollar amounts were statistically significant.

Table 2. Rejectable tile accuracy results.

Dependent variable	Group	Incorrect responses	Statistical significance
Tile values	Experimental	0	$p = .500$, Fisher's exact test, one-tailed
	Control	1	
Quantities	Experimental	0	$p = .047$, Fisher's exact test, one-tailed
	Control	4	
Dollar amounts	Experimental	0	$p = .047$, Fisher's exact test, one-tailed
	Control	4	

Task Speed. With respect to task completion time, the experimental group averaged 25 min ($SD = 6$), while the control group averaged 26 min ($SD = 3$). This difference was not statistically significant, $F(1, 22) = .222$, p $= .642$, 95% CI of the difference $[-3, 5]$, $d = .21$.

Error Analysis. Accuracy differences can also be understood by analyzing the specific types of errors that occurred. Eleven types of errors were observed in the present study, all of which occurred in the control group (Table 3). The experimental group process design mitigated or prevented each error. Apart from miscounts, all errors were prevented in the experimental group. The experimental group setup min-imized miscounting, as evidenced by an overall reduction in quantity errors from seven (control group) to one (experimental group), but did not completely prevent miscounting.

The failure types in Table 3 can further be interpreted with respect to Reason's four categories of errors: slips, lapses, mistakes, and violations [21]. In the present study, violations or intentional inappropriate acts were not observed; however, the remaining three error types did occur, primarily in the control group. For example, slips took the form of incorrectly recording a calculated dollar amount of $49 as $4 on paper. Slips occur when an action is taken, but it is not the action the individual intended. Forgetting to record the date and lot number on the control group paper form is an example of a lapse, wherein individuals forget to perform an activity they meant to accomplish. Incorrect categorizations represent mistakes—people perform the act they intended, but the act itself is inappropriate. In particular, some control group participants incorrectly categorized a rejectable tile type as acceptable, due to work instruction ambiguity. The experimental group process design successfully mitigated or prevented the three cat-egories of errors that occurred in the control group.

Reductions in Variability. Impacts of human factors interventions in the experi-mental group can additionally be considered in terms of reductions in variability. First, the experimental group exhibited a smaller number and variety of errors during task completion as compared to the control group (Table 3). The experimental group work

Table 3. Control group errors and experimental group mitigations.

Observed error	Control group instantiation	Experimental group mitigation
Miscounts	Tiles placed into piles or groupings prone to miscounting	Sorting trays contained multiple slots, each holding five tiles, to minimize miscounting
Miscategorizations	Rejectable tiles incorrectly categorized as acceptable	Work instruction and sorting trays contained photos of acceptable and rejectable tile types
Incorrect tile values	Tile values entered incorrectly on paper form	Electronic spreadsheet was pre-populated with static information such as tile values
Miscalculations	Dollar amounts calculated and recorded incorrectly on paper	Electronic spreadsheet automatically calculated dollar amounts
Overturned tiles	Stacks of sorted tiles bumped while inspecting remaining tiles	Sorting trays contained inspected tiles separate from unsorted tiles
Missing entries	Study ID, date, and lot number fields left blank	Electronic spreadsheet was pre-populated with this information
Scratchouts	Incorrect entries scratched out or overwritten	Changes made in the electronic form replaced existing entries
Handwriting	Handwriting sometimes ambiguous and open to interpretation	Electronic spreadsheet used only legible typewritten entries
Space allocation	Amount of space required to sort 350 tiles not well planned	Sorting trays held all 350 tiles, requiring an identifiable amount of table space
Re-counting	Tiles in configurations that did not support re-counting	Numbered and divided slots facilitated re-counting and count verification
End state	Tile groupings not conducive to transfer for follow-on work	Sorting trays also served as a convenient mechanism to transfer tiles for next level of work

instruction and tools prompted a more consistent, uniform approach and mitigated or eliminated the types of errors observed in the control group. Second, standard deviations were lower in the experimental group for 10 of 11 key dependent variables, and 7 of the differences were statistically significant (Table 4). Reduced standard deviations signify the process design minimized individual differences that contribute to process variation and hinder consistency in manufacturing.

3.2   Workload

The average global weighted NASA-TLX score was 14.9 (SD = 7.9) in the experimental group and 26.4 (SD = 14.2) in the control group. The mean difference was statistically significant, $F(1, 22) = 6.03$, $p = .022$, 95% CI of the difference [1.6, 21.4], $d = 1.0$. Control group participant comments revealed that devising an efficient sorting

Table 4. Experimental and control group standard deviations for key dependent variables.

Dependent variable	Mean (SD)	
	Experimental	Control
Acceptable quantity recorded*	335.08 (.29)	335.50 (1.0)
Acceptable dollar amount recorded*	$433.42 ($1.44)	$419.25 ($37.27)
Acceptable percent correct*	100.0% (0.00%)	99.9% (.29%)
Rejectable quantity recorded*	15.00 (0.00)	14.33 (1.16)
Rejectable dollar amount recorded*	$58.00 ($0.00)	$53.00 ($9.32)
Rejectable percent correct*	100.0% (0.0%)	95.6% (7.7%)
Task duration	25 min (6 min)	26 min (3 min)
NASA-TLX global workload*	14.9 (7.9)	26.4 (14.2)
Ease of completion usability rating	6.7 (.65)	6.0 (1.09)
Amount of time usability rating	6.3 (.62)	5.8 (1.03)
Work instructions usability rating	6.8 (.39)	6.7 (.65)

*$p < .05$, Levene's test for equality of variances

and counting method (while striving to meet the 20-min completion time) and performing manual counts and calculations were two primary challenges that increased demand. Three control group participants indicated that sorting tiles without bins and using a calculator with small buttons also added to the demand. Indeed, ratings for all six NASA-TLX subscales were higher in the control group as compared to the experimental group (Fig. 3). However, one-way ANOVAs of each workload dimension indicated that none of the differences was statistically significant ($p > .05$).

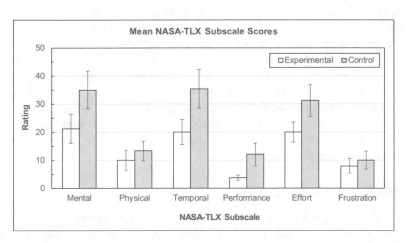

Fig. 3. Mean NASA-TLX subscale scores in the experimental and control groups. Error bars represent standard errors.

3.3 Usability Ratings

In accordance with the approach used by Walkenstein and Eisenberg, a criterion was established specifying the process would be deemed usable if 80% of participants provided ratings of 6 or 7 for all three usability items [14]. This criterion was met in the experimental group (92%), but not the control group (67%). In fact, experimental group participants did not have any ratings below 5. In contrast, control group participants assigned ratings as low as 4 and used more ratings of 5 than the experimental group.

4 Discussion

The value of human factors was demonstrated in a controlled between-groups experiment for a simple visual inspection process. The experimental group achieved greater performance accuracy, with reduced workload and more favorable usability ratings, as compared to the control group. Such improvements stemmed from applying a user-centered design approach for the experimental group that focused on thorough consideration of general human factors and specific visual inspection principles. The result was a more efficient and usable process for the experimental group that lends itself well to follow-on manufacturing steps and analysis. For example, use of electronic spreadsheets to enter inspection outcomes resulted in legible records for archiving and future analysis, with none of the scratchouts or writeovers observed in the control group (refer to Fig. 2). Further, the experimental group process design resulted in an end state configuration suitable for the next level of processing (refer to Fig. 1). By contrast, control group participants concluded the task with rows, columns, piles, or vertical stacks of tiles scattered across the table.

In a realistic manufacturing process, inspected parts must be organized to support subsequent processing, either installation in the next level of assembly (acceptable parts) or preparation for analysis and troubleshooting (rejectable parts). In the present study, this step was incorporated into the experimental group process via sorting bins and trays; however, this step was not specifically required in the control group. At a minimum, depositing tiles into separate bins after sorting was done would have not only prolonged completion time for the control group but also increased opportunities for error. In effect, although task completion time differences were not statistically significant, the experimental group was ultimately faster since the control group would still have to complete the final step.

4.1 Process Variation Reduction

Observed improvements in the present study were possible because the experimental group process design promoted more uniform task approaches among participants, effectively reducing process variation and mitigating or eliminating errors observed in the control group. Specifically, the observed number and variety of errors as well as standard deviations for most dependent variables were smaller in the experimental group as compared to the control group. Such differences indicate the process design, including the work instruction and tools, reduced individual differences in experimental group task approaches.

This reduction in process variation implies that including human factors during design can generate a consistent, repeatable manufacturing process by focusing on human factors for the visual inspection component of the manufacturing process. Understanding and reducing process variation in all components of the manufacturing process is critical. Excess variation at any point can increase scrap and rework, require additional inspections, impair functionality, and reduce reliability and durability [22].

4.2 Limitations and Directions for Future Research

The magnitude of effects in the present study was limited by inspection task simplicity, as evidenced in part by performance accuracy ceiling effects for acceptable tiles. This outcome was the result of applying a simple inspection criterion based on a single tile feature (character on the tile). In reality, receipt inspection typically involves simultaneous inspection for numerous defect types such as scratches, dents, and discolorations. Inspection only becomes more difficult as the number of different defect types increases, magnifying task demand and reducing performance accuracy [18]. Thus, additional differences between the experimental and control groups might have occurred in the current study with a more complex inspection task.

Future research might focus on designing a more complex process to increase task demands, while striving to minimize the amount of required participant training and preparation. Human error and workload tend to increase as task complexity increases, while usability can become degraded [23]. Therefore, a more complex task should generate more robust impacts in performance, workload, and usability and simultaneously enhance ecological validity. At the same time, however, any task should be simple enough to minimize requirements for specialized skills and training, accommodating the general population and reducing participant time burdens.

4.3 Conclusions

In summary, if the incorporation of human factors can make a difference in a simple task such as that used in the present study, even greater benefits might be expected to accrue for more complex products and processes. In effect, designing a task simply by using available tools, without true consideration of the human in the system, might yield a workable process, but not an optimal process that promotes effectiveness, reduces workload, and enhances usability. Systems engineers and other non-human factors practitioners may periodically require current, relevant evidence to help convince them that human factors issues must be expressly considered early and often throughout the lifecycle. To paraphrase Walkenstein and Eisenberg, experimental results such as these help demonstrate the value and need of involving human factors engineering in the design and development process and making it an integral part of that process [14].

References

1. Bailey, G.: Iterative methodology and designer training in human-computer interface design. In: Ashlund, S., Henderson, A., Hollnagel, E., Mullet, K., White, T. (eds.) Proceedings of INTERCHI 1993, INTERCHI 1993, pp. 198–205. IOS Press, Amsterdam (1993)
2. Bruseberg, A.: Presenting the value of human factors integration: guidance, arguments, and evidence. Cogn. Technol. Work **10**, 181–189 (2008)
3. Burgess-Limerick, R., Cotea, C., Pietrzak, E.: Human Systems Integration is Worth the Money and Effort! The Argument for the Implementation of Human Systems Integration Processes in Defence Capability Acquisition. Department of Defence, Commonwealth of Australia (2010)
4. Hendrick, H.W.: The ergonomics of economics is the economics of ergonomics. In: Proceedings of the Human Factors and Ergonomics Society 40th Annual Meeting, vol. 40, pp. 1–10 (1996)
5. Hendrick, H.W.: Applying ergonomics to systems: some documented "Lessons Learned". Appl. Ergon. **39**, 418–426 (2008)
6. Rouse, W., Kober, N., Mavor, A. (eds.): The Case for Human Factors in Industry and Government: Report of a Workshop. National Academy Press, Washington, D.C. (1997)
7. Sager, L., Grier, R.A.: Identifying and measuring the value of human factors to an acquisition project. In: Human Systems Integration Symposium, Arlington, VA (2005)
8. Shaver, E.F., Braun, C.C.: The Return on Investment (ROI) for Human Factors and Ergonomics Initiatives. Benchmark Research & Safety Inc., Moscow (2008)
9. Yousefi, P., Yousefi, P.: Cost Justifying Usability: A Case Study at Ericsson (Unpublished Master's Thesis). Blekinge Institute of Technology, Karlskrona (2011)
10. Steicklein, J.M., Dabney, J., Dick, B., Lovell, R., Moroney, G.: Error Cost Escalation Through the Project Life Cycle. Report JSC-CN-8435. NASA Johnson Space Center (2004)
11. Rauterberg, M., Strohm, O.: Work organization and software development. Ann. Rev. Autom. Program. **16**, 121–128 (1992)
12. Sen, R.N., Yeow, P.H.P.: Cost effectiveness of ergonomic redesign of electronic motherboard. Appl. Ergon. **34**, 453–463 (2003)
13. Yeow, P.H.P., Sen, R.N.: Ergonomics improvements of the visual inspection process in a printed circuit assembly factory. Int. J. Occup. Saf. Ergon. **10**, 369–385 (2004)
14. Walkenstein, M., Eisenberg, R.: Benefiting design even late in the development cycle: contributions by human factors engineers. In: Proceedings of the Human Factors and Ergonomics Society 40th Annual Meeting, vol. 40, pp. 318–322 (1996)
15. Lin, L., Isla, R., Doniz, K., Harkness, H., Vicente, K., Doyle, D.J.: Analysis, redesign, and evaluation of a patient-controlled analgesia machine interface. In: Proceedings of the Human Factors and Ergonomics Society 39th Annual Meeting, vol. 39, pp. 738–741 (1995)
16. Russ, A.L., Zillich, A.J., Melton, B.L., Russell, S.A., Chen, S., Spina, J.R., Weiner, M., Johnson, E.G., Daggy, J.K., McManus, M.S., Hawsey, J.M., Puleo, A.G., Doebbeling, B.N., Saleem, J.J.: Applying human factors principles to alert design increases efficiency and reduces prescribing errors in a scenario-based simulation. J. Am. Med. Inf. Assoc. **21**, 287–296 (2014)
17. Drury, C.G., Watson, J.: Good Practices in Visual Inspection (2002). http://www.faa.gov/about/initiatives/maintenance_hf/library/documents/media/human_factors_maintenance/good_practices_in_visual_inspection_-_drury.doc
18. See, J.E.: Visual Inspection: A Review of the Literature. Report SAND2012-8590. Sandia National Laboratories (2012)

19. Hart, S.G., Staveland, L.E.: Development of NASA-TLX (Task Load Index): results of empirical and theoretical research. Adv. Psychol. **52**, 139–183 (1988)
20. Lewis, J.R.: IBM computer usability satisfaction questionnaires: psychometric evaluation and instruction for user. Int. J. Hum.-Comput. Interact. **7**, 57–78 (1995)
21. Reason, J.T.: Human Error. Cambridge University Press, Cambridge, England (1992)
22. Steiner, S.H., MacKay, R.J.: Statistical engineering and variation reduction. Qual. Eng. **26**, 44–60 (2014)
23. Swain, A.D., Guttmann, H.E.: Handbook of Human Reliability Analysis with Emphasis on Nuclear Power Plant Application. Report NUREG/CR-1278-F SAND-0200. Sandia Corporation (1983)

Knowledge Management Model Based on the Enterprise Ontology for the KB DSS System of Enterprise Situation Assessment in the SME Sector

Jan Andreasik[✉]

University of Information Technology and Management, Sucharskiego Str. 2,
35-225 Rzeszów, Poland
jandreasik@wsiz.rzeszow.pl

Abstract. In the paper, the knowledge management model based on the original idea of the enterprise ontology is presented. This model is the basis of construction of the Knowledge Based Decision Support System (KB DSS) for evaluation of situation of enterprises in the SME sector. In the model, the SECI model of knowledge creation proposed by I. Nonaka and H. Takeuchi is applied. The model consists of a cycle of creating evaluation of situation of enterprises in the potential-risk space of activity. To design the enterprise ontology, ideas of Polish philosophers (J. Bochenski and R. Ingarden) are applied. Taxonomies of classes of the enterprise potential and risk are presented in the OWL language (the Protege editor). The KB DSS architecture is consistent with the Case Based Reasoning (CBR) methodology.

Keywords: Enterprise ontology · KB DSS system · CBR methodology
SECI model of knowledge creation

1 Introduction

Design of Knowledge Based Decision Support Systems (KB DSS) [1] requires a solution to the problem of knowledge representation including the process of knowledge creation. For this purpose, a properly selected concept structure (ontology) can be used. It enables us to cover the knowledge in the process of its formation and a respectively constructed cycle of knowledge creation. One of the knowledge creation models in organizations is the SECI model proposed Nonaka and Takeuchi [2]. A cycle of the knowledge conversion consists of four knowledge dimensions: socialization (tacit to tacit knowledge conversion), externalization (tacit to explicit knowledge conversion), combination (explicit to explicit knowledge conversion), internalization (explicit to tacit knowledge conversion). Bandera et al. [3] carried out research on application of particular processes of the SECI model in the SME enterprises. They have demonstrated a greater importance of the internalization process than an importance of the externalization and combination processes in the start-up enterprises category. This is related to the rapid process of incubation of such enterprises and

© Springer International Publishing AG, part of Springer Nature 2019
T. Z. Ahram (Ed.): AHFE 2018, AISC 787, pp. 146–156, 2019.
https://doi.org/10.1007/978-3-319-94229-2_15

intellectual property protection. In each process of the SECI model, the knowledge should be properly included and recorded. There is a lack of recommendations for operationalization of each of the processes in the SECI model. Figueiredo and Pereira [4] presented a process of knowledge creation from data in the following stages: data selection (data → target data), data preprocessing (target data → preprocessed data), data transformation (preprocessed data → target data), data mining (target data → patterns/models), interpretation/evaluation (patterns/models → knowledge).

Assessment of the proposed knowledge management model can be made on the basis of a set of requirements formulated by Ale et al. [5]. Those requirements correspond to a holistic view of knowledge management:

Requirement I: The knowledge management model should be convergent with the strategy of the organization.

Requirement II: The knowledge should be the strategic asset of the organization.

Requirement III: The model should include the process knowledge management in the organization.

Requirement IV: The model should include four categories of the knowledge management cycle: knowledge creation, knowledge communication, knowledge representation, and knowledge review.

Requirement V: The model should include the process of knowledge distribution in the organization (including creation of the knowledge repository).

Requirement VI: The model should take into consideration the social aspect (tacit knowledge transfer) and the technological aspect (declarative and procedural knowledge).

Requirement VII: The model should take into consideration the changes in the culture of the organization (using information systems).

The knowledge management model presented in the paper is created in terms of construction of the early-warning system for SME enterprises. This system is a knowledge repository as a key asset for enterprises. The vast majority of early-warning systems uses models related to prediction of bankruptcy. These models are based on research on statistical and machine learning methods. To build models using data in the form of indexes, mainly financial (profitability, fluency, capital structure, debt and others) are used. These indexes are calculated on the basis of historical financial reports. In SME enterprises, often, there is a lack of historical data, due to the short period of operation (the start-up enterprises category) as well as due to the simplified accounting of financial operations. In the proposed model, data are collected in the process of knowledge externalization (from tacit to explicit) by experts who assess business in certain ranges of strategic objectives.

To formalize the tacit knowledge, the model of knowledge representation, based on the original enterprise ontology, has been designed. A key aspect of creation of this ontology is a reference to conceptualization covered in the Ingarden's formal ontology as well as in the Bochenski's enterprise ontology [6]. To build the so-called applied ontologies, philosophical ideas are used as well. The most famous reference model is the BWW model proposed by Wand and Weber [7] using the Bunge's ontology. In the data model of information systems, Milton and Kazimierczak [8] use the conceptual apparatus of the Chisholm's philosophy. Dietz [9], in the enterprise ontology, uses the Wittgeinsten's ideas. In construction of the high-level ontology BFO [10], B. Smith uses ideas proposed by S. Lesniewski, the founder of mereology. The essence of ideas adopted in philosophical ontologies is to determine beings, which are subjects of cognitive processes, as a triple: subject, fact, object.

The presented knowledge management model generates the new knowledge in the form of recommendations for one of the four variants of enterprise competence assessment in competence potential – risk spaces: A1 (low potential/high risk), A2 (high potential/high risk), A3 (high potential/low risk), A4 (low potential/low risk).

Recommendations are automatically generated in the KM DSS system using aggregation of grades. For aggregation, multi-criteria decision making methods are used: AHP [11], EUCLID, Electre TRI [12]. To calculate threshold parameters in sorting models, the EUCLID method, proposed by Tavana [13], is used. This method requires data from an experiment conducted on a proper sample. Research on a sample of 220 Polish SME enterprises from Podkarpackie and Lubelskie voivodships was conducted by the author with a group of experts.

The so-called score trajectories according to metaphors defined by Argenti [14] are assigned to particular recommendations. Trajectories are characteristic patterns of particular results of the enterprise, for example, EBIDTA, EVA, etc., as functions of time. Recommendations related to assessment of enterprise competences and trajectory patterns are representations of cases in the CBR methodology [15]. This methodology is used in retrieving and modification of cases.

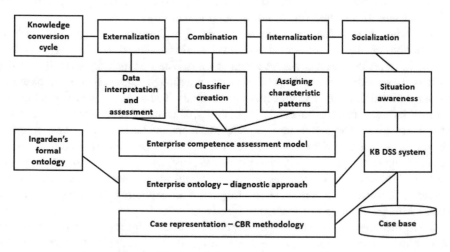

Fig. 1. The knowledge management model.

2 Knowledge Management Model Based on the Enterprise Ontology

In [29–35], we have presented an original view of the cycle of knowledge creation in the system for assessment of enterprise competences that is the Knowledge Based Decision Support System. The proposed knowledge management model is shown in Fig. 1.

The presented knowledge management model includes three main elements:

1. Enterprise ontology – a diagnostic approach.
2. Model for assessment of enterprise competences.
3. Representation of cases in the KB DSS system according to the CBR methodology.

Knowledge representation in the form of ontology is of a multi-aspect character. The first aspect concerns selection of theory which delivers the conceptual apparatus for defining particular beings. The second aspect concerns defining particular stages of the knowledge creation process with respect to concepts from domain ontology (the enterprise ontology). The third aspect is the adoption of such a representation of knowledge which leads to aggregation of assessment and expert opinion for indexation of cases in the CBR system.

The enterprise ontology proposed by the author is focused on a diagnostic aspect concerning assessment of enterprise competences. Enterprise competences are evaluated in two categories: the potential for activities and the risk of activities.

A base for distinguishing particular categories of the ontology is the enterprise model proposed by J.M. Bochenski. He considered an enterprise as a system consisting of internal and external elements. The internal elements include: capital, work and invention. The external elements include: customers, region and country. The author, according to the Bochenski's model, distinguishes five elements of an enterprise: capital potential, stakeholder potential, innovation and investment potential, relationship potential – neighborhood, relationship potential – surroundings. The capital potential is crucial in financing assets of enterprises, especially those of the nature of development projects (R&D), implementation projects and investment projects. In this kind of potential, the so-called structural capital, that is the component of intellectual capital, is included. Structural capital is formed by the ERP systems, the CRM systems, etc. The stakeholder potential is the result of synergy among different groups: the owners, managers, employees, experts, customers, sub-suppliers. The innovation and investment potential includes organizational assets concerning project management, process management and knowledge management. The relationship potential – neighborhood is the result of the influence of the policy of council organizations, non-governmental organizations (NGO), etc., on the sustainable development strategy. An enterprise should be also seen from the point of view of the risk of activities. A model, that is implemented by the author for defining the enterprise in terms of both the developmental potential and the risk of activities, is theory of object proposed by Ingarden [16]. An individual object is defined by the subject of properties as well as state of thing: positive and negative. State of thing is a function assigning properties to the object determined by the external subject of action. A positive state includes a statement of the real fact, where as a negative state concerns expression of what the object is not. The characteristic feature of the Ingarden's approach is the constitutive nature of the object. It is a matter which fully defines the object. The next element is the way of existence of the object (X is the self-contained being, X is the non-self-contained being, X is the primitive being, X is the secondary being, X is the independent being, X is the dependent being).

In analogy to concepts defining the individual object, the author proposes a set of concepts determining an enterprise:

```
object:=enterprise
subject:= competence system
positive state of thing:=assessment of potential
negative state of thing:=assessment of risk
property:=expert evaluation
constitutive nature:=pattern of enterprise according to Argenti
way of existence:=enterprise legal form   (charter)
```

The competence system is expressed by assessment of the enterprise potentials as well as assessment of the risk of activities. Figure 2 shows the taxonomy of the particular kinds of enterprise potentials and risks of activities.

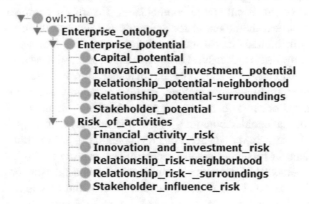

Fig. 2. A list of the main classes of the enterprise ontology in the Protege editor.

The model for assessment of enterprise competences is defined in a set of processes of externalization, combination and internalization (see Fig. 3). We assume that different experts can be authorized for each set of the enterprise competence analysis. Their task is to assess potentials and risks of activities. Experts can have data concerning the availability or unavailability of adequate resources (staff, machinery, technology, IT systems, capital). These resources should be recorded in a database system, e.g., ERP, CRM. Data should be gathered from the systems using proper queries formulated in the SQL or SPARQL language. A crucial system for gathering data is the BSC (Balanced Scorecard) system [17]. This system allows us to obtain data about indexes of strategic objectives.

Assessment of the enterprise potential is made by the expert on the basis of comparison of resource states (human, tangible, intangible) and effects of activities (profits) in the considered range of the potential with the target resource states determined by indexes of the strategy prepared by means of the BSC (Balanced Scorecard) method. The expert can use the AHP method and the EXPERT CHOICE software assuming a proper set of criteria for each range. Analogously, the expert makes assessment of the risk of activities on the basis of comparison of planned resource states and effects of activities with a pessimistic state caused by threats (losses).

Grades of assessment of the enterprise potential and risk are numbers from the interval [0, 1]. The task of the expert is to extract the tacit knowledge about the enterprise potential and about the degree of risk. The expert brings his/her own experience from identification of potentials and risks of different enterprises into the assessment process. In case of a large number of enterprises, assessment is made by a group of experts, exchanging knowledge and experiences and analyzing different cases.

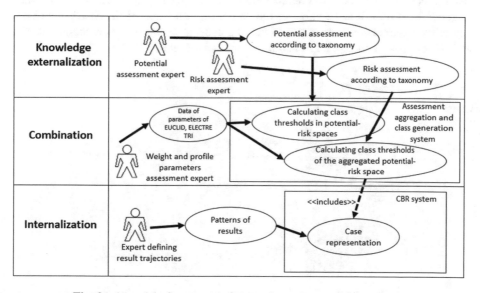

Fig. 3. A model of processes of enterprise competence assessment.

It is necessary to aggregate grades due to the fact that the presented ontology includes five main categories of the enterprise potential and five main categories of the enterprise risk as well as in each category there are distinguished lists of ranges for assessment. A convenient method of aggregation with the possibility of specifying the sorting classes is EUCLID. The author presented precise calculation procedures in [31]. On the basis of those procedures, four classes are determined:

I - threat class (low potential, high risk),
II - warning class due to the high risk,

III - good condition class (high potential, low risk),
IV - warning class due to the low potential.

Sorting parameters for these classes are simultaneously the knowledge extracted in the process of assessment generalization from the research sample. In the experiment carried out under the EQUAL Programme in the University of Management and Administration in Zamość, a group of 12 experts made assessment of 220 SME enterprises. Values of particular thresholds of identification of classes for five spaces were determined:

1. Capital potential – Financial risk.
2. Investment potential – Investment risk.
3. Stakeholder potential – Stakeholder cost risk.
4. Relationship potential - neighborhood – Relationship risk - neighborhood.
5. Relationship potential - surroundings – Relationship risk - surroundings.

The ELECTRE TRI method enables us to aggregate positions of an enterprise in each potential – risk space to one position in a generalized space: generalized potential – generalized risk. It is an effect of the combination process of the SECI model, i.e., knowledge conversion from explicit (expert grades) to explicit after processing (sorting model to classes of knowledge about the enterprise state).

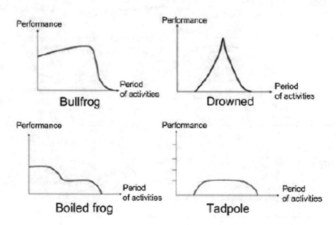

Fig. 4. The characteristics of results as a function of time according to [18].

The next process of the SECI model is the internalization process, i.e., knowledge conversion from explicit to tacit. It is the process connected to experience and a hidden form of information (symbols, metaphors, characteristic patterns of behavior, etc.). In the enterprise analysis, we can use characteristic patterns of enterprise results. Argenti in [14] presented an idea of three patterns of enterprise bankruptcy. Richardson et al. [18] extended those patterns and determined them using metaphors concerning frog behavior: bullfrog, drowned frog, boiled frog, tadpole (see Fig. 4). Figure 5 shows characteristics of results as a function of time according to Argenti's patterns. As a result in the case base of the presented system, selling income was selected. We can use other results, e.g., net profit, EBITDA, EBIT, EVA, CF, total assets. etc.

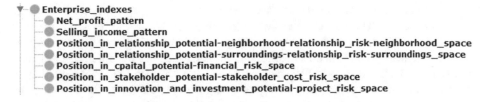

Fig. 5. Index of individual enterprise

In the case base of the KB DSS system based on the CBR methodology, an individual enterprise is represented by the set of indexes shown in Fig. 5.

In [29, 30], the author presented the architecture of the SOK-P1 system, which realizes the process of knowledge externalization, combination and internalization. The fourth element of the knowledge creation cycle according to the SECI model is the socialization process (tacit knowledge → tacit knowledge). This process is realized as retrieval and modification of cases recorded in the case base. The users can analyze a relationship between assessment of the enterprise state that is indexed by a position in particular potential – risk spaces and the Argenti's pattern that is of prognostic character.

3 Conclusions

In the paper, the knowledge management model based on the original enterprise ontology has been presented. This ontology is of a diagnostic character. It includes a structure of concepts necessary for assessment of an enterprise in the potential – risk space. The model fulfills requirements presented at the beginning.

Requirement I: The model is convergent with the organization strategy. In the assessment process, experts use data from strategic documents of an enterprise, especially from the balanced scorecard (BSC). In particular ranges of the enterprise potential and risk, discrepancies between assumed values of indexes of strategic objectives and actual values of indexes achieved during assessment are analyzed.

Requirement II: The presented model is a base for the design of the early warning system in an enterprise, which is a strategic asset of the organization.

Requirement III: The model includes data analysis process strategic for the organization, needed for particular stages of extraction of the knowledge about the enterprise state. Data are gathered from the ERP, CRM and BSC systems.

Requirement IV: The model consists of three categories of the knowledge conversion cycle: knowledge creation (generation of assessment of the enterprise potential and risk and aggregation of grades to indexes of the enterprise state), knowledge communication (presenting characteristic patterns of results according to the Argenti's models), knowledge representation and review (the designed system enables us to view and interpret results).

Requirement V: The model includes the case base which enables us to distribute the knowledge to particular users (stakeholders of an enterprise).

Requirement VI: The model is based on the SECI idea covering the knowledge conversion cycle. The presented enterprise ontology enables us to represent the tacit knowledge in the form of assessment of the potential and risk of enterprise activities.

Requirement VII: Particular taxonomies of assessment of the enterprise potential and risk are included in the OWL language and recorded using the Protege editor. It enables us to put corrections depending on changes in the culture of the organization.

The modern trend in the design of the KB DSS systems is to use semantic web analysis tools [19]. The key language for recording knowledge conceptualization is the OWL (Web Ontology Language) language [20]. A number of editors was designed to define domain and applied ontologies, e.g., Protege [21], Top Braid Composer [22]. The semantic web technologies allow us to design ontologically oriented decision support systems OB DSS (Ontology-based Decision Support Systems) [23] as well as ontologically oriented recommender systems OB RS (Ontology-based Recommender Systems) [24]. The author is developing the presented model towards the use of the myCBR and jCOLIBRI software, which are based on the CBR methodology.

The myCBR software was designed in the German Research Center for AI: DFKI. The myCBR web page [25] includes installation instructions and system documentation together with presentations. The software is developed by T. Roth-Berghofer. The myCBR software includes the editor for defining cases according to principles of creating taxonomy in a descriptive logic (DL) format. It also includes the editor for defining and analyzing similarity measures used to retrieve cases in the process of reasoning by analogy in the CBR cycle. Several recommender systems designed on the basis of the myCBR software is analyzed in the Sauer's Ph.D. thesis [26].

An alternative system to the reasoning process according to the CBR methodology is the jCOLIBRI2 [27] system developed in Facultad de Informatica, Universidad Complutense de Madrid [28]. The system is adopted to the text analysis in natural language. The engine of the system is based on the kNN algorithm. The system includes all of the elements of the CBR cycle: retrieve (review of cases), reuse (strategy of matching a case to the query or to the current feature signature of a new case), revision (introducing changes to the current case), retain (introducing a new case to the case base).

References

1. Zarate, P., Liu, S.: A new trend for knowledge-based decision support systems design. Int. J. Inf. Decis. Sci. 8(3), 305–324 (2016)
2. Nonaka, I., Takeuchi, H.: The Knowledge-Creating Company. How Japanese Companies Create the Dynamics of Innovation. Oxford University Press, New York, Oxford (1995)
3. Bandera, C., Keshtkar, F., Bartolacci, M.R., Neerudu, S., Passerini, K.: Knowledge management and the entrepreneur: insights from Ikujiro Nonaka's Dynamic Knowledge Creation model (SECI). Int. J. Innov. Stud. 1, 1163–1174 (2017)
4. Figueiredo, M.S.N., Pereira, A.M.: Managing knowledge – the importance of databases in the scientific production. Proc. Manuf. 12, 166–173 (2017)

5. Ale, M.A., Toledo, C.M., Chiotti, O., Galli, M.R.: A conceptual model and technological support for organizational knowledge management. Sci. Comput. Program. **95**, 73–92 (2014)
6. Bocheński, J.M.: Przyczynek do filozofii przedsiębiorstwa przemysłowego. Logika i filozofia. Wybór pism. Warszawa PWN, pp. 162–186 (1993)
7. Wand, Y., Weber, R.: An ontological model of an information system. IEEE Trans. Softw. Eng. **16**, 1282–1292 (1990)
8. Milton, S.K., Kazmierczak, E.: An ontology of data modeling languages: a study using a common-sense realistic ontology. J. Database Manag. **15**(2), 19–38 (2004)
9. Dietz, J.L.G.: Enterprise Ontology. Theory and Methodology. Springer, Heidelberger (2006)
10. Arp, R., Smith, B., Spear, D.: Building Ontologies with Basic Formal Ontology. Massachusetts Institute of Technology, Cambridge (2015)
11. Saaty, T.L.: Decision Making for Leaders. RWS Publications, Pittsburgh (2001)
12. Mousseau, V., Słowiński, R., Zielniewicz, P.: A user-oriented implementation of the ELECTRE TRI method integrating preference elicitation support. Comput. Oper. Res. **27**, 757–777 (2000)
13. Tavana, M.: Euclid: strategic alternative assessment matrix. J. Multi-Criteria Decis. Anal. **11**, 75–96 (2002)
14. Argenti, J.: Corporate Collapse: The Causes and Symptoms. McGraw-Hill, London (1976)
15. Aamodt, A., Plaza, E.: Case-based reasoning: foundations issues, methodological variations and system approaches. AI Commun. **7**(1), 39–59 (1994)
16. Ingarden, R.: Spór o istnienie świata. PWN, Warszawa (1987)
17. Kaplan, R.S., Norton, D.P.: Balanced Scorecard. Translating Strategy into Action. Harvard Business School Press, Boston (1996)
18. Richardson, B., Nwankwo, S., Richardson, S.: Understanding the causes of business failure crises: generic failure types: boiled frogs, drowned frogs, bullfrogs and tadpoles. Manag. Decis. **32**(4), 9–22 (1994)
19. Hitzler, P., Krotzsch, M., Rudolph, S.: Foundations of Semantic Web Technologies. Chapman & Hall/CRC, Boca Raton (2010)
20. OWL 2 Web Ontology Language Document Overview, 2nd edn. https://www.w3.org/TR/2012/REC-owl2-overview-20121211/
21. Protege platform. https://protege.stanford.edu/products.php
22. TopBraid Composer Maestro Edition. https://www.topquadrant.com/tools/ide-topbraid-composer-maestro-edition/
23. Giovannini, A., Aubry, A., Panetto, H., Dassisti, M., Haouzi, H.: Ontology-based system for supporting manufacturing sustainability. Ann. Rev. Control **36**, 309–317 (2012)
24. Middleton, S.E., Roure, D.D., Shadbolt, N.R.: Ontology-based recommender system. In: Stab, S., Studer, R. (eds.) Handbook on Ontologies, International Handbooks on Information Systems, pp. 779–795. Springer, Heidelberg (2009)
25. myCBR. http://mycbr-projekt.net
26. Sauer, S.: Knowledge elicitation and formalisation for context and explanation-aware computing with case-based recommender system. Doctoral thesis. University of West London (2016). https://repository.uwl.ac.uk/id/eprint/2226/
27. Recio-Garcia, J.A., Gonzalez-Calero, P.A., Diaz-Agudo, B.: jCOLIBRI: a framework for building case-based reasoning systems. Sci. Comput. Program. **79**, 126–145 (2014)
28. jCOLIBRI. www.gaia.fdi.ucm.es/research/colibri/jcolibri
29. Andreasik, J.: A case-based reasoning system for predicting the economic situation of enterprises – tacit knowledge capture process (externalization). In: Kurzyński, M., et al. (eds.) Computer Recognition Systems 2. Advances in Soft Computing, ASC, vol. 45, pp. 718–730. Springer, Heidelberg (2007)

30. Andreasik, J.: The knowledge generation about an enterprise in the KBS-AE (knowledge-based system – acts of explanation). In: Nguyen, N.T., et al. (eds.) New Challenges in Computational Collective Intelligence. Studies in Computational Intelligence, SCI, vol. 244, pp. 85–94. Springer, Heidelberg (2009)

31. Andreasik, J.: Decision support system for assessment of enterprise competence. In: Kurzynski, M., Wozniak, M. (eds.) Computer Recognition Systems 3. Advances in Intelligent and Soft Computing, AISC, vol. 57, pp. 559–567. Springer, Heidelberg (2009)

32. Andreasik J.: Enterprise ontology according to Roman Ingarden formal ontology. In: Cyran, K.A., et al. (eds.) Man-Machine Interactions. Advances in Intelligent and Soft Computing, AISC, vol. 59, pp. 85–94. Springer, Heidelberg (2009)

33. Andreasik, J.: Enterprise ontology for knowledge-based system. In: Hippe, Z.S., Kulikowski, J.L. (eds.) Human-Computer System Interactions. Advances in Intelligent and Soft Computing, AISC, vol. 60, pp. 443–458. Springer, Heidelberg (2009)

34. Andreasik, J.: Enterprise ontology-diagnostic approach. In: Proceedings of the Conference on Human System Interactions, HSI 2008, Book Series: Eurographics Technical Report Series, Krakow, Poland, pp. 503–509. IEEE (2008). https://doi.org/10.1109/hsi.2008.4581489

35. Andreasik, J.: Ontology of offers according to Ingarden's theory of individual objects. In: Proceedings of the 5th International Conference on Agents and Artificial Intelligence, ICAART 2013, pp. 429–432. SciTePress (2013). https://doi.org/10.5220/0004209304290432

The Human Side of Service Engineering: Advancing Smart Service Systems and the Contributions of AI and T-Shape Paradigm

Re-defining the Role of Artificial Intelligence (AI) in Wiser Service Systems

Sergio Barile[1], Paolo Piciocchi[2], Clara Bassano[3(✉)], Jim Spohrer[4], and Maria Cristina Pietronudo[3]

[1] Department of Management, Sapienza University of Rome, Rome, Italy
sergio.barile@uniroma1.it
[2] Department of Political, Social and Media Sciences, University of Salerno, Salerno, Italy
p.piciocchi@unisa.it
[3] Department of Management Studies and Quantitative Methods, University of Naples "Parthenope", Naples, Italy
{clara.bassano,mariacristina.pietronudo}
@uniparthenope.it
[4] IBM, Almaden Research Center, San Jose, CA, USA
spohrer@us.ibm.com

Abstract. Advances in Artificial Intelligence (AI) are raising important questions for companies, employees, consumers and policy makers. Researchers predict that intelligent machine will outperform humans in a wide range of tasks in the coming decade. Our purpose is to re-define the role of AI technologies and their relationship with people by re-thinking the concept of Intelligence Augmentation (IA), an interaction between AI technologies and people that, more than amplifying human capacities, produce a cognitive transformation. This transformation modifies the structure of humans thought, changing people's cognitive processes and providing new tools to optimize interpretative schemas, useful to analyze the real world. In line with the need to define new directions in Service Science, new rules should be formulated to create a synergic and collaborative processes between humans (people) and machines. In order to design a wiser service system, this paper proposes T-shaped professionals as especially well-adapted for augmented and collaborative intelligence.

Keywords: Service Science · Artificial Intelligence
Intelligence Augmentation · T-Shaped professional

1 Introduction

Over the last few years, an intensive debate on Artificial Intelligence (AI) technologies, such as - autonomous vehicle control, robots, speech recognition and other emerged applications [1] – has been taking place. Academics, especially economists [2, 3], are investigating effects that these technologies could produce on economic growth and on the future of work. AI has been considered the *general-purpose technology* [4] of our era [5], as it has altered the social and economic structure and introduced a series of *disruptive innovations*. However, this is not a new discussion; it emerged during the

© Springer International Publishing AG, part of Springer Nature 2019
T. Z. Ahram (Ed.): AHFE 2018, AISC 787, pp. 159–170, 2019.
https://doi.org/10.1007/978-3-319-94229-2_16

first industrial revolution - when the muscular energy produced by humans and animals was replaced by that of steam engines -and again during the computer revolution – when computers took on the work of book-keepers, telephone operators and cashiers [6–8]. However, contrary to previous revolutions, there is an especially deep fear associated with the emergence of AI in two regards. The first is the rapid pace at which all types of innovations are spreading; the time between an innovation and a successor to it is always shorter. The second is the computerization of cognitive and non-routine tasks such as composing music or writing a book. In particular, in healthcare, diagnostic tasks are already being computerized to decide the best treatment plan for a patient. A computer is able to compare each patient's individual symptoms, genetics, family and medication history, etc. formulating the most suitable and effective solution [9], in some case more quickly and accurately than the average doctor. Furthermore, nowadays, technologies are able to learn and solve technical problem alone, without more than a single initial human intervention (setting up the constraints and reward policies). Changes are also happening in a wide range of service-oriented roles and activities (e.g. customer care, HR assistants, secretaries, real estate brokers etc.) [10–12], where consequences are more complex to manage because they modify the value and the actors involved in the interaction. AI technologies are now catching up with and encapsulating more and more forms of human capabilities. They are no longer simple tools that facilitate interactions, but active actors capable of taking deliberate actions on their own [13]. Nevertheless, the future scenario is not so catastrophic. This is *the era of cognition as a service* [14]: machines learn and help humans make smarter decisions to obtain better outcomes. Maglio et al. (2010) already have considered integration between people, technology, organizations and information as a necessary condition to co-create both social and economic value in a service system [15, 16]. Now, however, organizations and managers should implement this integration avoiding a juxtaposition of tasks and functions and a complete automation of services with a high intensity of interaction. For that reason, in order to guarantee the co-existence of machines and humans, a system should be transformed from smart to wise, striking a balance between system's entities. This paper offers one possibility to consider as a first attempt to re-define the role of AI in wise systems characterized by an evolving concept of intelligent interaction called Intelligence Augmentation (IA) which is a specific type of augmented and collaborative intelligence. T-shaped professionals [17, 18] exhibit this type of augmented and collaborative intelligence. This type of professional exhibit deep problem-solving skills as well as broad communication talents, and these capabilities serve as facilitators of interactional relationships between entities of different natures (people and robots included).

The paper begins by a brief overview of AI, pointing out its limitations. After that, AI in service systems is investigated explaining why it occurs, re-defining its role, and presenting why it is appropriate to talk about Intelligence Augmentation. In the second section, the conditions for a wiser service system will be explained. In the third section, T-shaped professionals will be proposed as advocates to promote human-machine interactions. Implications and future research will be discussed in the fourth part. Some preliminary reflections are drawn in the final section.

2   Artificial Intelligence (AI) Overview

Artificial Intelligence (AI) studies are based on a very simple concept: *"If man's rational thoughts could be as systematic as geometry or physics, then it can be mechanized"* [19]. In 1950 Alan Turing, a theoretical computer scientist, considered the father of AI, designed a test called "The Imitation Game" to understand if a machine/computer could exhibit text-based conversational behavior indistinguishable from that of a human's. Turing's vision has become the cornerstone of a new inter-disciplinary science, artificial intelligence [20, 21] which views our brain as acting like a calculator. Put differently, AI takes the view that all human mental activities are reducible to algorithms and could therefore be implemented on a computer. From the perspective of AI, the aims of building intelligent machines can be summarized as in the seminal work of Russell (2009) as follows [22]:

- Acting like humans;
- Thinking like humans;
- Acting rationally;
- Thinking rationally.

Currently, for an ever-expanding range of tasks from computers learning to play chess to robots learning to run obstacle courses, a properly programmed machine can rapidly learn to perform tasks at the human or super-human levels with no further human intervention once the rules/constraints of the world and the rewards for ideal performance are defined [23]. Specifically, they can learn autonomously (deep learning), to analyze and produce a text in a common language starting from computer data (Natural language processing), and to optimize and automate human processes and activities (robotic process automation). However, challenges remain. Despite these efforts, even now, no intelligent machine can completely simulate human thought; in particular machines:

- have a low interpretability of decisions [5];
- are unable to have feelings and emotions;
- lack the commonsense reasoning abilities of a four-year-old child.

Concerning the first point, intelligent machines produce results without explaining how, since the output decisions are the result of complex network connections [5] based on mathematic models realized by software programs. Concerning the second point, even though recognizing facial expressions, body language, and speech prosody are correlated with feelings and emotions, machines still do not have emotional competence in responses and natural conversational interactions. Emotions are crucial to everyday human reasoning processes [24] since social-emotional learning skills are used in nearly every face-to-face interaction between people [25]. These machine limitations are strengths of most people, with the exceptions of individuals who suffer from autism and other related impairments. Intelligent machines still have to learn to share job's task and to augment human intellect [26] and naturally support people in complex, collaborative problem solving. The risk is not about creating an intelligent and emotive machine but creating an *intelligent machine without any emotion* [27].

However, only if they reach a complete autonomy, will we need to be concerned about the rights and responsibilities of a cognitive system as an independent service system entity. Because emotional and social intelligence are fundamental components in natural collaborative activities (i.e. decision-making process, during interactional process) the prospect of total replacement of human labor is in the distant future. Robotic managers or entrepreneurial robots have not yet been created, in part because these roles require social-emotional learning skills. In the near future, we can expect to only have hybrid teams, composed of humans and machines [28].

3   Understanding the Role of AI in Wise Service Systems

From the editorial published by Maglio in Service Science (INFORMS) (2017), the need for a revision of some fundamental notions of service science is perceived and beginning to be explored by the community [13]. Following the introduction of advanced interaction technologies, in particular AI technology there has been a perceived need to re-examine three basic principles of service science: service/ stakeholders, interaction and co-creation. Service, defined as the result of an interaction between different actors, who share skills and competences to create together, as joint stakeholders [18] who co-create [29] value [30]. The community is poised to re-examine these basic concepts in light of rapidly changing autonomous interaction capabilities of intelligent machines with social-emotional learning functions. As explored in the editorial, service, because of modern technologies, is increasingly *autonomous* in many sectors (transportation, financial services etc.) [13]. Given that the collaborative interaction is extended to intelligent machines using modern interaction technologies, the dilemma is how to analyze and re-think how "*the value produced by a human-machine interrelation can be defined as co-created?*" [13].

In our opinion, rather than attempting to re-define the already established concepts of service science, we try to clarify the role of AI technologies in service systems. Surely, compared to twenty years ago, intelligent machines have a more active role in the process of interaction and co-creation, because they have knowledge and skills sometimes greater than that of the human actors in the service situation. However, the value to be co-created remains a human value, to serve human purposes [31] for the benefit of people, organization, and certainly not machines. A service system is a configuration of elements, including people, technologies, information that work together via value propositions for mutual benefits [15]. But the technologies, although intelligent, do not have rights and responsibilities, and therefore maintain a partial role since they do not have the ability to propose, accept and evaluate the value from a legal justice perspective in case disputes arise [32]. They are more advanced, more autonomous tools that enhance people's capabilities and human purposes [33]. A machine will provide more appropriate schemes, will gather more information, sometimes interacting emotionally with the most active subjects of the interaction (i.e. humanoid robots or animated software agents that recognize human affective intentions and to produce also emotive facial expression like disgust or happiness [34]), but will not provide an interpretative key or introduce the values necessary for the co-finalization of the value proposition. In sum, intelligence machines do not have rights and responsibilities protected by a formal justice system, which is a type of service system that resolves disputes between entities with rights and responsibilities.

In analyzing the relationship between men and machines, Degani et al. (2017, p. 221) use the etymology of the term autonomy to outline three different concepts that originate from a single term [35]: (1) the authority to create and apply one's own laws; (2) the self-sufficiency, or better, the capacity of an entity to take care of itself without the help of others; (3) and the self-directedness, or freedom from outside control. The recent intelligent machines manage to self-govern their behaviors themselves by producing algorithms automatically. They can perform elaborations without anyone's intervention. But, at this stage they cannot be granted rights and responsivities and so must be redirected towards human purposes; they are not pro-actively requesting rights and responsibilities in pursuit of their own purposes, nor do they (yet) think outside of the box about better human purposes. Referring to the distinction made by Vargo and Lush (2006) between operant and operand resources [36], intelligent machines are still operand resources, whose usefulness is linked to the adequacy to generate a service. Their contribution is useful with the intervention of operant resources, or better human resources that amplify their usefulness. Service systems are complex systems; they need an interdisciplinary set of resources able to converge the interests of different actors. Therefore, the functions carried out by the AI technologies are not decisive for reaching a common goal, but they are certainly significantly increased compared to the past. Although we are in the age of cognition as a service, and machines can perform human-like functions (language, learnings, level of confidence), the machines/intelligence systems are used to grow and scale competences by increasing the productivity and creativity of humans [14]. So, we believe that in wise service systems, it is more appropriate refer to intelligence augmentation rather than artificial intelligence.

3.1 From Artificial Intelligence (AI) to Intelligence Augmentation (IA)

The difference between AI and IA has been addressed by the philosopher Peter Skagestad in his work "The Mind's Machines: The Turing Machine, the Memex, and the Personal Computer" [22] outlining a boundary between the two research domains. AI refers to computer programming of machines that know how to think better than men do. In contrast, IA refers to computer programming of machines that can increase the reliability of human thought by extending and assisting it. In addition to the two distinct software programming objectives, the distinction hides a different conception of the role of the person who uses the software systems. In the first, the human exists only for comparison purposes, and is a machine replaceable subject in a system; in the latter the human is a user with whom the machine can enhance. However, although at first AI and IA seem in conflict, they both come from theories on computational intelligence, and are, in fact, complementary [23, 37]. An intelligent machine can be a computer capable of mimicking the capabilities of the human brain – Artificial Intelligence (AI) – and at the same time be a support to the functionality of our brain – Intelligence Augmentation (IA). Using servant and maestro metaphors, we see that technologies, specifically AI, are at the same time a *servant*– replacing humans in complex activities (elaborate calculations, complicated mechanical activities) - and a *maestro*– empowering, instructing, and educating a human [33]. From both the AI servant and AI maestro perspectives, exchange of information and knowledge must be bi-directional, since people also have the ability to amplify the potential of a machine. Although machines are now able to acquire information and knowledge more efficiently, they lack an emotional and social intelligence and they are not yet able to put their knowledge at the service of the whole system, as can entities with right and responsibilities (authorities). Human interventions are therefore inevitable. Humans, in fact, possession social and emotional skills fundamental for the elaboration of thoughts, solutions, and decisions; they are able to use knowledge appropriately, addressing it towards common objectives that guarantee a balance between individual and collective value [38]. From our perspective, intelligence augmentation is based on a virtuous and circular interaction between human and machine, which guarantees smart and wise resolution of complex and systemic issues (Fig. 1, [39]). Particularly, an interaction between smart people, that have greater propensity for broad-communications and deep-technical skills, both optimize the process of augmented intelligence.

Thus, a service system should become wise (improving opportunities for emergence; beneficial surprises or unpredictable opportunities for future generations) as well as smart (improving productivity; trusted predictable opportunities for the current generation). The attribute "Smart" is a necessary but not sufficient condition for the viability of systems and for creating resonance and development in the context. A smart service system[1] creates value mediating and integrating interactions between (human) actors and efficient use of resources [42], but it is not able to ensure the complete exploration of opportunities for a better co-creation of value for future generations. A wise context is an evolution of a smart context across multiple generations; in fact, wisdom is referred to as virtuous relationships between smart components of today able to mediate human and society expectations and expand surprising, beneficial opportunities for future generations. Wise service systems need a sociological mediation approach, under which it is relevant to have a human component (problem solver and/or decision maker) who is able to ensure that the whole service system reaches the right decision in a specific situation and for a common welfare across multiple generations, as judged by future generations [43]. The human system component must be one that can appreciate the relativism of values, can understand the priorities of the individuals acting in the system, and be able to balance their own interests and those of others across generations [44] in a virtuous, harmonious process. Specifically, smart service systems characterized by AI technologies will be wiser if smart people cooperate and co-create in a choral manner with these technologies. From this interaction, more than an amplification of human capacities, a cognitive transformation emerges [25] that modifies the process of human thought. This is not utopia or dystopia. Unfortunate and unexpected things can always happen, but the resilience of the system increases so that bouncing-back from catastrophe is faster.

---

[1] A smart service system is a socio-technical system that amplifies human capabilities, routinizes interactions [40] controls, coordinates, and manages processes and goals of a system [41].

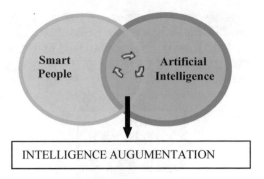

Fig. 1. Intelligence augmentation derives from a circular interaction between smart people and smart technologies (artificial intelligence), [39].

4   Intelligence Augmentation and T-Shaped Professionals

Although the human-machine interaction process is considered very similar to human-human interaction [45], to build a wise service system it is necessary to understand the characteristics that people must have to interact effectively with smart machines. Many studies have explored the design of machine interfaces (Human Computer Interaction (HCI) and Human-Factors and Ergonomics (HFE)) and seek to understand and interpret the various ways in which people communicate and interact with machines and technological systems [35]. Therefore, we focus on the skills that people must have to support human and machine interactions. This does not mean that people have to only develop technical skills to understand the complex algorithms of the machines, rather that they need to also develop creative abilities, relational and social skills, therefore, they need horizontal competencies to fill in for the machine's gaps so as to be able to operate for the benefit of the system. Furthermore, people will need to improve their analytical and critical senses and be able to understand the belief system in which the machine operates (its behavior, operational boundaries, and limitations) as well as its intentions [35]. In this way, the human partner will be able to adapt the profound skills of a machine to the social environment in which it operates, ensuring intelligent decisions that benefit the current generation and wise decisions that benefit future generations (as judged by future generations).

4.1   Why T-Shaped Skills

An analysis suggested by Service Science (SS) and the Viable Systems Approach (VSA) defines the best paradigm for considering and assessing professional competencies proposing T-Shaped Professionals (T-SP) [46–48]. T-SP are characterized by deep problem-solving knowledge in one disciplinary/system area, and broad communications skills across a range of disciplines/systems [48]. For example, an engineer who can communicate well with business and social science colleagues is a T-SP. In light of the discussion so far reported, we believe that to favor a Wise Service System and to avoid a future where intelligent machines replace workers, there is a need for

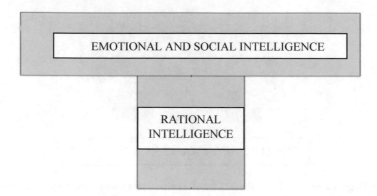

Fig. 2. People should develop more emotional and social intelligence and less rational intelligence to create a wise system based on a virtuous combination of machine and humans.

people to develop "T" – shaped intelligences. However, contrary to previous literature, in which an equitable distribution of horizontal and vertical skills was preferred to guarantee the ability across a wide range of disciplines and to cooperate with actors of different backgrounds, the proposed ideal configuration of T-shaped skills proposes a less deep and wider configuration, as machines help more with depth than breadth [48]. The proposed configuration favors the development of the human's social and emotional skills, therefore horizontal skills (Fig. 2), are more oriented to collaborative interactions and decision-making processes in IA systems designs.

While rational intelligence and technical/hard skills will be useful to verify the reliability of a result produced by intelligent machines, the social and emotional intelligences [49] will serve to verify the adaptability to the context of the solutions identified by the machines. In fact, they ensure the development of:

- intrapersonal (emotional) skills: understanding of one's own values, awareness of one's knowledge/awareness, flexibility, self-management;
- interpersonal skills (social): relationships with others (including intelligent machines), understanding of other people's values/empathy, active listening/ communication, cooperation;
- inter-generational skills for thinking long-term about the implications of today's decisions to future generation, especially decisions that impact the resilience of future generations (ability to rapidly rebuild from scratch after catastrophes).

Specifically, the T-SP optimized for the era of intelligent machines should possess a proactive attitude, creativity, change management orientation, understanding of complex situations, and negotiation skills. As technology races ahead, the demand for workers with social emotional learning skills also increases, because these skills are not yet susceptible to computerization [8] and can exceed the limits of intelligent machines, also known as the rapidly growing digital workforce each person can access [50]. The digital workforce can be seen today as the growing array of apps that people use on their smartphones. Overtime, apps will have episodic memories, commonsense reasoning, and conversational interfaces of a full-fledged digital workforce.

5 Implications and Future Challenges

As a result of the recent progress made in the field of artificial intelligent (AI) and other advanced interactions technologies (e.g., augmented reality), a transformation of our personal and intellectual habits is now gradually underway. Probably, the changes will not occur for all jobs, but in a wide range of cases new positions and greater flexibility will be required (freelance, self-employment, project-based work, temporary jobs at different levels of complexity and remuneration [51]). In any event, it will be necessary to adopt the right policies to support the current situation to prevent innovation from being destructive and generate higher unemployment rates than new jobs created [52]. Policy makers will have to adopt an information policy that makes people aware of the limits and potentials of IA technologies; an education policy aimed at preparing T-Shaped Professionals (T-SP) of the future; a collaborative policy to promote the virtuous and circular collaboration between humans and machines, which means favoring the development of intelligence augmentation (IA). At the same time, universities and researchers can play a key role by engaging with policy makers to design smarter and wiser human-centered service systems; driving new knowledge creation and stimulating qualify of life progress for everyone, including the weakest in society [53, 54]. The future challenges are numerous, the shift from smart to wise systems through intelligence augmentation process is still very difficult because there are many gaps and research questions, whose answer need to be obtained as a result of greater collaboration and multidisciplinary relationship between academia, industry and government institutions [55]. Certainly, it will be necessary to develop T-shaped skills and techniques for aligning human machine learning [14] and to design more intuitive machine interfaces. However, in our opinion, future research should investigate the new concept of Intelligence defined as the ability to approach a solution by changing our endowment of knowledge [39]. The question is: by modifying our knowledge endowment does a modification of intelligence result? Given that the endowment of knowledge consists of informative units, interpretative schemes and value categories from a Viable-Systems Approach perspective, it follows that if you can innovate the appropriate interpretative schemes, then you can advance intelligence augmentation (IA).

6 Conclusion

The paper highlights the growing awareness in the service science community that AI advances require a re-examination of fundamental concepts, especially the concepts of service/stakeholders, interaction, and value co-creation, for the purpose of intelligence augmentation (IA) in wiser service systems that benefit future stakeholders and future generations. From our perspective, the AI technologies are based not only on informative units and interpretative schemes but tends to an evolutionary dynamic characterized by a cognitive transformation. In this perspective, we can say that being aware of the shift from AI to IA means re-defining a Wise System from Smart Systems since AI progress changes the role of human-factors in the socio-technical systems. This means promoting a system in which the optimization process refers to the capability to improve the collaborative intelligence through intelligence augmentation processes. In short terms, what is changing is not the role of technologies, but the way we interact with them [25]. This transformation is at the base of the augmentation of intelligence (IA) since it modifies the processes of human thought, changing the thinking people way [25] through the optimization of new interpretative common schemes in which the effect of the application of the artificial intelligence into the problem-solving processes realizes the wise intelligence evolution system. Intelligence Augmentation - which cannot and does not intend to replace man - can revolutionize the concept of intelligence; it is not an artificial intelligence, but a support: it is up to man, with his competence, to decide how to use the collected evidence and how to modify it.

References

1. Jordan, M.I., Mitchell, T.M.: Machine learning: trends, perspectives, and prospects. Science **349**(6245), 255–260 (2015)
2. Frey, C.B., Osborne, M.: The future of employment. How susceptible are jobs to computerization (2013)
3. Brynjolfsson, E., McAfee, A.: Race Against the Machine: How the Digital Revolution is Accelerating Innovation, Driving Productivity and Irreversibly Transforming Employment and the Economy. Digital Frontier Press, Lexington (2011)
4. Rosenberg, N.: Inside the Black Box: Technology and Economics. Cambridge University Press, Cambridge (1982). ISBN 9780521273671
5. Brynjolfsson, E., Mcafee, A.: The Business of Artificial Intelligence. Harvard Business Review, Watertown (2017)
6. Bresnahan, T.F.: Computerisation and wage dispersion: an analytical reinterpretation. Econ. J. **109**(456), 390–415 (1999)
7. MGI: Disruptive technologies: advances that will transform life, business, and the global economy. Technical Report. McKinsey Global Institute (2013)
8. Frey, C.B., Osborne, M.A.: The future of employment: how susceptible are jobs to computerization? Tech. Forec. Soc. Chan. **114**, 254–280 (2017)
9. Cohn, J.: The Robot Will See You Now. The Atlantic, 20 February 2013
10. Maglio, P.P.: Editorial—smart service systems, human-centered service systems, and the mission of Service Science. Serv. Sci. **7**(2) (2015)
11. Medina-Borja, A.: Editorial column—smart things as service providers: a call for convergence of disciplines to build a research agenda for the service systems of the future. Serv. Sci. **7**(1), ii–v (2015)
12. Larson, R.C.: Smart service systems: bridging the silos. Serv. Sci. **8**(4), 359–367 (2016)
13. Maglio, P.P.: Editorial column—new directions in service science: value cocreation in the age of autonomous service systems. Ser. Sci **9**(1), 1–2 (2017)
14. Spohrer, J., Banavar, G.: Cognition as a service: an industry perspective. AI Mag. **36**(4), 71–86 (2015)
15. Maglio, P.P., Kieliszewski, C.A., Spohrer, J.C.: Handbook of Service Science (2015)
16. Barile, S., Saviano, M.: Complexity and sustainability in management: insights from a systems perspective. In: Social Dynamics in a Systems Perspective, pp. 39–63. Springer, Cham (2018)
17. Spohrer, J., Maglio, P.P., Bailey, J., Gruhl, D.: Steps toward a science of service systems. Computer **40**, 71–77 (2007)

18. Spohrer, J.C., Maglio, P.P.: Towards a science of service systems. Value and symbols. In: Handbook of Service Science (Service Science: Research and Innovations in the Service Economy), pp. 157–195 (2010)
19. Hendrickson, J.: Artificial Intelligence: Taking Over - How Will AI and Machine Learning Impact Your Life? (2017)
20. Ransdell, L.B.: Maximinzing response rate in questionnaire research. Am. J. Health Behav. **20**(2), 50–56 (1996)
21. Skagestad, P.: The mind's machines: the turing machine, the Memex, and the personal computer. Semiotica **111**(3–4), 217–244 (1996)
22. Russell, S., Norvig, P.: Artificial Intelligence: A Modern Approach. Prentice Hall, Upper Saddle River (2009)
23. Silver, D., Hubert, T., Schrittwieser, J., Antonoglou, I., Lai, M., Guez, A., Lanctot, M., Sifre, L., Kumaran, D., Graepel, T., Lillicrap, T.: Mastering Chess and Shogi by Self-Play with a General Reinforcement Learning Algorithm (2017)
24. Martınez-Miranda, J., Aldea, A.: Emotions in human and artificial intelligence. Comput. Hum. Behav. **21**(2), 323–341 (2005)
25. Picard, R.W.: Affective Computing. MIT Press, Cambridge (1997)
26. Carter, S., Nielsen, M.: Using artificial intelligence to augment human intelligence. Distill **2** (12), e9 (2017)
27. Minsky, M.L.: The Society of Mind. Simon and Schuster, New York (1986)
28. Rouse, W.B., Spohrer, J.C.: Automating versus augmenting intelligence. J. Enterp. Transform. 1–21 (2018)
29. Vargo, S.L., Maglio, P.P., Akaka, M.A.: On value and value co-creation: a service systems and service logic perspective. Eur. Manag. J. **26**(3), 145–152 (2008)
30. Maglio, P.P., Vargo, S.L., Caswell, N., Spohrer, J.: The service system is the basic abstraction of service science. Inf. Syst. e-Bus. Manag. **7**, 395–406 (2009)
31. Maglio, P.P., Lim, C.H.: Innovation and big data in smart service systems. J. Innov. Manag. **4**(1), 11–21 (2016)
32. Spohrer, J., Maglio, P.P.: The emergence of service science: toward systematic service innovations to accelerate co-creation of value. Prod. Oper. Manag. **17**(3), 1–9 (2008)
33. McGee, K., Hedborg, J.: Partner technologies: an alternative to technology masters & servants. In: Proceedings of the COSIGN (2004)
34. Breazeal, C.: Emotion and sociable humanoid robots. Int. J. Hum. Comput. Stud. **59**(1–2), 119–155 (2003)
35. Degani, A., Goldman, C.V., Deutsch, O., Tsimhoni, O.: On human–machine relations. Cogn. Technol. Work **19**(2–3), 211–231 (2017)
36. Vargo, S., Lush, R.: Development of new dominating logic of marketing. Russ. J. Manag. **2** (4), 73–106 (2006)
37. Ransdell, J.: The relevance of Peircean semiotic to computational intelligence augmentation (2002)
38. Spohrer, J., Bassano, C., Piciocchi, P., Siddike, M.A.K.: What makes a system smart? Wise? In: Advances in The Human Side of Service Engineering, pp. 23–34. Springer, Cham (2017)
39. Barile, S., Ferretti, M., Bassano, C., Piciocchi, P., Spohrer, J., Pietronudo, M.C.: From smart to wise systems: shifting from artificial intelligence (AI) to intelligence augmentation (IA). In: Poster in International Workshop on Opentech AI in Helsinki, 13–14 March 2018
40. Becker, S.O., Ekholm, K., Muendler, M.A.: Offshoring and the onshore composition of tasks and skills. J. Int. Econ. **90**(1), 91–106 (2013)
41. Barile, S., Polese, F.: Smart service systems and viable service systems: applying systems theory to service science. Serv. Sci. **2**(1–2), 21–40 (2010)
42. Beverungen, D., Matzner, M., Janiesch, C.: Information systems for smart services (2017)

43. Nonaka, I., Takeuchi, H.: The wise leader. Harv. Bus. Rev. **89**, 58–67 (2011)
44. Carr, A.: Positive Psychology: The Science of Happiness and Human Strengths. Routledge, New York (2011)
45. Reeves, B., Nass, C.: The Media Equation. Cambridge Press, New York (1996)
46. Freund, L.E., Spohrer, J.C.: The human side of service engineering. Hum. Factors Ergon. Manuf. Serv. Ind. **23**(1), 2–10 (2013)
47. Barile, S., Polese, F., Saviano, M., Carrubbo, L., Clarizia, F.: Service research contribution to healthcare networks' understanding. Innov. Serv. Perspect 71 (2012)
48. Piciocchi, P., Spohrer, J.C., Martuscelli, L., Pietronudo, M.C., Scocozza, M., Bassano, C.: T-Shape professionals co-working in smart contexts: VEGA (ST)–venice gateway for science and technology. In: International Conference on Applied Human Factors and Ergonomics. Springer, Cham (2017)
49. Goleman, D.: Emotional Intelligence. Bantam Books, New York (1995)
50. Piciocchi, P., Bassano, C., Pietronudo, M.C., Spohrer, J.: Digital workers in nested, networked holistic service systems: a new workplace culture from a Service Science perspective. In: Maglio, P.P., Spohrer, J., Kieliszewski, C.A. (eds.) Handbook of Service Science, vol. 2 (HOSS2), in press
51. Eurofound: New Forms of Employment. Luxembourg (2015)
52. Smith, A., Anderson, J.: AI, Robotics, and the Future of Jobs. Pew Research Internet Project, 6 August. Pew Research Center, Washington, DC (2014)
53. Spohrer, J., Piciocchi, P., Bassano, C.: Three frameworks for service research: exploring multilevel governance in nested, networked systems. Serv. Sci. **4**(2), 147–160 (2012)
54. Spohrer, J., Giuiusa, A., Demirkan, H., Ing, D.: Service science: reframing progress with universities. Syst. Res. Behav. Sci. **30**(5), 561–569 (2013)
55. Kline, S.J.: Conceptual Foundations for Multidisciplinary Thinking. Stanford University Press, Palo Alto (1995)

An Approach for a Quality-Based Test of Industrial Smart Service Concepts

Jens Neuhüttler[✉], Inka Woyke, Walter Ganz, and Dieter Spath

Fraunhofer-Institute for Industrial Engineering IAO, Nobelstr. 12,
70569 Stuttgart, Germany
{Jens.Neuhuettler, Inka.Woyke, Walter.Ganz,
Dieter.Spath}@iao.fraunhofer.de

Abstract. Using machine data for improving the service business offers new potentials for differentiation to manufacturing firms. However, the development of industrial Smart Service concepts is a rather complex task: Different elements (e.g. technologies, digital services and personal services) have to be considered together and customer requirements for these new offers are often unknown. In this context, testing the newly developed Smart Service concepts can help manufacturing companies to avoid development failures and to ensure their value. The focus of our paper is to introduce a new approach for testing industrial Smart Service concepts on their perceived quality for potential customers. The approach includes a process model, an evaluation framework that includes criteria derived from existing approaches and specific Smart Service items.

Keywords: Smart Services · Service testing · Industrial engineering
Perceived quality · Digitalization · Servitization

1 Industrial Smart Services as a Driver for Differentiation: Potentials and Challenges for Manufacturing Firms

Shifting focus from mainly producing and selling physical goods to providing services and solutions has been a popular strategy of manufacturing companies during the last decades [1]. This development, often referred to as servitization, has proven successful for differentiation against competition from lower cost economies in times of increasingly commoditized physical goods [2]. By adding a wide range of services to their products, manufacturing firms were able to provide more customer-centric solutions, leading to an increased value and thus to a higher customer retention [3]. However, since many services are moving towards commoditization as well, manufacturers are looking for new possibilities to achieve competitive advantages [4]. In this context, the ongoing digital transformation holds vast potentials for the increasingly important service business of manufacturing firms [5]. Due to the growing equipment of their physical goods with sensing and communication technologies, manufacturers receive a large amount of data regarding the condition and usage of their machines in the field. By analyzing, comparing and combining this data with data from additional sources (e.g. traditional enterprise data or social data [6]), manufacturers can provide

© Springer International Publishing AG, part of Springer Nature 2019
T. Z. Ahram (Ed.): AHFE 2018, AISC 787, pp. 171–182, 2019.
https://doi.org/10.1007/978-3-319-94229-2_17

individualized, contextualized and therefore "smart" services that lead to an increasing value for both, customer and provider. A well-known example of an industrial Smart Service is the usage of machine data for providing a more predictive maintenance. Based on the continuous collection and comparison of condition data from their installed base, manufacturers can identify conspicuous patterns for machine failures in advance and thus prevent breakdowns of their plants and machines. Consequently, these data-generated insights lead to a higher machine availability and optimized service processes, e.g. through more efficient route planning of service technicians.

Besides the undisputed potentials of enriching industrial Services with context sensitive data, many companies are also facing a number of challenges regarding the development and provision of these Smart Services [7]. Many manufacturers, for example, are lacking skills and capabilities to analyse large amounts of real-time data or to develop complex algorithms, since their core competences are still rather production-focused [4]. Other challenges arise from sharing machine data across company boarders to exploit further potentials, ensuring data security to customers or integrating Smart Services into existing processes and infrastructures [8]. However, one of the most important challenges is the development of Smart Services concepts that transfers the potentials of smart technologies and collected data into a substantial value for customers [9]. On the one hand, industrial Smart Service concepts are often linked to a number of uncertainties and risks in the perception of customers, such as passing on sensible data or a perceived loss of control, due to automated decision-making. Therefore, manufacturers have to design industrial Smart Service concepts in a way that the perceived value is exceeding the perceived risks of potential customers. This is rather challenging with only little knowledge about customer requirements in this field. On the other hand, the systematic development of Smart Services itself is a complex task, since technological, software and service elements have to be developed or integrated jointly [10]. In order to support a successful digitalization of industrial Smart Services, new procedures and methods are needed, which allow an integrated and holistic view on the development of all parts [11, 12].

In this context, our paper aims to support industrial companies by providing insights and a first approach for one of the most important but still neglected phase in the development of new services: Testing. Our first approach consists a procedure model that illustrates the integration of the different Smart Service elements and a refined evaluation framework that provides assessment criteria derived from perceived quality concepts. In the following paragraphs, we will first provide a short overview about the current state of the art. Secondly, we derive requirements based on occurring gaps. The third and final paragraph will introduce our first approach for a quality-based test of Smart Service concepts.

2   State of the Art

2.1   Smart Service Development in Manufacturing Industries

Although Smart Services have gained a growing popularity in research and practice, there is no coherent definition of the term Smart Service. Essential characteristics mentioned are the intensive use of data [10], the involvement of intelligent products [9] and the adaptability of the service offerings to a customer-specific need or situation [13]. Moreover, Smart Service offers can include digital as well as personally delivered service elements [14], as seen in the example of predictive maintenance. For our work, we define Smart Services as individually configurable bundles of digital and personally delivered services, which are based on data collected by intelligent products and additional sources. Consequently, Smart Services can include products, software and services. Unlike the closely related research stream of Product-Service-Systems, which deals with the enrichment of physical products with services in order to shift focus from selling to using the products, Smart Services research is much more focused on the context-specific combination of digital and personally delivered services. Physical products merely provide the basis for collecting necessary data of customer context and situation. However, technology and data elements are still part of the solution and therefore need to be considered during the development process. Nevertheless, we reviewed literature from a service perspective, since services play the major role in industrial Smart Service concepts.

A systematic and goal-oriented development of new services is one of the key suc-cess factors in increasingly dynamic markets, such as manufacturing [11]. Besides the growing relevance of new service development approaches [cf. 15, 16], the discipline of Service Engineering has gained popularity, especially in manufacturing industries [11, 17]. Service Engineering outlines a systematic development approach based on methods derived from software and product engineering, which are adapted and further developed on the specific character of services (e.g. value co-creation, intangibility etc.). During all stages, the integration of customers outlines a major impact factor for a successful development of new services. The main objectives of service engineering are an efficient development, the effective use of methods, a strong focus on customers and the achievement of high quality [18]. In order to advice companies during development tasks, different service engineering reference models were provided (we refer on this point to an overview from [16]). In most cases, these reference models are generic, phase-oriented models, with a strong link to product and traditional software development. Common phases of service engineering reference models cover stages like idea management, requirements analysis, service conceptualization, testing and implementation. The aspiration of service engineering is to provide service-specific approaches, methods and tools for each of the generic phases. However, this leads to following gaps with regard to our paper.

On the one hand, the provided approaches and methods are only partly suitable for Smart Service development because they do not sufficiently consider the physical and digital elements and their specifics. In reverse, the disciplines of product and software development neglect elements of personally delivered services. Therefore, new approaches that allow an integrated development of Smart Services are required [11]. Since the development logic of the different elements is often diverging (e.g. agile software development toward phase-oriented product engineering), establishing an overall development logic or reference model is a complex tasks. Moreover, a generic reference model only provides little value, if there are no appropriate approaches and methods to consider Smart Service specifics (e.g. integrated execution of the phases). On the other hand, the existence and maturity level of provided approaches and methods varies between the different phases of service engineering. One important but often neglected and rather complex phase in new service development and service engineering literature is testing [19]. However, testing new and highly innovative concepts, such as industrial Smart Service, supports manufacturers companies in increasing customer acceptance and preventing costly development failures by ensuring to meet relevant customer requirements.

2.2   Concept Testing in Service Engineering

Despite the central role of testing new service concepts and although testing is well established in product and software engineering, the testing phase of Smart Services is still widely neglected. It is mentioned in most of the development procedure models as central (cf. [20]), but there is little literature guidance about how to apply service concept testing and which methods are useful to do so [19]. The only considerable approaches, which provide deeper insights of service and product-service-system testing, are those of [19, 21]. Figure 1 displays the general testing process as part of new service development.

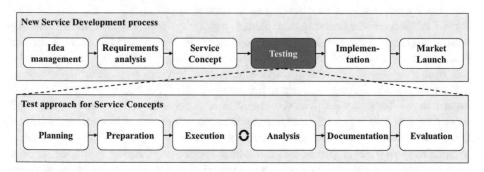

Fig. 1. Process model for testing as part of New Service Development (source: [19])

Moreover, [22] present results of two studies, which analysed the understanding, challenges and importance of testing service concepts in industrial companies. One of the results is that companies understand service testing as a verification of the service concept, a simulation of service provision and the evaluation of critical success factors from a customer perspective. Consequently, one important criterion for evaluating concept during testing is the fulfilment of customer needs. Furthermore, the study pointed out that an early evaluation of the desired customer benefits, the increase of service quality before market implementation and an early assessment of potential value are the essential objectives for testing service concepts in industrial companies [23]. According to the study, one of the main challenges of service tests is seen in the intangible character of services, which complicates an objective evaluation of service concepts with customers. Consequently, companies often test their service concepts only in final development stages, which reduces the initial benefits of testing (e.g. preventing costly development failures). Moreover, it prevents the application of already established methods from product or software testing leading to testing activities without a systematic approach [23].

The stated challenges are also transferrable to testing mainly intangible Smart Service concepts. In order to support a more systematic testing approach, a procedure model and a basis evaluation that allow an integrated view on physical, digital and immaterial elements as well as their interdependences is required. For our approach, we chose perceived quality as a basis for evaluation, since it addresses two of the main testing objectives. These are increasing service quality and assessing perceived value. The following paragraphs provide a short overview about the concept of perceived quality.

2.3 Perceived Quality in the Realm of Smart Services

For evaluating the quality of tangible products, objective criteria like the accuracy of sensor measurements or the durability of machines are appropriate. Regardless of this fact, it is much more complex to find objective criteria for predominantly intangible solutions. Therefore, service quality is a rather subjective construct, relying on the comparison between expected and perceived fulfilment of quality criteria from the customers' point of view [24]. Although the concept of perceived quality originally descends from service research, it also becomes more relevant in the evaluation of physical products, since customers are lacking the knowledge to evaluate complex products objectively. Due to the high complexity and the mainly intangible character of industrial Smart Service concepts, perceived quality seems to be a promising basis for evaluation. Furthermore, service quality is one of the main drivers for the perceived value of customers and thus supports above-mentioned objectives of service testing.

In our literature review, we found a lack of knowledge regarding the perception of Smart Service quality [25]. For each of the relevant elements of Smart Service concepts monolithic models and approaches can be found, such as SERVQUAL for personally delivered services [24], E-S-QUAL for digital services [26] or the Technology Acceptance Model [27] for perceptions about sensing technologies. However, customers are likely to evaluate Smart Services as a joint and integrated offering and consequently it takes an integrated understanding of Smart Service quality perceptions. In this context, [28] introduced a first framework to integrate evaluation criteria of existing approaches of perceived quality. The framework represents a matrix of the three elements of a Smart Service (technology, digital service and personally delivered service) on the one axis and the three service quality dimensions (resources, process and outcome), introduced by [29], on the other axis. For each of the nine occurring Smart Service quality fields, fitting criteria were derived from existing quality approaches and models leading to a sum of more than 180 different evaluation criteria. Although this framework might be used as basis for evaluation during the concept testing, it also possesses some gaps. On the one hand, it includes criteria from models that were developed a long time ago and thus might not be appropriate in the context of complex, configurable and data-based solutions. On the other hand, the framework does not integrate specific new quality items regarding Smart Services, such as for evaluating the integration of different elements or the perceived embeddedness of sensor technologies [9], so far. In order to provide a more substantial framework, we carried out a qualitative survey among 20 international experts in order to validate the framework and add new, relevant items. Section 4.2 will introduce an overview of results as part of our approach.

3 Requirements for Developing a New Approach

Previous paragraphs revealed several gaps in currently existing literature. Accordingly, our paper aims to fill these gaps with a new approach for quality-based testing of Smart Services. In order to do so, we can derive following key requirements for developing our approach:

- It needs to allow an integrated view on testing Smart Service concepts, including all present elements: Smart Technologies and data, digital services and personally delivered services (cf. Sects. 2.1, 2.2 and 2.3).
- It should include testing different scenarios, since Smart Services are context-specific, individually configurable (cf. Sect. 2.1).
- Since co-designing builds a main success factor in service engineering (cf. Sect. 2.1), the testing process needs to include potential customers.
- The approach has to be applicable for different maturity levels of the Smart Service concepts, since testing in early stages increases its benefits (cf. Sect. 2.2).
- An appropriate evaluation basis that includes specific criteria for Smart Services and adapts to the respective concept shall be part of the approach (cf. Sect. 2.3).

In the following paragraphs, we develop a first approach for testing Smart Services concepts on their quality, using derived requirements as a guideline. The approach is constituted by two elements: a process model and a revised framework that includes existing as well as new items for evaluation perceived quality.

4 An Approach for Quality-Based Testing of Industrial Smart Service Concepts

4.1 Process Model for Testing Industrial Smart Service Concepts

For developing a process model of our approach, we use existing service testing approaches (cf. [19, 21]) as a basis, but adapt them to meet stated requirements of testing industrial Smart Service concepts. Figure 2 illustrates a first version of our process model. Besides the three possible elements of a Smart Service concept "Technology and Data", "Digital Services" and "Personal Services", the element "Integration" completes the testing concept. In the context of predictive maintenance, for example, a reliable collection and analysis of data by sensing technologies as well as the predictive identification and alarming of an upcoming machine breakdown by the system leads to a good perceived quality of the two elements "Technology and Data" and "Digital Services". However, if no service technician is available to react upon this information and to fix the technical problem, the whole concept is failing. The overall perceived quality in this example might be low, although the personally delivered service might have had a good quality too and only the process integration between the different elements was bad. Thus, we consider the integration and coordination of the elements as another main contributor to overall perceived value. As already mentioned, our process model is an adapted version of the approach of [19] that includes six phases: planning, preparation, execution, analysis, evaluation and documentation. In the following lines, we explain the phases and annotate the necessary adaptions to Smart Services.

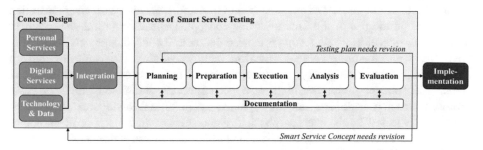

Fig. 2. Process model for testing Smart Service concepts (own illustration, referring to [19, 21]).

In the first phase, activities regarding the **planning** of the test take place. Planning is one of the most crucial phases for the overall performance of testing, since it enables a structured preparation as well as a reliable and smooth execution of all following activities. The first step addresses fundamental aspects, such as setting the overall testing objective (in our case: concept testing on perceived quality), nominating the testing personnel and defining their roles and responsibilities as well as setting the time and cost frame for the testing phases in coordination with the overall Smart Service development management team. One crucial factor of planning is to create a common understanding for the overall objective, since members of the testing team are likely to have different backgrounds in the realm of Smart Services (e.g. mechanical engineering, computer sciences or service management). Based on this joint understanding, subordinated objectives and further decisions regarding the testing concepts are to be made. This includes setting the test subjects that shall be included during the execution. Firstly, the testing team needs to decide, whether they want to include real customers or internal persons, which are acting as customers. A crucial factor is the inclusion of customers throughout the whole development and especially during the testing phase. However, the right decision also depends on the maturity level of the Smart Service concept and if it is possible to evaluate it for a real customer. Secondly, the testing team needs to define the customer role during the test. This is necessary, since customers in industrial settings are often represented by buying-centers, including different roles and persons (e.g. internal maintenance department, operations management or purchasing departments) that focus diverging objectives. Depending on the defined testing goals, relevant actors have to be selected and integrated in an appropriate way to receive reasonable results. Furthermore, the testing concept defines the testing objects, referring to the concepts elements and their delivery stage (e.g. potential, process or outcome) based on their maturity level.

During the **preparation** phase, the testing execution should be prepared and planned in detail. It starts with defining the testing scenarios, which include setting the application context (e.g. case of a machine failure) and the testing environment (e.g. virtually in a laboratory or in a real-life testing environment). Testing scenarios are an appropriate measure to consider the high degree of individuality of Smart Service concepts and to allow a realistic evaluation from a customer´s point of view (referring to the idea of value-in-use and value-in-context). Another major task is the preparation of relevant prototypes of the Smart Service concept. Depending on the development logic and the stage, in which testing is carried out, not all concept elements necessarily have the same maturity degree. Preparing and integrating the elements into an appropriate, assessable concept prototype, is one of the most complex yet important steps. The concept prototype might also be a mixture of different kinds of basic prototypes, such as scribbles, user stories, mock-ups or Virtual Reality models, and also more sophisticated kinds, such as physical demonstrators (e.g. sensing technologies implemented in machines) or a service theater that simulates the service process. However, the testing team needs to assure a logical consistency as well as the transferability of evaluation results from prototype to planned real-life concepts. Besides the test objects, the testing team needs to brief the included stakeholders, e.g. by providing them context-specific knowledge and information about the procedure. For finalizing the preparation, the evaluation process and criteria need to be defined.

During the **execution** phase, the testing team carries out all planned and prepared activities under conditions as realistic as possible. The testing scenarios and integrated prototypes are presented to or respectively performed with selected test subjects. One of the most important tasks during the phase is to gather relevant data and detailed observations for following phases. Data can be collected in many different ways, e.g. by including smart testing technologies, such as smart glasses in combination with fitness trackers for measuring heart rates at different activities or stages, by observing the test subjects personally or by a questionnaire-based evaluation after the execution.

In the **analysis** phase, the testing team examines collected testing data and feedback thoroughly. During the whole testing process, all activities and results are **documented** completely and conscientiously in order to provide a profound basis for the following evaluation. In the last phase, the testing team has to evaluate the test results with regard to the Smart Service concepts as well as the performance of the test itself.

The **evaluation** can lead to three decisions. In the first case, the results of testing the perceived quality of the Smart Service concept are satisfying and meet the requirements for implementation. The second case occurs, if the test performance was not good enough and therefore achieved results are not meaningful enough to make a decision. Consequently, the test has to be repeated and origins for the bad performance need to be corrected. In the third case, the team identifies potentials for improving quality during evaluation and decides that the Smart Service concept needs revision. Depending on the scope of affected elements (e.g. personal service, digital service, technology and data or the integration) respective information and recommendations are handed back to the Smart Service development process.

In order to provide a basis for evaluating industrial Smart Service concepts, we will introduce an adapted framework for Smart Service Quality as second part of our approach.

4.2 Framework for an Integrated View on Perceived Smart Service Quality

Since there is a lack of knowledge about appropriate dimensions and criteria for evaluating Smart Service concepts [25], we further developed the integrated framework presented by [28] for our test approach. Therefore, we presented a former version of the framework to 20 international experts (Germany: 9; USA & Switzerland: 3 each; Netherlands, China, Japan, Sweden, UK: 1 each) in a qualitative study. During the research, we asked the experts about their opinion regarding the framework's structure, the relevance of already derived quality dimension from existing approaches and about new and more specific evaluation criteria for Smart Services. A separate paper will publish the detailed results of our study, but some insights about the framework are presented in the following lines.

The advanced framework (see Fig. 3) consists 12 fields to describe Smart Service concepts. It is structured by the four relevant elements of industrial Smart Service concepts "Technology and Data", "Digital Services", "Personal Services", "Integration" and the three delivery stages "Resource", "Process" and "Outcome". The structure helps companies to identify relevant criteria for a specific test object that might comprise only certain parts of the Smart Service concept. For each of the 12 fields, numerous quality dimensions (e.g. "reliability") and respective criteria are assigned. During the testing process, manufacturing companies can use these criteria to evaluate the concept prototypes together with customers.

Most experts agreed that dimensions and items of existing perceived-quality models from personal service, digital service or technology management literature are still valid in the realm of Smart Service concepts. The quality of personal interaction between a service technician and a machine operator, for example, still contributes to the overall perceived quality, if there is a personal interaction. Therefore, quality dimension like "empathy", "courtesy" or "appreciation" are still valid for evaluating this specific part. However, for evaluating Smart Service concepts new and more specific items are needed. In the following lines, we provide some examples along the delivery stages.

The **resource** stage addresses quality aspects of the Smart Service prerequisites. Besides traditional quality dimensions (e.g. appearance and structure of digital applications, competences and equipment of employees or physical characteristics of technology), following new dimensions become important (selection):

- Data privacy issues (e.g. the perceived connection between collected data and their necessity for providing promised value)
- Perceived embeddedness of sensor technologies in the working environment of users, e.g. in regard of perceived surveillance
- Collaboration possibilities with other Smart Service platforms (e.g. by providing relevant API and data formats).

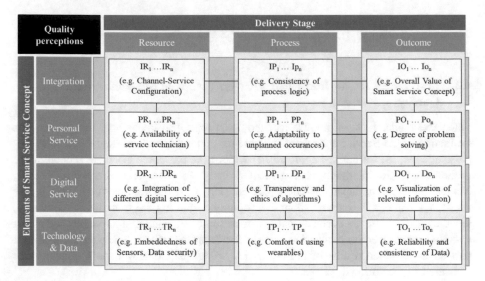

Fig. 3. Adapted framework for evaluating perceived quality of industrial Smart Service concepts (own illustration, referring to [28]).

- Projected size of installed machine base and additional data sources for providing intelligent solutions.

The **process** stage addresses personal and digital activities for providing the Smart Service. This considers the integration of intelligent products like human-to-human, human-to-machine as well as machine-to-machine interactions. Exemplary new aspects that influence perceived quality are:

- The integration of physical, digital and personally delivered activities
- Automatic adaption of processes to the context and situation of customers
- New forms of collaboration between customer and provider and the depth of integration into customer processes
- Perceived control options for intangible activities and automated decisions
- Empathy of systems and contextualized information provision
- Transparency, comprehensibility and ethics of algorithms and decision making logic

The **outcome** stage addresses the value provided by each of the Smart Service elements as well as its contribution to the overall Smart Service value. In many cases, the outcome of the sensing technology (e.g. perceived data consistency) builds the basis for the outcome of the digital (e.g. information visualization) or personal (e.g. solving a machine failure) service. The integration of the different outcomes was said to be of high importance as well as their individual adaption to the customer situation. Moreover, new value dimensions (e.g. emotional value or the joy of using an adaptable solution) were mentioned during our study.

5 Conclusion and Further Steps

In our paper, we presented a new approach for a quality-based test during the development of industrial Smart Service concepts. It consists of two elements: A process model that provides insights about how to carry out a test of Smart Service concepts and an adapted framework of perceived Smart Service quality that helps to find appropriate criteria for evaluation. However, our presented approach is only a first version that needs further development in order to support manufacturing companies in testing their Smart Service concepts and to provide a substantial value added. During the next steps, we plan to include appropriate methods for performing the six testing phases and to interlink identified evaluation criteria to generic maturity levels of prototypes in order to provide an adaptable framework for different development phases. Moreover, we are planning to test the approach with manufacturing companies.

References

1. Neely, A.: Exploring the financial consequences of the servitization of manufacturing. Oper. Manag. Res. **1**(2), 1–50 (2008)
2. Vandermerwe, S., Rada, J.: Servitization of business: adding value by adding services. Eur. Manag. J. **6**(4), 314–324 (1988)
3. Baines, T.S., Lightfoot, H.W., Benedettini, O., Kay, J.M.: The servitization of manufacturing: a review of literature and reflection on future challenges. J. Manuf. Tech. Manag. **20** (5), 547–567 (2009)
4. Schüritz, R., Seebacher, S., Satzger, G., Schwarz, L.: Datatization as the next frontier of servitization – challenges of organizational transformation. In: Proceedings of the 38th International Conference on Information Systems, South Korea (2017)
5. Neuhüttler, J., Woyke, I.C., Ganz, W.: Applying value proposition design for developing smart service business models in manufacturing firms. In: Freund, L.E., Wojciech, C. (eds.) Advances in the Human Side of Service Engineering, pp. 103–114 (2018)
6. Opresnik, D., Taisch, M.: The value of Big Data in servitization. Int. J. Prod. Econ. **165**, 174–184 (2015)
7. Bullinger, H.-J., Ganz, W., Neuhüttler, J.: Smart Services – Chancen und Herausforderungen digitalisierter Dienstleistungssysteme für Unternehmen. In: Bruhn, M., Hadwich, K. (eds.) Forum Dienstleistungsmanagement: Dienstleistungen 4.0. Springer, Berlin (2017)
8. Husman, M., Fabry, C.: Smart Services – Neue Chance für Services "Made in Europe" (2014)
9. Wünderlich, N., Heinonen, K., Ostrom, A., Patricio, L., Sousa, R., Voss, C., Lemmink, J.: "Futurizing" smart service - implications for service researchers and mana-gers. J. Serv. Manag. **29**(6), 442–447 (2015)
10. Lim, C., Kim, M.-J., Kim, K.-H., Kim, H.-J., Maglio, P.P.: Using data to advance service: managerial issues and theoretical implications from action research. J. Serv. Theo. Pract. **28** (1), 99–128 (2018)
11. Spath, D., Ganz, W., Meiren, T.: Dienstleistungen in der digitalen Gesellschaft – Chancen und Herausforderungen der Digitalisierung für Lösungsanbieter. In: Boes, A. (ed.) Dienstleistungen in der digitalen Gesellschaft. Campus Verlag, Frankfurt, New York (2014)
12. Genennig, et al.: Smart services. In: Heuberger, A., Möslein, K.M. (eds.) Open Service Lab Notes (2017)

13. Smart Service Welt Working Group, acatech (eds.): Smart Service Welt – Recommendations for the Strategic Initiative Web-based Services for Businesses, Berlin (2015)
14. Bullinger, H.-J., Neuhüttler, J., Nägele, R., Woyke, I.: Collaborative development of busines models in smart service eco-systems. In: Kocaoglu, D. (ed.) Proceedings of PICMET 2017, Technology Management for an Interconnected World, pp. 130–139 (2017)
15. Edvardsson, B., Olsson, J.: Key concepts for new service development. Serv. Ind. J. **16**(2), 140–164 (1996)
16. Kim, K.J., Meiren, T.: New service development process. In: Salvendy, G., Karwowski, W. (eds.) Introduction to Service Engineering, pp. 253–267. Wiley, Hoboken (2010)
17. Bullinger, H.-J., Meiren, T., Nägele, R.: Smart services in manufacturing companies. In: Proceedings of the International Conference on Production Research, vol. 23 (2015)
18. Schuh, G., Gudergan, G., Kampker, A. (eds.): Management Industrieller Dienstleistungen. Springer, Berlin (2016)
19. Burger, T., Kim, K.-J., Meiren, T.: A structured test approach for service concepts. Int. J. Serv. Sci. Manag. Eng. Technol. **1**(4), 12–21 (2010)
20. Ojasalo, J., Ojasalo, K.: Using service logic business model canvas in lean service development. In: Gummesson, E., Mele, C., Polese, F. (eds.) Service Dominant Logic, Network and Systems Theory and Service Science: Integrating three Perspectives for a New Service Agenda. Conference Proceedings 5th Naples Forum on Service, Naples (2016)
21. Freitag, M., Schiller, C.: Approach to test a product-service system during service engineering. Procedia CIRP **2**(64), 336–339 (2017)
22. Spath, D., Burger, T., Ganz, W.: Herausforderungen, Lösungsansätze und Entwicklungspfade für das Testen produktionsbegleitender Dienstleistungen. In: Schuh, G., Stich, V. (eds.) Enterprise-Integration. Springer, Berlin (2014)
23. Burger, T., Schultz, C.: Testen neuer Dienstleistungen Ergebnisse einer empirischen Breitenerhebung bei Anbietern technischer Dienstleistungen. Fraunhofer-Verlag, Stuttgart (2014)
24. Parasuraman, A., Zeithaml, V.A., Berry, L.L.: A conceptual model of service quality and its implications for future research. J. Mark. **49**, 41–50 (1985)
25. Maglio, P.P., Lim, C.H.: Innovation and big data in smart service systems. J. Innov. Manag. **4**(1), 11–21 (2016)
26. Parasuraman, A., Zeithaml, V.A., Malhotra, A.: E-S-QUAL, a multiple-item scale for assessing electronic service quality. J. Serv. Res. **7**(5), 1–21 (2005)
27. Venkatesh, V., Davis, F.: A theoretical extension of the technology acceptance model: four longitudinal field studies. Manag. Sci. **46**(2), 186–204 (2000)
28. Neuhüttler, J., Ganz, W., Liu, J.: An integrated approach for measuring and managing quality of smart senior care services. In: Ahram, T.Z., Karkowski, W. (eds.) Advances in the Human Side of Service Engineering, pp. 309–318. Springer International (2017)
29. Donabedian, A.: The Definition of Quality and Approaches to its Assessment. Health Administration Press, Ann Arbor (1980)

Smart Services Conditions and Preferences – An Analysis of Chinese and German Manufacturing Markets

Wenjuan Zhang[1(✉)], Jens Neuhüttler[2], Ming Chen[1],
and Walter Ganz[2]

[1] Sino-German College of Applied Sciences, Tongji University,
Shanghai, People's Republic of China
{zhangwenjuan, chenming}@tongji.edu.cn
[2] Fraunhofer-Institute for Industrial Engineering IAO, Stuttgart, Germany
{jens.neuhuettler, walter.ganz}@iao.fraunhofer.de

Abstract. Smart Services are individually configurable bundles of physically delivered services and digital services, based on data collected in the Internet of Things. Due to the increasing equipment of products with sensing technologies and communication modules, Smart Services become more and more important to manufacturing companies. Since German and Chinese manufacturing firms possess a strong trading relationship, it is important to understand the market conditions and customer requirements of the two country markets in order to develop and provide Smart Services successfully. In this context, our paper provides a first overview about these aspects, based on literature analysis and a small qualitative survey among four Chinese and four German experts.

Keywords: Smart Services · Industrial service management
Predictive maintenance · Customer requirements · Sino-German collaboration

1 Introduction

The manufacturing industry is highly important to the economies of China and Germany, but is currently on the move. The ongoing digital transformation, including the interconnection between machines and their equipment with sensing technologies and web-enabling modules, offers great opportunities for a more intelligent production. Besides that, large amounts of data collected by these smart products and machines can also be used to provide so-called Smart Services [1]. By using the data to provide individually adapted service offers to customers, manufacturing companies can unlock potentials for quality and productivity improvements as well as developing new services [2]. A common example for a Smart Service in manufacturing is predictive maintenance, where machine data is utilized to improve the maintenance processes of service technicians. Based on the continuous collection and analysis of data collected by machines and products for conspicuous patterns, manufacturers can plan maintenance operations demand-oriented and identify potential failures in advance. Consequently, machine availability and service operations can be improved, leading to benefits for providers and customers [3].

© Springer International Publishing AG, part of Springer Nature 2019
T. Z. Ahram (Ed.): AHFE 2018, AISC 787, pp. 183–194, 2019.
https://doi.org/10.1007/978-3-319-94229-2_18

Despite these potentials, the maturity level using machine data to provide Smart Services (e.g. maintenance, repair or operations) is rather low. One reason for that could be current market conditions that might prevent the provision of such Smart Services (e.g. missing regulations for data ownership). Another reason might be that the current concepts are not meeting customers' requirements and preferences and therefore aren't accepted by them. However, little is known so far about the market conditions and customer requirement regarding Smart Service [4].

Chinese and German manufacturing firms possess a long and strong trading relationship. For example, China has become the second most important export market for the German manufacturing industry during the last years [5]. In both countries, the development of smart products is understood as a starting point for the digital transformation and a springboard for upgrading manufacturing. Consequently, servicing smart products with data-based concepts in home and foreign markets becomes more and more important to these companies. However, to establish Smart Services in different country markets successfully, companies need to understand the similarities and differences between market conditions and requirements. Due to the strong trading relationship between the countries, the presumably high variances in technical, cultural and regulatory market conditions as well as customer requirements and the need of Smart Service providers for deep insights, we decided to examine these topics in a joint research project. The following chapters should be understood as a preliminary analysis, in which we tried to find interesting research aspects for upcoming activities. In our paper, we give a short introduction about Smart Service concepts and characteristics firstly and then provide a short overview about market conditions in China and Germany. In a final step, we present results from a qualitative survey among four German and four Chinese experts about customer requirements for Smart Services.

2 The Concept of Smart Services in Manufacturing Markets

Smart Services are data-based, individually configurable bundles of physically delivered services, digital services and intelligent products, that are organized and performed on integrated platforms [6]. Figure 1 shows a schematic layout of an integrated Smart Service platform, which typically includes three different layers [7].

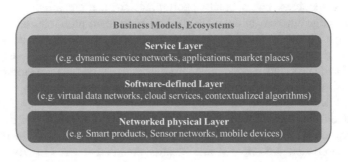

Fig. 1. Schematic layout of an integrated Smart Service platform (source: referring to [6]).

The first layer builds the technological basis for data acquisition and transmission, representing physical objects that are equipped with sensors, microprocessors, software, and communication modules. These so-called smart products are not connected to the internet only but also to other products, building complex a networked physical layer [8]. Examples for smart products in industrial settings are machines, technical equipment or mobile devices (e.g. smartphones) used by service technicians during their operations in production halls, which collect information about the state and usage of machine [3]. Besides smart products, sensors networks built another major pillar of this layer, since they provide context information about the conditions and situations of product usage and service delivery detached from specific products. Examples are independent sensors that measure the conditions in a production hall (e.g. energy consumption, heat or fine particulars) and therefore reveal information about the usage. In addition to that, other information sources, which allow the integration of user-generated content is relevant in the field of Smart Services. The prevalence of smart products lead to changes in the degree of collaboration between customers and providers. Due to the continuous stream of data and the possibility to access smart products remotely, the provider is deeply integrated into the customer processes.

Collected data is transferred to the software-defined layer, where data is merged and stored. Moreover, different data sets are combined and interpreted by context-specific and self-learning algorithms. Integrating data from different sources helps to gain new and extensive insights and builds the basis for new and innovative service offers. In an industrial context, schedules of service technicians can be adapted dynamically, based on the state of a machine and the order situation of the customer as well as on the specific competencies of the service technician, his current location and traffic forecasts [9]. On the service layer, structured data is used for developing and delivering smart service offers. Thereby, digital services (delivered by information systems) and physically delivered services (delivered by people) are combined individually and context-related in order to provide a solution that meets the requirements of customers best possible. The proportion between digital and physically delivered services can vary heavily and reaches from digital interactions between information systems of the provider and the machine control of the customer, to digitally mediated personal interactions between people [4].

Besides combining digital and physically delivered services of one provider, integrated Smart Service platforms can also function as fully-automated market places for trading capacities, data and services of different providers [6]. Consequently, platform providers can serve as intermediary between different providers and customers and thus gain market power and occupy the customer interface.

3 Manufacturing Industries in China and Germany

3.1 Current State and Developments in the Realm of Smart Services

Despite the trend of advanced economies towards becoming service economies, manufacturing remains a significantly important industry in China as well as in Germany. In 2016, manufacturing accounted for 30.45% of GDP in Germany and

39.81% of GDP in China [10]. In both countries, roughly 30% of the overall work force are employed in the industrial sector, which underlines its great economic importance [11]. Moreover, Chinese and German manufacturing companies possess a strong trading relationship with each other. German companies, for example, exported machinery and equipment worth 14.6 billion Euros to China in 2016, making China the second biggest export market for German mechanical engineering companies. At the same time, Germany imported mechanical goods worth 4.6 billion Euros from China, the highest amount of import nations outside the EU [5]. Figure 1 shows the traded product categories between Germany and China (Fig. 2).

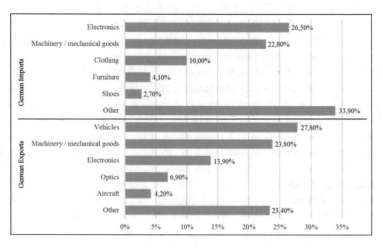

Fig. 2. Top product categories traded between China and Germany in 2014 (source: [12]).

Despite the state of Germany's manufacturing industry, companies are facing challenges regarding the future market development. Germany is well known for producing high quality goods and developing leading-edge technology, but international competition is growing more tense. The growing importance of providing solution (bundles of services and products) rather than selling products helped German manufacturers to stay competitive and to justify higher prices. Since the additional service offers also become increasingly commoditized, manufacturers are looking for new sources for innovation. However, China also faces similar challenges. For many years, Chinese manufacturing was seen as an extended workbench for more industrialized countries. Since the upgrading of Chinese manufacturing and an increasing salary level, the end of the low-wage era seems near. Consequently, China needs to enhance its efficiency and quality to draw level with countries like the USA or Germany [13].

In the context of these challenges, digital transformation holds manifold potentials for the two countries, and also for the cooperation between them. In Germany, the term "Industrie 4.0" has gained great popularity as strategic initiative of the Federal government to foster the digital transformation of the manufacturing industry: The concept describes a state of intelligent and interconnected machines, enabling autonomous decision making of machines and a more flexible, efficient and individualized production [14]. The goal of this strategic initiative is to secure Germany´s leading position in manufacturing, automation and factory equipment. Besides the rather technology-oriented initiative "Industrie 4.0", the German government has taken another important step to tap the potential of this new form of industrialization. The second initiative, entitled "Smart Service Welt", is focusing on establishing new value streams and business models for the service business of manufacturing companies, once their intelligent products have left the production line [6]. By using the potentials of digitalization and big data not only for optimizing production processes but also for the service business, Germany aims to secure long-term competitiveness and underlines the important role of industrial services.

In 2015, China launched a similar strategic initiative, entitled "Made in China 2025". The objective of the strategy is to use intelligent manufacturing to upgrade China´s manufacturing industry into a leading super power and climb up the value chain at once [11]. By making use of intelligent manufacturing and smart products, China aims to become an industry, driven by innovation, efficiency and quality rather than low costs. Moreover, developing a service-oriented manufacturing is another strategic goal, mentioned in "Made in China 2025". Often neglected during the past, the role of service is starting to change in China´s manufacturing industries from being a complement to products to become a differentiator with a distinct value [15]. Although service innovation and developing new customer-centred service offers have gained a strong momentum in China, they are still at an early stage of service transformation.

The close ties of cooperation between German and Chinese manufacturers will become even closer in the context of intelligent manufacturing. On the one hand, China views Germany as a preferred partner during its digital transformation, which offers opportunities for German manufacturers to sell their intelligent products and machines. On the other hand, the growing importance of services will lead to a close collaboration in developing new Smart Services and thus offer potentials to Chinese manufacturers to differentiate from their international competitors. To develop Smart Services with a high level of acceptance, conditions and customer requirements on the country market have to be taken into account. In the next chapter, we will present a short overview about framework conditions in China and Germany that might influence local requirements of customers.

3.2 Market Conditions for Industrial Smart Services in China and Germany

Technical Basis. The development and provision of industrial Smart Services is based on the prevalence of intelligent products, machines or equipment. Regarding this basis, German industry has taken on a particularly strong position [16]. According to an Accenture study, German companies have a high digital competitiveness compared to international competitors, especially in the fields of mechanical engineering, automotive, logistics, energy and chemical industries [6]. However, in other important industries, such as medical equipment or pharma, Germany´s leading suppliers are already slightly at the back of globally leading competitors. Nevertheless, Germany seems to have a good starting point for becoming a leading provider of industrial Smart Services, based on the excellence of their smart products combined with a skilled workforce. However, future development also depends on other factors, such as widespread high speed internet connection, that allows real-time data transfer. In this area, the performance of Germany is only mixed. Although the performance of cable based broadband and ability for real-time data transfer are sufficient, the mobile connection and LTE-coverage are rather poor in comparison to other countries, such as the USA [11]. Moreover, investments in ICT services are relatively low.

China's starting point is at a considerably lower technological level. However, China is working with full power on the digital transformation and the establishment of intelligent manufacturing in its industry [13]. Thereby, the Chinese manufacturing industry landscape is extremely heterogeneous in nature: On the one hand, major global corporations, such as Huawei, Sany or Haier, possess advanced and highly automated factories that built the technical basis for Smart Services [14]. On the other hand, many SMEs are not automatized or digitalized so far. Nevertheless, China is catching up fast and digital transformation enables them to leave out development steps. A recent study of Fraunhofer IAO showed that China has already overtaken Germany and the USA in the number and quality of patent applications for intelligent manufacturing basis technologies [17]. Moreover, China is on the way to provide a reasonable ICT infrastructure. Although, the broadband coverage is low, the mobile broadband coverage and the investments in ICT services are comparably high and illustrate the dynamics of China´s digital transformation. A good example for this dynamic and the speed of technological penetration is the development of mobile payment in China. In 2017, the penetration of mobile payment among Chinese Internet users has risen rapidly from 25% in 2013 to 68% in 2016 [18]. The Digital Evolution Index 2017 underlines this estimation by stating that China possess a fast rate of digital evolution but only scores mediocre in its current state of digital evolution. In comparison to that, Germany is already on a higher level, but possesses only mediocre rates of change [19].

Regulatory Framework. Another major influencing factor on the design and development of Smart Services is the regulatory framework. On the one hand, this refers to the general possibilities of providing services internationally and the necessary adaption to local regulations. On the other hand, regulatory aspects, such as data security or protection of IP rights, are central for high-tech companies and influence the possibilities of collecting and using sensible data from products and machines for new services heavily. The regulatory framework between China and Germany is highly divergent. Chinese economical regulation is based on a strong governmental supervision, providing only little room for control for private or individual initiatives. State-owned companies, for example, get a strong comprehensive support from the government. This means that government is smoothing the way for a rapid industrial modernization and digital transformation with its industrial policy, leading to a high speed of innovation, e.g. for the establishment of Smart Services. In Germany, on the contrary, private regulation is the dominant means of regulation. Companies and industries have more freedom, but governmental support for innovative topics can be rather reserved. In the realm of Smart Services for example, this could mean lacking investments in necessary infrastructures or regulation gaps regarding data ownership of smart products.

At the same time, the EU, frequently accuses China of import tariffs, largely closed service markets, technical barriers to trade, violation of property rights, transfer of technologies and discrimination by the authorities. Such disagreements on regulatory principles are a major barrier for joint innovations for Smart Services, especially in increasingly interdependent manufacturing industries. Regarding the service sector in China, for example, a fair competition environment is still missing, and development is restricted heavily by complex institutional arrangements and mechanisms. According to a survey conducted by the Development Research Center of China in 2010, the top three indicators that affect the development of the service sector are the institutional environment (including the legal system and property protection), government functions, and industrial regulations.

Other major barriers for providing Smart Services internationally can be seen in data security and privacy issues, since Smart Services are based on sensitive data. Although Chinese as well as German customers are paying high attention to security and privacy of their data in private as well as in industrial applications, data security is not guaranteed universally in China. The new Cyber Security Law established in 2017 forces many companies to save their data locally in China and only transfer data to other countries after an official security check, providing access to the data for governmental institutions. However, the lack of data security is seen as a major challenge for cooperation on setting up joint Smart Service platforms for the Chinese and German market. At the same time, data regulations in Germany are also not prepared for the diffusion of Industrial Smart Services. With the upcoming General Data Protection Regulation some relevant aspects are addressed, but there are still gaps, e.g. regarding data sovereignty of product-generated data or the collaborative usage of personal data across company boarders.

Cultural Aspects. Behavioral models vary across different cultures. Consequently, cultural aspects are believed to influence the requirements and acceptance of Smart Services on different country markets. In order to analyze the cultural differences between China and Germany, we rely on Hofstede´s cultural dimensions [20]. Hofstede defines culture as "the collective programming of the mind which distinguishes the members of one human group from another". He proposes six dimensions of national culture, which are displayed and described in Table 1. Based on these six cultural dimensions, Hofstede conducted several quantitative studies to identify cultural differences between countries.

According to the studies of Hofstede, China and Germany show similar values for the Masculinity-Femininity and the Long-term orientation dimensions and only little differences in the Indulgence-restraint dimension [20]. The biggest difference can be observed in the dimension "Power distance". In China, power differences and inequalities are much more accepted among people than in Germany. In Germany, the participation of employees in decision making process are much more common and required regardless of their rank in organization or society. Germans also dislike control, and leadership is challenged to expertise rather than rank. Another major difference can be found in "Uncertainty avoidance". While Chinese people often rely

Table 1. Description of the six cultural dimensions according to Hofstede [20].

Cultural dimension	Description
Individualism-collectivism	Degree to which an individual prioritizes its own benefit over the group's benefit and acts as an individual rather than as a member of a group
Power distance	Extent to which individuals accept large differences in power and inequality
Uncertainty avoidance	Degree of emphasizing risk, rule obedience, ritual behavior, and safety measures by an individual
Masculinity-femininity	Extent to which an individual accepts traditional role patterns and prefers different dominant values (e.g. firmness, achievement, success vs. finding consensus, carrying for others, quality of life)
Long-term orientation	The extent of long-term orientation versus short-term orientation toward the future
Indulgence-restraint	The extent to which members of a society control their own desires and impulses

on formal rules, strict control and clear instructions from superior levels, Germans prefer more freedom and would rather manage themselves independently. In Germany, differing ideas and innovative thoughts are much more appreciated that in cultures with a greater uncertainty avoidance. However, there is a certain ambiguity for sticking to the rules in China. In certain situation, rules are seen as flexible in order to better adapt to them. Moreover, Chinese and German culture differ in the dimension of "Individualism – Collectivism". German society is very individualistic one. Loyalty is based on personal preferences for people as well as a sense for duty and responsibility. German people like to communicate directly, even if mistakes were made. On the contrary, Chinese people have a strong collective culture, acting in the interest of a group rather than their own.

4   Customer Requirements in China and Germany

4.1   Approach

Based on the insights gained about the characteristics of Smart Services (Chap. 2) and the market conditions (Chap. 3), we decided to conduct a short qualitative survey among four Chinese and four German experts. All experts were members of manufacturing firms that possess knowledge about the customer requirements for Smart Services from experiences in introducing them or preparing the introduction. The interviews took 30 to 40 min and were based on a half-standardized questionnaire. In our survey, we asked the experts to state their opinions about customer requirements and preferences with respect to seven different factors that we found particularly relevant to the realm of Smart Services and their characteristics. The aspects are illustrated in Table 2.

Table 2. Aspects of interest regarding the requirements and preferences for Smart Services in Chinese and German manufacturing markets.

Aspects	Description
Value added	The potentials of Smart Services stated in literature are manifold. They reach from product to service focused potentials and address quality as well as efficiency.
Individual adaptation	One of the key characteristic of Smart Services is the individual configuration of services to the specific needs and situations.
Collaboration	The continuous collection of data from products located at the customer, remote access to these products and automated decision making for service operations have the potentials to shape new forms of collaboration.
Superordinated platforms	Consuming Smart Services from a platform that works as an intermediary between manufacturing firms and customers might lead to a strong dependency and a new balance of power.
Transparency and traceability	Smart Service processes are often highly invisible, underlying algorithms too complex to understand and decision-making automated.
Trustworthiness	Sharing sensitive data with a Smart Service provider requires trust. The aspect customers use to evaluate trustworthiness in this context are so far unclear.
Data privacy and security	Data privacy and security play a key role in the concept of Smart Services without a doubt. What are relevant aspects and requirements?

In the paragraphs below, an overview of the interview results is given.

4.2   Results

Expected Value Added. The experts of both countries agreed that the main value propositions of Smart Services are closely related to the machine or product, such as a higher machine availability or lower repair costs and downtimes. Moreover, the German experts stated even broader customer expectations, including a higher transparency and optimization regarding the whole production system, better preparation of service technicians and better time and capacity planning for themselves.

Individual Adaption of Service Offering. All experts agreed that an individual configuration of Smart Services is important, but needs to be balanced with a modularized service portfolio in order to reach economies of scale and provide the service at a reasonable price. One of the German experts stated that there will always be individualization to a certain degree, to meet the specific requirements and situation of customers. However, a Chinese expert pointed out that a high degree of individualization might also lead to different data standards and a missing comparison between machines of the installed base. In this context, another German expert proposed that individualization on a hardware layer (sensors and products) should be kept on a lower level and the individual adaption should be implemented on the software layer. In doing so, providers can reach a rather standardized hardware and data formats and customers receive an adapted service.

Collaboration Between Customer and Provider. The Chinese experts agreed that the degree of cooperation becomes deeper with the implementation of Smart Services. However, in their opinion the customer should profit from the new form of collaboration in order to be willing to provide data. Moreover, one expert is sure that customers still want to have the final decision in their hands and not so much by an autonomous system or the provider.

The German experts provided a differentiated view on collaboration. On the one hand, they stated that there are very advanced industries with highly sensible data. Those industries (e.g. car manufacturers or pharma industries) are not likely to share data with the provider on a regular basis but only in specific situations, in which they require help from the service provider. In these cases, the collaboration is not changing a lot. On the other hand, smaller and less advanced companies are likely to share the data in order to receive the full benefit (e.g. higher availability, higher efficiency and transparency) of Smart Service offers. Such companies are willing to step up the collaboration and provide more freedom of decision to the service providers. However, trust and legal rules for this cooperation are mandatory requirements for the collaboration.

Acceptance of a Superordinated Smart Service Platform. Experts of both countries showed skepticism regarding the current willingness to source data based Smart Services from a superordinated platform that comprises, bundles and sells services of different service providers. According to the experts, the main concern is seen in enabling access to a wide range of data from different machines and a resulting dominance of one central provider. However, there is a general willingness to accept integrated platforms, if the benefits are high enough and the data security issues are solved. German interviewees also indicated the big role of a direct one-to-one relationship between German manufacturing SMEs, which might lower an initial acceptance for a superordinated solution. Moreover, one German expert also stated that the tipping point for Smart Service platform is not reached yet. Therefore, the installed base of products and machines that are providing data is not big enough to exhaust all potentials of Smart Services.

Transparency and Traceability. The Chinese experts agree that transparency and traceability of automated decision logic is very important for business-to-business customers, but less important to those on business-to-consumer markets, since they are not familiar with the underlying technology. One expert stated that enabling transparency is a very complex task.

However, the German experts were divided in their opinion. Two of them stated that during the introduction of a Smart Service a clear transparency about algorithms and decision-making rules help to gain acceptance and trust for the solution. German customers want to understand what is provided and how decisions are made. The traceability of mainly intangible processes (e.g. by providing monthly reports about activities of data analytics) can contribute towards this objective too. After the implementation and piloting phase of the service, transparency and traceability become less important, because customers starting to trust the providers. Moreover, the decision basis (e.g. neural networks or deep learning algorithms) becomes more and more complex and thus less transparent and hard to understand. On the contrary, two other experts stated that in the realm of Smart Services, manufacturers are shifting the focus from selling products and services towards providing machine availability or selling results. Consequently, the measure and methods to reach this goal become less interesting to the customer, as long as the provider can deliver the promised value proposition.

Trustworthiness of Smart Service providers. Experts of both countries agree that the brand and reputation of the provider are still important factors influencing the choice of a provider. Moreover, a detailed description of the concept helps to establish trust. In addition to that, one Chinese expert highlighted the growing importance of qualification and skills of service technicians to increase trust among customers.

One German expert pointed out that industry-specific reputation, experiences and competences are still among the most important criteria, preventing independent software companies to establish platforms in the business-to-business markets. Moreover, the German interviewees mentioned localization (physical and legal) of data, a clear and transparent data security concept and certifications as additional factors to gain trust.

Data Privacy and Security. All respondents highlighted the vast importance of data privacy and security for the provision of Smart Services. The Chinese experts pointed out that not all data has to be secured to the same extent, which was also confirmed by one of the German interviewees. Moreover, data security is not seen as a technical but more of a conceptual challenge.

In addition to that, the German expert stated that local hosting of data within Germany or at least within the European Union is one of the key requirements of German customers. Furthermore, German customers require a clear and comprehensible data security concept. In this regard, a German expert stated that certifications from independent organizations could help to meet this requirement especially for smaller companies without specific knowledge about data security.

5 Conclusion and Limitations

The objective of our paper was to provide first insights about relevant differences and similarities of market conditions and customer requirements in the realm of Smart Services for manufacturing industries. The previous paragraphs have shown different market conditions for providing Smart Services. While the technical infrastructure in Germany might be more advanced today, China has a momentum in digital transforming their industries and seems to be more open for new technological developments. In both countries, the regulatory framework exhibit gaps in one of the most important precondition and requirement: Providing a clear and comprehensible basis for data security and privacy. Although there are cultural differences between China and Germany, the interviewees stated very similar requirements and preferences of customers regarding Smart Services. Of course, these results have to be understood in the limited depth and width of our survey and the rather low level of experiences and little literature provided in the realm of Smart Services.

References

1. Allmendinger, G., Lombreglia, R.: Four strategies for the age of smart services. Harvard Bus. Rev. **83**(10), 131–145 (2005)
2. Herterich, M.M., Uebernickel, F., Brenner, W.: Industrielle Dienstleistungen 4.0. Springer, Wiesbaden (2016)
3. Neuhüttler, J., Woyke, I.C., Ganz, W.: Applying value proposition design for developing smart service business models in manufacturing firms. In: Freund, L.E., Wojciech, C. (eds.) Advances in the Human Side of Service Engineering, pp. 103–114. (2018)
4. Wünderlich, N., et al.: "Futurizing" smart service - implications for service researchers and managers. J. Serv. Mark. **29**(6), 442–447 (2015)
5. Wiechers, R., Hell-Radke, S.: Mechanical Engineering – Figures and Charts 2017. VDMA publications, Mühlheim am Main (2017)
6. Smart Service Welt Working Group, acatech (eds.): Smart Service Welt – Recommendations for the Strategic Initiative Web-based Services for Businesses. Berlin (2015)
7. Bullinger, H.-J., Ganz, W., Neuhüttler, J.: Smart Services – Chancen und Herausforderungen digitalisierter Dienstleistungssysteme für Unternehmen. In: Bruhn, M., Hadwich, K. (eds.) Forum Dienstleistungsmanagement: Dienstleistungen 4.0, pp. 97–120. Springer, Wiesbaden (2017)
8. Porter, M.E., Heppelmann, J.E.: How smart, connected products are transforming competition. Harvard Bus. Rev. **92**(11), 64–88 (2014)
9. Bullinger H.-J., Neuhüttler, J., Nägele, R., Woyke, I.: Collaborative development of business models in smart service eco-systems. In: Kocaoglu, D. (ed.) Proceedings of PICMET 2017: Technology Management for an Interconnected World, pp. 130–139 (2017)
10. World Bank national account data. https://data.worldbank.org
11. Heilmann, D., Eickemeyer, L., Kleibrink, J.: Industrie 4.0 im internationalen Vergleich. Handelsblatt Research Institute, Düsseldorf (2017)
12. German Chamber of Commerce in China: German Business in China - Business Confidence Survey 2015 (2015)
13. Wübbeke, J., Conrad, B.: 'Industrie 4.0': will German technology help China catch up with the west? China Monitor **23**, 1–10 (2015). Mercator Institute for China Studies (ed.)
14. Kagermann, H., Anderl, R., Gausemeier, J., Schuh, G., Wahlster, W. (eds.): Industrie 4.0 in a Global Context: Strategies for Cooperating with International Partners (acatech STUDY). Herbert Utz Verlag, Munich (2016)
15. Wübbeke, J., Meissner, M., Zenglein, M.J., Ives, J., Conrad, B.: Made in China 2025: The making of a high-tech superpower and consequences for industrial countries. MERICS Papers on China 2 (2016)
16. Staufen AG: China – Industrie 4.0 Index (2015)
17. Le, N.T., Fischer, T.: Chinese Industry 4.0 Patents. Fraunhofer Verlag, Stuttgart (2015)
18. McKinsey Global Institute: China's digital transformation: the Internet's impact on productivity and growth (2014)
19. Chakravorti, B., Bhalla, A., Chaturvedi, R.S.: 60 Countries' Digital Competitiveness, Indexed. https://hbr.org/
20. Hofstede, G.H., Hofstede, G.J., Minkov, M.: Cultures and Organizations: Software of the Mind: Intercultural Cooperation and Its Importance for Survival. McGraw-Hill, New York (2010)

Using Digital Trace Analytics to Understand and Enhance Scientific Collaboration

Laura C. Anderson[✉] and Cheryl A. Kieliszewski

IBM Research – Almaden, San Jose, CA, USA
{lca, cher}@us.ibm.com

Abstract. Social interaction and idea flow have been shown to be important factors in the collaboration work of scientific and technical teams. This paper describes a study to investigate scientific team collaboration and activity through digital trace data. Using a 27-month electronic mail data corpus from a scientific research project, we analyze team member participation and topics of discussion as a proxy for interaction and idea flow. Our results illustrate the progression of participation and conversational themes over the project lifecycle. We identify temporal evolution of work activities, influential roles and formation of communities throughout the project, and conversational aspects in the project lifecycle. This work is the first step of a larger research program analyzing multiple sources of digital trace data to understand team activity through organic products and byproducts of work.

Keywords: Science teams · Collaboration · Discovery · Digital trace data

1 Introduction

Scientific projects today are composed of many complex parts that include people, technology, data, and process. The central hypothesis for our work is that we can identify the emergence of scientific discovery by examining scientific collaboration through digital trace data analysis. Our program of focus is on the examination of the social and collaborative aspects of the team that include discussion and social interaction, tools and technology usage, and information sharing. In this paper, we discuss a study using electronic mail (email) data to discern project activity.

The research question for this study is: are we able to identify meaningful research team activity and progress through conversation and participant interaction? Our research utilizes a combination of analytical methods, including social network analysis and concept analysis, on structured and unstructured digital trace data sources to study scientific activity. We are most interested in understanding teams composed of a heterogeneous mix of experts working together to leverage each other's skills and knowledge to accelerate time-to-discovery for the phenomena of interest.

© Springer International Publishing AG, part of Springer Nature 2019
T. Z. Ahram (Ed.): AHFE 2018, AISC 787, pp. 195–205, 2019.
https://doi.org/10.1007/978-3-319-94229-2_19

2 Measuring Scientific Work

2.1 Publications

Traditional bibliometrics is useful for a retrospective understanding of the impact of scientific and technical work within the larger research community and how scientific outcomes build upon previous work. This technique requires a long timeframe for recognition of peer-reviewed scientific results and discoveries in journals, patents, and conferences [15, 25]. In addition, groundbreaking discoveries, a subset of the overall scientific and technical work, are rare events [15].

Altmetrics seeks to shorten the timeline to assess discovery and scientific impact. This is done by collecting nearer to real-time informal reactions from emerging web sources such as social media, blogs, and other internet sources [13]. This approach captures electronic evidence of collegial activity in the invisible college and shows promise for reducing the time to recognize scientific contribution. The foundation of this analysis is still the publication, but additional information and metadata is gathered that clusters around the basic publication and author for an expanded representation of the scientific landscape.

2.2 Studies of Active Scientific Teamwork

Building an understanding of scientific activity and accomplishment prior to publication is also of interest to researchers. The body of multi-disciplinary literature examining the work of scientific and engineering teams explores a breadth of human and technological factors of teamwork using a range of research methods. Common research methods include interview, case study, ethnography, artifact analysis, participant feedback, and survey, informed by a range of theoretical frameworks. The human and technological factors of interest examined using these methods range widely from trust to coordination to collaboration among team members [1, 4, 9, 11, 12, 21].

Email is a rich data source for this purpose, and studies of email communication provide insights into both social interaction and the topics of discussion [26]. Tang and colleagues [23] conducted a literature survey of the extensive email mining techniques, including classification, content analysis, and visualization. Email has been found to be a communication mechanism providing affordances not available in other data [22, 23]. The public availability of the Enron email data corpus has been a rich source for mining and communication studies [5, 16, 19].

Text analytics methods on unstructured data are also employed to gain insights on larger and longitudinal data sources [20] and has been used on the full texts of published papers in the Web of Science [17]. Analytics of unstructured text and social networks are well-established data analysis methods to assess scientific impact [6, 13], [24] and to examine citation networks [24]. There is a growing acknowledgment of the importance of social relationships in scientific and technical endeavors and the use of social network analysis to gain insights in this area [6, 12].

Digital traces from a variety of communication, interaction, and other digital data sources are being utilized to understand human computer interaction behavior [7], and these can also augment conventional tracking of team progress. The digital trace data is a by-product of both human and computer system activity and a central component of the emergence of trace ethnography. As such, we believe that digital trace data can be used to better understand the idea flow, collaboration, and socio-technical aspects of scientific and technical teams [18, 25].

3 Research Design

This research uses email conversations to look for patterns of data-driven research activity as insight into scientific team collaboration. Structured metadata and unstructured message content covering a 27-month period (July 2012 to September 2014) was obtained from eight core members affiliated with a research project. The corpus contained email from a large circle of correspondents related to the project and provided a meaningful representation of both central and peripheral project discussions. Cultural-historical activity theory (CHAT) [8] was utilized as the theoretical framework. In focus were the elements of an activity system: role (subject), community, artifacts, and division of labor. Validation of our analysis and findings was obtained through discussions with the core project team.

3.1 Data Collection

This work was conducted at IBM Research - Almaden in the Accelerated Discovery Laboratory (henceforth "Lab")[1]. The Lab supports science and business engagement to solve challenges through the intersection of expertise, data, and analytics using data-driven research and discovery [10, 14, 2, 3]. The Lab "users" are categorized in three categories: *Maker,* the people charged with providing Lab services; *Partner*, all other Company employees; and *Client,* Company customers. The research purpose of our subject project team was to create a water cost index from global financial records with a client partner[2].

We started our investigation by obtaining the email messages from project participants near the conclusion of the project. Duplicate and non-project messages were removed, leaving 658 email messages. Each email message was treated as one autonomous and equal data record. The distribution of email messages by month is a bell-shaped curve showing a buildup from just a few at the project start, to the bulk of the email in the middle as the project is being fully executed, and then a decrease into the completion phase.

Our approach with both the data and tools was to keep the analysis as simple as possible – using digital trace data that was easily obtained and representative of at least one aspect of interaction along with generally available analysis tools. Three sets of analyses were performed on different segments of the same data. We used Microsoft Excel to code the metadata extracted from the email to gain a sense of what participant

---

[1] http://www-03.ibm.com/press/us/en/pressrelease/42169.wss.

[2] http://worldswaterfund.com/?accepted-notice=1.

roles and activities were represented in the data. We then used Leximancer[3] and SPSS[4] to perform concept analyses using both the structured and unstructured text in the body of the email messages; and Gephi[5] to perform a network analysis using the structured email fields.

3.2   Data Preparation

To prepare the data for analysis, we coded each email message based on information extracted from the structured data fields included in the email metadata. The metadata contained 10 structured data fields from which four were identified as useful for our analyses: 'To', 'From', 'DateTime', and 'Subject'. Each email message was then coded and supplemented with four additional fields of 'Role', 'Activity', 'Date', and 'Time'. The additional fields of role and activity were included to anonymize, aggregate, and explore the data (Table 1). Overall 128 individuals and three automated message systems were identified as either senders or recipients of the email messages and were grouped into nine role categories.

Table 1. Coded structured email data attributes.

Structured email data field	Description
To	Receiver or who the email was sent to
From	Sender or who initiated the email
Subject	The subject line of the email header information
Date time	Date as MM/DD/YY and Time as HH:MM
Additional coded category	Description
Role	Based on job title, position, responsibility, and/or affiliation of the Sender ("To" field) in the project: client, client proxy, partner management, maker management, partner staff, maker staff, provider executive, business development, provider general staff
Activity	Action or operation being addressed by the email in the Subject line as: event (e.g., meeting), reporting (e.g., materials needed for a review), project work (e.g., tasks related to the project scope), announcement (e.g., public relations, newsletter), or notice (e.g., lab shutdown, maintenance)
Date	Separated date stamp created from "Date" field
Time	Separated time stamp created from "Date" field

---

[3] http://info.leximancer.com.

[4] https://www.ibm.com/analytics/us/en/technology/spss/.

[5] https://gephi.org.

3.3   Analyses

Next, we performed three sets of analyses. First was an exploratory analysis to gain a general understanding, followed by concept and network analyses to more closely examine communication types, idea flow, and interaction.

3.4   Exploratory Analysis

We expected that the roles most highly represented would be Partner Staff, Partner Management, Maker Staff, and Maker Management. We performed a general numerical analysis of the email corpus by plotting the data across time, based on role and activity categories. Overall, 78 percent (516 email messages) of the data was sent during 2013 across all roles. This is compared to 13 percent in 2014 (85 email messages) and 9 percent in 2012 (57 email messages). To explore general activity, we plotted the individual email records across time based on the sender's role (Fig. 1). We could see clusters of messages by role. Of interest was the regularity and clustering of email messages during the same period within multiple roles.

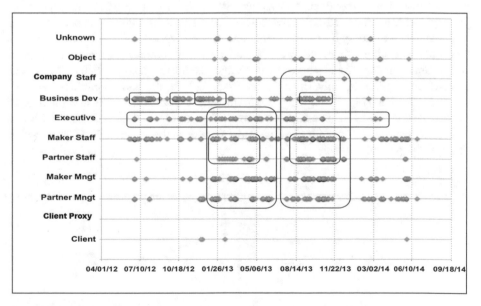

Fig. 1. Individual email messages plotted across time by sender role. Each point represents one email message.

3.5   Concept Analysis

Three software tools were used to perform this analysis: Leximancer, Microsoft Excel, and SPSS. Leximancer was chosen, and used first, because of the strong language analysis it provides, and the ability for the user to easily explore visualizations of the data. Deeper exploration was easily performed by iterative adjustment of the tuning parameters.

The Leximancer analysis yielded a total 235 concepts from the email corpus through the monthly segmentation of the data. These persistent concepts were divided into three categories: (1) those central to the focus of the project (e.g., 'data', 'water', 'index'); (2) general, non-specific concepts; and (3) technical concepts (e.g., 'project (s)', 'document(s)', 'table(s)', 'code'). Through this range of concept categories, we could start to see the breadth and depth of team discussion, from regular and mundane project management topics to more specific research aspects of the project. In addition, we found that important concepts that were *not* persistent could also be fruitful candidates for additional analysis. This is due to their unique emergence in the discussions, the timeframe of their appearance in the lifecycle of the project, and the co-occurrence with other concepts.

The last stage of the concept analysis, using SPSS, focused on the specific concepts of interest and deeper analysis of the semantic word relationships. For example, the Leximancer analysis shows the concept 'test/testing' first appearing in July 2013 and repeating in August, September, and October 2013, but not appearing in any other months. The SPSS analysis provides much more detail around the concept of 'test'. "Unit testing" has the highest relative strength compared to the other 'test' concept phrases in this first month of appearance, which is often one of the first steps for prototype testing. August 2013 shows fewer variations of 'test/testing' concepts, with a uniform relative strength among them, and the emergence of some new concepts ("completed testing", "current test"). There are multiple concepts related to 'regression', 'arguments testing', 'junit test cases', and 'exhaustive testing'. September shows a stark contraction to only three terms ('test set', 'test', and 'regression testing', whereas October (Fig. 2) expands again to nine concepts and the introduction of several new concepts. This analysis confirms the utility of email data used with this three-step analysis to provide insights into project activity, as well as progression through the project lifecycle.

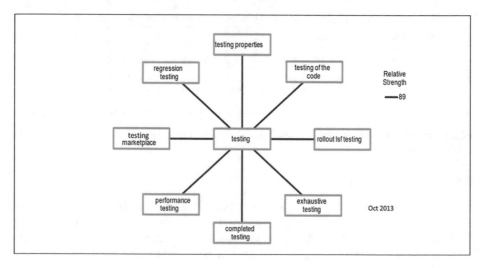

Fig. 2. Concept phrase networks for October 2013.

3.6 Network Analysis

The third examination that we performed was a network analysis using the structured email message metadata (Table 1) based on role affiliation. As with the concept analysis, sociograms were created based on a monthly timeframe. This analysis produced diagrams and social networking measurement output. Figure 3 illustrates the month-to-month change in the number of active roles and communication paths within the project team. The interaction ranged from two to 11 roles producing a range of one to 44 communication paths. We saw a similarity in the pattern of communication paths by month with the general email pattern. However, the number of communication paths illustrated in Fig. 3 provides a more precise view into the extremes between months of lighter and heavier email communication. For example, there is an increase in role participants and communication paths as the project starts, followed by a steep decline from June 2013 to August 2013, and a rebound in September 2013.

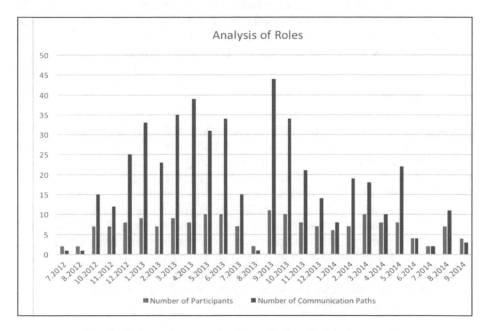

Fig. 3. Roles and communication paths for role-based networks.

To identify the most connected roles we looked at their ranking value by month, a measure based on the number of links to every node. We saw a slow progression at the start of the project, with Partner Management and Partner Staff membership being the only connected roles. Partner Staff, Maker Staff, and Partner Management were the most highly connected roles throughout the lifecycle of the project. In addition to these three roles, the Client, Maker Management, Executive, and Unknown roles were also regularly ranked as roles connected in the network.

The most influential roles were determined by using the Gephi betweeness centrality metric. This metric measures how often a node appears on the shortest paths between nodes within the network. The highest betweeness centrality value indicates the most influential node(s), or role(s) in our case. We found that Partner Staff and Partner Management were the two most influential roles. They tied as being the most influential role, each having the highest betweeness centrality value in eight of the 27-month project lifecycle. Second to these roles was Maker Staff, as the most influential role for three of the months. This indicates that these three roles had the most impact in affecting the project.

We also examined communities of engagement formed over time to see if these communities were rigid (remaining the same) or flexible (shifting role membership) over time. The community strength grew and became more prominent as the project progressed through its lifecycle. The number of communities ranged from one to three for each month, and the most common number of communities was two. Eight of the 27 months (30%) had only one community. These instances appeared mostly at the beginning and end of the project lifecycle, but also in August 2013 (Month 14) when there was a noticeable decrease in email activity. Eighteen months (67%) had two communities. Only one month (October 2012) expanded to three communities.

4   Discussion

The unique contribution of this study is the capture and analysis of email messages from a heterogeneous, multi-role scientific research team as a basis for indications of social patterns and configurations, information sharing, work activity focus, and project progress. As noted earlier, it is quite common to survey or interview team members to gain their perspectives and assessments about a project. In contrast, our approach depends on the digital footprints – organic byproducts – of the project work itself.

We saw indicators and patterns, of varying strength, of a range of project activities, focus, team engagement, and participant interaction using the analytical methods on the email corpus. These indicators were evident in the individual analyses (exploratory, concept, network), as well as in combination. We could also see evidence of differentiated participant interaction over the project lifecycle and detailed indications of meaningful team activity through email conversation.

4.1   Indications of the Social Dimension of Project Work

The analysis provides a view of the engagement and participation, varying influence and impact of different roles, and indications of who was active in the team interaction via the email communications (Fig. 3). We saw dramatic differences between months in the number of participants engaging in communication, number of communication paths, and in network ranking. Overall, the number of communication paths followed a general project lifecycle curve with the bulk of communications being created while the project work was being executed.

4.2   Project and Organizational Reflection

We were also able to gain insights into the trajectory of project progression and organizational involvement through the ebb and flow of topics over time. The analyses of communications provide an alternative lens to see elements of activity organically, and to provide feedback to a team to aid in identifying potential gaps. Providing these insights to the project participants and their organizations as the project is underway enables project "self-reflection" and has the potential to positively impact project practices. The use of a feedback loop with project participants enables the early warning of issues in a project, as well as novel insights from a different analytical perspective that contrast with standard project management and project status mechanisms.

4.3   Implications for the Method

This method provides flexibility in choosing the unit of analysis (e.g., individual contributor or role-based contribution) and varying levels of temporal granularity (e.g., day to years) based on the data being examined. We were limited by the size of our data corpus in being able to perform analyses more granular than monthly. However, with a larger corpus of the same type of data or combination of multiple data sources, shorter timeframes could be examined for a more real-time examination of interaction and progress.

A second limitation is that email is only one communication channel among many within a research project. This was not a major concern for this study because of the exploratory nature. However, in going forward with our research agenda, additional interactional data sources would add a depth of information sharing and idea exchange that may not appear via email, and reveal an additional dimension to scientific team collaboration and project progression.

5   Conclusions

Results of this method are promising to understand team progress, for both science and technical teams, using digital trace data to identify patterns of project activity. We were able to identify temporal evolution of work activities, influential roles and formation of communities throughout the project, variation in the conversational aspects as the project begins, matures, and then ends, and the ebb and flow of participation over time. We show the utility of using new data sources (digital trace data) to track and assess project progress, results, and contributions to the field through the examination of communication and interaction. The ability to begin this evaluation as a project is underway is a novel method.

The insights gained from the examination of the social and conceptual aspects of the scientific teamwork could be used to improve two dimensions of project work: (a) practices affecting team operations, such as information sharing; and (b) improvements to the computational environment, such as supported software configurations and data sources. This "organic" analysis of project focus, activity, and contribution is a new vantage point that could enable reflection by the project team members on a wide set of work dimensions, from identifying activity, gauging expected progress, and ultimately resulting in enhanced team performance.

Our success with three analyses on one data source provides only a partial picture of science team activity. Building on our current findings, future plans include: (a) adding additional data sources (e.g., such as system logs, meeting transcripts, and project artifacts) to deepen understanding and establish repeatability; (b) continued feedback and validation of our findings through interviews and participatory activities with the science teams; and (c) the identification and development of appropriate metrics to gauge the pace of science team activity.

Acknowledgments. Special thanks go to the Partner and Maker members of the project used for this investigation and to all our colleagues for contributions through critical research discussions and review of this effort.

References

1. Al-Ani, B., Bietz, M.J., Wang, Y., Trainer, E., Koehne, B., Marczak, S., Redmiles, D., Prikladnicki, R.: Globally distributed system developers: their trust expectations and processes. In: Proceedings of the 2013 CSCW Conference, pp. 563–574. ACM (2013)
2. Anderson, L.C., Kieliszewski, C.A. Co-creative aspects of data-driven discovery. In: Proceedings of 6th International Conference on Applied Human Factors and Ergonomics (AHFE 2015) and the Affiliated Conferences, pp. 3440–3447 (2015)
3. Anya, O., Moore, B., Kieliszewski, C., Maglio, P., Anderson, L.: Understanding the practice of discovery in enterprise big data science: an agent-based approach. In: proceedings of 6th International Conference on Applied Human Factors and Ergonomics (AHFE 2015) and the Affiliated Conferences, pp. 4389–4396 (2015)
4. Dabbish, L., Stuart, C., Tsay, J., Herbsleb, J.: Transparency and coordination in peer production. arXiv Preprint (2014). arXiv:1407.0377
5. Diesner, J. Carley, K.M.: Exploration of communication networks from the enron email corpus. In: SIAM International Conference on Data Mining: Workshop on Link Analysis, Counterterrorism and Security, Newport Beach, CA (2005)
6. Dong, W., Ehrlich, K., Macy, M.M., Muller, M.: Embracing cultural diversity: online social ties in distributed workgroups. In: Proceedings of the 19th ACM CSCW and Social Computing, pp. 274–287. ACM (2016)
7. Dumais, S., Jeffries, R., Russell, D.M., Tang, D., Teevan, J.: Understanding user behavior through log data and analysis. Ways of Knowing in HCI, pp. 349–372. Springer, New York (2014)
8. Engeström, Y.: From Teams to Knots: Activity-theoretical Studies of Collaboration and Learning at Work. Cambridge University Press, Cambridge (2008)
9. Floricel, S., Bonneau, C., Aubry, C.M., Sergi, V.: Extending project management research: insights from social theories. Int. J. Proj. Manag. **32**(7), 1091–1107 (2014)
10. Haas, L., Cefkin, M., Kieliszewski, C., Plouffe, W., Roth, M.: The IBM research accelerated discovery lab. ACM SIGMOD Rec. **43**(2), 41–48 (2014)
11. Herbsleb, J.D.: Global software engineering: the future of socio-technical coordination. Presented at the Future of Software Engineering FOSE 2007, pp. 188–198 (2007). https://doi.org/10.1109/FOSE.2007.11

12. Hoegl, M., Gemuenden, H.G.: Teamwork quality and the success of innovative projects: a theoretical concept and empirical evidence. Organ. Sci. **12**(4), 435–449 (2001)
13. Hoffmann, C.P., Lutz, C., Meckel, M.: A relational altmetric? network centrality on ResearchGate as an indicator of scientific impact. JASIST **67**(4), 765–777 (2016)
14. Kieliszewski, C.A., Anderson, L.C., Stucky, S.U.: A case study: designing the service experience for big data discovery. Adv. Hum. Side Serv. Eng. **1**, 460 (2014)
15. Li, J., Shi, D.: Sleeping beauties in genius work: when were they awakened? JASIST **67**(2), 432–440 (2016)
16. McCallum, A., Wang, X., Corrada-Emmanuel, A.: Topic and role discovery in social networks with experiments on enron and academic email. J. Artif. Intell. Res. **30**, 249–272 (2007)
17. McKeown, K., Daume, H., Chaturvedi, S., Paparrizos, J., Thadani, K., Barrio, P., Biran, O., Bothe, S., Collins, M., Fleischmann, K.R., Gravano, L.: Predicting the impact of scientific concepts using full-text features. JASIST **67**(11), 2684–2696 (2016)
18. Pentland, A.: Social physics: how Good Ideas Spread - the Lessons from a New Science. Penguin, New York (2014)
19. Rowe, R., Creamer, G., Hershkop, S. Stolfo, S.J.: Automated social hierarchy detection through email network analysis. In: Proceedings of the 9th WebKDD and 1st SNA-KDD 2007 workshop on Web mining and social network analysis, pp. 109–117. ACM (2007)
20. Song, J., Huang, Y., Qi, X., Li, Y., Li, F., Fu, K., Huang, T.: Discovering hierarchical topic evolution in time-stamped documents. JASIST **67**(4), 915–927 (2016)
21. Sonnenwald, D.H.: Scientific collaboration. Ann. Rev. Inf. Sci. Technol. **41**(1), 643–681 (2007)
22. Sproull, L., Kiesler, S.: Reducing social context cues: electronic mail in organizational communication. Manag. Sci. **32**(11), 1492–1512 (1986)
23. Tang, G., Pei, J., Luk, W.S.: Email mining: tasks, common techniques, and tools. Knowl. Inf. Syst. **41**(1), 1–31 (2014)
24. Velden, T., Yan, S., Yu, K., Lagoze, C.: Mapping the evolution of scientific community structures in time. In: Proceedings of the 24th International Conference on World Wide Web, pp. 1039–1044. New York, NY, USA ACM (2015)
25. Winnink, J.J., Tijssen, R.J., van Raan, A.F.: Theory-changing breakthroughs in science: the impact of research teamwork on scientific discoveries. JASIST **67**(5), 1210–1223 (2016)
26. Zehnalova, S., Horak, Z., Kudelka, M.: Email conversation network analysis: work groups and teams in organizations. In: Proceedings of the 2015 IEEE/ACM International Conference on Advances in Social Networks Analysis and Mining, pp. 1262–1268. IEEE (2015)

Smart University for Sustainable Governance in Smart Local Service Systems

Clara Bassano[1], Alberto Carotenuto[2], Marco Ferretti[1],
Maria Cristina Pietronudo[1(✉)], and Huseyin Emre Coskun[3]

[1] Department of Management Studies and Quantitative Methods,
University of Naples "Parthenope", Naples, Italy
{clara.bassano,marco.ferretti,
mariacristina.pietronudo}@uniparthenope.it
[2] Department of Engineering, University of Naples "Parthenope", Naples, Italy
alberto.carotenuto@uniparthenope.it
[3] Department of Economics and Management, University of Trento,
Trento, Italy
huseyinemre.coskun@unitn.it

Abstract. From the Service Science perspective, our work tends to provide a conceptual and methodological contribution to affirm the role of the university as the responsible agent for the growth and development of a local area. In this sense, the purpose is to promote a harmonious growth of the whole local service system, focused on academic quality and accountability through the approach to Social Responsibility, also involving government, business and society. The University as a place for higher education and research, at the same time, represents a privileged space of convergence of different growth perspectives of the local actors. This convergence should be seen as a "place" to share and develop a common sense of value that is smart, ethically, socially, and economically sustainable, i.e. a Smart Local Service System (S-LSS).

Keywords: University · Sustainable governance · Service science
Smart local service systems

1 Introduction

The concept of the entrepreneurial university emerged at the end of the 20th century following the footprints of 'research university' approach. Indeed, the emergence of the entrepreneurial university theme is based on a fundamental socio-economic change called knowledge-based society [1]. When these changes are examined, it is seen that industry's expectations of the university have changed correspondingly over time. Traditionally, enterprises in the market were expecting universities to train their future employees with the basic knowledge they would use during their work life [2]. Yet, while innovation has become the key concept of the competitive world [3], it shaped the relations of actors in the market as expected, along with the rapid change of the competition based on technological progress from "closed innovation" to the "open

© Springer International Publishing AG, part of Springer Nature 2019
T. Z. Ahram (Ed.): AHFE 2018, AISC 787, pp. 206–216, 2019.
https://doi.org/10.1007/978-3-319-94229-2_20

innovation" [4]. Thereby the need for externalizing the research to create a competitive advantage in the marketplace turned the eyes on to the universities [2, 4].

One of the most influential writers in the field, Etzkowitz [1, 2] argues that the universities had passed through two fundamental steps. With the first evolution, universities became institutions that not only teach but also perform research. With the second evolution, research universities evolved into a new structure called the entrepreneurial university that aims to commercialize the research outputs. Thereby universities have begun to respond to market and community demands. The same authors briefly refer to this process as shifting from "extension of knowledge" to the "commercialization of knowledge" or "capitalization of knowledge" [1, 2].

Moreover, when we look at the brief history of universities rather than market expectations that shape the university we see the efficient policy-making processes that transform the universities. When universities in Europe was surrounded by the influence of Church, firstly they gained freedom from the church [5] due to perform research under the economic and political protection of the sovereign [6]. Especially in the German case, Humboldt brothers ensued the German Idealism following Kant and asked the assurance from the King of Prussia when Europe was enchanted with Newton's physics [6]. From that moment, research university emerged as an institution that is isolated from the society and politics, stuck in its ivory tower [1]. With Humboldt principals, the concept of "performing research for science" got into the university without the abandonment of education [7]. And the Humboldt Ecole was highly successful from "basic sciences" to "applied research" and "spillover mechanisms" [8].

In following years, the land-granted universities in the USA that aim to boost agriculture production began to evolve into a new structure slowly. With the transformative power of the Patent Acts of 1790, 1836, 1922, 1952 and Morrill Land-Grant Acts at 1862 [1, 9] the USA became the world leading agriculture country [5, 8, 10]. Since American Universities became the world-leading research institutions, Senator Bayh opened the road for the academics to take copyrights, licenses, and patents of their research [8] with the Bayh Dole Act (1980) and universities became world-leading research and entrepreneurship institutions [1, 11].

And finally, in late 20th Century when entrepreneurial universities improving significantly on research and commercialization due to the rising R&D costs and requirement of the expertise from the firms led them to co-operate with the universities [12, 13]. Thus, the co-operation between firms and universities pushed the university to descend through the ivory tower and contribute to social welfare [1, 2]. When we look at the development of the university in history over time, we see that the university is transformed by both internal and external elements [2]. Furthermore, all these elements show us that the university as an institution evolved with the social, philosophical and economic enrichments. These fundamental changes have transformed the universities into service institutions that fulfill the requirements of firms and local society due to contribute economic prosperity. With this motive, regarding the fragmented structure of entrepreneurial university [14], the new concepts of the topic can be examined from a Service Science perspective (in short SS) [15], taking into account that in what dimensions co-operation between actors would take place to raise economic growth.

Following this objective, the role of University should be interpreted in a systemic view that can help local governments with technical ability, and intellectual and institutional resources [16], and in particular with applied programs for interpreting the territory as smart service systems.

In the first part, we propose a literature background concerning main lines on the concept of research on the role of the university. Then, through a service science framework, an extension of research role and its functions are suggested especially in the governance of the local system. In the final part, the framework is discussed, and implications are presented.

2   Literature Background

In 1998, Burthon Clark used the term "entrepreneurial university" in the title of his book *"Creating entrepreneurial universities: organizational pathways of transformation. Issues in Higher Education"*, explaining how universities had modernized strategies, management and structures [17]. Etzkowitz et al. [1] clarified the term referring to the new task of the university to innovate and promote the regional or national economic growth.

Two dominant approaches support this role: the "triple helix model" [18] and the "engaged university" [19].

The triple helix of university-industry-government emerges with the appearance of the university in the economic and political scenario, expanding his mission in a holistic perspective acting together with industry and government [20]. However, this model occurred a static vision of the triple relationship: the helixes are vertically independent, develop independently and only in a horizontal dimension form and interactive circulatory system [20]. During the non-linear interaction, each can take the role of the other creating hybrid organizations – spin-off, universities enterprises, and incubators – where roles overlap [21] and partially interact. The triple helix circulation creates a spiral, characterized by the evolution and circulation of vertical axes and the rotation of horizontal axes. The vertical evolution produces micro effects concerning single helix; and the horizontal evolution produces macro effects concerning collaborative programs, networks, and shared policies. The territorial impact is indirect since university pursue an aim focused on its profit saved by knowledge commercialization, not focused on local development [22] and neglecting interventions and role of institutions. However, transfer activities success does not depend only on organizational and strategic agreements of universities, but also on other institutional and structural aspects of the regions where a university operates [23]. Therefore, the triple helix approach does not give a systemic view. It discusses the overlapping role, but it does not devote any specific attention on the alignment of university products and regional demand [24] and neither on systemic links between local actors. For these reasons, the effect on regional development appears indirect, and the relationship between industries and government seem weak. In any case, the approach leads to new observations concerning University as:

- it is a catalyst for interactions and negotiations tensions between universities, industry and government [1, 25];
- its activities could produce an impact on the region;

however, there is no guarantee that it catches value for the region [26].

In addition to "regenerative" roles (universities and capital of knowledge), universities can play more "development" roles [23]. Gunasekara refers to the formation of human capital, the associative government e culture [27]. This kind of approach is characterized by more commitment and, it is defined as "engaged university" [28–30]. It assumes a broader and more adaptive role for the university [23] since it reacts to specific regional needs and extends his knowledge commercialization beyond the industrial innovations (patents, spin-off, etc.), versus civic tasks. In this way, it became an institutionalized actor acting on learning, innovation and also of governance process [31] of the territory.

For more than a decade, governments are involving Universities in policy formulation and production, conservation and evolution national and regional identities [32], posing a new challenge for academics who should maintain at the same time independence and a high level of knowledge production [33]. Tsipouri suggests three levels of action, through which university operates for the benefit of society: individual, institutional and collective [34]. At the individual level, it contributes to educate the future political elite and to involve them as an academic expert in a series of governance networks, supporting many managers and legislators in the exercise of their functions within modern governance structures.

At the institutional level, in many contexts, universities are recipients of decisions taken elsewhere. Governments delegate decisions to universities, but they were not the real institutional decision makers.

At the collective level-no as single entities but with other collective bodies - they are organizing collectively to define scientific policies and have greater influence and a greater power of initiative [33].

Nevertheless, according to Arbo and Benneworth [33], it is widespread a lack of recognition by governments of the ability of universities to intervene directly on local policies. Meanwhile, their role can no longer be ignored: they have extended their interests into new areas, they need to achieve concrete things, they are skilled consultants able to support regional governance networks [35, 36] and particularly they have a crucial role in disseminating global/local knowledge [26, 37, 38]. These elements could leverage the competitive advantage of the territory, but the primary barrier would be creating a complementarity of role (rather than an overlapping) between local actors.

The engaged university approach offers a dynamic and multifunctional view of the university, Kerr uses the term "multiversity" to indicate several functions of the modern university [39]. It has enriched its core activities (teaching, innovation/research/ technology transfer) with others residual (health and wellbeing, culture and sport, sustainable development and regeneration [40]; entrepreneurship promotion, building consortia, cultural networks, telematics networks, regional promotion, [31]). The latter constitutes the so-called third mission, i.e., the contribution that the universities make to the social and economic development of the territories [41, 42]. It could have a large-scale effect and more benefit on collectivity producing a social and academic utility; however, they are not enough for the success [33].

The difference between an entrepreneurial and engaged university is based on new forms of formal and informal collaborations with the local actors [43] that extend the academic commitment and the individual benefit derived from the commercialization of their activities [44, 45]. The authors D'Este and Patel [43] specify the commercialization of activities represent a measurable tool and immediate for the economic and social development; but formal agreement, conferences, meetings represent further convergence spaces to produce, share and apply knowledge. Therefore, the university could contribute to the local buzz of stimulating global/local interactions to improve the economic trajectories of these places [26]. The third mission reinforces university-government link, but at the same time could rise a difficulty of integrating universities into regional innovation strategies [46]. The solution is acting in an integrative way and combine influences and resources [26, 47], developing multi-layer governance [33]. Service Science [15] – SS – could provide a new perspective for the analysis and resolution of problems concerning expert thinking [49] or complex thinking capable of approaching to the variability of contexts by optimizing relational synergies.

3   Our Framework

SS is transdisciplinary and focused on the understanding of the evolution of value co-creation interactions between service systems [50], i.e., between a dynamic configuration of resources (people, technologies, organizations and information) that create and provide a service [51]. Universities are typical, but smallest entities of service system able to change the value of knowledge in the global service system ecology [52]. They have an essential role in terms of accelerating societal progress [50] since they serve as "glue" for a system in which operate. Having different areas/departments, they pursue strategies on many fronts, involving industries, government, citizens, students and other entities on several issues. For that reason, the key is guessing how to manage relationship and shared goals. Compared with previous approaches the SS clarifies the concept of governance, the result of political, economic and technologies characterized by network cooperation and collaboration at all levels of the organizations and/or companies [53]. In fact, SS enriches the university function as *knowledge applicator* since University practices are not the only commercial, but also partly linked to the governance (e.g. rules), [50]. Therefore, a further function could be the promotion of a smart, sustainable and multilevel governance to manage innovations and social progress in a system in which different actors operate. Such governance is typical of a Smart Local Service System, S-LSS [53, 54]. From our perspective based on the SS framework, the role of the university is interpreted as S-LSS, which means a local area in which different disciplinary perspectives coexist and converge for sharing innovation, competitiveness and improving quality of life through value co-creation process [55, 56]. This convergence should develop a common sense of smart value ethically, socially and economically sustainable.

4   Discussion

Even though University seems more active and aware of its strategic role on the territory, it appears to be weakened in situations in which it tries to integrate with other actors' concerning local policies, struggling to emerge as an independent actor [57, 58]. To finalize its third mission, University must be able to:

- build and maintain a network of advantageous relationships based more on competences and knowledge, on comprehension, trust, right and responsibility to obtain a non-zero-sum results in a service ecology [50];
- align the expectations of the various stakeholders [59];
- involve participants to share the process of definition and co-creation of the service [59].

Consequently, the qualification of an S-LSS becomes decisive to activate the relational components and determine the government structure. Explicitly, a S-LSS favors a smart and multilevel style of governance, characterized by a bottom-up approach [60–62], decentralizing the power of government and promoting a participatory leadership to manage projects and intervention policies. This governance approach avoids the overlapping of responsibility and moderate interaction between institutions, industries and university [63–65] taking into consideration the social and ecological diversity [66, 67]. Engaging efficiently on multiple scales is crucial for local systems that are invariably subject to powerful external influences, including changes in regulations, investment and the environment [63, 64] and internal influences as interests, profits, aims. Therefore, it is necessary to adequate rules and schemes dynamically to ensure a collaborative structure of governance in high variability and uncertain scenarios. In a similar context, University can create a convergence space, based on a proximity of intents, in which territorial proximity, or geographical proximity [68] become only a pre-condition that favors "the meeting" between actors. Cognitive, organizational, social and institutional proximity are elements, which compose the convergence (1), and that university can create.

So, we can assume that:

$$Convergence_{space} = \text{Geog.}_{prox} \sum \left(\text{Cog.}_{prox} + \text{Org.}_{prox} + \text{Soc.}_{prox} + \text{Institut.}_{prox} \right) \quad (1)$$

Where:

- *cognitive proximity* ensures the capacity to absorb new knowledge, communicate and learn knowledge from each other [68];
- *organizational proximity* represents the same space of relations [69] and favors an interactive learning process;
- *social proximity* is defined as micro level relations based on trust, friendship, kinship and shared experiences which facilitate the exchange of tacit knowledge [68];
- *institutional proximity* represents the macro- level framework: habits, routines, practices and lows [68].

By this means, University could create this condition educating, researching solutions, acting and transferring his international experience.

5 Implications and Future Challenges

An active and engaged role of the University produces several implications. The university becomes an institution that can fulfill the requirements of the industry not just as sustaining the human capital but also contributing to the discovery, design and production processes as a service institution. Industries benefit through the transfer of specific knowledge (skilled employees/T-shaped professionals) and from the reduction of innovation costs through patents or start-ups. Notably, some companies agreed with research centers and universities to innovate their product and services to externalize R&D and start process of open innovation. Policymakers could benefit from universities relationships and their knowledge transfer to implement national and international policies. Competition has increased at a global level since territory are competing for obtaining incentives and investments useful for their development [68], and the university is becoming ever more skilled to project initiatives and adopt them. All part of the local system could benefit from the convergence spaces created by university favoring local culture, regional policies and enhancing local identity (i.e. promoting local excellences).

At present, the study is merely descriptive and presents several limits.

In particular, future challenges are needed both at an academic and practical level. At an academic level, we refer to the analysis of exploratory survey to perform study the local network and examine the link between 'service science' perspective and, the five dimensions of "proximity": cognitive, organizational, social, institutional and geographical [68]. These dimensions could be read as the integrative condition to reach the systems purpose and co-created value. At a practical level, we refer to the opportunity to give a chance to university to operate actively, but not independently from territory. This is what happens especially where universities are public and need for an institutional and industrial relations to ensure the growth and the development of local areas. Unfortunately, not all communities have the capacity to implement smart growth strategies and collaborative policies to guarantee the territorial viability.

6 Conclusion

The main purpose of this paper is to interpret the University as a Smart Local Service System (S-LSS) which means as a "place" to share and develop a common sense of value that is smart, ethically, socially, and economically sustainable for higher education and research. This "place" represents what we call a "privileged space of convergence" in which many different actors collaborate according to our conceptual and methodological contribution to affirm the role of the university as the responsible agent for the growth and development of a local area.

Our perspective is coherent with the concepts of territory as a system in which systems dynamics prevail over single structural components, with a shift in focus from the traditional mechanisms of governance to shared and collaborative governance processes [59]. In this favorable context, the university becomes smart which means educating and transferring specific and deep knowledge to actors supporting the whole system to develop an absorptive capacity. At the same time, the different expectations of stakeholders are mediated ensuring the success of complex but cohesive and co-administered governance. The university's knowledge heritage not only aims at the development and sustainability of innovation in a local/regional context [50]. A reflection cannot be detached from the concept of responsibility, i.e. the ability to generate impacts on its territory of reference, to protect human resources and to promote and share knowledge, and sustainable research not only at the economic level but also at the environmental and social level. This means a transition from a social responsibility of an organization in a certain territory to the social responsibility of a territory as a whole system. The applicability of our conceptual and methodological perspective implies the synthesis of three important ingredients through the coordination of the University as a center of basic and specialized knowledge and democratic coexistence:

(1) the quality of the involved actors;
(2) the coordination in shared processes;
(3) the sustainability of common results.

Therefore, according to our view, the SS's contribution is to re-examine the university as an active actor in the system with a specific role of *knowledge applicator*, moving from a more indirect contribution for the local economic development and innovation to a more formal role, institutionalized and proactive [23].

References

1. Etzkowitz, H., Webster, A., Gebhardt, C., Terra, B.R.C.: The future of the university and the university of the future: evolution of ivory tower to entrepreneurial paradigm. Res. Policy **29** (2), 313–330 (2000)
2. Etzkowitz, H.: The norms of entrepreneurial science: cognitive effects of the new university–industry linkages. Res. Policy **27**(8), 823–833 (1998)
3. Mian, S.A.: University's involvement in technology business incubation: what theory and practice tell us? Int. J. Entrep. Innov. Manage. **13**(2), 113–121 (2011)
4. Chesbrough, H.W.: Open Innovation: The New Imperative for Creating and Profiting from Technology. Harvard Business Press, Boston (2006)
5. Audretsch, D.B.: The Entrepreneurial society. Oxford University Press, New York (2007)
6. Nybom, T.: The Humboldt legacy: reflections on the past, present, and future of the European university. High. Educ. Policy **16**(2), 141–159 (2003)
7. Schimank, U., Markus, W.: Beyond Humboldt? The relationship between teaching and research in European university systems. Sci. Public Policy **27**(6), 397–408 (2000)
8. Audretsch, D.B.: From the entrepreneurial university to the university for the entrepreneurial society. J. Technol. Transf. **39**(3), 313–321 (2014)
9. Etzkowitz, H., Stevens, A. J.: Inching toward industrial policy: the university's role in government initiatives to assist small, innovative companies in the U. Sci. Technol. Stud. (1995)

10. Audretsch, D.B.: The entrepreneurial society. In: New Frontiers in Entrepreneurship, pp. 95–105. Springer, New York (2009)
11. Mowery, D.C.: The Bayh-Dole Act and high-technology entrepreneurship in US universities: chicken, egg, or something else? In: University Entrepreneurship and Technology Transfer, pp. 39–68. Emerald Group Publishing Limited (2005)
12. Chesbrough, H.W.: Open innovation: a new paradigm for understanding industrial innovation. In: Chesbrough, H.W., Vanhaverbeke, W., West, J. (eds.) Open Innovation: Researching a New Paradigm. Oxford University Press, Oxford (2006)
13. Perkmann, M., Walsh, K.: University–industry relationships and open innovation: towards a research agenda. Int. J. Manage. Rev. 9(4), 259–280 (2007)
14. Rothaermel, F.T., Agung, S.D., Jiang, L.: University entrepreneurship: a taxonomy of the literature. Ind. Corp. chang. 16(4), 691–791 (2007)
15. Spohrer, J., Kwan, S.K., Fisk, R.P.: Marketing: a service science and arts perspective. In: Rust, R.T., Huang, M.H. (eds.) Handbook of Service Marketing Research, pp. 489–526. Edward Elgar, New York (2014)
16. Spohrer, J.C., Freund L.: Measuring T-shapes for ISSIP professional development. In: AHFE 2014 2nd Conference on Human Side of Service Engineering (2014)
17. Gulbrandsen, M., Slipersaeter, S.: The third mission and the entrepreneurial university model.Univ. Strat. Knowl. Creat. 112–143 (2007). In: Bonaccorsi, A., Daraio, C. (eds.) Universities and Strategic Knowledge Creation: Specialization and Performance in Europe. Edward Elgar Publishing
18. Etzkowitz, H., Leydesdorff, L.: The future location of research and technology transfer. J. Technol. Transf. 24, 111–123 (1999)
19. Ferretti, M., Parmentola, A.: The Creation of Local Innovation Systems in Emerging Countries: The Role of Governments, Firms and Universities. Springer, Heidelberg (2015)
20. Etzkowitz, H., Zhou, C.: Triple helix twins: innovation and sustainability. Sci. Public Policy 33(1), 77–83 (2006)
21. Etzkowitz, H., Leydesdorff, L.: The dynamics of innovation: from national systems and 'Mode 2' to a triple helix of university-industry-government relations. Res. Policy 29, 109–123 (2000)
22. Mowery, D.C., Sampat, B.N.: The Bayh-Dole Act of 1980 and university–industry technology transfer: a model for other OECD governments? J. Technol. Transf. 30(1), 115–127 (2004)
23. Uyarra, E.: Conceptualizing the regional roles of universities: implications and contradictions. Eur. Plan. Stud. 18(8), 1227–1246 (2010)
24. Cooke, P.: Regionally asymmetric knowledge capabilities and open innovation: exploring 'Globalisation 2' – a new model of industry organisation. Res. Policy 34, 1128–1149 (2005)
25. Etzkowitz, H., Leydesdorff, L.: Universities and the Global Knowledge Economy: A Triple Helix of University-Industry-Government Relations. Printer, London (1997)
26. Benneworth, P., Hospers, G.: The new economic geography of old industrial regions: universities as global/local pipelines. Environ. Plan. C 25(5), 779–802 (2007)
27. Gunasekara, C.: The generative and developmental roles of universities in regional innovation systems. Sci. Public Policy 33(2), 137–150 (2006)
28. Chatterton, P., Goddard, J.: The response of higher education institutions to regional needs. Eur. J. Educ. 35(4), 475–496 (2000)
29. Gunasekara, C.: Reframing the role of universities in the development of regional innovation systems. J. Technol. Transf. 31(1), 101–113 (2006)
30. OECD: Higher Education and Regions: Globally Competitive, Locally Engaged. OECD, Paris (2007)

31. Boucher, G., Conway, C., Van Der Meer, E.: Tiers of engagement by universities in their region's development. Reg. Stud. **37**(9), 887–897 (2003)
32. Keating, M., Loughlin, J., Deschouwer, K.: Culture, Institutions and Economic Development: A Study of Eight European Regions. Edward Elgar, Cheltenham (2003)
33. Arbo, P., Benneworth, P.: Understanding the regional contribution of higher education institutions: a literature review. OECD Educ. Work. Pap. (9) (2007)
34. Tsipouri, L.: Perceived and actual roles of academics in science policy. In: Siune, K. (ed.) Science Policy: Setting the Agenda for Research. The Danish Institute for Studies in Research and Research Policy (2001)
35. Bryson, J.: Spreading the message: management consultants and the shaping of economic geographies in time and space. In: Bryson, J.R., et al. (eds.) Knowledge, Space, Economy. Routledge, London (2000)
36. Muller, E., Zenker, A.: Business services as actors of knowledge transformation: the role of KIBS in regional and national innovation systems. Res. Policy **30**, 1501–1516 (2001)
37. Beaverstock, J., Smith, N., Taylor, P.: World city networks: a new meta geography. Ann. Assoc. Am. Geograph. **90**(1), 123–134 (2000)
38. Hospers, G.J.: Slimmestreken. Actuele Onderwerpen, Lelystad (2006)
39. Kerr, C.: The uses of the multiversity (2001)
40. Benneworth, P.S., Charles, D.R.: Bridging cluster theory and practice: learning from the cluster policy cycle. In: Bergman, E.M., et al. (eds.) Innovative Clusters: Drivers of National Innovation Systems. OECD, Paris (2001)
41. Göransson, B., Maharajh, R., Schmoch, U.: New activities of universities in transfer and extension: multiple requirements and manifold solutions. Sci. Public Policy **36**(2), 157–164 (2009)
42. Saad, M., Zawdie, G.: Introduction to special issue: the emerging role of universities in socio-economic development through knowledge networking. Sci. Public Policy **38**(1), 3–6 (2011)
43. D'Este, P., Patel, P.: University-industry linkages in the UK: what are the factors underlying the variety of interactions with industry? Res. Policy **36**(9), 1295–1313 (2007)
44. Bramwell, A., Wolfe, D.A.: Universities and regional economic development: the entrepreneurial University of Waterloo. Res. Policy **37**(8), 1175–1187 (2008)
45. Freitas, I.M.B., Geuna, A., Rossi, F.: Finding the right partners: institutional and personal modes of governance of university–industry interactions. Res. Policy **42**(1), 50–62 (2013)
46. Lagendijk, A., Rutten, R.: Associational dilemmas in regional innovation strategy development: regional innovation support organisations and the RIS/RITTS programmes. In: Bakkers, S., Boekema, F. (eds.) Universities, Knowledge Infrastructure and the Learning Region. Edward Elgar, Aldershot (2001)
47. Bathelt, H., Malmberg, A., Maskell, P.: Clusters and knowledge: local buzz, global pipelines and the process of knowledge creation. Prog. Hum. Geograph. **28**(1), 31–56 (2004)
48. Lampert, M., Clark, C.M.: Expert knowledge and expert thinking in teaching: a response to Floden and Klinzing. Educ. Res. **19**(5), 21–23 (1990)
49. Spohrer, J., Giuiusa, A., Demirkan, H., Ing, D.: Service science: reframing progress with universities. Syst. Res. Behav. Sci. **30**(5), 561–569 (2013). https://doi.org/10.1002/sres.2213
50. IfM, I.B.M.: Succeeding Through Service Innovation: A Service Perspective for Education, Research, Business and Government. University of Cambridge Institute for Manufacturing, Cambridge (2008)
51. Spohrer, J.C., Maglio, P.P.: Towards a science of service systems, value and symbols. In: Handbook of Service Science (Service Science: Research and Innovations in the Service Economy), pp. 157–195 (2010)

52. Piciocchi, P., Siano, A., Bassano,C., Conte, F.: Smart local service system. "Governamentalità intelligente" per la competitività del territorio. Atti del XXIV Convegno annuale di Sinergie (2012)
53. Piciocchi, P., Siano, A., Confetto, M.G., Paduano, E.: Driving co-created value through local tourism service systems (LTSS) in tourism sector. In: Gummesson, E., Mele, C., Polese, F. (eds.) Service Dominant Logic, Network and Systems Theory and Service Science. Giannini, Napoli (2011)
54. Spohrer, J., Maglio, P.P., Bailey, J., Gruhl, D.: Steps toward a science of service systems. Computer **40**, 71–77 (2007)
55. Spohrer, J.: Service science: progress and directions. In: AMA SERVSIG, Porto Portugal at FEUP Engineering School, June 2010. http://www.slideshare.net/spohrer/service-science-progress-and-directions-20100620
56. Balducci A.: La città come campo di riflessione e di pratiche per le università milanesi. In: Balducci, A., Cognetti, F., Fedeli, V. (a cura di) Milano città degli studi. Storia, geografia e politiche delle università milanesi, Associazione Interessi Metropolitani, Editrice Segesta, Milano, pp. 197–198 (2010)
57. Cognetti, F.: La third mission dell'università. Lo spazio di soglia tra città e accademia. Territorio (2013)
58. Piciocchi, P., Spohrer, J., Bassano, C., Giuiusa, A.: Smart governance to mediate human expectations and systems context interactions. In: Spohrer, J.C., Freund, L.E. (ed.) vol. 18, pp. 319–328 (2012)
59. Triantafillou, P.: Conceiving "network governance": the potential of the concepts of governmentality and normalization. In: Working paper 2004/4, Centre for Democratic Network Governance, Roskilde, May 2004
60. Trunfio, M.: Governance turistica e sistemi turistici locali: Modelli teorici ed evidenze empiriche in Italia. Giappichelli, Torino (2008)
61. Piciocchi P., Bassano C.: Governance and viability of franchising networks from a viable systems approach (VSA). In: The 2009 Naples Forum on Service, Service Dominant Logic, Service Science and Network Theory, Proceedings of the International Conference in Capri. Giannini Editore, Napoli, 16–19 June 2009
62. Berkes, F.: Cross-scale institutional linkages for commons management: perspectives from the bottom up. In: Ostrom, E., Dietz, T., Dolsak, N., Stern, P.C., Stonich, S., Weber, E.U. (eds.) The Drama of the Commons, pp. 293–321. National Academy Press, Washington (2002)
63. Young, O.R.: The Institutional Dimensions of Environmental Change: Fit, Interplay and Scale. MIT Press, Cambridge (2002)
64. Lebel, L.: Institutional dynamics and interplay: critical processes for forest governance and sustainability in the mountain regions of northern Thailand. In: Huber, U.M., Bugmann, H. K.M., Reasoner, M.A. (eds.) Global Change and Mountain Regions: An Overview of Current Knowledge, pp. 531–540. Springer, Berlin (2005)
65. Peterson, G.D.: Political ecology and ecological resilience: an integration of human and ecological dynamics. Ecol. Econ. **35**, 323–336 (2000)
66. Ostrom, E.: Understanding Institutional Diversity. Princeton University Press, Princeton (2005)
67. Boschma, R.: Proximity and innovation: a critical assessment. Reg. Stud. **39**(1), 61–74 (2005)
68. Shaw, A.T., Gilly, J.P.: On the analytical dimension of proximity dynamics. Reg. Stud. **34** (2), 169–180 (2000)
69. Golinelli, C.M.: Il territorio sistema vitale: verso un modello di analisi. G. Giappichelli, Torino (2002)

The Human Side of Service Engineering: Innovations in Service Delivery and Assessment

Using Augmented Reality and Gamification to Empower Rehabilitation Activities and Elderly Persons. A Study Applying Design Thinking

Oliver Korn[1(✉)], Lea Buchweitz[1], Adrian Rees[1], Gerald Bieber[2], Christian Werner[3], and Klaus Hauer[3]

[1] Offenburg University, Badstr. 24, 77652 Offenburg, Germany
`oliver.korn@acm.org`,
`{lea.buchweitz,adrian.rees}@hs-offenburg.de`
[2] Fraunhofer Institute for Computer Graphics Research,
Joachim-Jungius-Street 11, 18059 Rostock, Germany
`gerald.bieber@igd.r.fraunhofer.de`
[3] Agaplesion Bethanien Krankenhaus Heidelberg gGmbH,
Rohrbacher Street 149, 69126 Heidelberg, Germany
`{christian.werner,khauer}@bethanien-heidelberg.de`

Abstract. We present the design of a system combining augmented reality (AR) and gamification to support elderly persons' rehabilitation activities. The system is attached to the waist; it collects detailed movement data and at the same time augments the user's path by projections. The projected AR-elements can provide location-based information or incite movement games. The collected data can be observed by therapists. Based on this data, the challenge level can be more frequently adapted, keeping up the patient's motivation. The exercises can involve cognitive elements (for mild cognitive impairments), physiological elements (rehabilitation), or both. The overall vision is an individualized and gamified therapy. Thus, the system also offers application scenarios beyond rehabilitation in sports. In accordance with the methodology of design thinking, we present a first specification and a design vision based on inputs from business experts, gerontologists, physiologists, psychologists, game designers, cognitive scientists and computer scientists.

Keywords: Augmented reality (AR) · Human-centered rehabilitation
Personalized rehabilitation · Gamification · Human factors
Human-systems integration

1 Introduction

The demographic change is a serious challenge for most European countries and especially for Germany. The percentage of elderly persons increases, because of both improving medical conditions and comparatively low birthrates. Besides other areas, an ageing society challenges especially the healthcare sector. Elderly people are more susceptible to injuries and diseases [1]. If the percentage of seniors in society increases,

© Springer International Publishing AG, part of Springer Nature 2019
T. Z. Ahram (Ed.): AHFE 2018, AISC 787, pp. 219–229, 2019.
https://doi.org/10.1007/978-3-319-94229-2_21

more people are likely to suffer from health problems. The health sector will be at the edge of capacity, if 25% of the European population are aged 60 or older, with an increased risk of diseases and injuries [2].

Keeping elderly people autonomous and at good health for as long as possible does not only reduce pressure for the healthcare system but is a benefit for society as a whole [3]. "Active ageing" or "autonomous ageing" are key aims, for example in the prominent European program with the fitting title "Active and Assisted Living" (AAL) [4]. Considering the enormous diversity and number of medical histories and conditions, this seems to be only achievable by treating everyone individually. However, in the age of fast technical development, especially in the area of information and communication technology (ICT) new technical devices allow new application scenarios, creating combinatory innovations in a multidisciplinary approach.

Virtual reality (VR) and augmented reality (AR) offer promising features for autonomous health activities, as they allow to integrate digital objects and distant guiding into health-related activities. In this article, we present the design vision of a technical device using AR, which allows individualized therapy for everyone. It incorporates inputs from inputs from business experts, gerontologists, physiologists, psychologists, game designers, cognitive scientists and computer scientists. However, individualized therapy is just one application: the universally adaptable design also allows preventive training and could even support regular sports activities.

2   Related Work

In this section, we start by shortly presenting the medical background. We then discuss the state of the art development in AR technologies and gamification in the healthcare sector. Finally, we introduce the concept of design thinking.

2.1   Medical Background

After injuries (e.g., femoral neck fracture) or longer periods of illness, the first phase of rehabilitation is generally characterized by fast progress and improvements in mobility and flexibility. However, after the intensive training period in a rehabilitation center, progress often stagnates once the patient is responsible for regular and autonomous training. Repeating the same exercises every day is boring, so the training intervals become longer, which leads to slower progress or even no progress at all. The result: a lot of elderly patients never regain their previous mobility and flexibility. Unfortunately for those patients, walking and balancing are basic skills essential for autonomy and a high quality of life. Several studies associate gait disorders with cognitive deficits [5], reduced mobility [6], an increased risk of falling [7], a lower overall health [8] and a general decreased functionality [9]. Therefore, the preservation of a high gait quality is a fundamental preventative and rehabilitative approach.

In the medical context, several performance-based tests help to evaluate the gait capabilities of patients. Computer-aided systems (e.g., digital insoles or sensor-based 3D-motion analysis) and acceleration sensors are not yet fully established but offer promising features. Therefore a system collecting detailed sensor data outside the lab is very beneficial: it allows a better analysis and better adaptation of exercises. It this system also helps to motivate the patients, two major challenges are addressed in one solution.

2.2   Therapeutic Applications of Augmented Reality

In order to improve and preserve the mobility of patients with multiple sclerosis [10], a therapeutic system combines force feedback with a virtual environment (Fig. 1). The training exercises are embedded in an individually adjustable team game, which is designed to increase fun during training sessions. The system is stationary and especially created for arm movements, which limits the scope of applications to a limited field of rehabilitation training.

Fig. 1. Gamified AR-system for patients with multiple sclerosis.

Another AR-using system for upper limb rehabilitation is "SleeveAR" [11] (Fig. 2). Its main aim is creating an incentive for autonomous training without the need of regular physiological guiding.

Projections on the floor offer instructions of the motion sequence and the direction of the movement, whereas projections on the arm give feedback about incorrect joint positions and corrections if needed. After the patient has completed a movement, the system gives feedback on how close the patient's movement was to a reference movement previously saved by a therapist. The stationary system for rehabilitation of the upper limbs and requires a special floor covering and color. These constraints restrict the aim of autonomous training to specific training locations.

Fig. 2. SleeveAR supports training without instructions by a therapist.

However, there are not only stationary AR-systems used for rehabilitation purposes.

Researchers at the University of Washington developed "TrainAR" (Fig. 3), a portable training solution for the rehabilitation after sports injuries [12].

In this solution, the patient wears a head-mounted display (HMD) which projects training instructions and training elements (e.g. obstacles or a soccer goal) into the visual field of the patient. This system offers a totally autonomous training with the restriction that a HMD can be obstructive or uncomfortable (e.g. for elderly people, or people with glasses).

The targeted solution avoids such problems by projecting directly on the surfaces of the natural environment. However, it is much more susceptible too lighting conditions.

Fig. 3. TrainAR uses a headmounted display (HMD) for virtual training.

2.3   Gamification

Gamification describes the enrichment of non-playful tasks with elements from video games, to give a normal real-world task a playful character. This can create motivation, foster fun and increase the overall user experience. Ultimately, this leads to a better (or at least a faster) performance of the gamified task.

In the area of education, gamified solutions called "serious games" have already proven to work well [13]. When game consoles with movement controllers like the Nintendo Wii (released in 2007) and Microsoft Kinect (released in 2010) came up, gamification also was applied more frequently in the health sector as application like "VI-Bowling" [14], or "motivation60+" [15] show. A recent study shows that gamification approaches are now also applied more frequently in medical rehabilitation [16].

This is not surprising, as rehabilitation exercises can profit especially from gamification. As pointed out in the medical background, the process of regaining flexibility and mobility after injuries and accidents often needs a lot of discipline and motivation, which is sometimes lacking due to convenience or even emotional traumata [17]. Studies show that more than 65% of the patients do not follow the rehabilitation programs [18]. If the trainings are to be continued at home without supervision, this percentage gets even worse.

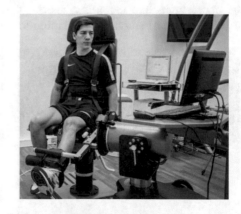

Fig. 4. Humac Norm, a stationary system which uses gamification for rehabilitation.

In a comparative analysis of a gamified and a non-gamified balance exercises, the gamified versions surpassed the traditional ones [19]. For instance, the rehabilitation system "Humac Norm" (Fig. 4) combines training exercises with classic Arcade games like Tetris. Another rehabilitation system, the "Lokomat", was gamified with the game "Gabarello" [20]. As stated in the previous section, all of these systems are stationary, which restricts autonomous training. Also, these systems are quite expensive, extending the budget for home use.

2.4 Design Thinking

Design thinking is often viewed as a new way of generating ideas and products, combining creativity, interdisciplinarity and a user-centered approach. It is also regarded as a step away from the more traditional mindset where engineers develop a product technologically and designers make it "look nice" in the end. Instead, "design thinking can be seen as an integrative approach which considers both form and function, takes both functional aspects and emotional involvement into account" [21]. This usually requires to involve experts and practitioners from several disciplines in the process of generating and evaluating ideas and forming new concepts.

Design thinking not only requires to survey and analyze the users' needs but also to envision what products and technologies users may want – without yet being aware of that desire [22]. Therefore, design thinking often involves the creation of prototypes which are tested, discarded and reviewed again. This method, accordingly called rapid prototyping, acknowledges the fact that defining "ideal" requirements or specifications by asking users at the beginning of a process is impossible. The process can go so far as integrating users in the development process, as seen in the concept of "living labs". In these labs, users can interact with products and technologies which are not yet on the market – not just for minutes but for hours or even days and weeks. This allows to determine the long-term acceptance of an innovation.

Instead of a linear process, design thinking follows a looser structure and consists of overlapping phases of developing, implementing and testing ideas. The collaborators from various disciplines agree in an open innovation process, accepting that "design is messy, needs probing, and that showing and testing unfinished work is part of the process" [21].

3 A Wearable AR-System for Autonomous Rehabilitation Training

We started a design thinking process to envision a technical device based on AR, which allows individualized therapy, preventive training and even support for normal sports activities. It is based on inputs from business experts, gerontologists, physiologists, psychologists, game designers, cognitive scientists and computer scientists. In the following we will describe the system's basic underlying technologies, as well as its features and capabilities.

3.1 System Overview

The system is a small, portable assistive device. It allows patients to do rehabilitation or preventive training sessions autonomously, at any time, in any location, without further equipment.

By combining training exercises with gamification elements, the solution is designed to improve not only flexibility, mobility and cognitive skills: it will also foster fun to keep the training motivation up, ideally even beyond the scheduled rehabilitation program.

Fig. 5. Gamified AR-device for autonomous rehabilitation training.

The assistive device itself consists of a micro-projector, several sensors and a computing device (typically a smartphone). The gamified exercises and instructions are projected on the floor in front of the user (Fig. 5) and can be adjusted, depending on the level of fitness, the soil condition and the rehabilitation program.

During the training, data from different sensors is collected and sent to the computing device for further analysis or interpretation by a therapist. In the following, each element of the system is presented in more detail.

3.2 Micro-projector

To provide an easy and comfortable use, the system's projector should be as small, light and bright as possible. Keeping a critical eye one the outward appearance and specifications of the system is not only relevant from a design perspective: The system's immediate

Fig. 6. Exemplary micro-projector.

appeal will be key to its successful application in rehabilitation and medical environments. As the projector is worn at the user's belt, this restricts weight and size, to not disturb the training, and to be conceived as unobtrusive as possible (Fig. 6).

As rehabilitation exercises are often performed during day-time and in luminous surroundings, brightness is a critical requirement. As a strong laser cannot be used in training contexts for safety reasons, the projector is based on LEDs which can project visible objects in daylight. Even with a projector with limited brightness, a wearable projection on several different organic surfaces in urban conditions and some outdoor conditions is feasible [23]. However, in bright sunlight the system will soon reach its limits. Under suitable lighting conditions, the brightness of the projection is automatically scaled to correspond to the illumination of the environment.

Another requirement for the micro-projector is the compensation of the user's movements. In order to keep the motivation for training up, the projections need to be stable and robust. This can be achieved by using data from inertial sensors, which will be described in the next section.

3.3 Sensors and Data Collection

The system not only provides movement instructions and training exercises, but also feedback for the user and the therapist. By combining sensors at various positions (e.g. at the leg, foot or knee), information on movement directions, speed and coordination can be collected. Mobile inertial measurement units (IMU) are available in a small package size and can be integrated directly into the wearable assistive device next to the micro-projector.

An inertial unit consists of 9D-sensors (acceleration, gyroscope, magnetic field) and detects horizontal and vertical deviation or a possible twisting of the micro-projector during usage. This allows an adequate recalculation of the viewpoint to secure a precise projection. Additional external sensors at the limbs could provide information about the movement accuracy with an expected position error of less than 1 cm [24]. Depending on the application and the exercise, different sensors can be attached to the body and controlled by the computing unit.

The therapist can access the collected data via a server or directly at the device. By interpreting the pre-analyzed sensor data, the therapist can deduce the training condition of the patient and adjust the exercise scheme accordingly. By incorporating feedback from both therapists and patients in the design process, we will adapt the level of feedback as well as the exercise types according to the individual users' needs. Furthermore, to allow for the "feeling of control" essential to interactions [25], we will enable them to adjust exercises, feedback or instructions on their own.

3.4 Outdoor Capabilities

A special feature of the AR training device is the possibility to do outdoor training (Fig. 7). The LED projector allows visible projections in suitable daylight conditions. Especially elderly persons highly profit from the training effect outside: after accidents or injuries with long rehabilitation phases, they often are scared to go outdoors again. This guided training approach allows outdoor training at an early stage of the rehabilitation, guided by the device and with frequent data control by the therapist. In the beginning, the training can also be accompanied by a physically present therapist, who reduces hesitation and worries. At later stages, the outdoor training can be done autonomously.

The advantages of outdoor rehabilitation training are the diverse and demanding soil conditions and the re-familiarization with outdoor settings. Different soil conditions offer different training stimuli, which makes the exercises more effective and the patients more self-confident in their own capabilities. In combination with small cognitive and motoric games, the familiarization with outdoor settings is reducing worries and preparing the patient for everyday life. The realistic training reduces the risk of injuries, resulting in long-lasting rehabilitation effects.

Fig. 7. Exemplary use of the AR-device outdoors.

The device also allows indoor training sessions, if the patient needs a more controlled environment. The combination of autonomous training, the possibility to get professional feedback on the health level, the option to adjust the exercise level and the opportunity to train either inside or outside makes the solution universally usable and suitable for everyday life.

4   Future Applications

Although the AR-device is primarily designed for outdoor exercises, its applications are manifold. First and foremost, the solution aims to support and motivate elderly persons to regain a high level of mobility and gait quality. However, the gamified training exercises can easily be combined with cognitive tasks to improve also light cognitive impairments. Foer example, a user can be asked to just touch one specific color of triangles with the right foot and ignore any other color (Fig. 8).

Another application area, which requires a 90-degree rotation of the system, is rehabilitation training for the upper limbs. If the exercises and instructions are projected on a wall, arm coordination and upper body movements can be trained, and more advanced cognition-specific tasks like mazes can be integrated as well.

Fig. 8. An elderly person exercising indoors with the AR-device.

Another application area is preventive training. If people use the system for keeping up their health level and balancing capabilities by doing individualized an motivating exercises, this training device can become an optimal solution to maintain autonomy as long as possible. Regarding the challenges of the demographic change, this device thus could work from both directions: rehabilitation and prevention.

Fig. 9. AR-device used for sports activities

As mentioned before, the applications of the AR-device are not limited to the medical area, to rehabilitation or even to prevention: collecting detailed movement data and accessing them at a later stage is an interesting feature for both amateur and professional athletes (Fig. 9). While amateur athletes could get fun information combined with some strengthening or gaming activities during the training, professional athletes and their trainers can improve motion cycles and movement techniques based on very detailed information. By adhering to the design thinking process and keeping an open mindset, other application areas will likely emerge wherever projection is preferred over HMDs.

5   Conclusion

We started a design thinking process to envision a technical device based on AR, which allows individualized and gamified therapy, preventive training and even support for normal sports activities. This first specification is based on inputs from business experts, gerontologists, physiologists, psychologists, game designers, cognitive scientists and computer scientists.

The device's primary function is to augment familiar surroundings for autonomous rehabilitation training, especially in outdoor settings. By increasing both the autonomy and the motivation of elderly persons, the systems can contribute to address the demographic change.

In Sect. 2, we described the underlying concepts and methods: the medical background, therapeutic applications using AR, gamification, and design thinking. In the following section we explained the underlying technological components focusing on projection and sensors. We also pointed out the special requirements and advantages of outdoor training. In Sect. 4, a range of future applications is described.

Indeed, we envision the AR-device as a universal tool. No matter what age the users are, once they attach the small wearable to their belt, there is activity for everyone: rehabilitation exercises, strengthening and balance training, training for the upper limbs, games involving movement and cognition and of course detailed movement data for therapeutic or sports-oriented motion analysis.

A main contribution of this AR-based training device is that it uses gamification to motivate the users by permanent feedback and adequate challenges. Thus, they ideally keep training because it is fun – and then progress is just a natural consequence.

References

1. Risk factors of ill health among older people, 05 February 2018. http://www.euro.who.int/en/health-topics/Life-stages/healthy-ageing/data-and-statistics/risk-factors-of-ill-health-among-older-people. Accessed 05 Feb 2018
2. Department of Economic and Social Affairs United Nations: World Population Prospects: The 2017 Revision, Key Findings and Advance Tables. Working Paper No. ESA/P/WP/248 (2017)
3. Siegel, C., Hochgatterer, A., Dorner, T.E.: Contributions of ambient assisted living for health and quality of life in the elderly and care services - a qualitative analysis from the experts' perspective of care service professionals. BMC Geriatr. **14**, 112 (2014)
4. Active and Assisted Living Programme | ICT for ageing well. http://www.aal-europe.eu/. Accessed 28 Feb 2018
5. Alfaro-Acha, A., Al Snih, S., Raji, M.A., Markides, K.S., Ottenbacher, K.J.: Does 8-foot walk time predict cognitive decline in older Mexicans Americans? J. Am. Geriatr. Soc. **55** (2), 245–251 (2007)
6. Cesari, M., et al.: Prognostic value of usual gait speed in well-functioning older people– results from the Health, Aging and Body Composition Study. J. Am. Geriatr. Soc. **53**(10), 1675–1680 (2005)
7. Montero-Odasso, M., et al.: Gait velocity as a single predictor of adverse events in healthy seniors aged 75 years and older. J. Gerontol. Ser. A: Biol. Sci. Med. Sci. **60**(10), 1304–1309 (2005)
8. Studenski, S.: Gait speed and survival in older adults. JAMA **305**(1), 50 (2011)
9. Verghese, J., Wang, C., Holtzer, R.: Relationship of clinic-based gait speed measurement to limitations in community-based activities in older adults. Arch. Phys. Med. Rehabil. **92**(5), 844–846 (2011)
10. Vanacken, L., et al.: Game-based collaborative training for arm rehabilitation of MS patients: a proof-of-concept game, March 2011

11. Sousa, M., Vieira, J., Medeiros, D., Arsenio, A., Jorge, J.: SleeveAR: augmented reality for rehabilitation using realtime feedback, pp. 175–185 (2016)
12. UW Design 2016 | TrainAR
13. Backlund, P., Hendrix, M.: Educational games - are they worth the effort? A literature survey of the effectiveness of serious games. In: 2013 5th International Conference on Games and Virtual Worlds for Serious Applications (VS-GAMES), pp. 1–8 (2013)
14. Morelli, T., Foley, J., Folmer, E.: Vi-bowling: a tactile spatial exergame for individuals with visual impairments. In: Proceedings of the 12th International ACM SIGACCESS Conference on Computers and Accessibility, New York, NY, USA, pp. 179–186 (2010)
15. Brach, M., et al.: Modern principles of training in exergames for sedentary seniors: requirements and approaches for sport and exercise sciences, vol. 11 (2011)
16. Korn, O., Tietz, S.: Strategies for playful design when gamifying rehabilitation: a study on user experience. In: Proceedings of the 10th International Conference on PErvasive Technologies Related to Assistive Environments, New York, NY, USA, pp. 209–214 (2017)
17. McNevin, N.H., Wulf, G., Carlson, C.: Effects of attentional focus, self-control, and dyad training on motor learning: implications for physical rehabilitation. Phys. Ther. **80**(4), 373–385 (2000)
18. Bassett, S.: The assessment of patient adherence to physiotherapy rehabilitation, vol. 31 (2003)
19. Brumels, K.A., Blasius, T., Cortright, T., Oumedian, D., Solberg, B.: Comparison of efficacy between traditional and video game based balance programs, vol. 62 (2008)
20. Martin, A.L., Götz, U., Müller, C., Bauer, R.: 'Gabarello v.1.0' and 'Gabarello v.2.0': development of motivating rehabilitation games for robot-assisted locomotion therapy in childhood. In: Games for Health 2014, pp. 101–104. Springer Vieweg, Wiesbaden (2014)
21. van Reine, P.P.: The culture of design thinking for innovation. J. Innov. Manag. **5**(2), 56–80 (2017)
22. Serrat, O.: "Design Thinking", in Knowledge Solutions, pp. 129–134. Springer, Singapore (2017)
23. McFarlane, D.C., Wilder, S.M.: Interactive dirt: increasing mobile work performance with a wearable projector-camera system. In: Proceedings of the 11th International Conference on Ubiquitous Computing, New York, NY, USA, pp. 205–214 (2009)
24. Zhou, H., Hu, H.: Human motion tracking for rehabilitation—a survey. Biomed. Sig. Process. Control **3**(1), 1–18 (2008)
25. Shneiderman, B.: Designing the User Interface: Strategies for Effective Human-Computer Interaction, 5th edn. Addison-Wesley, Boston (2010)

Method Cards – A New Concept for Teaching in Academia and to Innovate in SMEs

Christian Zagel$^{(\boxtimes)}$, Lena Grimm, and Xun Luo

University of Applied Sciences Coburg,
Friedrich-Streib Strasse 2, 96450 Coburg, Germany
{Christian.Zagel,Xun.Luo}@hs-cobrug.de,
Lena.Grimm@stud.hs-coburg.de

Abstract. The world of academic teaching is currently characterized by learning material in the form of books and lecture notes. Constantly renewing learning content, new concepts, and innovations require a new and more flexible knowledge base. Numerous initiatives in the e-learning area approach the issue through renewable digital content. Nevertheless, students in the fast changing VUCA (versatile, uncertain, complex, ambiguous) world demand for innovative approaches that focus on imparting competencies in addition to traditional knowledge. This paper presents the concept and prototype of a new blended-learning approach to foster creativity and innovation: the "Method Cards". We use the well-known format of traditional playing cards to create learning modules, e.g. representing trends, technologies, or methods. Additional content is linked through integrated NFC tags and QR codes. We additionally present the first results of two user studies conducted amongst Master students as well as in a business environment.

Keywords: Gamification · Innovation · Blended learning

1 Introduction

Teaching academia nowadays focuses more and more on delivering competencies. With the internet, allowing to easily access knowledge in huge databases it is now important to be able to process the information in the right way [1]. This especially is true for applications in innovative academic classes that teach, e.g., various forms of innovation management. The students are taught how to apply trends, technologies, and innovation methods that, afterwards should find application in the companies they work in.

Nevertheless, especially small and medium enterprises (SMEs) of the manufacturing sector face enormous challenges in strategically applying these methods. Also, research in the area of innovation management by now mainly focuses on large enterprises [2]. These issues can be seen in Upper Franconia, Germany which represents Europe's second largest industrialization density [3]. This area is characterized by an enormous amount of traditional, family-owned SMEs, often also known as hidden champions in their profession. Unfortunately, it is these industries that believe in their core values and the processes performed since hundreds of years. Hence, they might in

© Springer International Publishing AG, part of Springer Nature 2019
T. Z. Ahram (Ed.): AHFE 2018, AISC 787, pp. 230–241, 2019.
https://doi.org/10.1007/978-3-319-94229-2_22

future also be intimidated by the faster moving world and topics like digitalization and globalization. Even though driving innovations is regarded as very important, innovation management - if existing at all – is matter of the boss, who in many cases also represents the owner of the enterprise [4]. This offers a huge potential for developing not only a new teaching method, but also a solution to start implementing innovation methods in small enterprises.

This paper presents a new and innovative method that brings together academic research and the needs of regional small and medium enterprises in a playful learning concept. The so called "Method Cards" represent a physical card game that can be used for both teaching in academia as well as for innovation workshops in small businesses. We present thee initial card decks, representing topics also taught in innovation management courses at the university: innovation methods, technologies, and trends. After a theoretic introduction, we present the development of the first design prototypes as well as a concept evaluation conducted using the User Experience Questionnaire (UEQ) [5, 6] as a theoretical basis. The goal is to combine the idea of traditional playing cards with knowledge transfer and a motivation to collect. The proposed method will also allow to consequently update the knowledge base by adding new cards or even complete card decks. While this paper focuses on the basic idea of the method and the general design of the cards, future research will also include gaming structure and an evaluation of the learning effects in more detail.

2   Theoretic Background

2.1   Gamification

A method gaining attention since several years, especially in relation to its positive effects on user motivation and efficiency, is gamification [7]. Gamification describes the application of game-type elements in non-game contexts with the goal of increasing user motivation and the creation of an enhanced user experience [8]. Hence, applying the concept promises an increase in productivity while simultaneously improving user satisfaction [9].

The method is applied in many ways in organizational contexts. Nevertheless, literature and practice still demand additional practical examples that prove its efficiency [9, 10]. While previous studies could prove a positive influence of gamification towards motivation and engagement in everyday processes [11], the payment process at supermarket self-service checkouts has not yet been investigated. Most attempts to gamify the shopping experience have been done in online shopping or during the shopping process inside shops. The website Groupon, for example, implemented the so-called "SOS mechanic", where a certain number of people needed to order a service to get a cheaper price [12]. This way the company motivated their customers to additionally market their promotions and to motivate others in also ordering them. Lounis et al. [13] examined the use of gamification to promote a sustainable buying behavior for fast moving consumer goods. They found that customers, in general, are willing to participate in gamified services. Nevertheless, the results cannot be generalized, as different personality traits demand different ways of implementing gamification elements. The authors even state that a gamification scheme needs to be customizable and personalized to become efficient.

Gamification has also been found to enhance the retail experience in online shopping. Insley and Nunan [14] proposed to adjust the way of integrating game elements into the shopping experience in dependence of the customers' shopping task (e.g., recreational vs. functional shopping). They found that gamification can influence consumer behavior, e.g., reducing undesirable actions as the misuse of policies. Niels and Zagel [15] also share an example of gamified self-service checkouts and clarify how the method influences user motivation.

A concept linked to gamification but focusing more on a utilitarian value regarding education and training is the one called Serious Games [16]. Studies show a significant increase in the positive attitude towards learning and better learning results if content is taught through Serious Games techniques. Nevertheless, Korn [17] also mentions the challenges of integrating valuable educational content into game scenarios.

2.2   Blended Learning

The term "Blended Learning" describes a combination of physical presence learning with additional digital/electronic content [18]. This might, for example, include computer-based trainings in labs. Currently more and more different teaching forms are mixed to mutually compensate the drawbacks of single teaching forms. An important component of Blended Learning is self-studying [19]. The strategy of combining physically available information with additional digital content plays an important role in the subsequent development of the Method Card concept presented in this paper. While the paper at hand focuses on the design and layout of the physical cards, the idea of linking to digital content needs to be considered in the prototyping phase.

2.3   Existing Card Games and Methodologies

An evaluation in form of a literature review as well as a laboratory study was conducted regarding existing card games and their potentials for knowledge delivery, gaming, and if they might be collected or not. For the use case at hand, especially the more traditional card games are of interest. Therefore, we identified the most well-known card types and rated them regarding the before-mentioned aspects (see Table 1).

Table 1. Card game research (average rating, + existing, − not existing, o partly existing).

Card game	Gaming incentive	Knowledge transfer	Collectible
Happy Families	+	+	o
Poker	+	−	−
Pokémon	+	o	+
Trump	+	+	o
Learning Quiz	+	+	o

The evaluation was conducted amongst eight subjects, each providing their opinion on the implementation of each aspect in the respective game types. The goal of this evaluation was to identify an existing card game as a potential starting point for the development of our method cards. Obviously, none of the card games (see Fig. 1) can address all three investigated aspects. This leads to the conclusion that a new card game needs to combine the elements of multiple existing systems.

Fig. 1. Investigated Card Deck Types. From left to right: Trump [20], Pokémon [21], Poker [22], Learning Quiz [23], Happy Families [24]

3 Method Cards Design Prototype

The previously discussed background research serves as a basis for the development of our method cards. The goal is to use them in academic teaching as well as for conducting innovation workshops in small companies. A brainstorming session amongst 20 students and four professors conducted in December 2017 lead to several aspects that need to be considered for creating the first card decks. These are:

- **Design:** This aspect includes not only arrangement of the print on the cards but also their physical shape and size. Aspects like logos, color, or text size need to be considered as well.
- **Information:** Starting off with a blank card, space available for including information is rare. Hence, it needs to be decided, which gaming elements and which information on the respective method, trend, or technology should be included. Also, it needs to be defined, if content should be included only on one or on both sides.
- **Game type:** The content shown on the cards should on the one side provide enough information on the methods, technologies, or trends. On the other hand specific elements (KPIs, card numbers) need to allow the development of a game structure.
- **Digital content:** To overcome the size limitation of a playing card, the potentials of blended learning should be used by leveraging the functionalities of QR codes or NFC tags. These would link the user to additional and digital content.
- **Topics:** The core idea of the method cards is to provide a broad overview about topics relevant for innovation management. These might include innovation methods, trends, and technologies, but also more specific topics around, e.g., ethics,

team dynamics, or leadership. These topics may also focus on the specific needs of the local SMEs.

For the first prototypes, it was decided to develop three more general card decks that can be used in innovation workshops independent of a respective industry, representing basic knowledge taught in an innovation management: innovation methods, technologies, and trends. The gaming type applied in our method card system is still undefined. Nevertheless, the existing "Happy Families" card design is found to be a good example and well suited as an initial starting point. It represents the best realization of the investigated mixture of gaming incentive, knowledge transfer, and the possibility to collect which basically represents the general goal of our concept. Hence, our card decks will (comparable to "Happy Families") consist of 52 individual cards, each representing one individual method, trend, or technology. As many teaching guides mention the importance of visual aspects when it comes to a potential learning success, we initially focus on the general design of our method cards.

3.1   Shape

The shape of playing cards is limited by several factors. It is important to focus on a high level of (physical) usability. This can be done by users associating the final shape with existing game cards they might know. The size also defines the space available for content and images. Also, too much information limits clearness and consequently the learning effects.

Most of the card types have a quite similar format when it comes to size and shape. They have been analyzed and compared (see Fig. 2). All of them show rounded corners. While the Trump cards are the biggest, they also provide most information.

Fig. 2. Comparison of Card Formats (Trump 6.2 cm × 10 cm, Happy Families 6 cm × 9 cm, Poker 5.6 cm × 10 cm)

The narrow-elongated shape of the cards reviewed allow the user to hold many cards in one hand simultaneously. Their rectangular form allows keeping an overview about parts of the card content without completely covering the motifs. The rounded corners protect from damages and make it easier to shuffle, if needed.

To decide for the best suitable format for the method cards, paper prototypes in different sizes have been created and tested by 10 independent subjects in a laboratory study. The goal was to evaluate general appearance and handling. All the prototypes had an identical color and thickness and did not include any content. In addition to sizes of popular cards as shown above, we added a new format of 7 cm to 10.5 cm. The reason behind was that a larger card format would allow the integration of extended content. Results show that while a format of 7 cm to 10.5 cm is preferred by 9 of 10 subjects, no differences were mentioned regarding the handling of the cards. This format is used in the further proceeding of the card creation.

3.2   Color

Extensive research has been conducted in the area of color psychology. The learning platform "Diplomero", for example, mentions that colors arouse emotions and associations through which learning effects might be enhanced and intensified [25]. It is furthermore proposed to refrain from using neon colors, bright colors, as well as the arbitrary application of too many colors. In addition, aspects like differences in color perception amongst different cultures may also play an important role. More than 180 million people are color blind [26]. Hence, a newly developed card game should be created in a way that sparsely uses colors in an intelligent way.

3.3   Design and Layout

Using the aforementioned results, additional prototypes were created including card content and coloring. The goal was to cover the following requirements: (1) Multiple categories should be covered that are also represented in the course outline of the innovation methods of the university. (2) For each category multiple subtopics/methods should be realized and briefly described on the cards. (3) To arouse interest in various topics and methods and to verify suitability for playing, each card should also include certain KPIs to value, for example, innovational strength, previous knowledge required, complexity or manpower needed for a successful application of the method in a business environment. (4) In addition, it should be possible to individually identify each of the cards. The given ID represents the category and a collective number.

These aspects were applied in the prototypic design of the cards that, after several iterations, lead to the visualization shown in Fig. 3. This design was also used for a further evaluation.

4   Prototype Evaluation and Results

After finalizing the general layout of the method cards, their design and the overall concept was evaluated. To conduct the evaluation, multiple small card decks covering the categories "Innovation Methods", "Trends & Evolutions", "Technologies", "Communication", and "Management" were developed. The goal of developing an

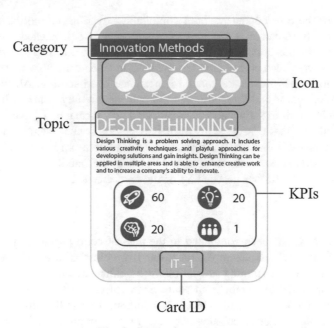

Fig. 3. Method card layout

initial set of different card decks was to show the easy extensibility of the concept by adding additional cards to one card deck or even adding completely new categories (see Fig. 4). The addition of digital content through integrated NFC tags inside or printed QR-codes on the back of the card was only verbally mentioned.

Fig. 4. Card decks used for the evaluation

Multiple methods were applied in the evaluation among 30 subjects (students, professors, externals). To receive as much feedback on the concept as possible, qualitative and quantitative research methods were combined. As an explorative evaluation method, we applied the feedback-capture-grid [27], a method often used in Design Thinking research. In this method subjects have the possibility to openly provide feedback in the categories "Things that I liked most", "Things that could be improved", "Things that I don't understand", and "New ideas to consider".

Feedback shows that the subjects liked the small format of the cards as well as the combination of a valuable knowledge base with a game approach. Also, the modern design including the chosen icons were positively mentioned along with text design (black text on white background). Nevertheless, and while some of the respondents positively noted the color coding of the different categories, others were asking for brighter colors. Interestingly this aspect stands in contrast with state of the art research [25]. Respondents criticized font size and the proposed given categories of the KPIs (which is also the only aspect responded as not understood). They need to be identical along the different card decks for being able to mix them. The subjects also demanded for using thicker paper for increasing the perceived quality and durability of the cards. Some also asked for more detailed content printed on the cards. New ideas were limited to the content provided. While some asked for less text to increase clearness, others demanded for additional information.

Following the qualitative evaluation, a quantitative test was conducted using the User Experience Questionnaire by Laugwitz et al. [5, 6] as a basis. In its original version, it is used to measure user experience, focusing on perceived attractiveness, quality of use, and design quality. Originally being used for the evaluation of software systems we transfer the general idea of its application to our physical use case by adding aspects of interest and our own categorization to capture feedback on the subjects' first impression, design, content & text, as well as the shape of the cards. The results are shown in Fig. 5.

The questionnaire consists of word pairs of contrasting attributes that may apply to the tested process, system, or software. The items are arranged in the form of a seven-stage semantic differential. We intentionally did not sort e.g. positive expressions to one specific side of the table as for the given use case some aspects need to be interpreted individually. In general, the results are comparable to the qualitative ones. The categories "first impression" and "design" show a slight positive rating over all elements, detailed analyses of the "content and text" subject show divided opinions. This verifies the verbal feedback already gathered through the feedback capture grid. Consequently, the amount of text as well as the general information provided need to be reworked. Positively, the shape and general form of the cards is well accepted, as many of the subjects compare the cards to the card games they already know.

5 Conclusions and Future Research

The development and evaluation process of the method card design showed possibilities for a further improvement of their design and structure. The feedback provided through the qualitative and quantitative evaluations was consequently used to update the design of the cards. This not only included modifications in font size, but also a re-arrangement of the content. Coloring was not modified, as it was not mentioned as a clearly negative aspect. In addition, the category KPIs are harmonized, keeping the respective logos. This resulted in the new design shown in Fig. 6.

First Impression	1	2	3	4	5	6	7	
1 conventional				●				original
2 outdated					●			modern
3 unobtrusive				●				obtrusive
4 refreshing		●						narcotic
5 boring					●			exciting
6 not interesting					●			interesting
7 diversified			●					monotonous
8 pleasant		●						unpleasant
9 of low quality					●			of high quality
10 tidy			●					messy
11 good		●						bad
12 attractive		●						unattractive
13 intuitive		●						counter-intuitive
Design								
14 obstruictive						●		supportive
15 conventional				●				inventive
16 creative		●						dull
17 confusing					●			clear
18 conservative				●				innovative
19 good	●							bad
20 non-supportive content		●						supportive content
21 neutral				●				playful
22 appropriate		●						inappropriate
Content & Text								
23 good to read			●					bad to read
24 complicated				●				easy
25 clear				●				confusing
26 interesting	●							not interesting
27 easy to learn			●					hard to learn
28 good		●						bad
Shape								
29 convenient	●							impractical
30 conservative			●					inventive
31 meets expectations			●					does not meet expectations
32 good	●							bad
33 handy	●							bulky

Fig. 5. Questionnaire results

The resulting design will be used for our future research. It will be used to develop the gaming method as well as to evaluate learning effects. The results might lead to another revision of the design. After the development of three complete card decks (innovation methods, trends, technologies), their application will be tested in three different scenarios: a master course for innovation management covering 70 students, an innovation workshop in form of a hackathon, as well as in a business setting in a

Fig. 6. Design prototype II

SME. The application in form of a card game will be tested separately. By then, each of the cards will furthermore act as the physical "portal" to additional digital content by integrating the beforementioned QR-codes and NFC tags, linking to a digital platform. These upcoming evaluations will therefore act as a proof of concept for a future rollout and commercialization.

In future studies, we will also be able to evaluate learning effects in comparison to traditional methods in education as well as in business settings. The integration of gamification elements into the platform offering the digital content might further enhance these effects.

References

1. Hoffmann, M., Loeffl, J., Luo, X., Thar, W., Valeva, M., Zagel, C.: zukunftsdesign - offen. innovativ.machen. In: Krahl, J., Loeffl, J. (eds.) Zwischen den Welten 11, Cuvillier Verlag Göttingen (2017)
2. Terziovski, M.: Innovation practice and its performance implications in small and medium enterprises (SMEs) in the manufacturing sector: a resource-based view. Strateg. Manag. J. **31** (8), 892–902 (2010)
3. Hunger, J.: BF/M – 25 Jahre Partner des Mittelstandes. In: Mittelstand im Fokus. Deutscher Universitätsverlag, pp. 13–14 (2004)
4. IHK für Oberfranken, Bayreuth: Studie 2015. Betriebliches Innovationsmanagement in der Region Oberfranken (2015)

5. Laugwitz, B., Held, T., Schrepp, M.: Construction and evaluation of a user experience questionnaire. In: Proceedings of the HCI and Usability for Education and Work, pp. 63–76 (2008)
6. Schrepp, M., Hinderks, A., Thomaschewski, J.: Applying the user experience questionnaire (UEQ) in Different evaluation scenarios. In: Marcus, A. (ed.) Design, User Experience, and Usability. Theories, Methods, and Tools for Designing the User Experience, pp. 383–392. Springer, Switzerland (2014)
7. Gamification. http://www.gamification.co
8. Deterding, S., Sicart, M., Nacke, L., O'Hara, K., Dixon, D.: Gamification: using game-design elements in non-game contexts. In: CHI 2011 Extended Abstracts on Human Factors in Computing Systems, pp. 2425–2428. ACM, New York (2011)
9. Gonzales-Schaller, P.: Trendthema Gamification: Was steckt hinter diesem Begriff? In: Diercks, J., Kupka, K. (eds.) Recrutainment, pp. 33–51. Springer, Wiesbaden (2013)
10. Hamari, J., Koivisto, J., Sarsa, H.: Does gamification work? – a literature review of empirical studies on gamification. In: Proceedings of the 47th Hawaii International Conference on System Science, pp. 3025–3034 (2014)
11. Neeli, B.K.: A method to engage employees using gamification in BPO industry. In: Third International Conference on Services in Emerging Markets Services in Emerging Markets (ICSEM), pp. 142–146. IEEE (2012)
12. Zicherman, G., Cunningham, C.: Gamification by Design: Implementing Game Mechanics in Web and Mobile Apps. O'Reilly, Cambridge (2011)
13. Lounis, S., Neratzouli, X., Pramatari, K.: Can gamification increase consumer engagement? A qualitative approach on a green case. In: Douligeris, C., Polemi, N., Karantjias, A., Lamersdord, W. (eds.) Collaborative, Trusted and Privacy-Aware E/M-Services, pp. 200–212. Springer, Heidelberg (2013)
14. Insley, V., Nunan, D.: Gamification and the online retail experience. Int. J. Retail Distrib. Manag. **42**, 340–351 (2014)
15. Niels, A., Zagel, C.: Gamified self-service checkouts: the influence of computer-related causal attributions on user experience and motivation. In: Ahram, T., Karwowski, W. (eds.) Advances in The Human Side of Service Engineering, pp. 24–36. Springer, Heidelberg (2017)
16. Michael, D.R., Chen, S.L.: Serious Games: Games That Educate, Train, and Inform. Course Technology PTR, Manson (2006)
17. Korn, O.: Serious game design: Potenziale und Fallstricke bei der spielerischen Kontextualisierung von Lernangeboten. In: Metz, M., Theis, F. (eds.) Digitale Lernwelt – Serious Games: Einsatz in der beruflichen Weiterbildung, Bielefeld, pp. 15–26 (2011)
18. Zehetmaier, S.: Blended Learning – Eine Lernmethode mit Erfolgsgarantie, p. 2. GRIN Publishing (2007)
19. Moriz, W.: Blended-Learning: Entwicklung, Gestaltung, Betreuung und Evaluation von E-Learninggestütztem Unterricht, pp. 21–22. Books on Demand (2008)
20. Liddell, T., van Straefen, J., Tschiggerl, M.: Top Trumps. Raubtiere. Winning Moves Deutschland GmbH, Duesseldorf (2011)
21. Nintendo: Pokèmon. Schnapp sie dir alle. Nintendo of America: GAMEFREAK (1995)
22. Altenburger: Rommé Canasta Bridge. Die echten Altenburger Spielkarten. Französisches Clubbild. Spielefabrik Altenburg GmbH
23. Moses: Pocket Quiz. Politik und Geschichte. Moses. Verlag GmbH, Kempen (2007)
24. Ravensburger: Supertrumpf. Die Legenden. Ravensburger Spielverlag, Ravensburg (1999)
25. Diplomero: Der korrekte Einsatz von Farben im Unterricht. https://www.diplomero.com/de/ratgeber/der-korrekte-einsatz-von-farben-im-unterricht.html. Accessed 10 Nov 2017

26. Eyebizz.de: 180 Millionen Menschen sind farbenblind. Ursachen und Wirkungen der Rot-Grün Schwäche. Ebner Verlag GmbH & Co KG, Ulm (2016)
27. Lewrick, M., Link, P., Leifer, L.: Das Design Thinking Playbook: Mit traditionellen, aktuellen und zukünftigen Erfolgsfaktoren. Vahlen, pp. 35–36 (2017)

Correlations Between Computer-Related Causal Attributions and User Persistence

Adelka Niels[1], Sophie Jent[1], Monique Janneck[1],
and Christian Zagel[2(✉)]

[1] Faculty of Electrical Engineering and Computer Science, Lübeck University of
Applied Sciences, Mönkhofer Weg 239, 23562 Lübeck, Germany
{Adelka.Niels,Sophie.Jent,
Monique.Janneck}@fh-luebeck.de
[2] ZukunftsDesign, Coburg University of Applied Sciences,
Friedrich-Streib-Strasse 2, 96450 Coburg, Germany
Christian.Zagel@hs-coburg.de

Abstract. This study used data collected from 2270 participants to investigate the impact of computer-related causal attributions on users' persistence. Attribution theory deals with subjectively perceived causes of events and is commonly used for explaining and predicting human behavior, emotion, and motivation. Individual attributions may either positively or negatively influence one's learning behavior, confidence levels, effort, or motivation. Results indicate that attributions indeed influence users' persistence in computer situations. Users with favorable attribution styles exhibit greater levels of persistence than users with unfavorable attribution styles. The findings can be used in HCI research and practice to understand better why users think, feel, or behave in a certain way. It is argued that an understanding of users' attributional characteristics is valuable for developing and improving existing computer learning training strategies and methods, as well as support and assistance mechanisms.

Keywords: Human factors · Applied cognitive psychology
Computer-related causal attributions · User persistence · User motivation

1 Introduction

Attribution theory deals with causal explanations people find for successful and unsuccessful outcomes and how they influence individuals' behavior, emotion, and motivation [1, 2]. This paper contributes to HCI research and practice by applying attribution theory, which to date has not received ample attention in the HCI community (e.g., [3]), although it is one of the most influential bodies of research of social psychology in the last 50 years [4]. For example, prior research has shown, that the users' computer-related problem-solving motivation depends on the attribution style. Users with favorable attribution styles exhibit greater levels of motivation in problem handling than users with unfavorable attribution styles [5].

To our knowledge, the impact of attributions on the users' persistence in computer situations, in particular in success situations, has not been investigated yet. However, a

© Springer International Publishing AG, part of Springer Nature 2019
T. Z. Ahram (Ed.): AHFE 2018, AISC 787, pp. 242–251, 2019.
https://doi.org/10.1007/978-3-319-94229-2_23

deeper understanding of how Causal Attributions impact users' persistence can help to design systems that fit their users' need better.

Persistence, as a personality trait, is often associated with stubbornness and perfectionism. It is often seen to be the prime ingredient in success in many pursuits such as athletics, academics, business, etc. Especially in a technologized world like ours, it is important to exhibit a sufficient level of persistence, for example, to handle even complex computer-related tasks. But why are some people more persistent in computer-related activities than others? Are personality traits like causal attributions responsible for persistence? These are the questions this paper will attempt to answer.

Persistence was often studied in terms of cultural differences. For instance, in a study conducted by Blinco (1998), it was found that American students are less persistent in learning than their Japanese counterparts. However, gender and school type could be excluded as influence factors [6]. Another study by Heine (2001) tested cultural differences between American and Japanese subjects on responses after success or failure situations regarding their task persistence [7]. The study confirmed that the Japanese subjects were more persistent in post-failure situations than their American counterparts. As a reason for the result, it was speculated that the Japanese subjects are more likely to attribute the cause of the failure to themselves, while Americans were more likely to believe that external factors caused the failure. Because Japanese tend to see themselves as the cause of the failure, they rather believe they could also solve the problem themselves. A further study among Western European and Chinese participants has shown that Chinese users (57%) are more likely to attribute computer-related success to external circumstances than Western Europeans (24%) [8]. The authors explain the external attribution patterns in success situations as a form of modesty that gives credit to others for their impact on individual success. These cultural studies hinted that computer-related task persistence may be predictable based on attribution style.

In this paper, persistence refers to the willingness to exert large amounts of effort over long periods of time to achieve a computer-related a goal. The goal of this study is to explore the relationship between different computer-related attribution patterns and the persistence of computer users. Strategies for avoiding and reframing negative computer experiences are also considered.

2 Theoretical Background and Related Work

In attribution research, a distinction is made between internal and external causes (locus) perceived by the individual. For example, a person may either feel responsible for a positive or negative outcome (internal) or relate it to external circumstances [9]. Three further dimensions are distinguished: Stability, controllability, and globality [1]. Causes are considered as stable, i.e. persistent over time or as unstable and singular. Furthermore, causes can be perceived as controllable or uncontrollable, as well as generally taking effect (global) or only applicable to a certain (specific) situation [10].

Stable attribution patterns which are present in a wide range of situations are called Attribution Styles. Originally, attribution styles derive from clinical psychology to explain and predict depression. Persons with a pessimistic attribution style tend to blame themselves when things don't go right (e.g., "it was my fault") and will not take credit for success, (e.g., "I was just lucky"). Contrary, persons with an optimistic style rather take credit for success and do not put the blame on themselves for things that go wrong (cf. [11–13]). Overall, people with an optimistic attribution style are more likely to succeed [14].

Research has shown that attributions are domain specific (e.g., [1, 15]) and therefore attribution patterns reported in other research areas may not represent the perceptions of computer users. They may even be completely unsuitable for an application in the HCI domain. Moreover, the application and theoretical testing of attribution theory is fairly young in the field of HCI research, compared to other disciplines. Nevertheless, it has already received some recognition and was found to be relevant in some HCI research issues. For example, it has been applied to explain computer system adoption [16], effects on users' evaluations of system quality [17], post-training reactions to and performance of computer systems [18], development of strategies to overcome computer anxiety [19], course performance [14], and satisfaction [20]. For a detailed review of attribution theory in the context of HCI, see [3]. Interestingly, the effects of causal attributions on users' persistence in computer situations have not been researched in detail yet.

Current research on attribution theory in the field of HCI tends to cluster people with regard to their computer-related attributions and developed a typology of six central computer-related attribution styles, three styles each for situations of success and failure. Similar to clinical psychology, optimistic styles (characterized by a feeling of control toward the technical systems) and pessimistic styles (marked by feelings of helplessness and resignation), as well as more 'neutral' styles, were found [21]. For situations of success, the Confident, the Realistic, and the Humble style were identified. Persons with a Confident style may explain their computer-related successes as "I am competent and responsible for my own success". They tend to attribute success to internal, stable, controllable, and global causes. Persons with a Realistic style expect that "Sometimes I am successful, sometimes not". They attribute the reasons for success rather temporally unstable and situation-related. For persons with a Humble style, the explanation is "This time I was lucky". They attribute success to external factors and experience only low levels of control when using computers [21]. For situations of failure, the Confident, the Realistic, and the Resigned styles were found. Persons with a Confident style reckon "I know it was my fault, but next time I will do better". They have high internality values and feel responsible for their failures, but also feel in control of the situation. For persons with a Realistic style, the explanation is "This time I failed, but I don't worry about it". They see internal as well as external reasons for failures and believe that they change over time and depend on a specific situation. Finally, if a failure occurs, persons with a Resigned style might feel "I never understand what computers do". They see external and temporally stable reasons for their failure and feel they have little control over the situation [21]. We will build on this typology of computer-related attribution styles in our study.

3   Methodology

3.1   Sample

A total of 2270 persons participated in this study (50.7% female and 49.3% male). The mean age was 42.21 years (Median = 42, SD = 13.13 years, range: 18–79). The general level of education was quite balanced, ranging from "without a school-leaving qualification" up to "university degree". They subjectively self-assessed their computer skills on a 7-point Likert-type scale ranging from 1 (low) to 7 (expert) on average at 5.25 (Median = 5, SD = 1.378, range: 1–7). In order to provide a well-balanced sample, participants were paid and recruited via an online research panel.

3.2   Measures

Attribution Questionnaire. The Attribution Questionnaire is an established and validated questionnaire to determine users' causal attributions in the field of HCI [22, 23]. The instrument includes hypothetical depictions of events, five addressing positive outcomes (success) and five addressing negative outcomes (failure). Sample events included, "Imagine you are working on a foreign computer. It is very easy for you to adapt to the new and unknown user interface." (success) and "Imagine while creating a document with the computer, you delete a text page. You are not able to recover this page." (failure). The perception of each event is rated on the four attributional dimensions of locus, stability, controllability, and globality. These dimensional items are answered on a 7-point Likert-type scale. Table 1 shows an excerpt from the English version of the dimensional items for failure situations. The items measuring attributions in situations of success are worded analogously. The construct allows to examine attributional dimensions separately, but also to determine overall attribution styles by using cluster analyses.

Table 1. Excerpt from the attribution questionnaire to measure the attributional dimensions in failure situations [22, 23].

What caused the breakdown?		
I would locate the cause of the breakdown…		
internally (I am to blame)	1 2 3 4 5 6 7	externally (the system is to blame)
The cause of the breakdown is…		
a singular event	1 2 3 4 5 6 7	recurring
The cause of the breakdown is…		
controllable	1 2 3 4 5 6 7	uncontrollable
The cause of the breakdown is likely to promote other breakdowns…		
just in this situation	1 2 3 4 5 6 7	in other situations as well

Users' Persistence Questionnaire. The *Achievement Motivation Inventory* (AMI) is an established personality inventory designed to measure the key dimensions that are addressed in different motivation theories. It is founded on the theoretical work related to the German *Leistungsmotivationsinventar* (LMI) [24] and contains 17 dimensions or "performance orientations" measured with 10 items each: compensatory effort, competitiveness, confidence in success, dominance, eagerness to learn, engagement, fearlessness, flexibility, flow, goal setting, independence, internality, persistence, preference for difficult tasks, pride in productivity, self-control, and status orientation. The items to be responded by participants on a 7-point-Likert scale ranging from "strongly disagree" to "strongly agree". The evaluation has a total value or score, as well as dimension-specific scores. It is possible to consider both, the values for each dimension separately or the total value across all 17 dimensions. However, in this work, only the dimension "Persistence" was considered and the items were slightly modified to fit the topic of HCI. For example, "When I'm working on the computer, it's hard for me to keep up my efforts for a long time" instead of the more general statement "It's hard for me to keep up my efforts for a long time". Moreover, to improve reliability, three items with poor selectivity ($r_{it} < .30$) were eliminated. Table 2 shows the final English version of the questionnaire items.

Table 2. Adapted persistence questionnaire items and results. Items denoted by * are inversely coded. Mean values and standard deviations for items and overall scale.

Item	Mean	SD
When I am determined to do something on the computer, and I don't succeed, then I do everything I can to still accomplish it.	5.75	1.26
When I'm working on the computer, it's hard for me to keep up my efforts for a long time.*	4.83	1.68
I could accomplish more on the computer if I did not fatigue that fast.*	4.80	1.74
When working on the computer, there is hardly anything that could distract me.	4.11	1.59
When working on the computer, it's hard for me to concentrate for a long time.	4.69	1.72
If something goes wrong, while working on the computer, I give up quickly. *	5.46	1.50
When working on the computer I find it hard to focus my attention on what I am doing.*	4.86	1.63
Overall scale	4.73	0.75

3.3   Procedure

In the online survey, the participants were presented with ten hypothetical events (five success and five failure situations) to measure attributions (See Sect. 3.2 – Attribution Questionnaire). The participants were instructed to imagine the respective situations as realistically as possible and to assess the cause of each situation on the four attributional dimensions of locus, stability, controllability, and globality (Table 1). Furthermore, they were asked to fill out the Persistence questionnaire. In addition, measures of demographics and computer experience were administered.

4 Results

K-means clustering was used to classify the attribution data (attributional dimensions) into existing clusters and to determine the attribution styles for each participant. Clusters identified in prior studies [21] served as the basis for classification (cf. Sect. 2).

The distribution of the individual clusters is relatively balanced for success situations. Cluster analysis revealed 736 with a Confident, 882 with a Humble, and 652 with a Realistic attribution style. In failure situations, merely 405 attributed in a Confident style, while 962 attributed in a Resigned, and 903 in a Realistic style. Table 3 shows the mean values for the six clusters. ANOVAs were calculated showing significant differences between clusters.

Table 3. ANOVA results for success and failure clusters.

Success	Confident	Realistic	Humble	F value	p	η^2
Locus	1.96	2.43	4.16	150.133	<0.001***	0.117
Stability	6.23	3.99	4.59	84.731	<0.001***	0.070
Controllability	1.71	2.54	3.99	173.687	<0.001***	0.133
Globality	5.71	3.25	4.22	69.004	<0.001***	0.057
Failure	Confident	Realistic	Resigned	F value	p	η^2
Locus	2.61	4.23	4.37	34.061	<0.001***	0.029
Stability	2.81	3.45	4.80	76.784	<0.001***	0.063
Controllability	2.34	3.29	4.21	65.270	<0.001***	0.054
Globality	2.68	3.14	4.63	113.152	<0.001***	0.091

*** = $p \leq 0.001$

4.1 Persistence Questionnaire

In a first step, inversely coded variables were inverted. Higher values on the overall scale as well as the subscales indicate a higher willingness to exert large amounts of effort over long periods in order to reach a computer-related goal. The results show that the overall persistence of the participants is slightly above average (Table 2). Reliability (Cronbach`s alpha) for the total score is $\alpha = .822$.

4.2 Correlation Analysis

Attribution styles and user persistence were tested globally for differences followed by post-hoc tests (LSD) for pairwise comparison. Because of non-normally distributed data the Kruskal-Wallis-Test was used instead of analyses of variance.

Kruskal-Wallis tests revealed significant differences concerning situations of success and failure (Table 4). Post-hoc tests results show that users with favorable attribution styles exhibit more persistence in computer-related tasks. Table 5 shows the results of the post-hoc test and the user persistence mean values for each attribution style.

Table 4. Relations between attribution styles and user persistence in situations of success and failure - results Kruskal-Wallis test.

	Chi^2	df	p
Success	257.782	2	<0.001***
Failure	125.350	2	<0.001***

*** = p $\leq$ 0.001

Table 5. Relations between attribution styles and user persistence - post-hoc test (LSD).

Success	Mean persistence		p
Confident	5.06	Realistic	<0.001***
Realist	4.74	Humble	<0.001***
Humble	4.46	Confident	<0.001***
Failure			
Confident	4.89	Realistic	0.772**
Realist	4.88	Resigned	<0.001***
Resigned	4.52	Confident	<0.001***

** = p $\leq$ 0.01; *** = p $\leq$ 0.001

In situations of success, the analysis showed significant differences between persons with the Confident and the Realistic styles (p < 0.001; M = 5.06 vs. M = 4.74), between persons with the Realistic and the Humble styles (p < 0.001; M = 4.74 vs. M = 4.46), as well as between persons with the Humble and the Confident styles (p < 0.001; M = 4.74 vs. M = 5.06). Persons with the more favorable attribution styles exhibit greater levels of persistence than users with the more unfavorable styles.

In situations of failure, the analysis showed significant differences between persons with the Resigned and the Confident (p < 0.001; M = 4.52 vs. M = 4.89) styles, as well as between persons with the Resigned and the Realistic (p < 0.001; M = 4.52 vs. M = 4.88) styles. Persons with the favorable Confident and the more neutral Realistic attribution styles exhibit greater levels of persistence than users with the unfavorable Resigned style.

5   Discussion

This study aimed to examine the relationship between computer-related attributions and users' computer-related task persistence. This section discusses the findings of the present study, its limitations, and offers suggestions for future research and practice. The results show that attribution styles indeed impact users' persistence. Users with favorable attribution styles are significantly more persistent in achieving a computer-related goal than persons with the more unfavorable styles.

The findings can be used in HCI research and practice to understand better why users think, feel, or behave in a certain way. Thus, design principles could be developed to support different types of users in a specific way. To our knowledge, this is the first study that directly examines the impact of computer-related causal attributions on users' task persistence. Therefore, this study contributes to a more complete and detailed knowledge of users' computer-behavior. The results encourage further research on causal attributions as personality traits in HCI research.

There are also implications for practitioners who develop and design computer systems. This study sheds light on different types of computer users regarding their explanations for successful and unsuccessful outcomes when working on computer-related tasks. In order to motivate people to put forth high effort to achieve a goal, several measures might be explored. For example, attributional retraining [25], which suggests that individuals' performance will increase when they learn to ascribe causes to more favorable attributions, could be a promising approach. Thus, our results are valuable for developing and improving existing computer learning training strategies and methods, as well as support and assistance mechanisms for users. Practitioners should attempt to adapt these findings and design specified systems by, for example, including attributional retraining strategies. This could be done, for example, by providing feedback that changes the beliefs of the users about the cause of computer-related outcomes (e.g., comments that contain the desired attributions). A first approach in this direction was made by [26]. They investigated the effect of different attributional wordings of error messages. System developers and designers should bear this in mind and future research should take this into consideration.

The present study also faces some limitations. First, the research design of this study carried certain limitations. Standardized hypothetical use situations were chosen to create a similar experience for all participants. Moreover, this method enabled a high number of participants. However, a drawback is that the situations were somewhat artificial and unrelated to the participants' normal use habits, which might result in reduced intensity and significance of the imagined situation (cf. [21]). Future research should bear this in mind and investigate these relations in real use situations.

Furthermore, participants are from Germany only and as mentioned above there is some evidence that people from other countries differ in their attributions [8]. In this regard, future studies should investigate cultural differences by expanding into a more international context.

Finally, the results presented here give insights regarding the relations between computer-related attributions and users' persistence. However, to make the findings more usable in practice, future studies should investigate the effects of reattribution training methods on users' persistence.

References

1. Weiner, B.: Achievement Motivation and Attribution Theory. General Learning Press, Morristown (1974)
2. Weiner, B.: An attributional theory of achievement motivation and emotion. Psychol. Rev. **92**(4), 548–573 (1985)
3. Kelley, H., Compeau, D., Higgins, C.A., Parent, M.: Advancing theory through the conceptualization and development of causal attributions for computer performance histories. ACM SIGMIS Database **44**(3), 8 (2013)

4. Martinko, M.J., Harvey, P., Dasborough, M.T.: Attribution theory in the organizational sciences: a case of unrealized potential. J. Organ. Behav. **32**(1), 144–149 (2011)
5. Niels, A., Jenneck, M.: The influence of causal attributions on users' problem-solving motivation. In: Mensch Und Computer 2017 – Tagungsband, pp. 127–136. Gesellschaft für Informatik e.V, Regensburg (2017)
6. Blinco, P.M.A.: A cross-cultural study of task persistence of young children in Japan and the United States. J. Cross Cult. Psychol. **23**(3), 407–415 (1992)
7. Heine, S.J., Kitayama, S., Lehman, D.R., Takata, T., Ide, E., Leung, C., Matsumoto, H.: Divergent consequences of success and failure in Japan and North America: an investigation of self-improving motivations and malleable selves. J. Pers. Soc. Psychol. **81**(4), 599–615 (2001)
8. Janneck, M., Xiao, J., Niels, A.: Computer-related attributions: an intercultural comparison. In: Tareq, Z.A., Karwowski, W. (eds.) Advances in the Human Side of Service Engineering, Proceedings of the AHFE 2016 International Conference on the Human Side of Service Engineering, Walt Disney World®, Florida, USA, 27–31 July 2016, pp. 161–172. Springer (2016)
9. Heider, F.: The Psychology of Interpersonal Relations, vol. 56. Lawrence Erlbaum Associates, London (1958)
10. Stiensmeier-Pelster, J., Schürmann, M., Eckert, C., Pelster, A.: Der Attributionsstil-Fragebogen für Kinder und Jugendliche (ASF-KJ) Untersuchungen zu seinen psychometrischen Eigenschaften. Diagnostica **40**, 329–343 (1994)
11. Abramson, L.Y., Seligman, M.E.P., Teasdale, J.D.: Learned helplessness in humans: critique and reformulation. J. Abnorm. Psychol. **87**(1), 49–74 (1978)
12. Kelley, H., Compeau, D., Higgins, C.: Attribution analysis of computer self-efficacy. In: AMCIS 1999 Proceedings, Americas Conference on Information Systems (AMCIS), Milwaukee, USA, pp. 782–784 (1999)
13. Seligman, M.E.P.: Learned Optimism: How to Change Your Mind and Your Life, vol. 9. Vintage, New York (2006)
14. Henry, J.W., Martinko, M.J., Pierce, M.A.: Attributional style as a predictor of success in a first computer science course. Comput. Hum. Behav. **9**(4), 341–352 (1993)
15. Anderson, C.A., Jennings, D.L., Arnoult, L.H.: Validity and utility of the attributional style construct at a moderate level of specificity. J. Pers. Soc. Psychol. **55**(6), 979–990 (1988)
16. Henry, J.W., Martinko, M.J.: An attributional analysis of the rejection of information technology. J. End User Comput. **9**(4), 3–18 (1997)
17. Niels, A., Guczka, S.R., Janneck, M.: The impact of causal attribution s on system evaluation in usability tests. In: Proceedings of the 2016 CHI Conference on Human Factors in Computing Systems - CHI 2016, pp. 3115–3125 (2016)
18. Rozell, E.J., Gardner, W.L.: Computer-related success and failure: a longitudinal field study of the factors influencing computer-related performance. Comput. Hum. Behav. **15**(1), 1–10 (1999)
19. Phelps, R., Ellis, A.: Overcoming computer anxiety through reflection on attribution. In: Williamson, A., Gunn, C., Young, A., Clear, T. (eds.) Winds of Change in the Sea of Learning: Charting the Course of Digital Education: Proceedings of the 19th Annual Conference of the Australasian Society for Computers in Learning in Tertiary Education (ASCILITE), UNITEC Institute of Technology, Auckland, NZ, 8–11 December 2002, vol. 2, pp. 515–524 (2002)
20. Barki, H.: Determinants of user satisfaction judgements in information systems. In: Proceedings of the Twenty-Third Annual Hawaii International Conference on System Sciences, pp. 408–417 (1990)

21. Niels, A., Janneck, M.: Computer-related attribution styles: typology and data collection methods. In: Proceedings of 15th IFIP TC 13 International Conference on Human-Computer Interaction – INTERACT 2015, Part II, Bamberg, Germany, 14–18 September 2015, pp. 274–291. Springer, New York (2015)
22. Dickhäuser, O., Stiensmeier-Pelster, J.: Entwicklung eines Fragebogens zur Erfassung computerspezifischer Attributionen. Diagnostica **46**(2), 103–111 (2000)
23. Guczka, S.R., Janneck, M.: Erfassung von Attributionsstilen in der MCI - eine empirische Annäherung. In: Reiterer, H., Deussen, O. (eds.) Mensch & Computer 2012: Interaktiv Informiert – Allgegenwärtig Und Allumfassend!?, pp. 223–232. Oldenbourg Verlag, München (2012)
24. Schuler, H., Prochaska, M.: Leistungsmotivationsinventar (LMI). Hogrefe, Göttingen (2001)
25. Försterling, F.: Attributional retraining: a review. Psychol. Bull. **98**(3), 495–512 (1985)
26. Niels, A., Lesser, T., Krüger, T.: The impact of causal attributions on the user experience of error messages. In: Tareq Z.A., Karwowski, W. (eds.) Advances in the Human Side of Service Engineering Springer, pp. 173–184 (2016)

Emotionalizing e-Commerce Pages: Empirical Evaluation of Design Strategies for Increasing the Affective Customer Response

Alexander Piazza[1]([⊠]), Corinna Lutz[1], Daniela Schuckay[1], Christian Zagel[2], and Freimut Bodendorf[1]

[1] Friedrich-Alexander-University Erlangen-Nuremberg, Lange Gasse 20, 90403 Nuremberg, Germany
{alexander.piazza,corinna.lutz,daniela.schuckay, freimut.bodendorf}@fau.de
[2] Coburg University of Applied Sciences and Arts, Friedrich-Streib-Straße 2, 96450 Coburg, Germany
christian.zagel@hs-coburg.de

Abstract. The interdisciplinary research of neuromarketing shows that the conscious and rational consumer is only an illusion, whereas emotions have a significant influence on consumer behavior. Therefore, this study examines the effect of emotionalized e-com pages on visitors' emotions as well as on their behavioral intention in hedonic situations. Three landing pages are conceptualized using diverse techniques of emotional boosting along with different procedures of triggering distinct levels of neuronal activity. The impact of these landing pages is examined in an online survey, generating a sample of 391 participants. The resulting dataset is analyzed by using structural equation modeling to test the proposed hypotheses. The results confirm that emotions can be triggered only by seeing a landing page of an e-com store and that these emotions influence the behavioral intentions. Additionally, the study shows a moderating effect of long-term involvement and mood and provides recommendations for appropriate and well-designed websites.

Keywords: Emotion · Human factors · e-Commerce · Landing pages
Design strategies

1 Introduction

The latest research in the interdisciplinary field of neuromarketing has proven that the conscious and rational consumer is only an illusion, whereas emotions are a critical factor in consumer behavior. Studies for example by Häusel [1] and Pispers [2] indicate that only 5% of decisions are made consciously while the remaining 95% are made by an unconscious autopilot, which is triggered by emotions. Therefore, examining emotions in the context of consumer research is crucial. One field, where emotions are especially important, is apparel shopping. Because of the emotional character of clothes [3, 4], it is necessary to create an emotional offering and atmosphere. This is especially important for online shopping because products cannot be touched and therefore no

© Springer International Publishing AG, part of Springer Nature 2019
T. Z. Ahram (Ed.): AHFE 2018, AISC 787, pp. 252–263, 2019.
https://doi.org/10.1007/978-3-319-94229-2_24

emotions can emerge from physical contact. One special category of apparel is sportswear, which combines the emotional character of clothes and sports. The importance of an appealing website design is underlined by the fact of on average 30% of visitors leaving websites due to a lack of rich media content [2]. In addition, 90% leave apparel websites after the third click, regardless of how good the functional features of the website are [5].

In general, the share of the German population shopping online has increased rapidly from 9.7% in 2000 to 67.6% in 2016 [6]. However, despite this impact for the German market, there seems to be a need for improvement. Visiting the e-com pages of the two most important sportswear brands in Germany, adidas and Nike [7], shows how similar and basically interchangeable the stores appear on the first impression. To get the online store visitors to buy the specific brand, sportswear provider need to trigger their customers in the right manner (Fig. 1).

Fig. 1. Screenshots of the adidas and Nike online stores (www.adidas.de; www.nike.de)

In this paper, we examine how emotions can be evoked by website design and how online shoppers of sportswear can be influenced by emotionalizing the online store. Therefore, we extract different design guidelines to arouse emotions from literature. Because shoppers only need seconds to build their opinion on an online store [2], the focus of this study is placed on the landing page.

2 Literature Background

Various definitions of emotions emerge in the academic literature [8]. While some define emotions as a static condition, others approach it from a more dynamic point of view. One reason for this might be that emotions are important in different scientific areas such as psychology or business studies [9]. Analyzing various established definitions [10–14] shows that behavior is an important component of all definitions, revealing the importance of the behavioral influence of emotions. Another important aspect is the object relation, which means that emotions refer to an object or are triggered by an object [9]. This is especially important for this study because emotions shall be triggered by websites that can be seen as external stimuli. In this study, we defined emotions as a subjective mental state that is related to a specific object and influences the behavior or at least the motivation of a person.

The basis of this study is placed in environmental psychology, which assumes that the environment influences the way individuals feel and behave. This connection is mostly seen as a stimulus-organism-response (S-O-R)-connection, in which the environment acts as a stimulus that triggers reactions of the organism and leads to behavioral reactions [15]. In accordance with this connection the pleasure-arousal-dominance (PAD) model by Mehrabian and Russell [15], which defines the organism as the emotional response of a person and measures emotions by three bipolar dimensions is used for this study. The three dimensions of the PAD model are: pleasure – displeasure, arousal – nonarousal and dominance – submissiveness.

3   Derived Hypotheses and Design Strategies

As emotion becomes stronger when there is a distraction or cognitive overload [16], we choose to conduct a field experiment. Following the approach of Eroglu et al. [17], we expanded the PAD model by Mehrabian and Russell by including attitude as a mediating variable and satisfaction as a response. The response comprises satisfaction regarding the online store as well as the degree of approach behavior of the individual, for instance, to show willingness to explore the online store [17]. Although it is assumed that the first impression of an online store influences the whole shopping behavior [18], this procedure is chosen for the study as it might be hard for the participants to imagine how they would behave in that online store without being able to get to know more about it. Therefore, these two more unconsciously made responses are included to measure the effect of the landing page. Based on the original model and the expansion of Eroglu et al. [17], the following hypotheses are tested:

H1: Landing pages can trigger emotions.
H2: Higher pleasure and arousal lead to higher attitude, more satisfaction, and greater willingness to show approach behaviors.
H3: Dominance does not influence the response in the form of satisfaction and approach behavior.
H4: Attitude partly mediates the influence of emotions on satisfaction and willingness to approach.

Another included variable is the involvement, meaning the degree of subjective relevance of an object [19]. Based on the insights from Eroglu et al. [17] as well as Ha and Lennon [20] low task-relevant cues seem to significantly influence emotions only in low involvement situations. Considering that situational involvement is an important moderator [17, 20] and no study has examined the (potential) effect of long-term involvement for internet and sports, it is included in this investigation.

H5: The influence of the environment on emotions is moderated by long-term involvement.

The actual mood of the participant is also included as a moderator between the environment and the organism because a different mood can change the emotional reaction to any environment [21]. In the research of Kim and Lennon [22], mood is found to be positively related to pleasure and arousal, stating that the influence of a stimulus on the emotional response is moderated by mood.

H6: The influence of the environment on emotions is stronger for a good mood.

It is assumed that the more neurons that are addressed, the higher is the resulting emotional response. The lowest level of addressing neurons is in the case where only an object is observed. More neurons are triggered when mirror neurons are addressed, e.g. when observing an actor. According to Capa et al. [23] and Kiesling [24], even more neurons are addressed if the participant is put into the situation himself and not only observes an actor. Based on these considerations, the landing pages represents the three levels of neuronal activity, having in mind that every situation triggers emotions and that by mirroring a situation, the participant experiences the same emotions as if being in the situation himself [15, 25].

H7: Addressing more neurons leads to a greater emotional response.

The expanded PAD model including the derived hypotheses, which are tested in this study are shown in Fig. 2.

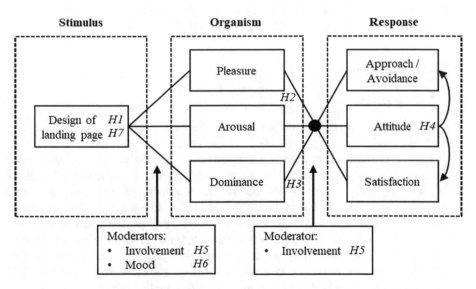

Fig. 2. Adaption of the Mehrabian Russel model including the tested hypotheses

For testing the proposed hypotheses, we designed three different landing pages, using the simulation theory as the basis. Therefore, 27 research articles focusing on website design are analyzed. Coupled with the insights from the mirror neuron research and neuromarketing [1], the three landing pages, which trigger different emotions, were conceptualized. The first page addressed canonical neurons only, by showing sports-wear items, which trigger the emotions normally felt while wearing them. The second one included acting people to address mirror neurons, leading to the simulation of the person's feeling in that situation. The third version managed to include the visitor to make it easier to imagine being in that situation himself. While the first page has a static design showing only pictures, the second and the third are characterized through a dynamic design by showing video sequences. To choose images, pictures, or video files and to design the website, referential, social, mythical, distinctive and recognition boosting can be used. For each design, different boosting techniques are utilized to trigger dominance, pleasure and arousal [13]. Table 1 shows the selected boosting technique for each website design to trigger specific emotions.

Table 1. Emotions triggered by different boosting techniques

Design	Utilized boosting technique	Emotion most effectively triggered
1	Referential and distinctive	Dominance
2	Social and mythical	Pleasure
3	Social and referential	Arousal

To make the effect of the different design options measurable, it is important that the foundation of the diverse landing pages is the same and that distortion factors such as audio or usability features of the website are not considered.

4 Data Collection

4.1 Survey Design

The questionnaire consists of three major parts: the measurement of the environment, of emotions, and of the response. Six bipolar items for each construct ensure an unbiased measurement. These items are the same as those introduced by Mehrabian and Russell [15]. They were measured on a seven-point rating scale. Although the PAD model has frequently been utilized for the measurement of emotions, only one investigation has made use of it in the German language. In a paper published in 1993, Hamm and Vaitl [26] translate the items proposed by Mehrabian and Russel into German and ask a bilingual expert to retranslate them to ensure the quality of their translation [26]. Afterwards, they test their scale and compare the results to those of Mehrabian and Russel, proving a high reliability of their scale. However, as this scale has only been used once before by Hamm and Vaitl, a pretest is conducted to test the comprehensibility of the items.

Seven out of ten participants of the pretest perceive some items as being unclear or confusing. This might be due to the fact that between the first use of the terms in 1993 and today the meanings of the terms might have changed. Therefore, the mentioned items were translated again, retranslated by a bilingual psychologist, and again presented to the participants of the pretest. After this new translation, no further questions arose, thus the improved items were used for the survey.

Concerning the measurement of the emotional response, two issues have to be considered. The first is that emotions are mostly of short duration [27]. Therefore, the measurement of emotions has to take place directly after the presentation of the website. In addition, the participants might already have emotions before seeing the website, which means that not the absolute response to the website, but the relative emotional change has to be measured. Following Donovan, Rossiter, Marcoolyn and Nesdale [28] the present investigation measures the emotions before and after the presentation of the website. Due to pre-existing emotions, the results would be highly biased if one participant had to evaluate his emotional response to each of the three websites. Therefore, every participant is only presented with one website, necessitating three samples. To ensure that these samples do not differ systematically, the online survey software randomly allocates the websites.

4.2 Descriptive Analysis

In total, 925 participants started the questionnaire with n = 391 of them completing (42.27%). This aggregated sample was split into the three sub-samples, resulting in a sample size of 129 participants for design one, 140 for design two, and 122 participants that evaluate design three.

Most terminations took place on the first page, where the restrictions concerning the Internet Explorer and the use of mobile devices were mentioned (320 takeouts). Although the short form of the mood questionnaire was chosen, this part of the questionnaire led to the second most terminations (73). The majority of the remaining 132 terminations were equally spread on the first and second part of the emotional scale.

Regarding the gender, 52.9% of participants were female, 46.5% were male, and 0.5% did not state their gender. This fits the general online shopping population very well, as in general 52% of online shopper are female, while 48% are male [29]. The age range of participants was distributed widely from under 16 to older than 65 years old. The majority of participants were between 20 and 49 years. This outcome does match neither the general distribution of online shoppers [30] nor the distribution of shoppers of sportswear [31]. Therefore, it constitutes a restriction concerning the generalizability of the results on the target group.

Concerning the internet use, 40.4% of the sample stated that they used the internet for at least three hours per day, including 16.6% who reported using it for more than five hours per day. An additional 35% of participants reported using the internet for one to three hours per day. Only two participants stated that they did not utilize the internet daily. This showed that the majority of participants could be classified as active online users. Regarding the frequency of online shopping, 21% stated that they had not bought any apparel using an e-com store within the last 12 months. Nearly the same proportion of participants, 22.8%, said that they bought 10 or more pieces on the internet within the same timeframe. The interest in sports is relatively high in the sample; 73.1% of the participants stated that they are at least rather interested in sports; only 5.1% stated that they were not interested at all.

5 Data Analysis

The hypotheses defined in Sect. 3 are tested with the collected data and by using the tools SPSS and AMOS. First, the hypothesis **H1** is examined, stating that landing pages are sufficient to trigger emotions. First, the changes in emotion based on the 18 items of the PAD model were calculated, representing the relative emotional change after the stimuli (landing page) was shown to the participant. A t-test is applied to the difference of all items shows that ten of the 18 items have significantly changed after the exposure to the website. This includes five items representing pleasure, two items measuring arousal, and three items for the dominance dimension. Some of these significant differences are relatively small. The reason might be that the participants did not select extreme values on the seven-level scale used for the measurement what results in small differences. Overall, the hypothesis H1 is considered to be maintained.

The hypotheses **H2** to **H4** were tested using the AMOS software based on the structural equation model illustrated in Fig. 3.

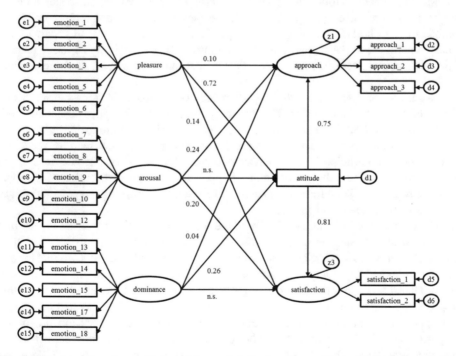

Fig. 3. Structural equation model used for testing the hypotheses H2 to H4 including the test results.

As the data does not show a multivariate normal distribution, the scale-free least square instead of the frequently used maximum likelihood approach was used in combination with Bayesian estimation and bootstrapping [32]. The results in Fig. 3 show that there are positive and significant correlations between all constructs except between the arousal to attitude and the dominance to satisfaction construct. The results indicate that emotions have the strongest influence on the attitude. The data supports the hypothesis **H2** as pleasure and arousal have a positive significant impact on attitude, approach, and satisfaction. The hypothesis **H3** is rejected, as the dominance dimension has an influence on the response constructs, even though the relationship is less strong than for the pleasure and arousal dimension. The hypothesis **H4** states that the attitude partly mediates the connection between emotion and response can be confirmed, as the direct influence of attitude on approach is 0.75 and on satisfaction 0.81. When calculating the total effect, approximately half of the influence of approach as well as on satisfaction emerges from attitude.

The influence of the moderator variables sports involvement and long-dated involvement on the different effects within the S-O-R scheme is tested by using the structural equation model in Fig. 4.

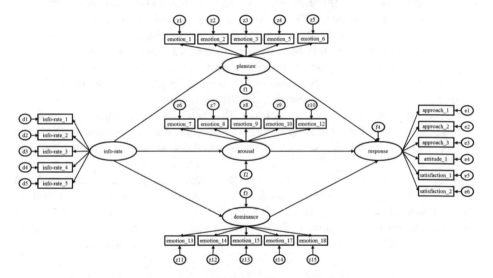

Fig. 4. Structural equation model for testing hypotheses H2 to H4

The impact of the moderator variables was tested by considering only the first percentile of the respective variables and comparing it with the model using only the fourth quantile of the variables. By doing this, only the difference between the extreme manifestations of the respective moderator were tested. As standardized regression coefficients cannot be compared due to their dependence on the variance of the dependent variable [33], unstandardized regression weights were used. Since there are no inference statistics telling if the difference between two coefficients was significant, the rule of Chin [34], which says that a path is meaningful if the regression weight at least accounts for 0.2, was utilized, This was adapted to the present question, only regarding those differences as meaningful that are at least as large as |0.2|.

The analysis regarding the impact of long-term involvement in sports and internet on the emotional response indicates that an increased involvement in internet and sports results in an increased emotional response. One exception is the pleasure dimension, which stays approximately stable for changing degrees of sport involvement. Another result of this analysis is that for participants with high involvement in the internet the pleasure dimension has an increasing impact. One explanation could be that people who are used to the internet have many web shops which they can choose from. Therefore, pleasurable websites are no longer sufficient as many websites are designed to be pleasurable. In summary, the analysis indicates that long-term involvement is a major moderator for the influence of the website on emotion as well as emotion to the response and therefore the hypothesis **H5** can be confirmed.

The analysis of the hypothesis **H6** was conducted with the same approach as conducted for H6. The results indicate that the mood was largely not affecting the relation between environment and emotion and was therefore rejected. In the last analysis, the hypothesis **H7** was tested, examining the relation between addressing higher number of neurons with the website design and the emotional response. Therefore, a Kruskal-Wallis test was conducted to test the differences between the three designs [35]. For this, separate tests were performed comparing each website design with each other. As the differences for all emotional dimensions were relatively small, only six items were found to be significantly different between all of the three website designs. The significant differences indicate that the second design was rated best and the first design was the least successful one. This indicates that emotions are triggered best when mirror neurons are addressed. As assumed, canonical neurons triggered the lowest emotional response. Contrary to the expectation, the third design triggers less emotions than the second design. Consequently, the hypothesis **H7** is only partly supported, meaning that that addressing mirror neurons leads to a higher emotional response than triggering canonical neurons, but including the participant into a scene is less successful than triggering mirror neurons. One explanation might be that it is easier for people to mirror the emotions of other persons than to develop emotions oneself by only seeing a web store. An overview of the results is given in Table 2.

Table 2. Overview of hypothesis testing

Hypothesis	Result
H1: Landing pages can trigger emotions	Supported
H2: Higher pleasure and arousal lead to higher attitude, more satisfaction, and greater willingness to show approach behaviors	Supported
H3: Dominance does not influence the response	Disproved
H4: Attitude partly mediates the influence of emotion on satisfaction and willingness to approach	Supported
H5: The influence of the environment on emotion is moderated by long-term involvement	Supported
H6: The influence of the environment on emotion is stronger for a good mood	Disproved
H7: Addressing more neurons leads to a greater emotional response	Partly supported

6   Conclusion and Discussion

The objective of this paper was to assess whether emotions can be triggered by emotionalizing landing pages of e-com stores and whether these emotions have an impact on the response. Therefore, the S-O-R paradigm of Mehrabian and Russell [15] was used. Based on the insights from the mirror neuron research and from neuro-marketing, three landing pages that are supposed to trigger different emotions are conceptualized. The impact of these landing pages was tested in an online survey, generating a sample of 391 participants.

The results show that landing pages can trigger each emotional dimension identified by Mehrabian and Russell [15], which in turn influence the attitude as well as the response. Further insights can be derived from this study. One example is the outcome that long-term involvement is a major moderator of the connection between stimulus and emotions, as well as emotions and response. Another is that mood moderates the connection between emotions and behavior. One further result is that triggering emotions is most successful when mirror neurons are addressed. Additionally, different emotionalizing techniques are shown to influence diverse emotional dimensions, thereby offering the possibility of directly addressing the emotional dimension, which is most important for the respective target group of an e-com store.

7   Limitations and Further Research

This study generates important insights for practice as well as for research. Nevertheless, there are also limitations and leverage points for further research that have to be mentioned.

The study only takes e-com stores for sportswear and hedonic motivation into account. There are no insights on the transferability on other branches or utilitarian motivation, thus both should be tested in future research. Furthermore, color and sound are excluded from this study. Since both tools are supposed to be emotionalizing, further research should combine neuroscientific knowledge with website design concerning those features.

Other limitations arise from the questionnaire. Since attitude is measured by using global items the attitude loads on the same factor as the response items. In addition to this, the situation chosen induced emotions, which reduced the possibility of finding great emotional changes. In future research using scale consisting on more manifestations should be considered since the used seven-point rating scale might not be convenient to measure emotional change. Due to the type of the investigation, only behavioral intentions are measured. As intentions do not always have to be implemented into practice, results that are more realistic might be possible when using a whole e-com store. Another limitation arises from the conceptualization of the website and that the sample was not normally distributed. The conceptualization of a landing page is a creative and highly subjective task, therefore other designs based on the results of this thesis should be conceptualized and tested in order to confirm the outcomes.

References

1. Häusel, H.-G.: Neuromarketing: Der direkte Weg ins Konsumentenhirn? In: Häusel, H.-G. (ed.) Neuromarketing. Erkenntnisse der Hirnforschung für Markenführung, Werbung und Verkauf, 2nd edn., pp. 7–15. Haufe Verlag, München (2013)
2. Pispers, R.: Neuromarketing im Internet: Von der Website zum interaktiven Kauferlebnis. Haufe Verlag, Freiburg (2013)
3. Moody, W., Kinderman, P., Sinha, P.: An exploratory study: Relationships between trying on clothing, mood, emotion, personality and clothing preference. J. Fashion Mark. Manag. 14(1), 161–179 (2010)
4. Andrée, P.: Marktsegmente im Onlinehandel der Bekleidungsbranche: Entwicklung eines Marketingkonzepts für den Onlinehandel stationärer Mehrmarkenhändler der Bekleidungsbranche in Deutschland zur Erschließung von Marktpotentialen online-affiner Kundengruppen. Hampp, München (2003)
5. Web Arts AG: Emotionale Aktivierung als Konversionstreiber. Web Arts AG, Bad Homburg v. d. Höhe (2009)
6. IfD Allensbach (2016). https://de.statista.com/statistik/daten/studie/2054/umfrage/anteil-der-online-kaeufer-in-deutschland/
7. Statista (2017). https://de.statista.com/statistik/daten/studie/150745/umfrage/groessten-sport artikelhersteller-nach-umsatz/
8. Gröppel-Klein, A.: No motion without emotion: getting started with hard facts of a soft topic. GfK Mark. Intell. Rev. 6(1), 8–15 (2014)
9. Franke, M.-K.: Hedonischer Konsum: Emotionen als Treiber im Konsumentenverhalten. Springer Gabler, Wiesbaden (2013)
10. Bagozzi, R.P., Gopinath, M., Nyer, P.U.: The role of emotions in marketing. J. Acad. Mark. Sci. 27(2), 184–206 (1999)
11. Frijda, N.H.: The Emotions. Cambridge University Press, Cambridge (1986)
12. Hupp, O., Gröppel-Klein, A., Dieckmann, A., Broeckelmann, P., Walter, K.: Beyond verbal scales: measurement of emotions in advertising effectiveness research. In: GfK Nürnberg e. V. (ed.) Yearbook of Marketing and Consumer Research, 6th edn., pp. 72–98. Duncker & Humblot, Berlin (2008)
13. Mau, G.: Die Bedeutung der Emotionen beim Besuch von Online-Shops: Messung, Determinanten und Wirkungen. Gabler Verlag, Wiesbaden (2009)
14. Trommsdorff, V., Teichert, T.: Konsumentenverhalten, 8th edn. Kohlhammer, Stuttgart (2011)
15. Mehrabian, A., Russell, J.A.: An Approach to Environmental Psychology. MIT Press, Cambridge (1974)
16. Kroeber-Riel, W., Weinberg, P.: Konsumentenverhalten, 8th edn. Vahlen, München
17. Eroglu, S.A., Machleit, K.A., Davis, L.M.: Empirical testing of a model of online store atmospherics and shopper responses. Psychol. Mark. 20(2), 139–150 (2003)
18. Häusel, H.-G.: Think Limbic!: Die Macht des Unbewussten verstehen und nutzen für Motivation, Marketing, Management, 4th edn. Haufe Verlag, München (2015)
19. Homburg, C.: Marketingmanagement: Strategie - Instrumente - Umsetzung - Unternehmensführung, 4th edn. Gabler Verlag, Wiesbaden (2012)
20. Ha, Y., Lennon, S.J.: Effects of site design on consumer emotions: role of product involvement. J. Res. Interact. Mark. 4(2), 80–96 (2010)
21. Scheier, C., Held, D.: Die Neuro-Logik erfolgreicher Markenkommunikation. In: Häusel, H.-G. (ed.) Neuromarketing. Erkenntnisse der Hirnforschung für Markenführung, Werbung und Verkauf, 2nd edn. pp. 97–134. Haufe Verlag, München (2013)

22. Kim, H., Lennon, S.J.: E-atmosphere, emotional, cognitive, and behavioral responses. J. Fashion Mark. Manag. **14**(3), 412–428 (2010)
23. Capa, R., Marshall, P., Shipley, T., Salesse, R., Bouquet, C.: Does motor interference arise from mirror system activation? The effect of prior visuo-motor practice on automatic imitation. Psychol. Res. **75**(2), 152–157 (2011)
24. Kiesling, L.L.: Mirror neuron research and Adam Smith's concept of sympathy: three points of correspondence. Rev. Austrian Econ. **25**(4), 299–313 (2012)
25. Gordon, R.M.: The simulation theory: objections and misconceptions. Mind Lang. **7**(1–2), 11–34 (1992)
26. Hamm, O.A., Vaitl, D.: Emotionsinduktion durch visuelle Reize: Validierung einer Stimulationsmethode auf drei Reaktionsebenen. Psychologische Rundschau **44**(3), 143–161 (1993)
27. Donovan, R.J., Rossiter, J.R.: Store atmosphere: an environmental psychology approach. J. Retail. **58**(1), 34–57 (1982)
28. Donovan, R.J., Rossiter, J.R., Marcoolyn, G., Nesdale, A.: Store atmosphere and purchasing behavior. J. Retail. **70**(3), 283–294 (1994)
29. Konzept & Markt (2015). http://de.statista.com/statistik/daten/studie/425499/umfrage/online-shopper-in-deutschland-nach-geschlecht/
30. VuMA (2017). https://de.statista.com/statistik/daten/studie/538490/umfrage/online-kaeufer-in-deutschland-nach-alter/
31. Arbeitsgemeinschaft Verbrauchs- und Medienanalyse (2014). http://de.statista.com/statistik/daten/studie/293376/umfrage/umfrage-in-deutschland-zum-alter-der-kaeufer-von-sportbekleidung/
32. Weiber, R., Mühlhaus, D.: Strukturgleichungsmodellierung: Eine anwendungsorientierte Einführung in die Kausalanalyse mit Hilfe von AMOS, SmartPLS und SPSS, 2nd edn. Springer Gabler, Berlin (2014)
33. Backhaus, K.: Multivariate Analysemethoden: Eine anwendungsorientierte Einführung, 13th edn. Springer, Berlin (2011)
34. Chin, W.W.: Issues and opinion on structural equation modeling. MIS Q. **22**(1), vii–xvi (1998)
35. Theodorsson-Norheim, E.: Kruskal-Wallis test: BASIC computer program to perform nonparametric one-way analysis of variance and multiple comparisons on ranks of several independent samples. Comput. Methods Programs Biomed. **23**(1), 57–62 (1986)

Patient-Centered Design of an e-Mental Health App

Leonhard Glomann[1]([⊠]), Viktoria Hager[1], Christian A. Lukas[2],
and Matthias Berking[2]

[1] LINC Interactionarchitects GmbH, Munich, Germany
{leo.glomann, viktoria.hager}@linc-interaction.de
[2] Chair of Clinical Psychology and Psychotherapy,
Friedrich-Alexander-Universität Erlangen-Nürnberg, Erlangen, Germany
{Christian.aljoscha.lukas, matthias.berking}@fau.de

Abstract. Clinically diagnosed patients who suffer psychological illnesses are usually well supported during ambulatory treatments. Once back in their personal environment, the likelihood increases for them to relapse into past conscious or subconscious habitual patterns. The key to a successful and long-lasting medical attendance is regarded as an individually tailored and constantly available support. In this respect, an e-Mental Health app, acting as a constant companion, is envisaged to support an ongoing personal treatment of a patient during or after an ambulatory treatment. As part of the "mindtastic" project, the app "mindtastic Phoenix" is being created in a cooperation between the University of Erlangen-Nürnberg's Chair of Clinical Psychology and Psychotherapy, the related information technology department and the service design company LINC Interactionarchitects. This paper describes the design process of this e-Mental Health app and highlights the deviations from the design process of conventional apps.

Keywords: eHealth · mHealth · Mental health · Interdisciplinary
Human-Centered Design · Patient-Centered Design

1 Introduction and Objective

Human-Centered Design [1] is one of the most-established approaches for service design. Similar to Design Thinking [2], it is especially useful when creating a new service from scratch, with a target group yet to be understood. However, service design teams are usually dealing with either B2C (Business-to-Consumer), B2B (Business-to-Business) or B2E (Business-to-Enterprise) projects. Target groups of eHealth services cannot be defined as a part of one of these, as their behavior is typically not economically motivated. When translating the principles from Human-Centered Design to the eHealth context, the term Patient-Centered Design is used [3].

This paper introduces the project *mindtastic* as an interdisciplinary collaboration between the Chair of Clinical Psychology and Psychotherapy of the University of Erlangen-Nürnberg, the related information technology department and the service design agency LINC Interactionarchitects, all based in Germany. The aim of *mindtastic*

© Springer International Publishing AG, part of Springer Nature 2019
T. Z. Ahram (Ed.): AHFE 2018, AISC 787, pp. 264–271, 2019.
https://doi.org/10.1007/978-3-319-94229-2_25

is to provide digital patient-centered psychological therapy, e-Mental Health [4], through mobile device-based services, with each service specifically designed to target a certain mental illness. In this paper, the project and the collaboration model is described in an introductory fashion, specifically targeting the following question:

How does the process of designing an e-Mental Health app differ from the process of designing a conventional B2C, B2B or B2E app?

2 Project Description

2.1 Project Background

eHealth. eHealth comprises all measures that can be taken to help patients to get better or to prevent them from getting ill in the first place. This can be accomplished through self-service computer programs, web applications, or mobile apps. In addition to self-service solutions eHealth measures which accompany classical treatment are also in place [5]. In both cases, gathering, analyzing and interpreting data about the patient's behavior is essential for the effectiveness of the digital service.

In this context, the technical term best suited is mHealth, which stands for health measures in the form of a mobile app. But as the health measures in question are of a *mental* health nature, the term e-Mental Health is most appropriate.

It is important to note that the purpose of eHealth is not to replace conventional treatment, but to offer a meaningful enhancement for future medical treatments. Where deficits in healthcare exist, e-Mental Health can play a beneficial role: Such deficits include long waiting periods for treatment [6], the social stigma associated with psychotherapeutic treatment, or simply the fact that, in order to permanently change negative forms of behavior, a patient needs to constantly confront his problems both during and after his treatment [7]. It has already been shown, that computer-based health interventions and training procedures are growing in popularity and relevance [8].

Project set-up. In 2017, the University of Erlangen-Nürnberg's Chair of Clinical Psychology and Psychotherapy, together with the related information technology department began their cooperation on *mindtastic Phoenix* with the service design agency LINC Interactionarchitects. As first part of the *mindtastic* service, *mindtastic Phoenix* is an e-Mental Health app that is designed to help patients suffering from depression by means of self-help intervention to reduce their depressive symptoms [9]. In order to help patients battle their depression, *mindtastic Phoenix* employs different kinds of training sessions, tasks and data collection. The Chair of Clinical Psychology and Psychotherapy has been doing all kinds of research in the field of eHealth, mHealth and e-Mental Health for years, for example app-based treatment programs for procrastination [10] or body dissatisfaction [11]. With the scientific psychological knowledge and a fundamental concept of the service in place, the team embarked in an interdisciplinary cooperation with the information technology department for development and LINC for information architecture, interaction design and visual design capabilities.

2.2   *mindtastic Phoenix* – Structure, Content and Implementation

The purpose of the *mindtastic Phoenix* app is to trigger and support a change in behavior of the patient. The core part of the app is the 'training' section, which is divided into three categories. The categories are 'knowledge', 'practice' and 'tasks'. With the help of these categories, the user trains to unlearn negative thinking patterns and establish new behavior [9]. In the following sections, these categories are described, each highlighting the purpose of a certain part of the service.

User Onboarding. For everyone working on *mindtastic Phoenix*, it was clear that, with regard to the target group, the app would have to communicate in a sensitive way and with positive tonality. The aim is that users feel comfortable while using the service. For that reason, there are a lot of options to personalize the user feed, similar to personalizing a social media profile. When the app is first started, an introduction appears in the form of a chat dialogue, that will help users learn about how *mindtastic Phoenix* works and how it can help them therapeutically. By using the format of a chat, there is immediately a kind of personal relationship established and the user can feel free to ask any question.

Knowledge Transfer. The same principle is employed when a training starts. The category 'knowledge' uses information, termed 'psychoeducation'. In this category, users learn about their personal condition. Studies have proven that by learning about depression, whether about the symptoms or possible treatments, patients are more willing to take action against their condition and thus guard against future recurrences [12]. The psychoeducation section always ends with a quiz, that allows the user immediately to test her newly achieved knowledge and gives her positive feedback when she chooses the correct answer.

Serious Games. The category 'practice' consists of all different kinds of so-called 'brain games', videos and audio tutorials. When playing, users will train to approve of positive messages such as 'I am worth just as much as others are' and avoid negative messages, whether through tossing negative messages into a trashcan, or zooming in on a picture with a positive connotation. Through these games, users will start to reinforce positive thoughts and behavior in a playful way. Because of positive feedback when winning the game, users will restructure thinking patterns with priority given to more helpful thinking [13]. It was important that these games are designed in a way that they are self-explanatory and not too hard to pass. To this aim, positive feedback is essential when the user successfully completes the game.

Implementation into Daily Routine. The category 'tasks' goes even further. Here, the user is challenged to apply what she has learned during psychoeducation and the 'brain games'. Small tasks are given to the users, which they have to fulfill in real life. Through these tasks, the patient leaves the digital context and learns to implement a changed kind of behavior in his everyday life. An example would be the following task: 'Go for a ten-minute bike ride'. Again, the actual purpose of this task is not to get people on bikes more often but to motivate users to get active and feel good about it. To prove the accomplishment of the task, *mindtastic Phoenix* asks the user to take a picture of the said task and congratulates the user for successfully finishing the task [14]. This is also an extremely important part of *mindtastic* in general: Motivating users to reflect on what they have accomplished and inviting them to enjoy this feeling.

3 Design Process

3.1 Conventional Service Design Process

In order to compare the design process of conventional B2C, B2B or B2E apps to that of an e-Mental Health app, the regular design process is described below as a benchmark. Conventional apps offer all different kinds of functions but in contrast to eHealth apps, they usually aim to increase sales or improve process efficiency. This means that, with economic interests in mind, conventional apps offer services to a consumer or person with a business interest. Looking at the design directive of such an app, the principles of Human-Centered Design or Design Thinking can be set as established standard. One of their key principles is to understand users, their context and their needs. Other include addressing the entire experience that a user has with the service and involving actual users throughout design and development [1]. Following these principles, design teams first need to fully comprehend the project's background, environment and aim and at the same time the user's background, actual context and underlying needs. Through research, e.g. target group surveys, contextual inquiries or observations, they find out what the users truly need. Afterwards, the user requirements and to-be scenarios are defined. Based on these definitions, they come up with information structures and design solutions. The solutions are evaluated with users and adjusted where necessary. In order to assure an ideal outcome, these steps are repeated until there are no more adjustments necessary [15].

3.2 E-Mental Health Design Process by the Example of *mindtastic Phoenix*

In contrast to a conventional service design process, the specific process steps of creating *mindtastic Phoenix* are described below. The Chair of Clinical Psychology and Psychotherapy has been working in the field of e-Mental Health for years and has been partnering with the University's information technology department when LINC Interactionarchitects came into play in early 2017 to support with conceptual, interaction and visual design. Together, they drew up first drafts and an initial design style. After a phase of early expert evaluations of an early pre-alpha version with psychologists who deal with the target group on a day-to-day basis, it has been found out that the service concept needed change content-wise as well as structurally. LINC reviewed and analyzed the outcome of the evaluation and came up with a new conceptual design. In order to be able to transfer this concept onto the whole service, the team has created a couple of key screen designs. Based on these screens, standards for interactions as well as design styles, such as colors, shapes, fonts etc. were defined. Subsequently, further screen designs have been adapted to the definitions as set forth in the preceding screens. During this process, which took a few weeks' time in mid 2017, the screen designs were iteratively revised for conceptual, usability and aesthetic reasons. As soon as the revised concept was in a good state, software development resumed in parallel. The designers and developers were now working simultaneously, continuously reviewing and checking results and clarifying ambiguities. With the result of that, the Chair organized and conducted a study with actual patients, using the alpha version of the app. This study has been carefully constructed and supervised by psychologists because of the sensitive evaluation context with mentally unstable users. With that, the team has tested the app in real conditions by giving it to people that were in treatment for depression. The tests have been accompanied by expert evaluations. This combined approach has proven very helpful in enhancing the usability of the app.

According to the study [9], "preliminary evidence of the effectiveness of the app-based self-help intervention *mindtastic [Phoenix]* in treatment of depression symptoms was found". The response to the tested version of *mindtastic Phoenix* has been predominantly positive. Nearly a third of the users assessed the app as absolutely useful and had practically no complaints. The study also showed that it is able to help reduce symptoms of depression. More specifics are still being tested. The evaluation results have been discussed in a full team workshop at the end of the year, resulting in clear operational intent to bring a first release of *mindtastic Phoenix* to completion.

While designing all aspects of *mindtastic Phoenix* with a psychological reference, the designers from LINC were always careful to realize the service in such a way, that there would be a well-rounded, in itself coherent outcome, which is easy to understand by actual users. The principles of Human-Centered Design generally stress the need to keep the user and her experience in the center of any project activity. However, for this kind of target group, mentally unstable or ill people, the user involvement proves to be a practical challenge.

3.3   Comparison of the Design Processes

Based on the empirical insights gained from designing *mindtastic Phoenix*, the deviations in the design process of a conventional app from that of an e-Mental Health app are explored below.

Special Target Group Preconditions. The main challenge, when creating e-Mental Health services, lies with the special characteristics of the target group and the dependent evaluation options. There are not many possibilities when it comes to test method selection for e-Mental Health services, as described in the following. Furthermore, there are many differences as opposed to conventional service evaluation during test execution.

Constrained User Involvement. A key difference is the difficulty with the practical realization of the Human-Centered Design principle "involve users throughout design and development". This is only possible in parts, because of the mentally constrained individuals of the target group. With them, early user research in the format of contextual inquiries or similar, prior to the actual design activities is in practice very limited. Also, the data gathered from representatives of that target group through regular lab user testing sessions might not be fully reliable and therefore useless in the design process.

Undeterminable Experience Duration. Another complexity is the practical realization of the Human-Centered Design principle "address the entire experience", as the entire experience begins even before diagnosis and ends even after the alleviation or removal of symptoms. The "experience" of a diagnosed patient does not end with being free of symptoms and he might have an increased risk to suffer a relapse. The design team will never be able to observe or analyze all parts of this process on equal terms.

Evaluations Only with Actual Software. Another difference is that evaluations need to be conducted with actual functioning software. With conventional service design, early user tests are typically executed with the help of click dummies. Click dummies consist of wireframes or screen designs that do not contain specific business logic. They are interactive but are not equipped to fully simulate the real application experience, for example when it comes to processing user input and user data. However, e-Mental Health apps rely on being able to truly interact with the individual over a longer period in order to help them get better. Due to the special preconditions of the target group and given the limited validity of regular user testing results, integrated data analytics are mandatory to track and evaluate user behavior.

Impractical Rapid Prototyping. Multiple iterations in the design process are not just recommended but are absolutely mandatory to learn about the actual user's behavior with the e-Mental Health service. Because of the above described necessity to test in real life environments with a projected user behavior impact, evaluation study periods should be planned to take a couple of weeks. As described above, the test can only be performed with actual functioning software with each evaluation using a considerable number of changed features. Therefore, incremental rapid prototyping with just slightly changed software features proves to be impractical. Studies are recommended to be set in real life environments with pre-alpha, alpha and beta testing versions and ongoing releases of the functioning software.

4 Conclusion

As we have seen, the differences described between the process of designing an e-Mental Health app and the process of designing a conventional B2C, B2B or B2E app are predominantly related to *user evaluation*: Owing to the special preconditions and the constrained involvement capabilities of target group representatives and uncontrollable factors with regards to the entire process experience, user evaluations are only possible in real life settings with actual working software. The result is that a rapid way of prototyping is unusable. The patient-centered approach needs to take the user's motivation into consideration while planning the whole design process. The experience that patients have with testing early software versions needs to be as real and therefore impactful as the experience when using the fully functioning service.

5 Further Research

During the collaboration on *mindtastic Phoenix*, several further questions about designing an e-Mental Health or eHealth service arose. An open question is, for example, whether visual design is able to improve the effectiveness of an eHealth service. It is interesting, for example, to consider how much and at which parts in the experience, visual design impacts the effectiveness. Another question is about the level of interplay between therapists, patients in ambulatory treatment and the e-Mental Health service. Lastly, there is the question whether a service interaction structure for one mental health issue, e.g. depression, is applicable also for other mental health issues, such as addiction or social anxiety. Studies in these fields are about to commence at the Chair of Clinical Psychology and Psychotherapy of the University of Erlangen-Nürnberg.

References

1. International Organization for Standardization: DIN EN ISO 9241-210 Human-centred design for interactive systems (2010)
2. Brown, T.: Design thinking. In: Harvard Business Review, pp. 87–92. Harvard Business Publishing, Boston, June 2008
3. Rodriguez, M., et al.: Patient-centered design: the potential of user-centered design in personal health records. J. AHIMA **78**(4), 44–46 (2007)
4. The Royal Australian College of General Practitioners Ltd. https://www.racgp.org.au/your-practice/guidelines/e-mental-health/the-guide-an-introduction/what-is-e-mental-health/. Accessed 28 Feb 2018
5. Innovatemedtec. https://innovatemedtec.com/digital-health/ehealth. Accessed 28 Feb 2018
6. Melchior, H., Schulz, H., Härter, M.: Faktencheck Depression Regionale Unterschiede in der Diagnostik und Behandlung von Depressionen. Bertelsmann Stiftung (2014)
7. Albrecht, U.: Chances and risks of mobile health apps (CHARIMSHA). Peter L. Reichertz Institut für Medizinische Informatik der TU Braunschweig und der Medizinischen Hochschule Hannover (2016)
8. Berking, M.: Entwicklung und Evaluation einer Emotionsregulations-App. http://www.psych1.phil.uni-erlangen.de/forschung/emotionsregulation/entwicklung-und-evaluation-einer-emotionsregulations-app.shtml. Accessed 28 Feb 2018
9. Benevolenskaya, K.: Pilotstudie zur Evaluation der Effektivität und Usability einer App-basierten Intervention zur Linderung von depressiver Symptomatik. Chair of Clinical Psychology and Psychotherapy of the University Erlangen-Nürnberg, pp. 2–3, 16–17, 35–36, 45 (2017)
10. Lukas, C.A., Berking, M.: Reducing procrastination using a smartphone-based treatment program: a randomized controlled pilot study. Internet Interv. **12**, 83–90 (2017)
11. Kollei, I., Lukas, C.A., Loeber, S., Berking, M.: An app-based blended intervention to reduce body dissatisfaction: a randomized controlled pilot study. J. Consult. Clin. Psychol. **85**(11), 1104–1108 (2017)
12. Cuijpers, P., Muñoz, R., Clarke, G., Lewinsohn, P.: Psychoeducational treatment and prevention of depression: The "Coping with Depression" course thirty years later. Clin. Psychol. Rev. **29**(5), 449–458 (2009)

13. Lukas, C.A., Bogun, S., Fahrendholz, L., Fels, A., Kremer, J., Berking, M.: Eine Smartphone-basierte Intervention zur Behandlung von Depression: Ergebnisse einer Pilot-studie: Smartphone-basierte Behandlung von Depression (2017)
14. Burns, D.D., Spangler, D.L.: Does psychotherapy homework lead to improvements in depression in cognitive-behavioral therapy or does improvement lead to increased home-work compliance? J. Consult. Clin. Psychol. **68**(1), 46–56 (2000)
15. Goodwin, K.: Designing for the digital age. Wiley Publishing Inc, Hoboken (2009)

Artificial Intelligence and Social Computing

New « Intelligence » Coming to the Cockpit…Again?

Sylvain Hourlier[✉]

Thales AVS France, Thales Campus, 33700 Merignac, France
sylvain.hourlier@fr.thalesgroup.com

Abstract. Adaptive automation/agent has been "a good idea" for 40 years now. Yet it's hardly used so far. Automations changing in accordance to internal rules are widely distributed and eventually fail to be understandable whenever their inner change can't be grasped by the operator supposedly "trained & in charge". All that could change because for technological reasons AI is back with the assumption is that it will fix it all. How can we build cooperative agents capable of helping Humans by building them with a techno-centered view? The epic fail of Ai in the 90' will just repeat itself. We need to analyze the root of our need when envisioning cooperation with agents. So, what is it that we want from adaptive agents? If you take the example of an assistant surgeon, you have your answer, we want that kind of adaptation. They facilitate the surgeon work without any (verbal) exchanges (not resource demanding to control). They know what to do and when to help. They can interpret any sign from the surgeon as a directive for help. They completely share the same references. They know so well the implicit that collective work seems like the work of a single entity. Alas that is the description of a human being. So definitely what we seek in a cooperative agent are qualities reserved to the living like the ability to adapt. We have misplaced assumptions of humanity on AI without giving it the potential for it: socializing for cooperation through proper communication. It's called articulation work and it's been around 30 years at least. It's the key to enable effective cooperation between agents. DARPA has just realized it and has launched in 2017 a massive research project so as to use AI to digitize the interaction level between an agent and an operator. Modeling with AI what makes a proper cooperation between agents and Human could be the answer.

Keywords: Human Factors · AI · Articulation work · Agent assistance

1 Introduction

Over two decades ago, adaptive "intelligent" assisting systems were the hype. Having played a part in the development of the Electronic Copilot for the Rafale Aircraft [1] it seemed like a good idea to share some thoughts about the latest AI comeback, that obviously no one could miss. AI based automation made a flop at that time, because of multiple weaknesses and over enthusiastic "R2D2" expectations. A renewed interest in AI induced assistance is again emerging, with major investments being made. We could be heading for another cruel deception or towards a true revolution in Human-Agent interactions. Maybe our problem lies with what we expect from an

© Springer International Publishing AG, part of Springer Nature 2019
T. Z. Ahram (Ed.): AHFE 2018, AISC 787, pp. 275–282, 2019.
https://doi.org/10.1007/978-3-319-94229-2_26

assisting agent. Maybe it is time to let Human operators do what they do best and let the technology, system, AI, whatever… do the rest. This means we have to focus, as we should have in the first place, on the Human part of that sort of collaboration. Technology will do our bidding, we just have to know what to ask for.

2 Why AI is Back

Why is it we are having this second wave of "AI is the key to all our problems"? Several enablers have matured since the mid 90'. First the computing capability has followed the predicted Moore's law [2] and risen since 1995 from 5M to 10B transistors on integrated circuit chips in 2016. Computers are definitely faster now. We have gone from gigaFlops[1] to zetaFlops to petaFlops. The second enabler is the abundance of data, because of the third enabler: cheap computer storage; From $500,000 per gigabyte in 1981 to less than $0.03 per gigabyte today. So, no data is lost ever and there is a lot of it out there. Hence the term Big Data. There is a wealth of information that only obnoxious obsessive computing frenzy could make sense of. That sense being the emergence of statistically meaningful patterns no Human could ever have the time to find. Then, the internet is providing now a network capable of interlacing all available sources of data. And those continue to thrive as the latest trends in terms of data collection is happening through novel sensors: bio or non-bio sensors. Everyone is complaining about big brother and at the same time sharing all their biodata on social media. It's a brave new world. Last but not least, machine/deep learning is starting to make sense of subliminal signals deeply hidden in all that data.

All of these combined, have given the opportunity for some sort of "intelligent process" to emerge. It's not brilliant it's just analyzing large pools of data and applying statistical analysis to "see" emerging patterns that can later be promoted as "rules" because they have been proven right by a great number of occurrences. That is today's AI, it's called "Weak AI" as opposed to "Strong AI". Siri and Alexa are often considered AI, but they are weak AI programs. Even advanced chess programs are considered weak AI, just try to have them play another game and see how they perform.

Let's be clear, if you're seeking movie AIs like HAL from *2001 a space odyssey*, that's Strong AI and it does not currently exist. Strong AI would be a type of machine intelligence that is equivalent to human intelligence. We are not there yet, nevertheless some prominent personalities[2] have voiced their fear of Strong AI as potential predators for the Human species we should design with extreme caution.

Today's Weak AI has limitations. Its application to car driving is often limited to a "guardian" paradigm: Lane keeping, Blind-spot monitoring, Adaptive cruise control (speed & spacing), Automated Emergency Braking, Forward Collision Warning, etc. All these are closed loops rule applications. Easy enough yet still tricky and surprising

---

[1] Flops are used to measure the numerical computing performance of a computer.

[2] Just the shortlist: Stephen Hawking, Elon Musk, and Bill Gates.

sometimes. The latest trend is the car as "chauffeur" paradigm: Autosteering (Tesla) and DrivePilot (Mercedes Benz). But the stats aren't good. As Vera [2] synthesized it:

Human drivers' reliability is quite high: ~ 1 injury accident per million miles driven, with only one percent of those leading to a fatality.

It isn't the same for self-driving cars. A disengage is when the control shifts from the system to the operator, quite often brutally. Most of those happen on the streets (more obstacle & events) than on the highway. One must acknowledge that a handover is hardly completed under 1 mn when it's initiated by the system and the operator isn't expecting it. The numbers speak for themselves:

- Google & Nissan: ~ 1 disengage per 5,000 miles
- Tesla; ~ 1 disengage per 3 miles

Indeed, for cars, the path will be a long one towards safe autonomy.

3   The Epic Fail of AI Assistances in the 90'

The ability to adapt is at the core of all living systems. Twenty and some years ago, AI was definitely not alive and proved it the hard way. Most "AI for pilots" programs across the Atlantic fell short for applying an inappropriate metaphor for the interaction between human and machine [4]. Ours was no different.

In 1994, I joined a research team devoted to the extraction and completion of a knowledge database that would serve as the basis for the "intelligence" behind The Electronic Copilot of the Rafale Aircraft. After 3 years of interviewing military pilots on a single reference mission and feeding it in the proper format into a gigantic (at the time) database, we eventually came across some limitations.

Our system, though able to deliver alternate solutions/plans, failed to do so in a "pilot" acceptable time frame (though proper computing power would overcome that eventually). More problematic, was that it also failed to provide "recognizable" solutions. Pilots failed to grasp the "Human coherence" behind the proposed solutions. The reason being that such a database emerged as an aggregate of the best possible pilot practices, thus losing its coherence (as compared to the coherence of a solution given by a single pilot). To circumvent that, the database was reduced around the knowledge of a singled-out pilot to preserve its Human coherence.

Pilot comments changed from:

- "I don't know anyone who would come up with that sort of solution, why would I do that?"

To:

- "Hey, I know a guy who does that, I flew with him once."

Pilots were reluctant to trust a solution that lacked recognizable human limitation; they were not too keen on applying "magical" solutions that would potentially endanger their situation. Understanding is what draws the line between a good solution and a magical one. Pilots do not trust magic.

That being dealt with, we moved on to the simulator runs. The first thing the invited pilots said was:

– "How do you turn the system off?"

We explained (again) that the system needed a continuous surveillance of past and present cues to be able to position an acceptable alternate proposition to the ongoing plan, and thus there were no on/off switch, it had run in the background… And their answer was:

– "Oh, ok, I see, … So how did you say you turn it off?"

Eventually we understood that there would be no compromise here: they wanted to be able to turn it off like any other system they eventually would not trust. They said their life depended on it. We added an on/off switch. Trust is the key; it does not come easy, especially with technology.

In the end, the project never overcame the awkward industrial consortium setup that kept undermining whatever brilliant solution the engineers developed. All that was left was a complete coherent piloting database that was ultimately delivered to each partner.

Exit the French in the pursuit of R2D2.

Several years later, on a test site, during a military exercise in the south of France, a colleague of mine came up quite distressed about some trials they were doing. They were testing UCAV management from the back seat of a fighter. The scenario was quite simple: the fighter and its four UCAVs were supposed to deliver armament on a primary target at high speed and fall back on a secondary in case of overshoot. On and on it went until a pilot did an over shoot and decided to proceed to the secondary target. At that moment, all its UCAVs brutally disappeared and demanded shooting clearance. Not understanding what was happening; the WPO in the back seat cancelled all authorization as fast as he could. What happened here? Well obviously, our piloting database was used after all. I recognized the algorithm: "until all possible manners are exhausted, primary target must be processed". That is the rule and all pilots know it. The UCAVs knew it, so they applied it, because given their performance envelope; they could easily perform a sharp 360 the fighter could not, so they did.

Sharing a common knowledge does not build shared understanding.

This collective lacked shared understanding of each other's performance and its consequences on the application of a simple rule. Until the Humans in that collective learn all about the agent's way to apply all the rules, there could be no efficient, trustworthy collective work. Cooperation between an "intelligent agent" and a Human being implies in-depth development of trust and knowledge about the "other one", and this needs time.

There are two ways to set up an efficient team:

1. always team together the same ones,
2. train everyone with cooperation techniques and normalize all the work.

The first option leads to higher immediate performance and potential complacency, while the second option evens at a lower performance but enables high collective turnover. Both imply a great deal of training: with one another, on the former, while on cooperative techniques and normalized procedures, on the latter.

Why can't we have an artificial assistance just helping without all the fuss?
Well, there is no such thing as a free lunch.

4 Human Assistant Paradigm

Our expectations on "intelligent assistance" are off the charts when compared to the
actual abilities of AI systems. We expect ideal Human like effectiveness. Why settle for
less?

The ideal teamwork is much sought after, for instance in surgery team. Though
head surgeons are the most capable operators, their assistant is indispensable to them
because:

- They facilitate the surgeon work without any (verbal) exchanges, they are not
  resource demanding to control,
- They know what to do, how and when to help
- They can interpret any sign from the surgeon as a directive for help
- They completely share the same references rules and processes
- They know so well the implicit that collective work seems like the work of a single
  entity

It is an extremely efficient Human collective work. Having such level of demand on
agent assistance only reveals the Humanity we seek in artificial aids. The efficient
Human assistant paradigm is extremely difficult to compete with as such Humanity
comes as a great challenge for AI.

Also, for an AI assistance, we drag along other *misplaced assumptions* about any
form of assistance:

- The "limited investment rule": we assume that a system supposed to help us *should
  not demand any investment from our part*, otherwise it isn't helping, it's becoming
  part of the problem.
- The "cognitive continuity respect rule": Any assistance intrusion is considered
  disruptive if not compliant with the operator's internal train of thoughts. So,
  inherently, *assistance should always be coherent with the operator's action plan
  without having to interfere*. It's also called intention detection, implicit recognition
  or even divination. It's quite the challenge actually.
- The "automatic sync rule". *Knowledge about the other should not be my problem*.
  Assistance legibility should suffice for collective synchronization.
- The "respect and recognition rule": *others should respect me before it's the other
  way around*.

Those objectives are already hard to meet in Human collectives; anyone has a
colleague with poor social skills?

There are other models to base an assistance on, beyond the Human like Paradigm.
In their paper [4], Schutte et al. mention three other sorts of collaboration: Domesti-
cated animal metaphor, Body metaphor and Tool metaphor. Their preference goes to
the Horse like paradigm. It's cooperative within a certain envelope of skilled based
abilities, obeys orders and the feedback make sense to the Human user. It reminds me

of how, 40 years ago, tank commanders (in their turret) used to kick the right or the left shoulder of their pilot (because it was all they could reach directly with their boot) to give them directions; effective but mostly unidirectional, lacking a bit of synergy, yet still demanding continuous command and control. They put forward that the rider does not need any "awareness of individual muscle movements or internal bodily activities" of the system (i.e. horse), that knowledge is superfluous. However, they also mention that intelligence is distributed between operator (high level of control, mid & long term) and System (low level of control, short term).

In the body metaphor the machine "enhances, empowers, and extends the human's abilities". The intelligence is only the operator's. It also means that the operator has developed an intimate awareness of the ability of the whole (operator + system), just as a reviewed version of their body schema. It's a model that will gain in application as robotics is developing much faster that Strong AI.

The tool metaphor is applied when assisting the operator through the enhancement of a limited task performer: for instance, a type writer for replacing a copyist, a word processor for replacing a type writer, etc. the tool/system is extremely efficient but also quite monomaniac. It is quite hard to draw with a typewriter.

All these models are legitimate within their potential application. But when developing AI based autonomous agents Human need to work with, are they still relevant?

5   What Do We Need to Be Able to Work with Agents?

Working with someone or something is not always a breeze… it is already a known challenge when only humans are involved. Computer supported Cooperative Work (CSCW) [5] has been around for quite some time now and has focused on that link between agents and how to facilitate it.

We already know what works and what does not. There is a huge amount of knowledge in the field of articulation work[3] [6] and meta-functional activity that explains how collaboration fails or succeeds. Such knowledge could benefit the design of artificial assistance.

> *"[we should] Help characterize those aspects of human performance that will allow the enabling capabilities of the human to function effectively when teamed with machine intelligence."* [2]

Articulation work is a kind of supra work. It's not the exact work of designing a system, building a building or producing a product. It's all the work around that cooperation that makes it possible. In a sense it can be viewed as a parallel work process.

Articulation work focuses on various components of cooperation management: dividing task, allocating, coordinating, scheduling, meshing, interrelating, etc.; All that

---

[3] "Articulation work is work to make work work" Or to be exact, "articulation work is cooperative work to make cooperative work work".

is performed by a number of actors with responsibilities and activities, in conceptual structures, with common or distributed resources.

An agent will have to comply with such articulation processes. An "adaptive" system/agent should follow "Human like" references: Predictability, legibility, trust-worthiness, … All these are part of our heritage as social beings. But how could a machine know that the action a Human is performing is for an immediate objective rather than just an anticipated move for a future objective. Intention detection is just one of the critical problem cooperating agents need to master.

We all learn it in kindergarten with peers. Maybe cooperative agent will have to learn it just the same, while training with young Human operators.

Indeed, the training needs to change hands. In many industries, we have already reached system complexities that impose impossible training challenges. When combined, agent complexity and training are aggravating factors.

Complexity in systems will increase due to the incorporation of self-learning, self-evolving, self-maintaining databases in an ever complexifying net-workplace putting the operator (here the pilot) in the quasi impossible mission of making sense of it all if a failure happens. Millions of lines of code, continuous access to open sources, learning algorithm, etc…., all in real time will beat Human ability to understand.

Training the Human isn't the answer anymore. It's a simple statistical fact.

First, the increased use of the same technology at the same level of integrity will increase the emergence of events beyond its certified level of integrity (10^{-9} for aviation, one failure every billion counts). If you have 10 billion uses, you have, potentially 10 failures that can happen. Failures only happening beyond the accepted level of integrity (thus never witnessed today) may appear. Now if you also increase the technology complexity, you make it even harder (costlier) to maintain such level of integrity.

Second, if you double the numbers of aircrafts in flight, you also double the potential occurrence of facing critical weather conditions (and I am not accounting for potential climate change in this equation). Combination of possible mishap either from technology or from weather will lead to an increased emergence of critical situations, some of which being incredible and unforeseeable.

Training is effective trough repetition. Training on an extraordinary rare event (for example high altitude stall) is costly and ineffective if not repeated or encountered often. Repeating such training to make it pertinent means it should take the place of other training that are referring to much more frequent events: QED. It would not be responsible to pursue in that direction. Answers should be coming from the system, not problems.

DARPA figured that out already and launched a massive research program called XAI for Explainable AI [7].

The idea is to focus AI, not on system internal efficiency, but on AI self-explanation. Using AI to explain its own complexity, to help complex systems "Speak Human", is just brilliant and may be the breakthrough we need.

In a parallel view, European project eXCockpit [8] is aiming at having an AI serve as an "interpreter" to translate into human plain lingo the environmental/technological complexity of today's Cockpits facing exceptional situations. The next five years should be extremely interesting in terms of Human Factors for AI.

6 Conclusion

As is, we are not ready to use the Human assistant paradigm. It would mean having the ability to create artificial humans and we are not there yet, we can all be reassured the terminator is not around the corner. In the meantime, we could take an ecological position and see how cooperation works between species. We live in harmony with the billions of bacteria thriving in our digestive system. They help us process food for a fraction of the energy, while we host them in a protective environment. Another example of cooperation is the plover bird cleaning the crocodile's teeth. The bird gets the food and the crocodile the dental hygiene: a win-win situation. Of course, there is always the possibility of a parasitic relationship and one could fear that would be the case with upcoming AI agents.

I prefer to choose a future where Humans and Artificial Intelligence work in symbiosis and XAI is paving the route towards that goal so symbiosis can be the model for implementing Human-AI cooperation.

References

1. Hourlier, S., Grau, J.Y., Amalberti, R.: The "Electronic Copilot", a human factors approach to pilot assistance. In: Proceedings of the 1999 World Aviation Congress, San Francisco, (USA), 19–21/10, pp. 210–215 (1999)
2. Vera, A.: What machines need to learn to support human problem-solving. In: Presented at Ames Machine Learning Workshop, Moffett Field, CA, 29–31 August 2017 (2017)
3. Brock, D.C. (ed.): Understanding Moore's Law: Four Decades of Innovation. Chemical Heritage Press, Philadelphia (2006). ISBN 0941901416
4. Schutte, P.C., Goodrich, K.H., Cox, D.E., Jackson, B., Palmer, M.T., Pope, A.T., Schlecht, R.W., Tedjojuwono, K.K., Trujillo, A.C., Williams, R.A., Kinney, J.B.: The naturalistic flight deck system: an integrated system concept (2007)
5. Bannon, L.J., Schmidt, K.: CSCW: four characters in search of a context. In: ECSCW 1989: Proceedings of the First European Conference on Computer Supported Cooperative Work. Computer Sciences Company, London (1989)
6. Schmidt, K.: Cooperative work and its articulation: requirements for computer support. Le travail humain **57**, 345–366 (1994)
7. DARPA project: Explainable Artificial Intelligence (XAI). www.darpa.mil/program/ explainable-artificial-intelligence
8. eXCockpit project n° 815045-1. https://ec.europa.eu/research/participants/portal/desktop/en/ opportunities/h2020/topics/mg-2-1-2018.html

Beyond the Chatbot: Enhancing Search with Cognitive Capabilities

Jon G. Temple[✉] and Claude J. Elie[✉]

CIO Design, IBM, Armonk, NY, USA
{jgtemple,eliec}@us.ibm.com

Abstract. Conversational chatbot interfaces face pragmatic challenges that must be overcome in order for the user to obtain a positive end user experience. For example, users must understand the rules associated with the chatbot (domain, context, understanding of natural language, etc.). A null result will often end with, "Let me search the web for you", which can lead to dissatisfaction in the technology or system. This paper discusses an enterprise solution, which turns the expectation around for the user. Starting with a traditional search, the cognitive agent can work with the search keywords in the background to determine if intervention will provide value. If the cognitive evaluation has sufficient confidence in a solution, it is provided in addition to the traditional search results, which will likely delight the user. If the cognitive evaluation has insufficient confidence, the user receives traditional search results with no loss in user expectation.

Keywords: Chatbot alternative · AI · Hybrid search · Cognitive

1 Introduction

People are gaining increasing experience with cognitive interfaces (e.g., Cortana, Alexa, Siri, etc.) and many have seen amazing demonstrations about the potential of cognitive computing (e.g., Watson on Jeopardy). These real-world experiences make some users eager to try a cognitive interface primarily to test their limits.

This trend has led to a proliferation of chatbot interfaces, where users ask natural language questions to an avatar to receive a natural language (or scripted) response to their query. In some cases, this may delight the user by returning a specific answer to a specific question. In other cases, it may frustrate the user by misunderstanding the input or providing a less than satisfactory answer.

While the associated cognitive technologies are rapidly evolving, it is clear that most commercial attempts at conversational interfaces come with some level of frustration. The natural language training required to properly anticipate a wide range of speech variability can be daunting and will often lead to some gaps in interpretation. Moreover, unless sufficiently trained in a particular knowledge domain, the user may receive a generic answer to a specific question, which may be less than satisfying. So the question is, how can we use cognitive technologies to facilitate user tasks while minimizing the frustrations associated with a conversational interface?

© Springer International Publishing AG, part of Springer Nature 2019
T. Z. Ahram (Ed.): AHFE 2018, AISC 787, pp. 283–290, 2019.
https://doi.org/10.1007/978-3-319-94229-2_27

In the area of chatbots for IT support, one thing that is being explored is a hybrid search approach where users are given familiar search patterns, which are then augmented with cognitive technologies. For example, a search interface can support both keyword and natural language input. If the natural language processing leads to high probability match/solution/answer, the user can be presented with an augmented response in addition to traditional search results.

This approach leverages existing search behavior and provides cognitive technologies a platform to wow the user without the overhead associated with user expectations with conversational interfaces.

To further increase the chances of success, rather than allowing open-ended input from the user, wherever possible, options are given to the user to increase the likelihood of success. For example, if a user enters that they are having a hardware problem, the response can be a list of options which have reliable solutions (Fan, Battery, Display) rather than an open-ended prompt for more information.

This paper discusses a hybrid approach between search and chatbots and the associated benefits and challenges.

A note on terminology: As a rapidly evolving technology, the terminology is in constant flux. For simplicity, we will use the term "chatbot" generically to refer to all related technologies, such as virtual assistants, intelligent assistants, and personal assistants. We are also focused on IT Support, although applicability may be far wider.

2   What Problem Were We Trying to Solve?

Our users have access to a large database of solution documents which is continually growing and evolving to address the current IT issues of the business. Users can search or browse the database for solutions to their issues.

The issue we are trying to address is that sometimes users have difficulty locating solutions even when a solution may exist. We'd like to apply a cognitive solution to see if we can facilitate the surfacing of applicable self-service solutions, providing a better end user experience while reducing costly help desk services.

2.1   Why Not a Chatbot?

Training to the Right Level of Specificity

There is significant effort required to train a chatbot to reliably respond to natural language queries. Depending on the scope/domain of your chatbot, this could be a fairly limited exercise or an extensive one. Not surprisingly, new chatbots or chatbots on a budget will not be as fully trained as a mature solution. They may, for example, return generic answers to specific questions as they are much easier to develop than dynamic answers. For example, a "show me jobs available in Florida" query may come back with 'here is a list of available jobs' (not scoped to Florida). The response is not wrong, per se, but if the user sees jobs that are not in Florida, or worse, doesn't realize that their answer wasn't properly scoped, the user will invariably be disappointed, so it is incumbent on the development team to anticipate the level of specificity of user questions and prepare accurate solutions.

A chatbot is only as useful as the amount of training it has received, and it is not possible to train a chatbot on every possible question for every domain. If the corpus for your chatbot is too shallow, it may feel more like an interactive FAQ; if the corpus is too deep, it may take too long to get to market. Finding that Goldilocks point can be quite challenging. One cost of getting this balance wrong is a lack of repeat visitors; you can only make a first impression once.

Chatbots Need to Provide Answers, Not Just Search Results

Commonly, chatbots degrade to search when the answer cannot be found. For example, one of the more common disappointing things Siri can say is, "here is what I found on the web." This is clearly better than nothing, but it also means the user will not be getting a direct answer to the asked question. Also, search results surfaced in a chatbot tend to be limited in scope and lack basic search functions such as filters or pagination. A chatbot is a poor replacement for dedicated search. Ideally, degrading to search should happen as little as possible, or you may have a training or scope issue. Chatbots may also compete with a search function (if an independent search exists), and users are left having to guess which path is more likely to produce a successful result.

Natural Language Processing

Natural language processing can be wonderful when everything works flawlessly, but there are significant limitations. Chatbots need a sufficient vocabulary, which may include phrases, idioms, and synonyms, as well as an adequate technical lexicon for the chosen knowledge domain. Users must guess what words work and what terms don't. In the worst case, users may need to memorize those keywords that have proven most effective in the past.

There are also challenges associated with non-native speakers and regional accents, illustrated by the popular comedy skit where two Scots become stuck in a voice-recognition controlled elevator which has been optimized to understand an "American accent" [1]. Even natural language chatbots that leverage typed input require non-native speakers to both read and type phrases in another language than their own. Translation of an entire chatbot corpus into additional languages requires a significant investment.

Communicating the Domain of the Chatbot

A chatbot's knowledge domain has its own challenges, as the user must puzzle out what can be asked, and how to boil those concepts down to simpler concrete relationships. Amazon sends out a newsletter that teaches users about the Echo's new capabilities and skills, all of which have to be recalled and spoken precisely like a magical invocation. Moreover, it is common for a chatbot to instruct the user with a format like, "here are the things you can ask me" [2], and have these terms or phrases conveyed in the chat log. Psychologists have long understood that recognition is superior to recall [3], yet chatbots play to this weakness by focusing on the role of recall.

The popularity of chatbots actually compounds some of the problems stated above. For each chatbot encountered, there is cognitive overhead, requiring that the user must learn its set of rules: what can I ask about?, what words can I use?, how do I correct any mistakes?, does the chatbot understand context or is each statement treated as independent? Users can and do learn these rules, but reducing this cognitive overhead should be a paramount goal of every chatbot designer.

2.2   Why Not Just Search?

Recognizing a Solution from a Search Result List

Regardless of whether or not the user is successful in getting a search result that will solve the user's problem, the user must ascertain from a list view (which contains limited information) whether or not the details of a document will address their particular concern. It is not difficult to overlook correct articles simply because the title might be too generic or was not the top result [4]. For example, if a user searched for '6 × 7' an article containing the right answer tiled, 'Common mathematical solutions' could leave the user unsure if they would find the answer within.

Constructing Good Search Queries is an Acquired Skill

There are some users that are very proficient in using keywords and adapting them to quickly narrow down a list of search results. Any given search could produce thousands of results and some users have grown quite adept at identifying unique words while other users have difficultly constructing adequate queries [5]. For some, a natural language interface may be better suited to phrasing a question or issue.

When Do You Know a Solution Doesn't Exist?

When does a user decide that the solution they are looking for does not exist? Will they give up prematurely or appropriately? One of the challenges of a search interface against an unknown domain is that there may or may not be a solution that you haven't located yet either because the user didn't recognize it, or they didn't refine their search properly.

3   Competing Solutions: Search and Chatbots

Two common ways to assist users obtain IT support are search and a chatbot. Increasingly, both ways are offered to end users as separate solutions, despite the functional high overlap between them. This puts users in a position where they must choose between interfaces, potentially wasting more of their time should the chosen method not contain the right answer, and then needing to take the solution not attempted with no guarantee it will work, either.

Another possibility is to integrate any planned chatbot capabilities directly in search, thereby eliminating the "two interfaces" problem and many of the other issues with chatbots already discussed. In our user experience research, we have noticed that our users preferred searching over browsing for their information. Because search was already high traffic, it is an excellent foundation to showcase new technologies such as AI, eliminating the need to educate users about it or even how to get started. We also asked 213 users whether they would prefer improving search with AI or a separate chatbot. We found that twice as many of our users preferred enhancing search with AI to a completely separate chatbot interface.

While standalone chatbots have become the norm, some exceptions exist. For many, Google has become a verb that means search [6]. Rather than potentially confuse users with a competing solution that would potentially undermine their most valuable asset, Google has invested heavily in "deep learning" technologies to improve the effectiveness of their existing search [7]. In addition, Google leverages this and natural language to surface interactive self-service Widgets, from calculators to foreign language translation and currency conversions, or to look up movie times or to book an airplane flight.

4   What is Hybrid Search?

Hybrid search leverages existing search behavior and provides cognitive technologies to provide additional assistance to the user without the overhead associated with user expectations with conversational interfaces.

Regular chatbots start with a natural language query, and if trained with the appropriate intent, will provide an answer; failing all else, it will gracefully degrade to perform a search. Hybrid search also starts with natural language, while still accepting traditional keyword terms. Unlike a chatbot, search results are always returned; if there happens to be a positive match against an intent, this is returned above the search results.

Not only does this approach allow both search and AI recommendation to be combined in a single interface, but it appropriately sets expectations. When a chatbot returns a search result instead of a specific answer it has been trained on, the user may feel disappointed. On the other hand, the hybrid search approach grounds the user in the familiar world of search results and the user interface can be optimized for that purpose, with filters and pagination. Building on that foundation, the AI component would feel like a bonus that was not promised in advance, appearing only if the AI was specifically trained to provide a better answer than search alone.

To illustrate this, consider a scenario where a user has been encountering difficulties resetting her password (see Fig. 1). As with a regular search engine, she may simply search for "I need to reset my password". If there were no intents trained specifically for this, a full set of search results would be returned, and the user would be no worse off than before. However, if a set of password intents had been trained, she may instead see a prompt asking for clarification on which password she wants to change, listing out the most common passwords (providing options with clear success paths instead of an open-ended conversation). Should other questions be needed from the AI, they would be handled with additional prompts as needed.

Moreover, hybrid search also leverages self-service widgets. In the search example, the AI may return a form directly to the search results page to be completed by the user to reset her password, or provide her with a simple 'Reset password' button, or offer to open a ticket with the Help Desk should it be a type of password the AI cannot handle. By displaying the form or prompt, you are indicating to the user that you understood their question and can provide an answer right now, without digging any deeper into an article.

Hybrid search also differs in its approach towards natural language. For experienced users, keyword search can be extremely precise, especially when combined with booleans and other advanced functions. This capability is retained. However, natural

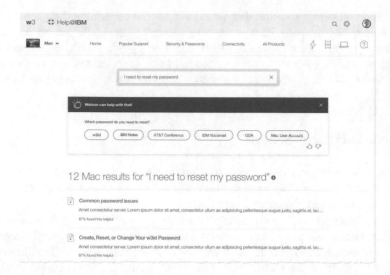

Fig. 1. Example of hybrid search, with cognitive recommendation on top, and search results below.

language may prove superior in areas where the search is more nuanced or where there is a need to establish context for the search, something that may be difficult with keywords. For example, if our user had searched for "I want to reset my Mac password", there would be no need to ask her which password she would like to reset since that ambiguity has been resolved through use of natural language.

Moreover, natural language is only used to initiate a new search. A conversation may start with natural language, but any additional responses from the user are elicited through additional menus prompts in order to reduce ambiguity and guess work. Should the user enter a new phrase in the search field, this is treated as the beginning of a new conversation if a matching intent is found. No attempt is made to store a retrospective history of the conversation.

5   Advantages of Hybrid Search

Single Point of Entry. Since search and the AI are combined, user does not need to decide between a chatbot and search.

Reduced Learning Curve. Due to the large variety of available chatbots, there is overhead in learning a new chatbot. However, search is pretty ubiquitous and there is much less to learn.

Support Natural Language Input. If users are more comfortable asking a question in natural language format, it is supported but not required.

Correcting Mistakes. It is much easier to refine a mistake as an input to a search field than to a submitted conversational element to a chatbot in a messaging platform which may or may not understand the context of the previous query.

Follow Up Refinements. With a full natural language chatbot, a refinement or clarification may involve either a prompt or the request for another natural language exchange. And each natural language exchange may entail the need for more clarification to deal with ambiguity. Hybrid search standardizes all such refinements as prompts, thereby eliminating a lot of guess work by human and machine.

Intent-triggered Advice. Hybrid search makes search paramount - but unlike your Uncle Ed who you only see at Thanksgiving and weddings, it only offers assistance when it actually knows something useful.

Graceful Ramp Up. Related to the last item is that chatbots do not ramp up gracefully. It may take time to write all of the intents you need, but you may not want to wait till it is perfect before making it publicly available. People who tried your chatbot on the first day may have a very different impression from those try it on Day 365, assuming they ever come back. The search engine sets expectations, not the AI; but over time, hybrid search will keep getting smarter.

Multiple Language Support. Training an AI on multiple languages may be very costly and time consuming. For nonsupported languages, hybrid search degrades to regular search. However, you probably should not consider deploying a chatbot for a population until it fully supports that language.

6   Disadvantages of Hybrid Search

Not Able to Reuse Existing Chatbots UIs. There may be existing chatbot technology/code that you can use as a base for reuse which can save some time in the development process. That said, the more significant component will be training up the chatbot, which can be reused.

Not Ideal Choice for Screening a VRU or Chat. Hybrid search is not appropriate for a voice response unit (VRU), which may benefit from either a conversational voice recognition system, a menu driven system, or a combination of the two. As a screener for human to human chat, a within-window switch from AI to human seems a more natural transition.

Not Personal. In essence, this is search. The 'personal' touch is less applicable with a search system as your base. That said, when the AI does provide an answer, there is room for personality in how the intent is presented.

Optimized for Typed Input, Not Speech Input. While suitable for a desktop or mobile device, hybrid search is not a good choice for a voice recognition appliance like the Amazon Echo or Google Home.

Search System is Required for the Base. Hybrid search assumes you already have an effective search engine as a base.

7 Considerations

Technical

Throughout this paper, we have stayed away from technical implementation details. This is something that could be addressed in a future paper. We have implemented hybrid search using Watson Conversations, married to an open-source search engine. In theory, hybrid search could be implemented using a variety of products and open source solutions, although we are understandably quite partial to the Watson line of products.

Interestingly, there is a large amount of technical overlap between hybrid search and a regular chatbot. The challenge to write the intents is largely the same; training requirements appear to be very similar as well. We also compared many of the key requirements and benefits around a proposed chatbot to hybrid search and found them to be similar as well.

Final Words

The most important thing to consider when deciding to implement an enhancement is the impact the enhancement will have on your users. If, after you've done an analysis of your particular situation, you feel like a chatbot is the correct solution to proceed, by no means should this paper discourage you. For our environment, however, our research and analysis gave us a clear path forward. Hybrid search openly embraces cognitive aspects of a chatbot without the associated cognitive overhead by leveraging and augmenting established patterns of behavior.

References

1. Voice Is the Next Big Platform, Unless You Have an Accent. https://www.wired.com/2017/03/voice-is-the-next-big-platform-unless-you-have-an-accent/
2. Questions for a Starter Chatbot Template: A Simple Case Study. https://chatbotslife.com/questions-for-a-starter-chat-bot-template-a-simple-case-study-fbb0a16ef84b
3. Recall (memory). https://en.wikipedia.org/wiki/Recall_(memory)
4. Clive Thompson on Why Kids Can't Search. https://www.wired.com/2011/11/st_thompson_searchresults/
5. Seven Web Search Habits You Should Know. https://www.lifewire.com/highly-effective-web-searchers-3482841
6. Google (verb). https://en.wikipedia.org/wiki/Google_(verb)
7. Google's vision for AI is the right one. https://mashable.com/2017/05/17/google-io-artificial-intelligence-plan/#lprL27eooqqx

A Comparative Analysis of Similarity Metrics on Sparse Data for Clustering in Recommender Systems

Rodolfo Bojorque[1,2(✉)], Remigio Hurtado[1,2], and Andrés Inga[1]

[1] Universidad Politécnica Salesiana, Cuenca, Ecuador
{rbojorque, rhurtadoo, cingac}@ups.edu.ec
[2] Universidad Politécnica de Madrid, Madrid, Spain

Abstract. This work shows similarity metrics behavior on sparse data for recommender systems (RS). Clustering in RS is an important technique to perform groups of users or items with the purpose of personalization and optimization recommendations. The majority of clustering techniques try to minimize the Euclidean distance between the samples and their centroid, but this technique has a drawback on sparse data because it considers the lack of value as zero. We propose a comparative analysis of similarity metrics like Pearson Correlation, Jaccard, Mean Square Difference, Jaccard Mean Square Difference and Mean Jaccard Difference as an alternative method to Euclidean distance, our work shows results for FilmTrust and MovieLens 100K datasets, these both free and public with high sparsity. We probe that using similarity measures is better for accuracy in terms of Mean Absolute Error and Within-Cluster on sparse data.

Keywords: Clustering · Recommender systems · Similarity measures

1 Introduction

Nowadays, the expansion of information and communication technologies (ICT), the mass use of the Internet and social networks; and the product customization techniques are causing the paradox of choice problem. There are a lot of same class products that enterprises offer to customers and the user cannot choose the product that best adapts to their necessities. In this context, the recommender systems (RS) have an important role to help users to choose what they search.

Recommender Systems are a type of information filter to overcome the information overload problem [1]. These systems predict the impact that an unknown item will have on a user. RS collect information on the preferences of its users for a set of items (e.g., movies, songs, books, jokes, gadgets, applications, websites, travel destinations and e-learning material) [2]. Hitherto, collaborative filtering (CF) is the most successful approach for personalized product or service recommendations [3]. Neighborhood based collaborative filtering is an important class of CF, which is simple, intuitive and efficient product recommender system widely used in commercial domain [4].

The most popular items to which RS have been applied are movies due to open datasets [5], which are released for research. However, RS have a drawback, datasets

© Springer International Publishing AG, part of Springer Nature 2019
T. Z. Ahram (Ed.): AHFE 2018, AISC 787, pp. 291–299, 2019.
https://doi.org/10.1007/978-3-319-94229-2_28

are sparse and difficult to clustering groups of users or items [6]. Recently, increasing relevance of groups of users in the social web has led to a significant expansion of Group Recommender Systems [7, 8]. There are certain scenarios in which recommending a set of items to a group of several users is more appropriate than providing several sets of recommendations to each individual user of the group (e.g. recommending a movie to a group of friends or recommending a tourist pack to group of travelers).

This work is organized as follows. In Sect. 2 we introduce basic concepts - such as clustering approaches, classical K-Means algorithm, and similarity measures - that form a basis for all subsequent exposition. Then we proceed to explain our methodology and evaluation metrics in Sect. 3. Section 4 shows results on public datasets MovieLens 100K and Filmtrust and finally in Sect. 5 we present our conclusions.

2 Related Work

2.1 Clustering

Clustering is an unsupervised classification approach for recognizing patterns, which is based on grouping similar objects together. This approach is useful for finding patterns in an unlabeled dataset. Machine learning, bioinformatics, image analysis, pattern recognition and outlier detection are few of many application areas of clustering [9]. There are two approaches to clustering are hierarchical and partitional clustering.

Hierarchical clustering produces nested series of partitions and produces a dendrogram showing pattern and different similarity levels of grouping [10]. Partitional clustering aims at finding the single partition [11, 12] rather than numerous as in hierarchical methods. The second one approach benefit is that it can be applied to large datasets for which dendrogram does not work.

K-Means clustering is one of the classical and most widely used clustering algorithm developed by Mac Queen in 1967. This approach is a partitional clustering algorithm which divides de datasets into k disjoint clusters. Let be C as the set of centroids and n the number of samples, (1) represents the objective function named as Within-Cluster. The K-Means objective J_c is minimized the sum of Euclidean distances between each sample X_i and its centroid C_j. Though K-Means is known for its intelligence to cluster large datasets its computation complexity is very expensive for massive data sets [13].

$$J_c = \sum_{j=1}^{c} \sum_{i=1}^{n} \left\| X_i - C_j \right\|^2 \tag{1}$$

2.2 Similarity Measures

The similarity measures between two users are calculated taking into account only the ratings made by these two users [14]. Traditional measures such as Pearson correlation (PC), Jaccard based metrics are frequently used in recommendation systems [15].

- Pearson Correlation: Measures like two user u and v are linearly correlated, where I' is the set of co-rated items, r_u is the rating average of user u and r_v is the rating average of user v.

$$S_{PC}(u,v) = \frac{\Sigma_{i \in I'}(r_{ui} - \bar{r}_u)(r_{vi} - \bar{r}_v)}{\sqrt{\Sigma_{i \in I'}(r_{ui} - \bar{r}_u)^2}\sqrt{\Sigma_{i \in I'}(r_{vi} - \bar{r}_v)^2}} \tag{2}$$

- Jaccard: Uses the rating information provided by a pair of users, where I_u is the set of items rated by user u and I_v is the set of items rated by user v.

$$S_{Jac}(u,v) = \frac{|I_u \cap I_v|}{|I_u \cup I_v|} \tag{3}$$

- JMSD: Combines Jaccard similarity and Mean Square Difference similarity defined in (4). $|I'|$ is the cardinality of set co-rated items between user u and user v. This metric uses the numerical information from the ratings (via mean squared differences) also uses the non-numerical information provided by the arrangement of these.

$$S_{MSD}(u,v) = 1 - \frac{\Sigma_{i \in I'}(r_{ui} - r_{vi})^2}{|I'|} \tag{4}$$

$$S_{JMSD}(u,v) = S_{MSD}(u,v) x S_{Jac}(u,v) \tag{5}$$

- MJD: Combines some similarity measures on a neural network, where S_v^0, S_v^1, S_v^2, S_v^3, and S_v^4, are weights defined in [16].

$$S_{MJD}(u,v) = S_{v^0}(u,v) + S_{v^1}(u,v) + S_{v^2}(u,v) + S_{v^3}(u,v) + S_{v^4}(u,v)$$
$$+ S_{MSD}(u,v) + S_{Jac}(u,v) \tag{6}$$

3 Methodology

We are using similarity instead of distance, which can be computed by measuring the similarity between a sample and centroid. As similarity and distance are inversely proportional to each other, distance function can be modeled as follows [17]:

$$\left.\begin{array}{l} \frac{1}{sim} \\ \\ MAX_{DIST} \end{array}\right\} \begin{array}{l} if \quad sim \neq 0, \\ otherwise \end{array} \tag{7}$$

Where $MAX_{DIS} = 1{,}000$ denotes the maximum distance between two points.

Unlike to minimize the distance between sample and centroid, we maximize the similarity between a user and centroid, then (1) is modified by (8) where *sim* define any similarity measure. The K-Means clustering algorithm converges when centroids no longer changes or a defined number of iterations is reached.

$$J_c = \sum_{j=1}^{c} \sum_{i=1}^{n} sim(X_i, C_j) \tag{8}$$

For evaluation methodology, we have randomly divided the dataset into training (D^{Train}) and testing (D^{Test}) by performing 5-fold cross-validation and reported average results. Since we are using movies domain (FilmTrust and MovieLens 100K) for each user, 80% randomly divides movies (rated by them) are chosen as training set and the rest (20%) as the test set.

Mean Absolute Error (MAE) has been extensively used in many research projects, such as [9]. It computes the average absolute deviation between predicted rating provided by a recommender system and a true rating assigned by the user. It is computed as:

$$MAE = \frac{1}{|D^{Test}|} \sum_{i=1}^{|D^{Test}|} |p_i - a_i| \tag{9}$$

Where $|D^{Test}|$ is the total number of ratings provided by the test set, p_i is predicted rating and a_i is the actual rating assigned by the user to item i. To obtain pi we use clusters defined by K-Means. Finally, we use a K-Nearest-Neighbour (K-NN) approach.

4   Results

In this section, we present the performance comparison of aforementioned approaches in terms of MAE and cluster quality (Within-Cluster). We make clusters from k = 2 to 10, because the best results in terms of accuracy are obtained with low k values.

Figures 1 and 2 show the results of MAE for Euclidean Distance (MAEwED), Pearson Correlation (MAEwPC), Mean Square Difference (MAEwMSD), Jaccard (MAEwJaccard), Jaccard Mean Square Difference (MAEwJMSD) and Mean Jaccard Difference (MAEwMJD) on MovieLens 100K and FilmTrust datasets respectively. MJD is the best metric in terms of accuracy (MAE). However, a good accuracy is not related to good Within-Cluster. Figures 3 and 4 show the result of Within-Cluster for MovieLens 100K and FilmTrust datasets for Euclidean Distance (W-CwED), Pearson Correlation (W-CwPC), Mean Square Difference (W-CwMSD), Jaccard (W-CwJaccard), Jaccard Mean Square Difference (W-CwJMSD) and Mean Jaccard Difference (W-CwMJD) on MovieLens 100K and FilmTrust datasets. For Within-Cluster Pearson Correlation is the best metric. Furthermore, MSD based metrics are better than Euclidean Distance.

Tables 1 and 2 show the numerical results for MAE and Tables 3 and 4 show numerical results for Within-Cluster on MovieLens 1M and FilmTrust datasets respectively.

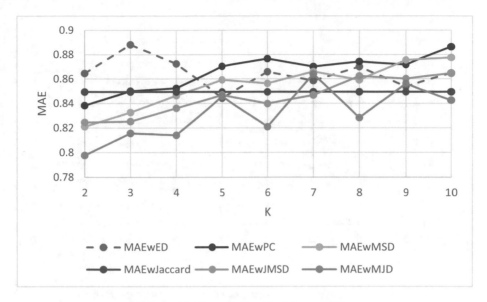

Fig. 1. MAE results for MovieLens 100K.

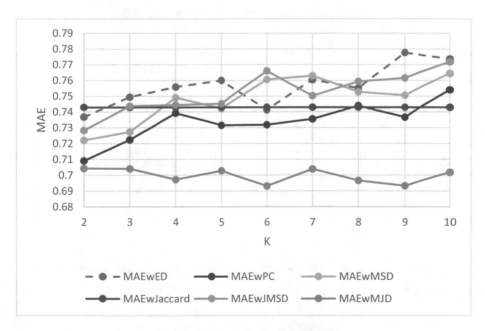

Fig. 2. MAE results for FilmTrust.

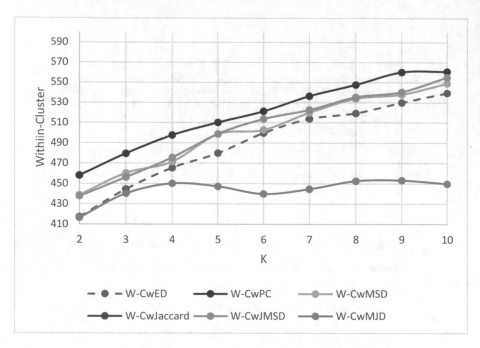

Fig. 3. Within-cluster results for MovieLens 100K.

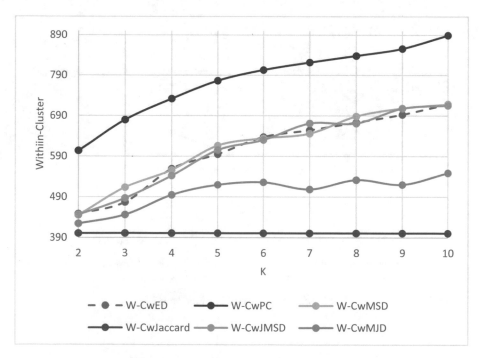

Fig. 4. Within-cluster results for FilmTrust.

Table 1. MAE results for K = 2 to K = 10 for MovieLens 100K

K	MAEwED	MAEwPC	MAEwMSD	MAEwJaccard	MAEwJMSD	MAEwMJD
2	0.865	0.838	0.821	0.849	0.824	0.798
3	0.888	0.850	0.833	0.849	0.825	0.816
4	0.872	0.852	0.846	0.849	0.836	0.814
5	0.844	0.870	0.859	0.849	0.847	0.846
6	0.866	0.877	0.856	0.849	0.840	0.821
7	0.859	0.870	0.866	0.849	0.847	0.864
8	0.870	0.874	0.859	0.849	0.862	0.828
9	0.854	0.872	0.876	0.849	0.860	0.856
10	0.865	0.886	0.878	0.849	0.865	0.843

Table 2. MAE results for K = 2 to K = 10 for FilmTrust

K	MAEwED	MAEwPC	MAEwMSD	MAEwJaccard	MAEwJMSD	MAEwMJD
2	0.737	0.709	0.722	0.743	0.728	0.704
3	0.749	0.722	0.727	0.743	0.744	0.704
4	0.756	0.739	0.749	0.743	0.744	0.697
5	0.760	0.732	0.742	0.743	0.745	0.703
6	0.741	0.732	0.761	0.743	0.766	0.693
7	0.760	0.736	0.763	0.743	0.750	0.704
8	0.755	0.744	0.753	0.743	0.759	0.697
9	0.778	0.737	0.751	0.743	0.762	0.693
10	0.774	0.754	0.764	0.743	0.772	0.702

Table 3. Within-cluster results for K = 2 to K = 10 for MovieLens 100K

K	W-CwED	W-CwPC	W-CwMSD	W-CwJaccard	W-CwJMSD	W-CwMJD
2	417.141	458.095	438.776	395.980	437.683	416.505
3	444.514	479.496	460.280	395.980	456.196	440.152
4	465.499	497.622	471.447	395.980	475.640	450.146
5	479.889	510.199	498.633	395.980	498.925	447.215
6	499.823	521.316	502.888	395.980	513.721	439.876
7	513.858	536.417	519.995	395.980	522.744	444.650
8	519.440	547.671	533.841	395.980	535.391	452.671
9	529.921	560.254	537.940	395.980	540.490	453.283
10	539.481	560.956	549.105	395.980	555.279	449.852

Table 4. Within-cluster results for K = 2 to K = 10 for FilmTrust

K	W-CwED	W-CwPC	W-CwMSD	W-CwJaccard	W-CwJMSD	W-CwMJD
2	448.564	603.729	445.587	400.768	446.480	424.527
3	477.065	679.989	513.676	400.768	487.031	446.325
4	559.580	732.070	556.583	400.768	542.545	494.800
5	595.320	776.493	616.568	400.768	605.462	519.867
6	637.567	803.212	634.591	400.768	630.463	526.233
7	654.481	821.785	646.173	400.768	671.152	508.671
8	673.496	838.468	688.699	400.768	671.434	532.139
9	693.024	856.079	708.443	400.768	708.069	520.502
10	718.611	889.260	718.445	400.768	715.447	549.540

5   Conclusions

We propose a novel method for clustering on sparse data, the results show that using Pearson correlation as similarity measure to train K-Means clustering algorithm obtains better results in terms of accuracy and cluster quality instead of classical approach that uses Euclidean distance. Our fundamental contribution consists in using similarity measures to clustering sparse datasets.

Future works include a comparison of different similarity metrics, specially Jaccard based metrics since Pearson correlation has some drawbacks when there are a few co-rated items.

References

1. Ortega, F., Hernando, A., Bobadilla, J., Kang, J. H.: Recommending items to group of users using matrix factorization based collaborative filtering. Inf. Sci. **345**, 313–324 (2016). ISSN 00200255, https://doi.org/10.1016/j.ins.2016.01.083
2. Arthur, D., Vassilvitskii, S.: K-Means++: the advantages of careful seeding. In: Proceedings of the Eighteenth Annual ACM-SIAM Symposium on Discrete Algorithms, SODA 2007, Society for Industrial and Applied Mathematics, Philadelphia, PA, USA, pp. 1027–1035 (2007). ISBN 978-0-898716-24-5
3. Meteren, R., Someren, M.: Using content-based filtering for recommendation. In: Proceedings of ECML 2000 Workshop on Maching Learning in Information Age, pp. 47–56 (2000)
4. Adomavicius, G., Tuzhilin, A.: Toward the next generation of recommender systems: a survey of the state-of-the-art and possible extensions. IEEE Trans. Knowl. Data Eng. **17**(6), 734–749 (2005)
5. Su, X., Khoshgoftaar, T.M.: A survey of collaborative filtering techniques. Adv. Artif. Intell. **4**, 2 (2009). ISSN 1687-7470, https://doi.org/10.1155/2009/421425
6. Lü, L., Medo, M., Yeung, C. H., Zhang, Y.-C., Zhang, Z.-K., Zhou, T.: Recommender systems. Phys. Rep. **519**(1), 1–49 (2012). ISSN 03701573, https://doi.org/10.1016/j.physrep.2012.02.006
7. Jameson, A., Smyth, B.: Recommendation to Groups, pp. 596–627. Springer, Heidelberg (2007). ISBN 978-3-540-72079-9, https://doi.org/10.1007/978-3-540-72079-920

8. Boratto, L., Carta, S.: State-of-the-Art in Group Recommendation and New Approaches for Automatic Identification of Groups, pp. 1–20. Springer, Heidelberg (2011). ISBN 978-3-642-16089-9, https://doi.org/10.1007/978-3-642-16089-91

9. Zahra, S., Ghazanfar, M.A., Khalid, A., Azam, M.A., Naeem, U., Prugel-Bennett, A.: Novel centroid selection approaches for KMeans-clustering based recommender systems. Inf. Sci. **320**, 156–189 (2015)

10. Bouguettaya, A., Yu, Q., Liu, X., Zhou, X., Song, A.: Efficient agglomerative hierarchical clustering. Expert Syst. Appl. **42**(5), 2785–2797 (2015). ISSN 0957-4174, http://dx.doi.org/10.1016/j.eswa.2014.09.054

11. Ghazanfar, M. A., Szedmak, S., Prugel-Bennett, A.: Incremental kernel mapping algorithms for scalable recommender systems. In: 2011 IEEE 23rd International Conference on Tools with Artificial Intelligence, pp. 1077–1084 (2011). ISSN 1082-3409, https://doi.org/10.1109/ictai.2011.183

12. Hruschka, E.R., Campello, R.J.G.B., Freitas, A.A., Ponce, A.C., de Carvalho, L.F.: A survey of evolutionary algorithms for clustering. IEEE Trans. Syst. Man, Cybern. **39**(2), 133–155 (2009)

13. Nazeer, K.A.A., Kumar, S.D.M., Sebastian, M.P.: Enhancing the K-Means clustering algorithm by using a o(n logn) heuristic method for finding better initial centroids. In: 2011 Second International Conference on Emerging Applications of Information Technology, pp. 261–264 (2011)

14. Bobadilla, J., Ortega, F., Hernando, A.: A collaborative filtering similarity measure based on singularities. Inf. Process. Manage. **48**, 204–217 (2012). ISSN 03064573, https://doi.org/10.1016/j.ipm.2011.03.007

15. Patra, B.K., Launonen, R., Ollikainen, V., Nandi, S.: A new similarity measure using Bhattacharyya coefficient for collaborative filtering in sparse data. Knowl.-Based Syst. **82**, 163–177 (2015)

16. Bobadilla, J., Ortega, F., Hernando, A., Bernal, J.: A collaborative filtering approach to mitigate the new user cold start problem. Knowl.-Based Syst. **26**, 225–238 (2012)

17. Ghazanfar, M.A., Prügel-Bennett, A.: Leveraging clustering approaches to solve the gray-sheep users problem in recommender systems. Expert Syst. Appl. **41**, 3261–3272 (2014)

Development of an Integrated AI Platform and an Ecosystem for Daily Life, Business and Social Problems

Kota Takaoka[✉], Keisuke Yamazaki, Eiichi Sakurai,
Kazuya Yamashita, and Yoichi Motomura

National Institute of Advanced Industrial Science and Technology Artificial
Intelligence Research Center, Tokyo, Japan
{kota.takaoka, k.yamazaki, e.sakurai, yamashita-kazuya,
y.motomura}@aist.go.jp

Abstract. Artificial intelligence (AI) has been making extraordinary progress. To keep developing the AI, reciprocal feedback-communication between AI and people in many use cases are important. Hence, this study aims to effectively build an AI platform from data collection to data analysis as well as the eco-systems in the field. The platform includes multiple applications for data collection, a cloud database to store data, probabilistic latent semantic analysis and Bayesian network as the AI by which people understand why predictions and recommendations are provided as a white-box. The platform can be easily customized and comfortably deployed for each use case depending on user needs. In the test phase, as part of this study, the system has been deployed in several fields, such as museum events, vending machines, and local Child Guidance Centers that respond to child maltreatment. As part of future studies, the systems should continue to be tested and developed more openly.

Keywords: Artificial intelligence · Eco system for AI · System engineering
Bayesian network · Probabilistic modeling

1 Introduction

Currently, Artificial intelligence (AI) has been broadly applied owing to the immense development of machine learning, big data and highly specialized computer environments. However, these factors cannot be improved if additional data is not provided. In other words, sustainable data growth is important for their development. Therefore, this data should be utilized in valuable use cases as part of a streamlined start to this process. Furthermore, it should be frequently deployed in diverse areas by a large number of people. The environment and reciprocal communication between AI and humans is called an ecosystem.

In order to develop the AI platform and ecosystem, three aspects, namely, transparency, accountability, and a positive impact on society, are among the key values [1].

To demonstrate transparency, data should be consistent and the algorithm should be well understood [2]. In addition, a good explanation of what AI represents is essential for accountability because AI will be utilized in communities, local administration, and

© Springer International Publishing AG, part of Springer Nature 2019
T. Z. Ahram (Ed.): AHFE 2018, AISC 787, pp. 300–309, 2019.
https://doi.org/10.1007/978-3-319-94229-2_29

social justice [3]. Furthermore, a positive impact on the economy and society is related to the generation of positive value chains in the daily life of a community. These are broadly referred to with frame paradigms, from individual decision making, user community issues and communication contexts to governmental policy making.

With respect to such open innovation platforms, it is important to develop an AI platform from data collection to data analysis effectively as well as a mature eco-system to enable it to learn efficiently. Therefore, transparency for an AI algorithm and data collection, accountability for the explanation and prediction by AI, and a positive impact on value chains of communities are required in the ecosystem of our society.

By following the theoretical framework of daily life behavior modeling [4], living support applications such as digital signage, smart phones, and wearable devices can build user models to compute user behavior patterns (See Fig. 1).

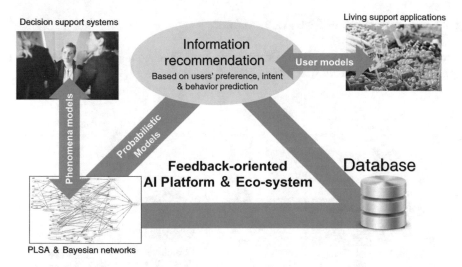

Fig. 1. Feedback-oriented AI platform and eco-system

Once the user-data is stored onto a cloud database, it will be analyzed by the AI which enables people to understand the algorithm. Then, when the AI has evaluated and made predictions based on the data, it can produce two different models. Firstly, phenomena models are built when users need to know what they should do. The models are able to detect the situations for which decision making support would be most highly requested. Depending on the situations (frames) which people are concerned about and do not have enough confidence to judge, the models can develop phenomena models of people's difficult situations related to their own decisions. Secondly, probabilistic models are generated and are computed to build mathematical models to provide information recommendations. Once a database stores user-model data appropriately, AI analyzes the data in step 1, collects each person's phenomena data in step 2, and provides a recommendation as a part of step 3. That is to say, AI can understand when, where, how, and what questions users have as a part of each frame, and optimize the best recommendation as much as possible. Then, the user can refer to the recommendation and consider his or her next behavioral response. The sequences of the information communication between the AI platform and users may lead to an improvement in their positive behavior experiences in the ecosystem. The feedback-oriented AI platform and ecosystem would be considered as a piece of evidence about the transparency, accountability, and positive impact on daily community life.

Therefore, this study aims to develop both an applicable AI platform and ecosystem related to daily life to employ the circular process shown in Fig. 1. As a part of open innovation, the system can be easily customized for the needs of each use case. As a concrete use case, the study introduces the core concepts and three use cases which can be comfortably deployed into community environments where people have social concerns such as business matters, daily life issues, and social problems.

2   Core Concept of the AI Platform

The AI platform contains four different modules. The concept map, shown as an application diagram is shown in Fig. 2.

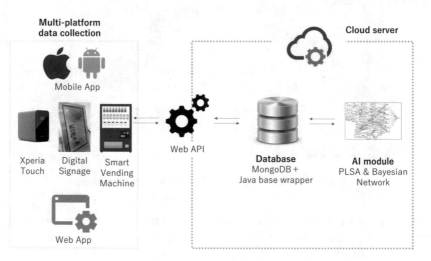

Fig. 2. Application diagram

First, multiple applications (c.f. iOS, Android, Web application, digital signage, XperiaTouch by Sony, a smart vending machine with display and Microsoft Kinect, and so on) are available for data collection. Depending on the situations, multi-platform data collection is beneficial because data should be collected from young children to elderly people. In the case of young children, almost all of them are familiar with smartphones and tablet applications for data collection. Conversely, elderly people might prefer a big display, such as digital signage and smart vending machines to answer questions. Furthermore, based on research questions, data collection environments are distinctly different. For instance, if users want to answer each question by themselves, smartphones and tablet apps will be preferred. However, if users want to discuss each question with friends and colleagues, digital signage and XperiaTouch might be useful because they have a big display and enable users to share the process of what they choose. To reflect many the user and research needs, the study considered the situation context for answering and developing a multi-platform for data collection.

Second, a cloud database for the storage of numerous data files (c.f. text, numerical, image, video files, and so on) was included. In each community, there could be several unplanned issues regarding data collection. For example, depending on the research progress or change of research purposes, additional data could be required. Therefore, the study needs to assume that the number of data variables could increase. Hence, the study employed a NoSQL database, such as MongoDB, because it is easy to add extra variables and respond in a flexible manner to diverse data.

Third, probabilistic latent semantic analysis (pLSA) for clustering and the Bayesian network were implemented on a server for probabilistic modeling and recommendations. Their advantages are that their results are easy to understand graphically and they take into account the uncertainty of the circumstances of daily life. Recommendations are also feasible for not only presenting information but also establishing effective questionnaires. The main component of the AI algorithm are explained in the following section.

Fourth, the foundation of the RESTful API is included and can be connected to each module and the cloud server. The system is flexible and able to be customized depending on the user needs and required functionally. In addition, the server has a login function and a supplemental database for data validation.

3   AI Algorithm

The study employed pLSA and a Bayesian network as the AI algorithm. This is because the Bayesian network is utilized in many fields, from marine conservation [5] to health care [6]. In addition, the study considers the importance of explaining AI. Basically, how current machine learning works is unknown and most explanations consider it as a black-box. However, if pLSA and a Bayesian network are combined, this could be one option for understanding why predictions and recommendations are provided as a white-box. The main algorithm is described in the following paragraphs.

pLSA [7] is a clustering method. In this method, the study assumes that the variable x in data d is generated via a latent variable z. As a part of the likelihood maximization of the Expectation–Maximization (EM) algorithm, the latent variable $z \in Z = \{z_1, \ldots, z_k\}$ is added to the co-occurring data. The co-occurrence frequency is shown as n(i, j). The joint probability is described by the following equation.

$$P(x_i, d_i) = \sum_k P(x_i|z_k)P(d_i|z_k)P(z_k) \tag{1}$$

Subsequently, $P(x|z), P(d|z), P(z)$ are calculated by the EM algorithm which maximizes the following log likelihood function.

$$L = \sum_i \sum_j n(i, j) log P(x_i, d_i) \qquad (2)$$

pLSA can maximize the information content of the EM algorithm. That is, it consists of dimension reduction, which has less information content loss and can be implemented by the producing segments. The latent classes extracted by pLSA are beneficial for big data analysis. However, it does not clearly explain what the latent class means. Therefore, it is important that relationships between latent classes be calculated graphically by the Bayesian network with other nodes. As pLSA often tends to fall into a locally optimal solution, multiple initial value settings are necessary.

The Bayesian network [8] enables the prediction of specific events and provides reasonable decision making presented by probabilistic modeling. The product of joint probabilities among the variables shows the simultaneous distribution of the model that the Bayesian network presents. The Bayesian network is structured as a graphical representation of probabilistic relationships between several random variables and represents their conditional dependencies via a directed acyclic graph. Each variable is called a node and nodes are related by strings called links. For example, the dependence relation between random variables X_i and X_j is represented by a directed link $X_i \rightarrow X_j$. In this case, X_i is the parent node and X_j is the child node. If it is assumed that the set of parent nodes $\pi(X_j) = \{X_1, \ldots, X_i\}$ with child node X_j exists, the dependency relation between X_j and $\pi(X_j)$ is quantitatively indicated by the following conditional probability.

$$P(X_j|\pi(X_j)) \qquad (3)$$

Furthermore, for each of the n random variables $X_1, \ldots, X_n$ similar to a child node, the joint probability distribution of all the random variables is shown below.

$$P(X_1, \ldots, X_n) = \prod_j P(X_j|\pi(X_j)) \qquad (4)$$

If the Bayesian network is applied to big data, the number of states for the discrete random variable becomes very large. Therefore, the size of the conditional probability table also becomes large and the frequency distribution becomes sparse. In order to solve the problem, it is necessary to cluster the state to an appropriate granularity before building the Bayesian network [9]. Consequently, the study operates pLSA as prior processing which can prevent the frequency distribution from becoming sparse [10].

4   Use Cases

In the test phase, we deployed the system into several fields and specifically collected numerical and categorical data. For example, we implemented the AI platform and ecosystem into museum events, vending machines, and local Child Guidance Centers in Japan which respond to child maltreatment cases.

4.1   Museums and Events

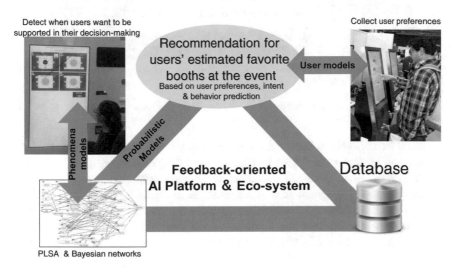

Fig. 3. AI platform and ecosystem for museum and events

First, in the museum, the AI platform and ecosystem collected visitors' data and suggested what the visitors should view (See Fig. 3). For example, when users in the event entered their information into a digital signage system and the AI provided personalized recommendations about which booths in the event might be preferred by each user. Once users received the recommendation, they can decide where they would like to go based on them. In addition, the AI platform can collect when and where users want recommendations. AI can learn the time and situation frames from the data and update the probabilistic modeling.

4.2   Vending Machine (Under Development)

In vending machines, the AI platform and ecosystem collected user preferences connected with their NFC cards to understand user context, such as time, occasion, place, weather, temperature, the number of people in their party at the moment, and so on. Subsequently, the system provides the estimated preferred beverages for each user and shows the reasons why the system makes a certain recommendation (See Fig. 4). For

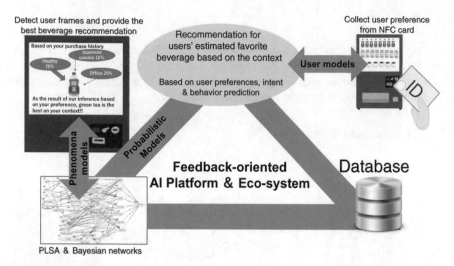

Fig. 4. AI platform and ecosystem in vending machines

example, a user is likely to buy coffee during the day based on his/her purchase history. However, the system may detect that he or she often buys beer at night during the summer season. Then, during the autumn, he or she might change his or her preferences and tend to buy green tea. The system keeps learning the users' profile and frames. If the user comes to a vending machine in which the system is embedded during the daytime and over the summer season, the system utilizes the user's past frames and recommends a bottle of green tea.

4.3   Child Abuse Cases

In Child Guidance Centers, the AI platform and ecosystem collect case data (See Fig. 5). It also calculates the risk of child abuse and provides decision making support for child protection. This system learns the risk of each child abuse case and the past decision-making history of social workers. Subsequently, it updates the probabilistic modeling and social workers can refer to the latest risk probability. Therefore, it enables social workers to promote their best practices. In fact, the AI platform and ecosystem is "practice as research" in the field in this case, which means local practitioners input risk assessment data as official logs in their practice into the system and the system provides a recommendation based on all previous child abuse cases that they and their predecessors have responded to. The system can learn each case and enhance their practice in the eco system.

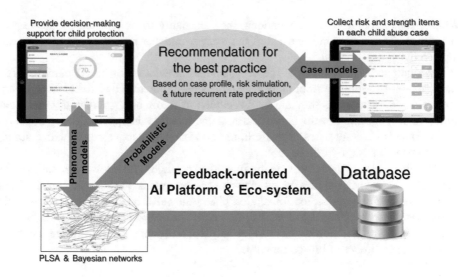

Fig. 5. AI platform and ecosystem for responding to child abuse cases

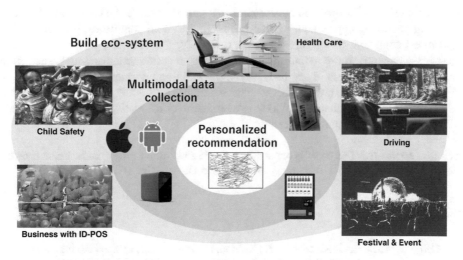

Fig. 6. Diverse opportunity to develop the AI platform and eco-system

5 Discussion

The study demonstrates an AI platform and ecosystem which supports multi-modal data collection and personalized recommendations (See Fig. 6). These functions have three unique points. Firstly, the AI platform and ecosystem can learn and utilize user and/or case frames, such as context, situation, timing, environmental issues, and so on. The Bayesian network can update prior information (user frames) according to actual user or case profile data (likelihood) into posterior information (an updated frame).

When people receive recommendations, they are likely to be affected by the recommendation. This situation means that the system updates people's own frames and supports good transparency.

Secondly, a consideration of the ecosystem is inevitable for an AI platform. The platform can generate an intuitively understandable prediction and the user can learn something from them, as they are well-regarded in terms of accountability. As mentioned at the beginning of the paper, learning and updating happens not only in the AI system but also among users. Therefore, the ecosystem generates continuous learning opportunities for both AI and users.

Thirdly, the system can provide the functional modules and a low-cost platform for different use cases. Given that the AI platform and ecosystem have already almost been completed and deployed in the foundation phase, they will be embedded into multiple fields to evaluate how users will react to the social collective impact to their daily lives in the next phase. In addition to the first and second points, the third point creates a positive value impact in our communities.

6 Future Work

The proposed AI platform and ecosystem have started to drive the next evaluation and updating phase. To apply them to different needs and field conditions, it is required to continue testing and developing them more openly. Furthermore, in order to ensure their dissemination, more attention should be paid to confidentiality issues, such as in health care, government data, and so on. In addition, a scientific implementation methodology is required; this is because each field has a different way of implementing AI, based on the industry methods, government conservativeness, and academic robustness of the computed AI result.

Acknowledgments. This paper is based on results obtained from a project commissioned by the New Energy and Industrial Technology Development Organization (NEDO).

References

1. Cath, C., Wachter, S., Mittelstadt, B., Taddeo, M., Floridi, L.: Artificial intelligence and the 'good society': the US, EU, and UK approach. Sci. Eng. Ethics **24**, 1–24 (2017)
2. National Science and Technology Council Networking and Information Technology. Networking and Information Technology Research and Development Subcommittee. The National Artificial Intelligence Research and Development Strategic Plan, Washington, DC, USA (2016)
3. Executive Office of the President National Science and Technology Council Committee on Technology: Preparing for the future of Artificial Intelligence, Washington, DC, USA (2016)
4. Motomura, Y.: Predictive modeling of everyday behavior from large-scale data. Synth. Engl. Ed. **2**(1), 1–12 (2009)
5. Trifonova, N., Maxwell, D., Pinnegar, J., Kenny, A., Tucker, A.: Predicting ecosystem responses to changes in fisheries catch, temperature, and primary productivity with a dynamic Bayesian network model. ICES J. Mar. Sci. **74**(5), 1334–1343 (2017)

6. Ide, A., Yamashita, K., Motomura, Y., Terano, T.: Analyzing regional characteristics of living activities of elderly people from large survey data with probabilistic latent spatial semantic structure modeling, Boston, MA (2018)

7. Hofmann, T.: Probabilistic latent semantic indexing. In: Proceedings of the 22nd Annual International ACM SIGIR Conference on Research and Development in Information Retrieval, pp. 50–57 (1999)

8. Pearl, J.: Bayesian networks: a model of self-activated memory for evidential reasoning. In: Proceedings of the 7th Conference Cognitive Science Society, pp. 329–334 (1985)

9. Ishigaki, T., Takenaka, T., Motomura, Y.: Category mining by heterogeneous data fusion using PdLSI model in a retail service. In: Proceedings - IEEE International Conference on Data Mining, ICDM, pp. 857–862 (2010)

10. Hirokawa, N., Murayama, K., Motomura, Y.: Probabilistic latent spatiotemporal semantic structure models based on travel history data. In: 29th Annual Conference Japanese Society for Artificial Intelligence, pp. 3–4 (2015)

Entropy and Algorithm of the Decision Tree for Approximated Natural Intelligence

Olga Popova[✉], Yury Shevtsov, Boris Popov, Vladimir Karandey,
Vladimir Klyuchko, and Alexander Gerashchenko

Kuban State Technological University, Krasnodar, Russia
{popova_ob,pbk47,alexander_gerashchenko}@mail.ru,
{shud48,epp_kvy}@rambler.ru, kluchko@kubstu.ru

Abstract. An actual task is the classification of knowledge of a specified subject area, where it's represented not as information coded in a certain manner, but in a way close to the natural intelligence, which structures obtained knowledge according to a different principle. The well-known answers to the questions should be classified so that the current task could be solved. Thus a new method of decision tree formation, which is approximated to the natural intelligence, is suitable for knowledge understanding. The article describes how entropy is connected to knowledge appearance, classification of previous knowledge and with definitions used in decision trees. The latter is necessary for comparing the traditional methods with the algorithm of the decision tree obtaining approximated to the natural intelligence. The dependency of entropy on the properties of element subsets of a set has been obtained.

Keywords: Entropy · Natural intelligence · Knowledge · Production rule
Decision tree · Question and answer system binary tree

1 Introduction

Nowadays numerous classification tasks [1–3] are being solved in various areas of science. They have become actual due to the fast development of information technology. The progress in the methods of data collection, storage and processing has made it possible to collect huge masses of data. That is why the methods of automated analysis are necessary for such masses of data in order to process such a great amount of knowledge. Decision trees are suitable for solving the tasks of classifications. These trees make it possible not only to classify all objects but also to obtain the tree structure with the optimal code [1, 2] having the minimal length. The demand for it grows with the increase in the number of classified objects, because the height of the tree increases as well. Here the resulting code is represented according to the Huffman tree type, by means of zeros and ones. For obtaining the optimal tree they use the information theory method suggesting the usage of the notions of information gain [3] and entropy [2]. The obtained Shannon's formula for the nodes of classification trees and the basic definitions taken from information theory provide a conventional way of building a decision tree. The way of choosing the attribute [4] and the rules for obtaining the nodes [3] is always a creative process, which has the peculiarities conventional for this

© Springer International Publishing AG, part of Springer Nature 2019
T. Z. Ahram (Ed.): AHFE 2018, AISC 787, pp. 310–321, 2019.
https://doi.org/10.1007/978-3-319-94229-2_30

theory. Due to the fact that classification trees suggest adding the objects in the process of getting new experience, the problem of reconfiguring the tree is actual here. It is necessary to develop the ways of teaching the decision support systems using such trees [5]. The resulting situation is as follows. There are no volume limits for describing the objects, their properties, facts, etc., which are to be classified and used for solving the complex scientific and applied tasks due to the modern advances in computer technology. The limits start to appear when the apparatus of analysis for these facts, objects, etc., is being arranged. These limits are the disadvantages of using the decision tree method, and they have to be corrected urgently. The authors suggest another way to eliminate the disadvantages of this method: they propose to change the conventional approaches to using it. The pre-existing approach to using the decision tree method did not make it possible to solve a number of tasks by the method of decision trees, which are especially vulnerable to using the equivalent structures of knowledge representation and the rules for obtaining the tree nodes [6–9]. That is why the effect produced by automating such tasks should be greater than or at least equal to the efficiency of solving them manually. In this article let us analyze one of such tasks and the way of solving it.

It is the task of the classification of knowledge for a given subject area. For solving such tasks, the authors suggest using a specific approach to obtaining decision trees and knowledge representation, which is approximated to the natural intelligence. Such an approach made it possible to obtain an automated system with the efficiency close to the work of the natural intelligence. For this purpose, it was necessary to analyze the efficient modern ways of structuring and working with information, which are used by the natural intelligence while obtaining the ideal and checked knowledge of a subject area. In this connection the modern ways of improving the work of the intelligence and the authors' own experience of working with scientific information. The readers will probably find the common features in the approach to learning and memorizing the checked information, which are used by them as well.

Any human being is accustomed to the idea of knowledge as possessing the checked information (answers to the questions) making it possible to solve a set task. So, the natural intelligence is accustomed to structuring the knowledge it obtains following the principle of classifying the existing and the known (to it) answers to the questions in such a way for the purpose of solving any task it receives. The question "What method should be used to solve the task or what tool should be applied?" is usually important for any scientist. And then the set of knowledge of a given subject area is the methods of solution, which are structured by the natural intelligence in a specific way making it possible to choose the proper solution method while identifying the task properties. In this approach the decision methods are classified, their numeric equivalent being the number many times smaller than the number of all the tasks that already exist and can appear later. Such an approach to identifying the aims for memorizing and further structuring is typical for the natural intelligence. It lets the intelligence omit the unnecessary information that can "overload the human brain" and "confuse the process of understanding" the information obtained and, therefore, the process of its memorizing and structuring for further use. A well-trained human brain can structure the knowledge at a high level, turning a human into an expert in a certain subject area. That's why the correct methods of teaching students, which represent the knowledge in a way convenient for understanding and memorizing, are sometimes of such great importance [8]. Such methods can be observed in the mathematics and physics textbooks, where the methods of solving the tasks and the laws of physics are presented in the order of their appearance or making the discoveries. As we see, the process of structuring and teaching, i.e. adding new ideal knowledge, is carried out gradually, from the simpler methods to the more complex ones. The reader has probably noticed that a definite amount of school and university knowledge to be studied in a certain subject area is limited to certain methods or laws, and the rest stays out of the curriculum. The other methods, which appeared later, are presented in the specialized literature, articles, scientific reports, etc. This situation reflects the natural intelligence quality to remember only a certain amount of information: not more than 7–9 positions, as it has been found out. It is clear that when the amount of information increases very fast, it is hard to remember it all and to structure it correctly. Those who want to have strong intelligence use modern methods of memory training and information structuring for the efficient planning of time, money and learning process. Those who have succeeded to develop their abilities in this area are considered high-class specialists – experts. Compared to others, their knowledge and experience is considerable, and the way of functioning of their intelligence is more efficient. As we see, the human intelligence could be much more efficient than a computer system if it had more "memory volume" and "task performance speed" reserves.

There are well-known efficient methods of structuring by the natural intelligence, which resemble decision trees by the form of presenting information. It was necessary to compose a modified algorithm of obtaining decision trees in a way, which is approximated to that of the natural intelligence, defining how the notion of entropy is connected to the appearance of new knowledge and the classification of previous knowledge. Let us connect entropy to the main definitions used in building decision trees, because using this notion makes it possible to obtain the efficient tree structure and reflect the way of knowledge structuring used by the natural intelligence. The latter is necessary for comparing the new method of building decision trees, approximated to the natural intelligence, with the traditional methods.

2   Materials and Methods

It is known that entropy depends on the proportion in which a set is divided, and it decreases symmetrically following the increase of this proportion from 0 to ½. In the case of dividing a set into several parts entropy is at its maximum when the parts have equal sizes and equals zero if one of the parts occupies all the set. In order to obtain an efficient division of a set into subsets it is necessary to select an attribute which gives the largest increase of information. According to the standard understanding of entropy, accepted in information theory, the leaves of decision trees are classes, and for the classification of an object it is necessary to descend the tree sequentially; then the way from the tree root to its leaves can be described as the explanation of placing this or that object within a certain class. Then entropy is an average quantity of bits required for coding the attribute S of the set element A. So, if the occurrence probability for S equals ½, entropy equals 1 and a full-valued bit is required; and if S occurs at no equal probability, the element consequence A can be encoded more efficiently. Such an idea of entropy makes it possible to use coding theory for obtaining an optimal code of binary decision tree in order to use it in intelligent information systems, e.g. in decision support systems, for solving various tasks. Such decision trees make it possible to apply for them the well-known ways of data representation, accepted in modern databases and knowledge bases. This idea of entropy, information, and knowledge is typical for a process which can be automated well with the help of computers, but it differs from the way of information structuring followed by the natural intelligence.

It is typical for the natural intelligence to operate not the bits, but the amounts of knowledge it has learnt, i.e. understood, and could therefore structure in a proper way. Then, according to the above-mentioned statements of information theory and the definitions of entropy, we can consider that the knowledge 3 increase of a certain size $\varDelta 3$ will enable the entropy S decrease of a certain size $\varDelta S$ (Fig. 1). It is possible if knowledge $\varDelta K$ differs from knowledge K by a certain property C. The property C can be considered an attribute.

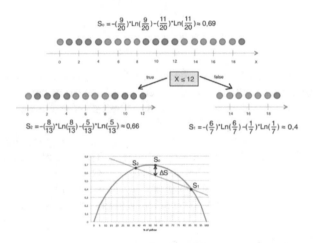

Fig. 1. Sample of calculating the entropy when choosing x

For the explanation of the method of decision tree building involving entropy they use a simplest example, which we shall use for comparing two approaches: the traditional one and the one suggested by the authors of the article. There is a set of two-coloured balls (Figs. 1 and 2), where the ball colour depends only on coordinate x. Obtaining the tree nodes, we shall perform calculations using Shannon's entropy

$$-\sum_{i=1}^{2} (p_i * \log_2 p_i) = -\frac{N_i}{N} * \log_2 \frac{N_i}{N} - \frac{N - N_i}{N} * \log_2 \frac{N - N_i}{N}, \tag{1}$$

where N is the number of balls belonging to the group to be split and N_i is the number of balls of the same colour in the group to be split, and the following learning algorithm: it is necessary to find the rules (predicated) as the basis for the training data set

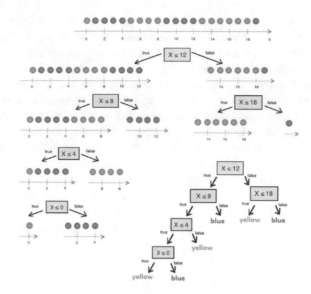

Fig. 2. Sample of solving the task in a traditional way.

so that the average value of entropy diminished. The process of dividing a data set into two parts according to a certain predicate leads to decrease in entropy. If the entropy value turns out to be less than the entropy of the initial set, it means that the predicate contains a certain generalizing information on the data.

Now let us demonstrate the suggested approach, which suggests the absence of the dependency of the ball colour on the coordinate x. The fact of the character of distributing the balls of two colours between themselves will not be taken into consideration. It is necessary to identify the property which will make it possible to divide the multitude of balls into two groups. The readers will notice that if they look at the given sequence they can see that

- all the figures within the sequence are balls;
- all the balls have the same size;
- all the balls within the sequence are of only two colours: yellow and blue.

The first two properties unite all the group of balls, while the third one divides them into two parts. So, the solution will be the property "Is the ball yellow?" which will divide the balls into two classes: the yellow ones and the blue ones (Fig. 3).

If a figure with a different property appears, it can be immediately distinguished from the already classified groups by that very property. If it is a square, the new property "Is this figure a circle?" will make it possible to divide the figures into two classes – the circles and the squares. And the new property will become the root of the tree. The new nodes can be added to the tree taking into account the probability for the other types of figures or the already existing figures different in colour or size to appear. They can become either new roots of the tree or its intermediate nodes. The more diverse the classified multitude is, the more efficient is the way of choosing the properties classifying this multitude. Comparing the decision trees presented in Figs. 2

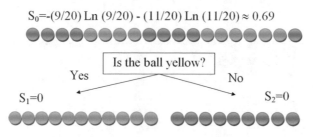

$$S_0 = -(9/20)\,Ln\,(9/20) - (11/20)\,Ln\,(11/20) \approx 0.69$$

Fig. 3. Sample of the solution using the new approach.

and 3, we come to the conclusion that the new approach provides a more efficient solution.

The analysed example characterizes a random non-meaningful formation of a multitude and adding elements to it. A subject area, on the other hand, is characterized by a well-grounded introduction of the new knowledge based on the already existing knowledge. The new knowledge can contribute to the already existing knowledge or discover new knowledge areas. This is what determines whether the new knowledge property will be an intermediate note of a decision tree or its new root.

Let the process of getting new knowledge of a subject area be infinite. The initial moment of this process is characterized by the fact that $3 = 0$ (where 3 stands for knowledge). When a number of tasks to be solved emerge, interdependent changes occur in the system, and they lead to the appearance of new solution methods having the properties C_1 and solving the tasks of the same class. The ongoing external influence leads to the appearance of new knowledge $\Delta 3$, which will influence the appearance of new solution methods. The well-known and new methods have the common property C_2, but the new methods do not have the property C_1. As it is seen in Fig. 4, with the appearance of new amount of knowledge the current entropy meaning, which characterizes the general condition of a subject area, diminishes in the course of time. The appearing new property, e.g. C_2, is the border between the already existing knowledge of a subject area and the non-knowledge, which will turn into knowledge in the course of time. It does not contradict to the idea of knowledge in philosophy, according to the definition given by Anaximenes. He presents knowledge as a circle with non-knowledge existing beyond its borders (green lines in Fig. 4).

The process of appearance of new knowledge and new properties corresponds to the learning algorithm for the new approach, which defines the rules for obtaining a new tree root. Popova, Romanov and Evseeva developed and published these rules in the scientific articles [7, 8] in regards to the question and answer system binary tree, which describes such a subject area as optimization methods [6, 9]. The initial data here are the optimization methods, which were to be classified. The result of transition along such a tree is the optimization method, which should be used for solving the task with the set of properties obtained in the solution process. The solution method itself pre-determines the class of tasks possessing these properties. The addition of a new solution method for an optimization task to the decision tree takes place after it is published and becomes well-known to the scientific community. The standard approach, on the other hand, would suggest the classification of all the existing

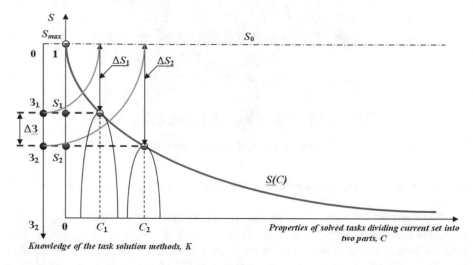

Fig. 4. Character of changing maximum entropy from the time, which depends on the appearance of new knowledge, i.e. of new methods of solving the tasks of a new type.

optimization tasks, including those of the same type, with gradual addition of new tasks to the tree with a similar solution method [6], which is very inefficient.

These rules were used for obtaining the decision tree classifying 64 well-known optimization methods. The rest 63 nodes are intermediate nodes, giving their unique from root to leaf paths and showing the course of finding a solution – selecting the most suitable method (see VIDEO1.mp4, binary tree of Q&A system.xlsx in [6]). In sum it gives 127 nodes in the decision tree. It is now possible to realize how effective this approach is, because the number of all the possible classification methods is considerably greater than 64.

3 Theory/Calculation

Now let us connect the well-known principles of information structuring performed by the natural intelligence for memorizing it with the definitions of entropy and the postulates accepted in information theory.

Following the first principle of structuring, information must be divided into groups and subgroups following a certain criterion meaningful for us. According to the second principle, the identified groups must be logically connected and arranged in a necessary order (following their importance, time, intensity, etc.).

It does not contradict the information theory statement that entropy depends on the proportion in which a set is divided. That is why it will diminish in the course of time as it is a necessary transition from chaos to order in case of obtaining a logical structure, which is easy to remember and visualize. This process will occur gradually due to the introduction of more and more new knowledge $\Delta 3$. The process of dividing a set of data into parts, leading to diminishing entropy, can be estimated as information production, where information is knowledge structured in a specific way.

Let us consider that, before the new method of obtaining decision trees emerged, the knowledge a certain subject area had't been classified properly, i.e. entropy had been at its maximum. Then all the set will have to be divided recursively into two parts in a proper way until only one element is left in the tree leaves. Such a way of dividing a set into parts makes it possible to take into account the main postulates obtained for entropy in information theory, which are listed above (in part 2 of the article), and to obtain an optimal and logical decision tree structure, which is easy to be implemented and memorized. It would be possible to divide a set into parts equal to the number of set elements, then at one division time entropy would diminish to zero. Such a situation could be possible if every element had its property. But, as a rule, for any set of knowledge in a certain subject area there is a series of properties, each corresponding to a certain number of the set elements. The properties that need to be selected for division first of all are the properties possessed by a larger number of elements (Fig. 5).

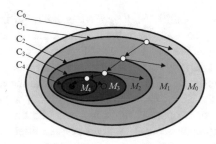

Fig. 5. Knowledge set of a certain subject area with its own series of properties.

It is seen in Fig. 5 that all the elements of the set M_0 possess a common property C_0. Let M_0 be the multitude of optimization methods to be classified, and the property set C_i is the properties, which first divide the optimization methods into groups and then separate each method from another as it is shown in Fig. 6. Then the subset M_1 has a property C_1, which unites a large number of elements of the initial set M_0. Therefore, this property will be the first to break the initial set M_0 into two parts, diminishing the maximum entropy S_{max} by the size ΔS_1. In the same way the entropy S_1 will be diminished by the size ΔS_2 due to the division of the subset M_1 by the property C_2 into two parts. This process will continue until the subset M_3 consisting of two elements (the black and the violet balls) is divided into two parts by the property C_4. As the subset M_4 consists of one element (the black ball), entropy is close to zero. Let us consider it close to 0 as there is a possibility that some more elements of the subset M_4 with the same properties C_0, C_1, C_2, C_3, and C_4 will appear in the course of time.

Recursively, the same actions are repeated with the remaining elements of the subsets M_0–M_1; M_1–M_2; M_2–M_3. As there is only one element (the lilac ball) left in M_3, partitioning is not performed.

Now the knowledge of a subject area structured in a specific way give the smallest size of entropy, but not zero, as knowledge will go on being generated by scientists. It

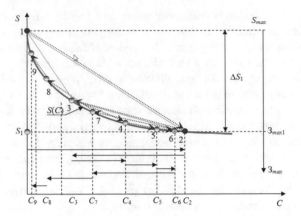

Fig. 6. Character of change of the dependency $S(C)$ in the process of the initial partitioning of the set recursively by the properties $C_1, C_2, C_3 \ldots, C_9$.

will continue until new scientific and applied tasks requiring their solution keep appearing. The tasks may have different levels of complexity and be connected to the topics referring to the already known and classified knowledge. That is why two more actions are added to breaking a set into parts: transition along the decision tree with searching the place of potential addition of the elements to the necessary subset and its further partitioning in the above-mentioned way; adding a new root of the tree if elements with property or propertied differing from the already existing ones have appeared. The latter action is similar to dividing an initial set into two parts by a property uniting the greatest number of subset elements. If several elements with the property C_{-1} have appeared, then the largest subset (Fig. 5) will be M_0 with the property C_0, which will break all the set into two parts. The first part will be the already classified elements, which will enter the tree without changes. The second part will be the subset of several elements with the property C_{-1}, which will be recursively divided in an analogous way. Therefore, in case of repeating this situation, the next property creating the new root of the tree will be C_{-1}. When the new root of the tree is obtained, entropy gradually diminishes and aims at zero.

Let us define the character of the dependency $S(C)$. In Fig. 6 it's seen that for the initial division of the set M_0 into two parts by the property C_2 the graph has the form of a line for the elements of the subset M_1 with the same property C_2 – it's the interval with points 1-2.

In the case of dividing this subset M_1 into two parts by the property C_3 a broken line with points 1-3-2 is obtained. During further division of the subsets into two parts the broken line gets a deflection and becomes smoother. For the example depicted in Fig. 6 it's the broken line with points 1-9-8-3-7-4-5-6-2. In the course of time new knowledge will make this broken line very smooth and close to the logarithmic function in form. Therefore, the dependency $S(C)$ can immediately be depicted in the form of such a graph (see the blue logarithmic curve)

$$-\sum_{i=1}^{2} (p_i * \log_2 p_i) = -\frac{N_i}{N} * \log_2 \frac{N_i}{N} - \frac{N - N_i}{N} * \log_2 \frac{N - N_i}{N}, \qquad (2)$$

where N_i is the number of set elements with the property C_i and N is the maximum possible number of the set elements.

4   Results and Discussion

The graph of the dependency $S(C)$ reflects not only the current points of dividing the initial set into parts, but also the potentially possible ones. Using entropy in the approach we suggest helps to obtain interesting results, whose analysis helps to connect the appearance stages for new knowledge of a subject area with the properties of new solution methods and with the subject area classification stages:

- Each new property, which divides knowledge from non-knowledge, diminishes the current value of entropy for the classified subject area.
- Each new property, which gets into the already discovered knowledge area, makes the classification of this area more detailed.
- The more elaborated the subject area is and the more classified the methods used in it are, the stronger the graph of entropy dependency $S(C)$ resembles the logarithmic curve (see Fig. 6).
- The properties identified for a subject area are clear for the natural intelligence and can be used to structure the subject area knowledge, are easy to remember, and can be used in teaching and learning as well as for building an efficient decision tree that can be used in the information system.
- The closer a property is to the non-knowledge area, the more complex and general it is; and it creates the "Scio me nihil scire" ("I know that I know nothing") effect.
- The closer the knowledge is to the axis beginning, the more properties it has, with those properties being simpler and characterizing the knowledge more accurately; and the way to that knowledge along the decision tree is longer.
- The properties must not be repeated.

The idea of entropy change represented in Fig. 6 makes it possible to have a different look at the representation of knowledge in the information system when it refers to the subject area. It's naturally reflected in the ways of decision tree implementation as the corresponding means of implementation (both for the data structure and the programming languages) already exist. That's why it is possible to make a conclusion that the possibility of solving the subject area classification task was first gradually prepared and developed, and then it took the form of a corresponding solution.

The formula (1) corresponds to Shannon's entropy. If for the standard application of entropy in the universally accepted algorithm of obtaining a decision tree this formula plays a certain role in the calculations of values for the predicates of the given attributed for obtaining the interim nodes and leaves of the tree, the new method, which is approximated to the natural intelligence, does not need the calculation of entropy. It is just necessary to provide the proper selection of properties, which structure the subject area knowledge, i.e. classify the task solution methods.

The obtained dependency $S(C)$ makes it possible to see the following connections, which can be implemented using recursion in the algorithms for obtaining the decision tree nodes [6]:

between the choice of the current root of the tree and the property, which unites the largest number of the subset elements;

between the choice of the current interim node of the tree and the property, which unites the largest number of the elements of the current interim subset.

Then for adding new elements into the set M_0 it is necessary to use the two programs indexed in the subprogram algorithm, which will help to find the place for adding the elements in the decision tree or to create a new tree root, preserving the remaining structure untouched. For the implementation of the algorithm described above it is most efficient to use indicators and recursion, i.e. the binary tree is presented as a dynamic structure.

This algorithm was used in the software Optimel v1.0.1 [6]. The software performs a visual choice of the optimization method for the current task from the subject area of optimization methods. The most suitable optimization method for the problem was found. The volume of an axial flux electric motor is to be minimized. The electric output parameters – power, voltage, current and shaft rotation speed – are set. The optimal geometric dimensions of the electric motor stator so that its volume was minimal are to be defined [6, 9].

Acknowledgments. The work was carried out with the financial support provided by the Russian Foundation for Humanities within the research projects No. 16-03-00382 within the theme "Monitoring the research activity of educational institutions in the conditions of information society" of 18.02.2016.

References

1. Correa Bahnsen, A., Aouada, D., Ottersten, B.: Example-dependent cost-sensitive decision trees. Expert Syst. Appl. **42**(19), 6609–6619 (2015)
2. Mantas Carlos, J., Abellán, J.: Analysis and extension of decision trees based on imprecise probabilities: application on noisy data. Expert Syst. Appl. **41**(5), 2514–2525 (2014)
3. Ferreira, D.R., Vasilyev, E.: Using logical decision trees to discover the cause of process delays from event logs. Comput. Ind. **70**, 194–207 (2015)
4. Kamadi, V., Allam, A.R., Thummala, S.M., Nageswara Rao, V.P.: A computational intelligence technique for the effective diagnosis of diabetic patients using principal component analysis (PCA) and modified fuzzy SLIQ decision tree approach. Appl. Soft Comput. **49**, 13–145 (2016)
5. Cui, Y.Q., Shi, J.Y., Wang, Z.L.: Analog circuit fault diagnosis based on quantum clustering based multi-valued quantum fuzzification decision tree (QC-MQFDT). Measurement **93**, 421–434 (2016)
6. Popova, O., Popov, B., Romanov, D., Evseeva, M.: Optimel: software for selecting the optimal method. SoftwareX **6**(C), 231–236 (2017). http://www.sciencedirect.com/science/article/pii/S2352711017300316?via%3Dihub

7. Popova, O., Popov, B., Karandey, V., Evseeva, M.: Intelligence amplification via language of choice description as a mathematical object (binary tree of question-answer system). Procedia – Soc. Behav. Sci. **214**, 897–905 (2015). http://www.sciencedirect.com/science/article/pii/S1877042815061030

8. Popova, O., Romanov, D., Evseeva, M.: A new approximated to the natural intelligence decision tree. Int. J. Eng. Comput. Sci. (IJECS) **5**(8), 17555–17561 (2016)

9. Popova, O., Shevtsov, Y., Popov, B., Karandey, V., Klyuchko, V.: Theoretical propositions and practical implementation of the formalization of structured knowledge of the subject area for exploratory research. In: Karwowski, W., Ahram, T. (eds.) Advances in Intelligent Systems and Computing 2018. IHSI2018, vol. 722, pp. 432–437. Springer, Dubai (2018)

Business Intelligence Analysis, the New Role of Enhancing and Complementing the Internship of Students from Information Technology Program

Shutchapol Chopvitayakun[✉]

Faculty of Science and Technology, Suan Sunandha Rajabhat University,
Bangkok 10300, Thailand
shutchapol.ch@ssru.ac.th

Abstract. Internship programs are a vital course for most study programs in terms of integrating and demonstrating the body of knowledge of the real professional work domain. Most schools adopt this program and give it a very high priority due to its significance and core value. However, internship program management consists of several components - especially the work performance of interns, which is one of the most crucial factors. Business Intelligence or BI can address this concern well with its analytical capability to provide multi-dimensional or multi-pivotal reports and highly responsive interaction to any query regarding an internship's work performance. Moreover, it features adaptive visualization and provides many insightful strategies for interns, advisors and stakeholders enabling them to manage the work of interns in the organization more effectively. One current study is employing Business Intelligence software called Microsoft Power BI. This desktop platform has been handling sets of cumulative data for the past 5 years (from 2013 to 2017) involving 470 students in Information Technology programs. It illustrates graphical outputs in a dashboard format that processes raw data and transforms it into meaningful information and insightful strategies, respectively.

Keywords: Business Intelligence · Information Technology
Internship program · Management

1 Introduction

Information systems play a crucial role in strategy planning and decision-making in the business world. There are many kinds of information systems available in the market such as Management Information System (MIS), Decision Support System (DSS), and Executive Information System (EIS). These systems can provide a lot of insightful information in the form of charts, graphs, and other statistical information to help management define their organizational strategies and plans [1]. However, information collection and distribution are very critical to the decision-making process. Information has to be facilitated by various channels of information and communication technology (ICT) to process large volumes of data using the enhanced features of information

© Springer International Publishing AG, part of Springer Nature 2019
T. Z. Ahram (Ed.): AHFE 2018, AISC 787, pp. 322–329, 2019.
https://doi.org/10.1007/978-3-319-94229-2_31

systems in order to obtain relevant information [2]. Eventually, the integration of advanced ICT and predecessor information systems such as MIS and DSS evolved into the advent of a new sophisticated information system called Enterprise Resource Planning (ERP). This system can handle all data, information, and knowledge which covers every level of enterprise users from top to bottom [3]. Recently, Business Intelligence (BI) has been adopted and widely implemented in medium and large size organizations. BI features complement its predecessors by providing insightful information in various dimensions. Business Intelligence is a domain term that consists of technologies such as online analytical processing, data mining, business performance management, data warehousing, business analytical tools, etc. [4]. This research studied analytical techniques and methods of Business Intelligence and deployed them to analyze a set of data regarding the internship work performance of students from the Information Technology program of Suan Sunandha Rajabhat University, Thailand. The results of this study are presented in interactive reports and a responsive dashboard monitor to aid in planning the IT internship program over the next following years.

2 Literature Review

Business Intelligence is defined as the process of gathering information in the field of business; the process of turning data into information and then into knowledge. Business demands always constantly evolve with the business target. There are 3 approaches proposed in this study [5];

1. IT-Centric. This approach initially focuses on data collection and analytical tool selection. It emphasizes making better business decisions through the analysis of historical data.
2. Information Management. This approach focuses on real-time decision-making and emphasizes integrating data from CRM and ERP applications.
3. Predictive Insight. This approach focuses on advanced analytics and predictive modeling to anticipate likely future events and capitalize on new trends or market opportunities. It emphasizes business outcome optimization.

The Business Intelligence System is very important in the decision-making process. It also plays an important role in corporate strategic planning to achieve effectiveness incorporate management. Business Intelligence Systems contain sets of methodologies, processes, architectures, and technologies. It can transform raw data into meaningful and useful information and help to enable more effective strategic, tactical, and operational insights and decision-making [6].

Business Intelligence implementations require business process knowledge and skills. These are 3 vital skills in order to implement a successful Business Intelligence system [7];

1. Business Skills: Organizations must be able to quickly analyze market changes and to adapt processes consistent with existing conditions. Also, according to the demands of multiple customers, they must be able to prioritize needs and expectations using different tools and have a strong organizational strategy.

2. IT Skills: Organizations should provide all the necessary infrastructure to updates by identifying, collecting, and receiving data and saving and maintaining it and be able to integrate the existing data. Also, they should be able to monitor the BI programs and assist this process effectively.
3. Organizational Skills: Organizations should be able to implement this new system and warn the appropriate organizational culture, implementing step by step. If they fail to adequately do this, staff and personnel may resist the new changes. On the other hand, the organization needs to have the strength to explore and describe data and to finally analyze and summarize it carefully.

Chopvitayakun [8] identified the relevant internship program as an essential core course for most undergraduate study programs. It involves real-setting workspace and real-life problem-solving skills integrated with knowledge and IT competencies. Internship programs benefit interns in several ways. Interns have to apply their knowledge which they have learned over the last four 4 years with assignments or tasks given by their host organization. In the Information Technology industry, many businesses allow interns to practice and work in their organizations as a way of recruiting new employees. Well qualified interns who can complete tasks with satisfaction may have a chance to be recruited as a full-time worker. Each organization will demand slightly different qualifications from interns. However, most organizations have common qualifications for their desired interns such as punctuality, manners, responsibility, and competence. According to this study, there are a few uncertain factors regarding job attainment of each intern and these uncertainties need to be studied.

3   Methodology

This study gathered data over 5 years from 2013–2017, from 470 interns in the Information Technology program of Suan Sunandha Rajabhat University, Thailand. It uses the CSV input format file for Microsoft Power BI software to import and analyze data to foresee trends that might happen and involve internship decision-making and strategy planning such as competitive rates for interns to attain a job within each organization and comparative scenarios among intern's work performance. Dashboard was generated by Microsoft Power BI. The visualizations on this dashboard come from several key reports and each report is based on one relevant dataset. Dashboard is similar to an entryway into the underlying reports and datasets. This study created important visualizations in the form of charts, graphs, and lines to provide nice and effective graphical interface.

3.1   Microsoft Power BI

Power BI is a product of Microsoft Corporation available on cloud-based, desktop-based, and mobile phone-based platforms. It has some interesting features such as a data warehouse, data preparation, data discovery, and interactive dashboards. In this study, Desktop-based Power BI was used to import datasheets from Microsoft Excel and migrated them in to the data entities. Each one of them was linked accordingly to their meaningful relationship and principles of data manipulation. Finally, these datasets were analyzed and generated a set of insightful reports pinned on a multi-window dashboard. At one time, this provides an interesting perspective, as a new visualization can be customized and generated efficiently.

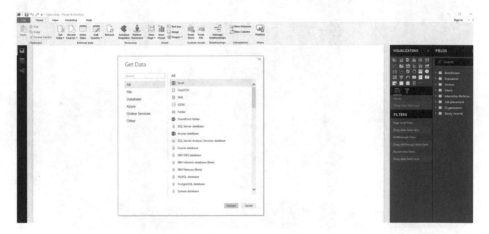

Fig. 1. Data source and import panel

Power BI can handle different types of data sources such as Text, Excel, and Database files, or retrieve data from other Database software real-time via Open Database Connectivity (ODBC). Moreover, it can retrieve data from a Cloud server such as a HTTP website. Data in this study was provided in Excel file format and imported using Power BI by column and work sheet.

Data entities and relationships were normalized by using internal and external keys according to the rule of data standardization. Microsoft Power BI has several tools to visualize information and filter findings for specific purposes. Microsoft Power BI also provides dynamic range of query to retrieve particular pieces of information effortlessly.

Power BI supports a wide range of data type from simple formats such as text or numerical to sophisticated formats such as date/time and currency. The data pane on the right bar shows table structures and data fields for data manipulation. Once data is imported into this project, it can be edited within Power BI workbook. Workbook looks like a table format in Excel with columns and rows.

4   Result

Power BI report created a multi-perspective view retrieved from internship datasets. It provided visualizations that constitute different findings and insights from the data of internship programs from 2013–2017. Figure 5, shows some example reports representing the job attainment of IT interns and overall competitive rate of organizational

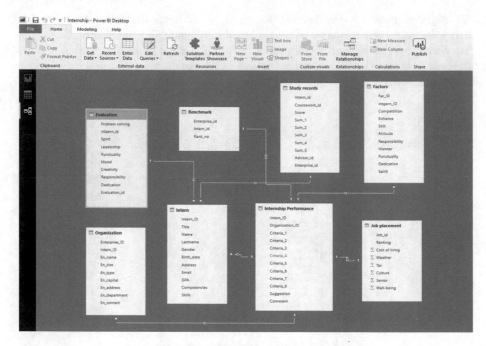

Fig. 2. Entities relationship in Microsoft Power BI

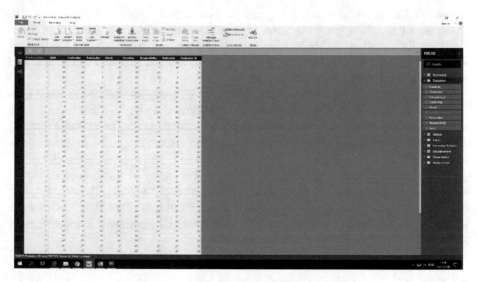

Fig. 3. Data sheet and pane in Power BI

employee recruitment. These reports were analyzed from internship datasets. Each visualization in these reports depict a nugget of information pivotal to interns and their advisors. Moreover, these visualizations are not static, they are very dynamic and ready

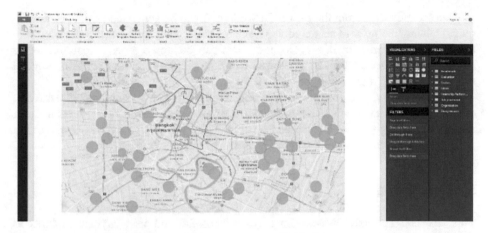

Fig. 4. Geolocation report of intern's work sites

to apply and customize new filters or criteria to retrieve new relevant information. Each type of visualization such as the geolocation, radar, and scattered charts, and the stacked bar or column chart are highly customizable and promptly updates visualizations as the underlying data changes. Moreover, an alerting feature can be deployed in Power BI. Sensitive information, such as the decrease in rate of acceptable intern job performance for 2 consecutive periods, this system can be set to notify you automatically by e-mail or message box.

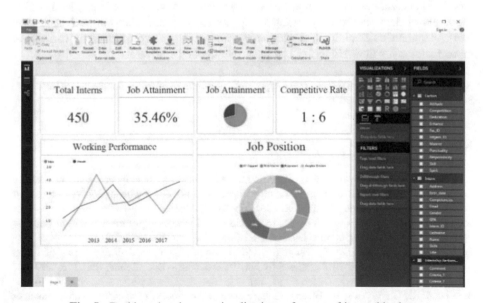

Fig. 5. Dashboard and some visualizations of report of internship data

Geolocation data from the Power BI workbook can be generated as pin points on a map, showing a cluster of scattered work sites of the organization where interns receive their training. Format of this geolocation data can be stored as X, Y coordinates or latitude and longitude format. A Power BI geolocation report can illustrate several types of maps such as bubble and shape maps.

This dashboard is highly responsive and adaptive to user's requirements. Different criteria from users will affect the visualization of graphical information that each report represents. This dynamic feature of Power BI allows proactive queries employing a wide range of information regarding the internship program.

5   Conclusion

This research applied Business Intelligence software and its analytical features to provide dynamic reports called dashboard. Its responsive and interactive features easily provide a wide range of queries regarding the internship program; specifically, the intern's work performance. There are many pivotal reports generated from this study and all of them serve the needs of the internship program management. Business Intelligence concepts and frameworks have been implemented in this research in order to stipulate strategies and plans proactively. However, it is stand-alone system, accessed by computer clients having Microsoft Power BI software installed. It cannot be accessed via network or Internet. Then, Cloud-based platform of Microsoft Power BI should be implemented for the next phase. Cloud-based platforms for Business Intelligence can enhance a lot of features such as ubiquitous access through mobile phone via the Internet or access of other mobile devices. For further study, other segments of data should be integrated into the cloud-based platform such as course work, program curriculum, and academic training. These considerations can utilize more features for analysis in the Business Intelligence software.

Acknowledgments. This research was supported by Research and Development Institute, Suan Sunandha Rajabhat University.

References

1. Silahtaroğlu, G., Alayoglu, N.: Using or not using business intelligence and big data for strategic management: an empirical study based on interviews with executives in various sectors. Procedia – Soc. Behav. Sci. **235**, 208–215 (2016)
2. Kubina, M., Koman, G., Kubinova, I.: Possibility of improving efficiency within business intelligence systems in companies. Procedia Econ. Fin. **26**, 300–305 (2015)
3. Antoniadis, I., Tsiakiris, T., Tsopogloy, S.: Business intelligence during times of crisis: adoption and usage of ERP systems by SMEs. Procedia – Soc. Behav. Sci. **175**, 299–307 (2015)
4. Vizgaitytė, G., Skyrius, R.: Business intelligence in the process of decision making: changes and trends. Ekonomika **91**(3), 147–157 (2012)

5. Gartner: A step-by-step approach to successful Business Intelligence: featuring research from Gartner. http://resources.idgenterprise.com/original/AST-0066459_YTW03194CAEN.pdf. Accessed 20 Oct 2017
6. Marín-Ortega, P.M., Dmitriyev, V., Abilov, M., Gómez, J.M.: ELTA: new approach in designing business intelligence solutions in era of big data. Procedia Technol. **16**, 667–674 (2014)
7. Azma, F., Mostafapour, M.A.: Business intelligence as a key strategy for development organizations. Procedia Technol. **1**, 102–106 (2012)
8. Chopvitayakun, S.: The study of internship performances: comparison of information technology interns towards students' types and background profiles. Int. J. Soc. Behav. Educ. Econ. Bus. Ind. Eng. **10**(7), 2503–2506 (2016)

Clustering-Based Recommender System: Bundle Recommendation Using Matrix Factorization to Single User and User Communities

Remigio Hurtado Ortiz[1,2(✉)], Rodolfo Bojorque Chasi[1,2],
and César Inga Chalco[1]

[1] Universidad Politécnica Salesiana, Cuenca, Ecuador
{rhurtadoo, rbojorque, cingac}@ups.edu.ec
[2] Universidad Politécnica de Madrid, Madrid, Spain

Abstract. This paper shows the results of a Recommender System (RS) that suggests bundles of items to a user or a community of users. Nowadays, there are several RS that realize suggestions of a unique item considering the preferences of a user. However, these RS are not scalable and sometimes the suggestions that make are far from a user's preferences. We propose an RS that suggests bundles of items to one user or a community of users with similar affinities. This RS uses an algorithm based on Matrix Factorization (MF). To execute the experiments, we use released databases with high dispersion. The results obtained are evaluated per the metrics Accuracy, Precision, Recall and F-measure. The results demonstrate that the proposed method improves significantly the quality of the suggestions.

Keywords: Recommender System · Bundles of items · Matrix Factorization

1 Introduction

The RS has been the object of study and development in recent years. Usually, the RS are used over the World Wide Web (WEB) where it acts as a filter on the abundant content that rests on the WEB. The RS use clustering algorithms that allow to analyze the preferences of a user and suggest items (e.g., movies, books, songs, e-commerce) that the user finds them interesting [1, 2].

Some of the most used clustering algorithms (K-means, K-neighbors) have limitations as [3]: (i) suggest one item at a time (ii) their behavior over dispersed data still is object of study because it can make imprecise recommendations. There are developed have been some improvements in the mentioned algorithms as: (i) realize a prior grouping to the start of the clustering process (ii) implementation of scalable filtering techniques to improve the quality of the predictions. However, no significant improvements have been obtained [4].

It is necessary to continue with the study of clustering algorithms for improve the quality of the suggestions and be able to suggest bundles of items instead of just one item. [5, 6]. For example, a user who will travel, could project a budget and based on

© Springer International Publishing AG, part of Springer Nature 2019
T. Z. Ahram (Ed.): AHFE 2018, AISC 787, pp. 330–338, 2019.
https://doi.org/10.1007/978-3-319-94229-2_32

this information the RS will suggest a tour package (bundles of items) that includes options for: mobilization, food, lodging, leisure and points of interest that the tourist will can visit [7].

Considering the previous example, it could be thought that there is already websites or travel agencies that already do this type of work. But these organisms only restrict the results without considering the user experience. A RS is a more complete and robust system. It would allow that once the trip is finished, the user can qualify each item that the RS suggested and could make a feedback about the items suggested by the RS. So the RS will consider these feedback to improve future suggestions [7, 8].

Therefore, we propose a scalable technique based on the MF model to represent both user and items in a set of latent probability factors [9]. In the second section we mention the previous works on which our proposal is based. The third section contain the design of the proposed method and the experimental tests in which released databases were used. In the fourth section shows the results obtained which demonstrate that the proposed method generates suggestions a better quality. Finally, in the fifth section we show our conclusions.

2   Related Work

2.1   Clustering

Clustering is the process that all SR perform to find patterns that allow grouping objects of similar characteristics [3, 6]. For the object of study of this document, the RS is oriented to the clustering of items with similar affinities. The algorithms most used in memory-based clustering is K-means. It was proposed by Mac Queen in 1967. Consists of partitioning a set of users u into a determined number of clusters k where each u is assigned at the k with the closest means value [3, 4]. K-neighbors is an algorithm similar to k-means. The unique difference is that it functioning consists of grouping the k nearest. [10]. Something common in these algorithms is that not are scalable, so, cannot process large datasets (users rating over the items). In addition, the predictions these algorithms they make are very imprecise (pour accuracy) and not are able to suggest bundles of items [4].

2.2   RS of Bundles

This RS makes suggestions of bundles of items that fit together, based on a series of preferences of a user or community of users [1]. Its implementation is mostly aimed at the tourist area, suggesting tourist packages similar to the example initially proposed [7]. This model is scalable and its clustering is oriented to users and items. The bundles can be re-used again suggesting it to a user or community of users with similar affinities. For this process, the RS considers the interaction and ranking that each user assigned to each item that makes up the bundle [9].

This SR analyzes one item at a time and continues with this process until to form the items bundles. Users choose the quantity of items that make up the bundle by through of restrictions for each item or for the bundle. It clustering process is to determine the probability that an item belongs to a bundle and the probability that this bundle is of interest to a user or community of users [9, 11]. Is necessary this RS use MF for realize suggestions clustering bundles of items.

2.3   Matrix Factorization

The MF is a model that decomposes the training data. Separates users from the items and represents them in a same space of latent factors which allow generating the recommendations. The latent factors provide information on relationships between users and items. This relation allows that the RS: (i) be scalable (ii) it make suggestions that are of most interest to each user (improves accuracy) (iii) be flexible and adapt to different scenarios (e.g., movies, tour packages, purchases) [12]. Using the example of the tourist packages proposed at the beginning, the latent factors can define the types of mobilization (e.g., air, land, maritime).

The general structure of an MF begins representing users and items in the a same space of latent factors with dimension f. All the interactions that a user performs with an item are represented as an internal product of this space. Each item i is associated with a vector $q_i \in \mathbb{R}^f$ and each user is associated with a vector $p_u \in \mathbb{R}^f$. For an item i the vector elements q_i measure the degree to which the item is related to these factors, which may be positive or negative. For a user u the vector elements p_u measure the level of interest that the user presents on the items i, which may be positive or negative. This product of vectors results $q_i^T p_u$, which represents the interaction between a user and an item [9, 12, 13]. The approximation of the suggestion to a user u with regard to an item i is shown in (1) and is represented by r_{ui}.

$$r_{ui} = q_i^T p_u \tag{1}$$

Of all the MF techniques we use the Bayesian non-negative Matrix Factorization (BNMF). The technique BNMF use the same approximation of the suggestion that is detailed in (1). BNMF is a probabilistic model that establishes that items can be grouped into k clusters according to their preferences [13, 14]. The parameter $\propto$ represents the degree of overlapping between the clusters [13, 14].

2.4   Evaluation of Quality

The performance of a RS is ideal when the majority of suggested articles are accepted by the user MF is a model that decomposes the training data [15]. The metrics of evaluation calculate the precision of the recommendations considering the suggestions accepted per the user on the total of realized suggestions by the RS [16]. To determine the Accuracy of the RS of bundles, we use the metric Mean Absolute error (MAE). The Metrics Precision and recall need the confusion matrix shown in Table 1 to evaluate every suggestion realized by the RS. The True Positive (TP) and False Positive (FP) represent the correct recommendations, whereas True Negative (TN) and False Negative (FN) represent incorrect recommendations [9].

Table 1. The confusion matrix.

	Correct	Not correct
Suggestion	TP	FP
Successful suggestion	FN	TN

Description. TP: True Positive. FP: False Positive. FN: False Negative. TN: True Negative.

The *Precision* metric (2) shows the proportion of suggested items that the user found relevant of the total of suggested items [9, 15].

$$Precision = \frac{TP}{TP + FP} \tag{2}$$

The *Recall* metric (3) shows the proportion of suggested items that the user found relevant of the total of suggested relevant items [15, 17].

$$Recall = \frac{TP}{TP + FN} \tag{3}$$

In order that MAE realizes an exact measure it is necessary that analyze one item at a time from the total of items that make up the bundle. In (4) $r_{u.i} \neq \cdot$ represents the absence of a user's vote on an item. O_u represents the set of items voted by a user u with predictive values O_u [9].

$$O_u = \left\{ i \in I | p_{u,i} \neq \cdot \wedge r_{u.i} \neq \cdot \right\} \tag{4}$$

The prediction of O_u is defined by (5) the metric *MAE* [9, 17].

$$MAE = \frac{1}{\# U} \sum_{u \in U} \left(\frac{1}{\# O_u} |p_{u,i} - r_{u.i}| \right) \tag{5}$$

A last metric *f-measure* is used that represents the harmonic mean between the Precision and Recall metrics. Its highest value 1 is obtained when the precision and recall metrics reach an optimal value and 0 when poor values are obtained [9].

3 Experimentation

3.1 Description of Method Proposed

For the execution of the experiments was used the dataset Filmtrust 100k y MovieLens 100k. Both dataset present a high dispersion of data. In the Fig. 1 shows the dataset represented by a matrix where the rows are the users ($u = u_1, u_2, \ldots, u_m$) and the columns the items ($i = i_1, i_2, \ldots, i_n$). BNMF divides the dataset into two matrices that represent users and items in a same space of latent factors f. The vector q_i represents the relationship between an item and a user. The vector P_u represents the interest a user has over an item. Finally, the recommendation r_{ui} is obtained [13].

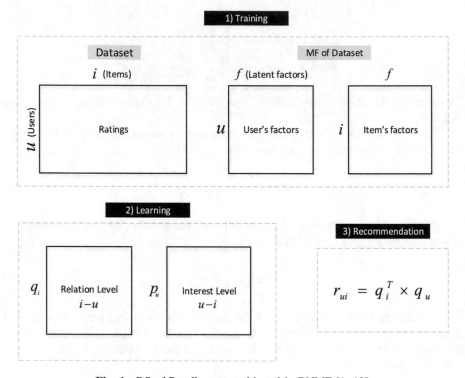

Fig. 1. RS of Bundle proposed based in BNMF [1, 13].

3.2   Program Code

The algorithm used for BNMF is detailed below:

Method proposed based BNMF.

Input: U, training users; itr, number of iteration.

Ouput: $Y_{i,Gi}$, probability that an item belongs to a bundle; $B_{Gi,u}$, probability that this bundle interests to an user; $B_{Gi,Gu}$ probability that this bundle interests to an user communitie; $Y_{u,Gu}$, probability that an user belongs to a bundle.

> *Procedure*
>> Initialize randomly parameters of model
>> Repeat until changes are not significant:
>>> Update parameters according to equations of model [13]
>> Output $Y_{i,Gi}$
>> Output $B_{Gi,u}$
>> Calculate the predictions of the user's tastes
>> With $B_{Gi,u}$ as input BNMF is applied again
>> Output $Y_{u,Gu}$
>> Output $B_{Gi,Gu}$
> *end procedure*

The algorithm used for K-means is detailed below:

K-Means clustering

> *Input:* U, users; k clusters
> *Output:* C centroids, idx index
>
> *Procedure*
> C=centroidselect
> Iter=0
> while (C no longer changes) OR (Iter=Iterations)
>   Assign user's U to Clusters k. Output idx.
>   Update Centroids. Output C.
>   iter=iter+1
> end while
>   Use Pearson correlation to find the similarity between an active user and k other centroids.
>   Find the neighbors of the active user, i.e. l (l <= k) most similar centroids.
>   Make prediction on target item using the weighted average of the ratings provided by neighbors.
> *end procedure*

4   Results Obtained

The Fig. 2(a) shows the MAE results obtained with the dataset Filmtrust 100k. Using BNMF with a value of alpha 0.8 is obtained the best results from MAE. Using the Movielens 100k dataset (Fig. 2b) the best MAE result is obtained with alpha of 0.8. Note that BNMF presents better results than K-means.

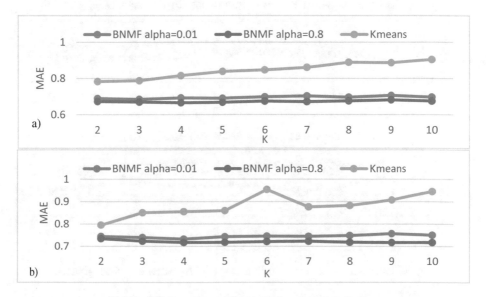

Fig. 2. MAE results using (a) Filmtrust (b) Movielens

The Fig. 3(a) shows the results of the dataset Filmtrust 100k with the Precision metric. The best results are obtained with an alpha of 0.01. With the dataset Movielens 100k (Fig. 3b) the best results are obtained with an alpha value of 0.8. In both datasets the SR suggests 5 items (N).

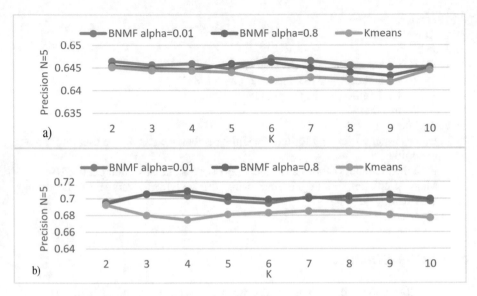

Fig. 3. Precision results using (a) Filmtrust (b) Movielens with 5 suggestions (N).

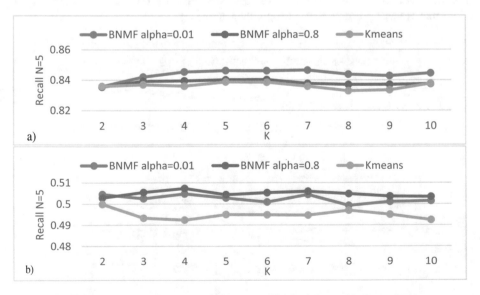

Fig. 4. Precision results using (a) Filmtrust (b) Movielens with 5 suggestions.

The Fig. 4(a) shows the results of the dataset Filmtrust 100k with the Recall metric. The best results are obtained with an alpha of 0.01. With the dataset Movielens 100k (Fig. 4b) the best results are obtained with an alpha value of 0.8. In both datasets the SR suggests 5 items (N).

The Fig. 5(a) shows the results of the dataset Filmtrust 100k with the F-measure metric. The best results are obtained with an alpha of 0.01. With the dataset Movielens 100k (Fig. 5b) the best results are obtained with an alpha value of 0.8. In both datasets the SR suggests 5 items (N).

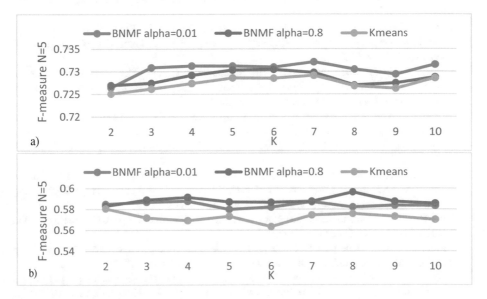

Fig. 5. (a) F-measure results using Filmtrust with 5 suggestions. (b) F-measure results using Movielens with 5 suggestions.

5   Conclusions and Future Lines of Research

We conclude that BNMF is a method that separates users of the items and represents them in a same field of latent factors. This allows to determine the relationship between an item-user and the interest that each user has on each item. With these relationships the RS to suggest bundles of items to a single user or user groups. The evaluation metrics used as: Accuracy (MAE), Precision, Recall and F-measure show that the proposed RS makes suggestions of better quality than K-means. BNMF uses alpha as a parameter that regulate cluster overlapping, it is necessary to carry out some tests until determining which value of alpha best fits our dataset and gives us better results.

The future lines of research related to this paper should focus on implementing this RS on the WEB. They must carry out experiments generating their own dataset based on the ranking assigned by each user that accesses the web. The evaluation metrics are a means that measures the quality of the SR suggestions and are efficient to obtain experimental results. But the Ideally, an RS already mounted on the web makes suggestions to a person and this is who judges the quality of the suggestions.

References

1. Ortega, F., Hernando, A., Bobadilla, J., Kang, J.H.: Recommending items to group of users using Matrix Factorization based Collaborative Filtering. Inf. Sci. **345**, 313–324 (2016)
2. Arthur, D., Vassilvitskii, S.: K-means ++: the advantages of careful seeding. In: Proceedings of the Eighteenth Annual ACM-SIAM Symposium Discrete Algorithms, pp. 1027–1035 (2007)
3. Zahra, S., Ghazanfar, M.A., Khalid, A., Azam, M.A., Naeem, U., Prugel-Bennett, A.: Novel centroid selection approaches for K-means-clustering based recommender systems. Inf. Sci. (Ny) **320**, 156–189 (2015)
4. Ortega, J.P., del Rocio, M., Rojas, B., Somodevilla Garcia, M.J.: Research issues on K-means algorithm: an experimental trial using matlab. In: Proceedings 2nd Workshop Semantic Web New Technology, pp. 83–96 (2009)
5. Boratto, L., Carta, S.: State-of-the-art in group recommendation and new approaches for automatic identification of groups. In: Soro, A., Vargiu, E., Armano, G., Paddeu, G. (eds.) Information Retrieval and Mining in Distributed Environments, pp. 1–20. Springer, Heidelberg (2011)
6. Mohankumar, R., Saravanan, D.: Design of quality-based recommender system for bundle purchases. In: National Conference on Recent Trends in Communication on Engineering Tamil Nadu College, Coimbatore-641 659, vol. 1, pp. 1–8 (2014)
7. Casimiro, K.H.: Diseño de un sistema de recomendación turístico para Cd. Juárez, Chih., México Karina Hernández Casimiro
8. Silvia, S.: 'Un enfoque para Sistemas de Recomendación de paquetes turísticos basado en restricciones de usuario.' Trabajo final de Ingeniería de Sistemas (2016)
9. Bojorque, R., Hurtado, R.: Técnicas híbridas en Sistemas de Recomendación para optimizar el Modelo Non Negative Matrix Factorization. Universidad Politécnica de Madrid (2017)
10. Hartigan, J.A., Wong, M.A.: Algorithm AS 136: a k-means clustering algorithm. J. R. Stat. Soc: Ser. C (Appl. Stat.) **28**(1), 100–108 (1979). http://www.jstor.org/stable/2346830
11. Zhu, T., Harrington, P., Li, J., Tang, L.: Bundle recommendation in eCommerce. In: Proceedings of the 37th International ACM SIGIR Conference on Research and Development in Information Retrieval, pp. 657–666 (2014)
12. Koren, Y., Bell, R., Volinsky, C.: Matrix factorization techniques for recommender systems. Computer **42**(8), 30–37 (2009)
13. Bobadilla, J., Bojorque, R., Hernando, A., Hurtado, R.: Recommender systems clustering using Bayesian non negative matrix factorization. IEEE Access **6**, 1 (2018)
14. Hernando, A., Bobadilla, J., Ortega, F.: A non negative matrix factorization for collaborative filtering recommender systems based on a Bayesian probabilistic model. Knowl.-Based Syst. **97**, 188–202 (2016)
15. Hernández del Olmo, F., Gaudioso, E.: Evaluation of recommender systems: a new approach. Expert Syst. Appl. **35**(3), 790–804 (2008)
16. Pham, M.C., Cao, Y., Klamma, R., Jarke, M.: A clustering approach for collaborative filtering recommendation using social network analysis. J. Univers. Comput. Sci. **17**(4), 1–21 (2011)
17. Ricci, F., Rokach, L., Shapira, B.: Recommender Systems Handbook, 1003 p. Springer +Business Media, New York (2015). ISBN 978-1-4899-7636-9

User Input-Based Construction of Personal Knowledge Graphs

Xiaohua Sun[✉] and Shengchen Zhang

Tongji University, Shanghai, China
{xsun, seanzhang}@tongji.edu.cn

Abstract. Personal knowledge plays a key role in the development of more intelligent applications. Applying knowledge representation techniques like knowledge graphs to the representation of personal knowledge is under active research. However, current knowledge graph construction methods are hindered by problems like absence of knowledge, ambiguity, conflicts and erroneous knowledge when applied to personal knowledge. This is largely due to its unique properties, such as its user-specific, volatile nature and limited data availability. We present in this paper a novel method supporting user input-based construction of personal knowledge graphs. We develop a new knowledge graph structure specifically to counter the said problems, and present a method that uses an iteration-specific subgraph as the intermediate layer between the user and the actual personal knowledge graph for better integration of user input. We also propose a deprecation mechanism to address the volatile nature of personal knowledge.

Keywords: Knowledge graph construction · Human-systems integration
User input-based · Personal knowledge graph

1 Introduction

Rapid development in the field of artificial intelligence has led to widespread attention on building more intelligent applications that has the ability to adaptively learn knowledge about users in the environment of use. Effective representation of personal knowledge of the user plays a key role in the functioning of such applications.

Knowledge graph (KG) is favored as a means of knowledge representation. It has been successfully applied to various intelligent applications like QA systems, search engines and expert systems, enabling such applications to take advantage of available large databases of general knowledge of various domains [1, 2]. However, research into effective KG construction has not yielded fully satisfactory results, due to difficulties such as erroneous information. [3–6] all have attempted to bring in user input as a means of overcoming said problems, and reached promising results.

Study has been carried out on utilizing KG to represent personal knowledge and aid adaptive learning. Zoliner et al. proposed a method for robots to learn procedural knowledge using a and-or-graph [7], and Kollar et al. successfully used KG to achieve conversation-based construction of environmental knowledge graphs [8]. However, the

© Springer International Publishing AG, part of Springer Nature 2019
T. Z. Ahram (Ed.): AHFE 2018, AISC 787, pp. 339–345, 2019.
https://doi.org/10.1007/978-3-319-94229-2_33

construction of personal knowledge graphs is faced with additional obstacles that arises from the particular properties of personal knowledge. Different from general, factual knowledge, personal knowledge is by nature volatile and has limited availability, therefore can only be acquired locally and is prone to deprecation and errors. This can cause significant problems during construction and utilization of personal knowledge graphs, namely:

1. Absence of knowledge. The information required to generate a response is not present in the knowledge graph. This is very probable, due to the limited availability of personal knowledge.
2. Ambiguity. During one query of knowledge, there exist multiple candidates that are all plausible. This can be common, for example, when querying possible reaction of user from historical knowledge about user's past behavior.
3. Conflicts. When integrating new knowledge into existing knowledge graph, there exist pieces of knowledge that conflict with the new knowledge, likely due to changes in highly volatile properties such as personal preference.
4. Erroneous knowledge. This category can be further divided into three cases, all characterized for being impossible to detect yet cause unexpected or undesired behavior for applications. The cases are further categorized as:
   a. *False knowledge*. Certain knowledge is present but is false, either due to the knowledge not having real-world meaning or that it being wrong.
   b. *Deprecation*. Certain knowledge is no longer valid and needs update.
   c. *Inaccuracy*. Data for certain knowledge is not sufficiently collected such that it fails to support the actual utilization of the knowledge.

Based on above analysis, especially (4)a–c where the error is not detectable by nature, we argue that it is necessary to integrate user input into the construction process of personal knowledge graphs. However, most of present KG structures are not designed with user input in mind, and fails to incorporate important features like validity. Therefore, the design of a new user input-based knowledge graph suitable for the representation of personal knowledge is needed for building a more effective personal knowledge graph.

In this paper, we aim at addressing issues encountered in utilizing knowledge graphs to represent personal knowledge by proposing a novel technique supporting user input based construction. We specifically develop a new KG structure to model the characteristics of personal knowledge, and present a method that uses an iteration-specific subgraph as the intermediate layer between user input and the actual personal knowledge graph. We also propose a deprecation mechanism to address the volatile nature of personal knowledge.

2   A KG Structure Representing Personal Knowledge

In this study, we concern factual knowledge that surrounds the users of a certain application. Such knowledge often consists of (1) information about individual users, (2) concepts specifically related to each user, (3) properties of said concepts and users, and (4) arbitrary semantic relations between (1)–(3). Also of interest is the validity of each piece of knowledge, to account for the volatile nature of personal knowledge.

We model personal knowledge and corresponding validity using a knowledge graph consisting of entities set E and relations set R as a weighted graph $G = \{E,R\}$. The weight on relation edges is added to represent the knowledge's validity. We further categorize entities into four sets: users U, concepts C, properties P and relational templates T. The categories are given detailed description with examples in Table 1 below.

Table 1. Entity categories.

Entity category	Description	Example
Users	An entity that represents a user of a certain application	`JohnSmith` as in `(cup, owned_by, JohnSmith)`
Concepts	Non-user entities that has a certain relationship with a user entity that is not present in general knowledge	`cup` as in `(cup, owned_by, JohnSmith)`
Properties	Descriptive entity that attach specific property to a user or concept. Takes the form of an set element in ordered pair *(user or concept, [properties])*	`hot` as in nested knowledge `(JohnSmith, likes, (drink, {hot}))`
Relational templates	Takes the form of a triple containing two sets and a learned arbitrary semantic relation	`(C, owned_by, U)`

Knowledge within the proposed personal KG is therefore represented with a 5-tuple that takes the form (`head, relation, tail, weight, depreca-tion_rate`). The addition of varying weight and depreciation rate enables automatic invalidation of deprecated knowledge and further improves choice accuracy in reasoning process. Head and tail can both be entities with no properties attached, or take the form of an ordered pair with the second element being a set of properties. This enables direct queries that has more complex structure, with (`JohnSmith, likes, (drink, {x?})`) as an example of querying a user's preference of a drink. Relations are regarded as instances of their corresponding relational templates. This approach supports the representation of inter-relational relationships (e.g. (`like, is_a, preference`)), which provides the possibility to adaptively learn new relationship patterns and enable reasoning of relations using knowledge graph embedding techniques. Typical structure can be seen in Fig. 1. Each edge is weighted and has specific deprecation rate.

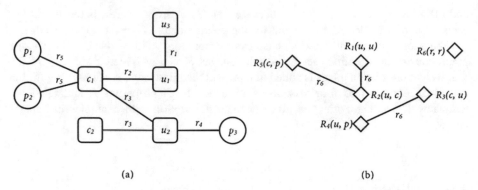

(a)   (b)

Fig. 1. Typical structure of personal knowledge graph. (a) Shows entities and relations. u, c, p, r denotes users, concepts properties and relations respectively. (b) Shows relational templates and their relations. $R(\cdot, \cdot)$ denotes relational templates that can be further encoded by inter-relational relations (e.g. r_6).

3   User Input Integration Using Iteration-Specific Subgraph

There are two key factors to consider when integrating user input into knowledge graph construction. One is the typical interaction process between a user and an application that uses knowledge graph. A user typically interacts with such an application in multiple iterations of queries or conversation, each consists of an initiation, a response, and a reflection to that response. Each iteration may concern different knowledge and have different purpose. Errors can be observed by the user only after the response is given. Another key factor to consider is the nature of personal knowledge and the requirements of the application. While volatile and prone to deprecation, personal knowledge like preferences is often recurring or has multiple alternatives all being plausible. Therefore, it is desirable to be able to keep record of historical information and multiple possibilities of a piece of knowledge. Methods like direct modification from the user, while being natural and easy to understand, lacks the ability to support this functionality. It is often not plausible to assume that users understand the difference between overwriting an existing value and creating a new entry, but the construction of personal knowledge graph requires correct selection and implementation of both cre- ation and modification. Based on consideration of these two key factors, we present a method that adds an intermediate layer between the user and the personal knowledge graph to support user-based input. We propose building an iteration-specific subgraph for each interaction iteration, and direct all user modification to the subgraph. Personal knowledge graph is then updated at the end of each iteration, and the subgraph is discarded after use. More specifically, we describe the interaction process utilizing the subgraph in detail in Table 2 below.

Table 2. Interaction process utilizing a iteration-specific subgraph.

Interaction stage	Description
Initiation	Input from the user (e.g. a query, a statement of fact, etc.) is parsed and understood. In this process, the application constructs a subgraph of the personal KG specific to this iteration. Errors in the construction is noted to the user. A correction session is initiated in this case, and the user inputs new knowledge or modifies existing knowledge. All modifications are made to the iteration-specific subgraph
Response	A response is formed and presented to the user
Reflection	User replies to the application's previous response, and point out erroneous information and undesired results. A correction session is initiated in this case as well
Update	Iteration-specific subgraph is used to update the personal knowledge graph. Errors such as conflicts are again noted to the user, and subsequently corrected in a correction session

Using a iteration-specific subgraph has several advantages:

1. User modifications are not direct to the knowledge graph, thus allowing further processing such as detection of conflicts when updating, which helps reducing the probability of human errors.
2. A subgraph specific to one iteration is much smaller in size compared to the personal knowledge graph, thus can lower the user's burden in understanding the structure and mechanism of a knowledge graph. It also provides a context that is highly related.
3. A smaller subgraph reduces computation time on analytical operations.
4. The use of a smaller subgraph opens up the possibility of visual analysis, which is yet to be explored for knowledge graphs.

4 Deprecation Mechanism

Probability of deprecation is a key property of personal knowledge. Certain types of knowledge may be short-lived, like a user's current objective or a device's current status, while others may last for a longer period of time, such as a user's personal information or interpersonal relationships. We reason that short-lived knowledge can be identified by its high rate of errors and modification, while long-term knowledge tends to be recalled and validated to be true for multiple times during its lifespan. This poses an opportunity to use knowledge deprecation to automatically achieve different continuous representation for different kinds of knowledge. Therefore, we propose a deprecation mechanism that simulates the forgetting processing present in human memory. Our method assigns varying deprecation rate to each relation present in the personal KG that is used on a regular interval to update the weight of the relation. Deprecation rate is adjusted upon each recall and modification of the knowledge. More formally, we adopt the exponential curve estimation of the forgetting curve proposed by Woźniak et al. with minor modifications [9]:

$$\omega = e^{-\frac{at}{d}} \tag{1}$$

Where t denotes time since last recall and d denotes the rate of deprecation. Value of d is doubled for each successful recall and respectively halved for each modification. Parameter a is present to scale t to its respective time unit. This update of d happens after each iteration of user interaction.

Adopting the deprecation mechanism has several desirable advantages. It's closed-form property does not require constant update of the entire knowledge graph. Calculation is only needed upon activation of a certain piece of knowledge. It auto-matically generates ranking among possible variations of the same relation, and translates well to knowledge graph embedding techniques that applies to weighted graphs. Also of significance is its support for pruning of the knowledge graph. This makes application of knowledge graph possible on devices with limited memory.

5   Conclusion and Future Work

The effective representation of personal knowledge is important to developing more intelligent applications, and the unique nature of personal knowledge raises problems such as absence of knowledge, ambiguity, conflicts and erroneous knowledge when applying common knowledge representation techniques like knowledge graphs. User input plays a key role overcoming the problems. In this paper, we proposed a novel method for user input-based construction of personal knowledge graphs. We presented a knowledge graph structure specifically designed for representing personal knowledge. Methods for using a iteration-specific subgraph to integrate user input and adopting a deprecation mechanism to address the volatile nature of personal knowledge is also discussed. With this research, we aim to provide a basis and effective method for developing personal knowledge based applications. We plan to extend our research further into the development and field testing of our methods, and other possible means of integrating user input in the construction of personal knowledge graphs.

Acknowledgements. This paper was supported by the Funds Project of Shanghai High Peak IV Program (Grant DA17003).

References

1. Bollacker, K., Evans, C., Paritosh, P., Sturge, T., Taylor, J.: Freebase: a collaboratively created graph database for structuring human knowledge. In: Proceedings of the 2008 ACM SIGMOD International Conference on Management of Data, pp. 1247–1250. ACM, June 2008
2. Auer, S., Bizer, C., Kobilarov, G., Lehmann, J., Cyganiak, R., Ives, Z.: DBpedia: a nucleus for a web of open data. In: The Semantic Web, pp. 722–735. Springer, Heidelberg (2007)

3. Acosta, M., Zaveri, A., Simperl, E., Kontokostas, D., Auer, S., Lehmann, J.: Crowdsourcing linked data quality assessment. In: International Semantic Web Conference, pp. 260–276. Springer, Heidelberg (2013)
4. Thaler, S., Simperl, E.P.B., Siorpaes, K.: SpotTheLink: a game for ontology alignment. Wissensmanagement **182**, 246–253 (2011)
5. Sarasua, C., Simperl, E., Noy, N.F.: CrowdMap: crowdsourcing ontology alignment with microtasks. In: International Semantic Web Conference, pp. 525–541. Springer, Heidelberg. (2012)
6. Wang, J., Kraska, T., Franklin, M.J., Feng, J.: Crowder: crowdsourcing entity resolution. Proc. VLDB Endow. **5**(11), 1483–1494 (2012)
7. Zoliner, R., Pardowitz, M., Knoop, S., Dillmann, R.: Towards cognitive robots: building hierarchical task representations of manipulations from human demonstration. In: Robotics and Automation 2005, ICRA 2005, April 2005
8. Kollar, T., Perera, V., Nardi, D., Veloso, M.: Learning environmental knowledge from task-based human-robot dialog. In: 2013 IEEE International Conference on Robotics and Automation (ICRA), pp. 4304–4309. IEEE, May 2013
9. Woźniak, P.A., Gorzelańczyk, E.J., Murakowski, J.A.: Two components of long-term memory. Acta Neurobiol. Exp. **55**(4), 301–305 (1995)

Hierarchical Clustering for Collaborative Filtering Recommender Systems

César Inga Chalco[1(✉)], Rodolfo Bojorque Chasi[1,2],
and Remigio Hurtado Ortiz[1,2]

[1] Carrera de Ingeniería de Sistemas,
Universidad Politécnica Salesiana del Ecuador, Cuenca, Ecuador
{cingac,rbojorque,rhurtadoo}@ups.edu.ec
[2] Universidad Politécnica de Madrid, Madrid, Spain

Abstract. Nowadays, the Recommender Systems (RS) that use Collaborative Filtering (CF) are objects of interest and development. CF allows RS to have a scalable filtering, vary metrics to determine the similarity between users and obtain very precise recommendations when using dispersed data. This paper proposes an RS based in Agglomerative Hierarchical Clustering (HAC) for CF. The databases used for the experiments are released and of high dispersion. We used five HAC methods in order to identify which method provides the best results, we also analyzed similarity metrics such as Pearson Correlation (PC) and Jaccard Mean Square Difference (JMSD) versus Euclidean distance. Finally, we evaluated the results of the proposed algorithm through precision, recall and accuracy.

Keywords: Recommender Systems · Collaborative Filtering
Agglomerative Hierarchical Clustering · Similarity metrics

1 Introduction

Before the Internet existed, most products and services were made known to the public through advertising. Over the years the reach of this medium became limited. Its main disadvantages were the scarce information provided by: (i) the products, (ii) the large offer available in the market [1].

The overcrowding of Internet access and the service known as the World Wide Web (WEB) allow us to experience a scene contrary to that of decades ago. Today it is possible to access more than a billion web pages that cover all areas of knowledge and that according to the InternetLiveStats® portal generate Internet traffic that exceeds two thousand terabytes a day [2, 3]. This has been a benefit for Internet users. But also, it has been the source of a problem, which we will define as overflow of content [1].

The overflow of content is generated due to the excess of information available on the web. This problem makes it difficult for users to extract content that is useful and of interest to them. The RS were devised in order to solve this problem. These systems act as a filter on the overflow of content and recommend items, that are related to a user, through a clustering algorithm (e.g., movies, music, sports, books, applications) [4, 5]. This prevents the user from analyzing all the content on the web, saving time and resources.

© Springer International Publishing AG, part of Springer Nature 2019
T. Z. Ahram (Ed.): AHFE 2018, AISC 787, pp. 346–356, 2019.
https://doi.org/10.1007/978-3-319-94229-2_34

Currently, the RS of greatest development and interest are those based on CF. Its operation consists in identifying affinities between new users and existing users in the RS through similarity metrics. That is, if a user A has similar affinities on an item to a user B, it is very likely that A has the same interest as B on items that A has not examined yet [6, 7].

K-means is one of the most used clustering algorithms to find similarities between users. Its objective is to generate groups of users, where the Euclidean distance between the user and the centroid (center of mass) of the cluster is minimized [8]. Some disadvantages of this algorithm are [9]: (i) at the end of the clustering algorithm, groups with a spherical geometry are obtained, which generates local minimums and outliers. (ii) it requires a high computing time. Although this algorithm is ideal for the grouping of bulky data sets, its behavior with scattered data is still being studied [8, 9]. K-neighbors is another popular algorithm; its operation is similar to K-means with the difference that it generates groups among the closest (most similar) users. The given recommendations are usually quite accurate, as long as they are not used in scattered data [10].

Dispersed data is generated when users have not defined their interest in an item through a ranking [11]. This lack of information directs CF-based RS to look for more scalable and precise techniques. To this end, they have adapted clustering techniques that allow [7, 12]: (i) defining the users' belonging to clusters. (ii) experimenting with metrics that provide a better similarity between users. (iii) comparing the distance between users in the clustering process. All mentioned techniques improve the filtering scalability compared to K-neighbors or K-means algorithms [13] and are considered in the development of this paper.

Our work proposes an Agglomerative method for Hierarchical Clustering (HAC) using CF. The main contribution is that we obtain groups of varied geometry, which improves the quality of the cluster [7]. By implementing CF on the HAC it is possible to obtain very precise suggestions despite using scattered data [6]. This document is structured as follows: Sect. 2 contains works related to our proposal; Sect. 3 shows the design of the experiments; in Sect. 4 the obtained results are shown and finally, conclusions are presented.

2 Related Work

2.1 Description of HAC

The HAC generates unified clusters in ascending and successive ways [14]. Figure 1(a) shows the clustering process where each user represents a cluster (# users = # clusters) at the beginning [11, 15]. The similarity between all the clusters is calculated and in each iteration two clusters with a greater similarity are merged (1st Iteration: merger cluster A–B, 2nd Iteration: merger cluster D–E). The agglomeration is carried out until all the users are unified (from user A … n) in a single cluster or until a previously defined cluster number is obtained. The clustering sequence constructs a dendogram or ascending hierarchical tree, as shown in Fig. 1(b). Each merger is represented by a horizontal line and the similarity between each cluster by a vertical line. The first mergers (cluster A–B, cluster D–E) correspond to the users with greater similarity and continue to the least similar (cluster AB - CDE).

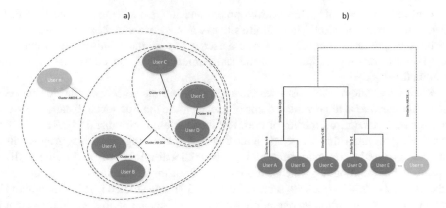

Fig. 1. HAC process and generation of the ascending dendogram from users A… n [11].

To avoid obtaining a single cluster where all the users are located at the end of the HAC process we propose two guidelines: (i) allow controlling the number of clusters generated by the HAC. (ii) avoid mergers between clusters with a distant similarity gap. (iii) improve the clustering process when outliers' users are presented. The last guideline corresponds to the evaluation of the quality of the HAC.

- Set the number of clusters *(k)* that you want to obtain from the HAC prior to the beginning of the clustering process.
- Condition the clustering process to stops once the number of clusters generated by the HAC equals the fixed number of k.
- Measure the quality of clustering through precision, recall and accuracy.

2.2 Methods for Fusion in HAC

By varying the criteria for merging clusters, it is possible to identify which method best fits our dataset and allows us to obtain better results. Five of the seven fusion methods of the HAC were used. The use of the Ward and Median methods is discarded because they are appropriate only for the Euclidean distance which would avoid comparing with the other similarity measures [16]. In all the methods the distance d defines the cluster.

Table 1 shows the *Single linkage* method that performs a cluster fusion at a local level. Clusters C_I, C_J are merged to present the smallest distance d measured between the closest users. This method is the simplest of HAC since it does not consider the most distant users nor the hierarchy of each merger [14].

The *Complete linkage* method merges the C_I, C_J clusters by presenting the largest d measured among its most distant users. This method provides very compact clusters. It is sensitive to the presence of outliers' users [17].

The *Weighted Method* use Weighted Pair Group Method with Arithmetic Mean (WPGMA). Initially the clusters (C_I, C_J) are merged because they are the closest. The next merger is made between the clusters (C_I, C_J) and the cluster C_k that is at the shortest arithmetic mean measured between the clusters (C_I, C_J) and C_k [14, 17].

The method *Average linkage* use Unweighted Pair Group Method using Arithmetic Averages (UPGMA). That is, merges the clusters C_I, C_J whenever they have the shortest half d between each user [17].

The *Centroid Method* use Unweighted Pair Group Method with Centroid Averaging (UPGMC) i.e. it merges the clusters C_I, C_J whenever they have the shortest d between the centroids, with each centroid being the center of mass of each cluster. It does not consider the order in which the clusters were formed [14].

Table 1. Methods of merging cluster in the HAC

Method	Equation of distance	Graphic
Single linkage	$d(C_I, C_J) = \min_{i \in I} \min_{j \in J} d(C_i, C_j)$	
Complete linkage	$d(C_I, C_J) = \max_{i \in I} \max_{j \in J} d(C_i, C_j)$	
Weighted method	$d(C_I \cup C_J), C_k = \dfrac{d(C_i, C_k) + d(C_j, C_k)}{2}$	
Average linkage	$d(C_I, C_J) = \dfrac{1}{|C_I||C_J|} \sum_{i \in I} \sum_{j \in J} d(C_i, C_j)$	
Centroid method	$d(C_I, C_J) = \dfrac{1}{|C_I||C_J|} \sum_{i \in I} \sum_{j \in J} d(C_i, C_j) \ldots$ $- \dfrac{1}{|C_I|^2} \sum_{i \in I} \sum_{\substack{i' \in I \\ i'>i}} d(C_i, C_j) \ldots$ $- \dfrac{1}{|C_J|^2} \sum_{j \in J} \sum_{\substack{j' \in J \\ j'>j}} d(C_i, C_j)$	

Description. d: distance between two clusters; C_I and C_J user clusters; C_k: cluster k from which the distance toward the clusters is measured C_I and C_J.

2.3 Similarity Measures

Similarity measures are metrics that determine the similarity or dissimilarity between users. They are calculated based on the rating that each user assigned on each item [18]. So that the HAC is able to cluster the users using CF it is necessary that the lack of vote (rating) is not considered as zero. The lack of rating should be considered as a vacuum and should not be considered in the process of clustering.

Table 2 shows the *Pearson Correlation* (PC) that is one of the most commonly used metrics in CF for memory-based algorithms. It determines the similarity ($S_{Pearson}$) between two users (u, v) through $\overline{r_u}$ which represents the average of all users' votes u y $\overline{r_v}$ representing the average of all the user's votes v. PC defines the similarity between users in a range from -1 to 1. Values close to -1 indicate absence of similarity and values close to 0 indicate low similarity but not the absence of it. Values close to 1 indicate the presence of maximum similarity [15, 18].

The *Jaccard* is a metric that defines the similarity (S_{Jac}) between the users (u, v) through the division of the cardinality of intersection between the votes $(I_u \cap I_v)$ and the cardinality of the union of the vows $(I_u \cup I_v)$. Is represent by $S_{Jac}(u, v) = \frac{I_u \cap I_v}{I_u \cup I_v}$ [11, 18].

The *Mean Square Difference* (MSD) is a metric that uses the Euclidean difference to define the similarity between the votes of the users (u, v). Is represent by $MSD(u, v) = \frac{\sum_{i \in I'} (r_{ui} - r_{vi})^2}{|I'|}$ [18].

The *Jaccard Mean Square Difference* (JMSD) is a metric whose product between the Jaccard and MSD metrics defines the similarity between users (u, v). JMSD defines the similarity (S_{JMSD}) between users in a range of 0–100%. The values which are close to 0% indicate absence of similarity and the values close to 100% shows the maximum similarity [11, 15].

The *Euclidean distance* is a metric that defines the similarity between the users (u, v) by calculating the distance between the Cartesian coordinates of each user $(u_{i,j}, v_{i,j})$ [15, 18].

Table 2. Metrics of similarity and distance.

Method	Equation of similarity or distance
Pearson correlation	$S_{Pearson}(u, v) = \dfrac{\sum_{i \in I'} (r_{ui} - \overline{r_u})(r_{vi} - \overline{r_v})}{\sqrt{\sum_{i \in I'} (r_{ui} - \overline{r_u})^2} \sqrt{\sum_{i \in I'} (r_{vi} - \overline{r_v})^2}}$
JSMD	$S_{JMSD}(u, v) = (1 - MSD(u, v)) \times S_{Jac}(u, v)$
Euclidean distance	$E_{distance}(u, v) = \sqrt{(u_i - v_i)^2 + (u_j - v_j)^2}$

Description. u and v users. $\overline{r_v}$ and $\overline{r_v}$ median of the users' votes. r_{ui} and r_{vi} votes of the users. I_u and I_v number of users' votes. I' : cardinality of the votes set. $u_{i,j}$ and $v_{i,j}$ cartesian coordinates of the users.

When using PC or JMSD similarity metrics, it is necessary to convert users' similarities to distances between users. This conversion is necessary since the five HAC methods need a distance value to merge the clusters. The conversion for the two methods is the following: if you have a value of $S_{JMSD} = 0$ or $S_{Pearson} = 0$ the distance value is $d = 1000$ which indicates a distance gap between users. For all other values that S_{JMSD} or $S_{Pearson}$ may take, the distance value is $d = 1/S_{JMSD}$ or $d = 1/S_{Pearson}$ as appropriate [19].

2.4 Evaluation of Quality

The objective of the evaluation metrics is to measure the quality of the recommendations made by the RS [20]. By having an evaluation metric, it can be determined [13]: (i) what is the optimal k number for our RS. (ii) which method of the HAC and which similarity metric gives us better results. Figure 2 shows the confusion matrix composed of the recommendations made by the RS and the recommendations accepted by the RS user. TP and TF represent the correct recommendations, while TN and FN represent incorrect recommendations [21].

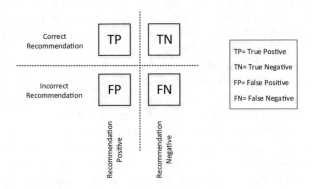

Fig. 2. Confusion matrix of retrieval information for RS [18].

Accuracy measures the difference between the prediction made by the RS and the value of the real ranking on an item. Table 3 shows the metric *Mean Absolute Error* (MAE) which is the most used to determine Accuracy. $r_{u,i} \neq \blacksquare$ It means that a user u has not voted on an item i. O_u represents the set of items voted by u with predictive values $O_u = \{i \in I | p_{u,i} \neq \blacksquare \wedge r_{u,i} \neq \blacksquare\}$ [18, 20].

The *Precision* metric shows the proportion of correctly recommended items of the total number of items that entered the RS [20, 21]. The *Recall* metric shows the proportion of relevant items of the total of items correctly recommended [18, 21]. The metric *f-measure* represents the harmonic mean between the Precision and Recall metrics. A value close to 1 must be obtained, this indicates that the quality of the RS recommendations is excellent.

Table 3. Metrics of evaluation.

Method	Equation
MAE	$= \frac{1}{\#U} \sum_{u \in U} \left(\frac{1}{\#O_u} |p_{u,i} - r_{u,i}| \right)$
Precision	$= \frac{TP}{TP+FP}$
Recall	$= \frac{TP}{TP+FN}$

Description. U: number of users. O_u: set of voted items.$p_{u,i}$: absolute difference of the prediction. r_{ui}: users' votes. TP: True Positive. FP: False Positive. FN: False Negative.

3   Developed Experiments

3.1   Description of the HAC with FC

The dataset used for the experiments is Filmtrust 100k. This dataset presents a high dispersion of data and correspond to user's rankings on movies (items). In the Fig. 3 shows the dataset that contains the votes that each user assigns to each item. The dataset is a matrix where the users ($u = u_1, u_2, \ldots, u_m$) are represented in the rows and the items ($i = i_1, i_2, \ldots, i_n$) in the columns.

When using the Euclidean Distance metric, the *user distance matrix* is calculated directly with the votes of the dataset of each user $d_{(u1-u2)}$ up to $d_{(u_n-u_m)}$. Only the upper triangular matrix is calculated because it is a mirror matrix and it would obtain the same data in the lower triangular as in the upper triangular matrix.

When using the PC or JMSD metrics, the *user similarity matrix* is first calculated from $S_{(u1-u2)}$ to $S_{(u_n-u_m)}$. Then the similarities are converted to distances applying the considerations of Sect. 2.2. *Note:* The lack of a vote (rating) must not be considered as a zero so HAC can use FC to calculate the *user similarity matrix* or the *user distance matrix*. The lack of rating should be considered as a gap and should not be considered in the calculation of the matrices.

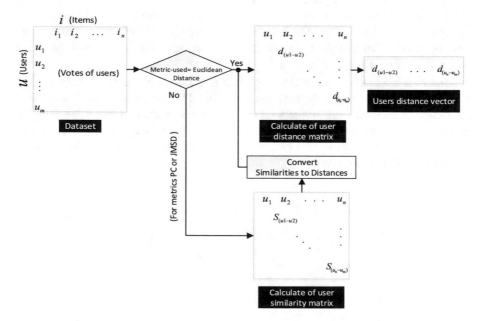

Fig. 3. Flowchart of HAC using CF to obtain the users distances vector

The upper triangular of the *user distance matrix* is extracted to obtain the *users distances vector* from $d_{(u1-u2)}$ - $d_{(u_n-u_m)}$. To start the HAC, each distance of the **users'** *distances vector* represents a cluster, as mentioned in Sect. 2.1. It varies among the five fusion methods described in Sect. 2.2 for the HAC process. Finally, the clusters (k) obtained from HAC are evaluated using the evaluation metrics of Sect. 2.3.

3.2 Results Obtained

For the experiments we use the Single and Complete Linkage fusion methods. We choose these methods due to the size of the Filmtrust 100k dataset and the restrictions of the magnitude of this work. Figure 4 shows the results of the fusion methods with different metrics using the dataset Fimltrust that presents a high data spreading and 4

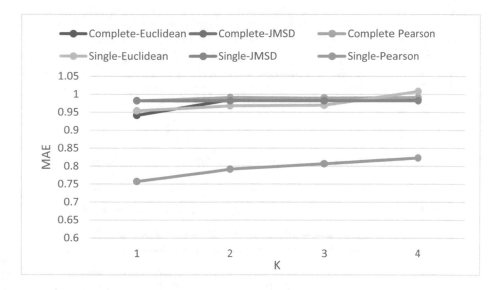

Fig. 4. Results of MAE with the metrics: Euclidean distance, JMSD and PC

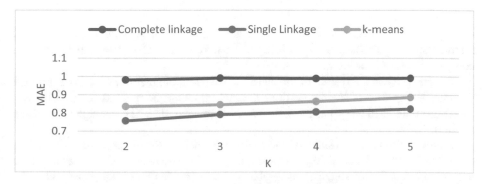

Fig. 5. Results of MAE with: Complete linkage, Single linkage and k-means.

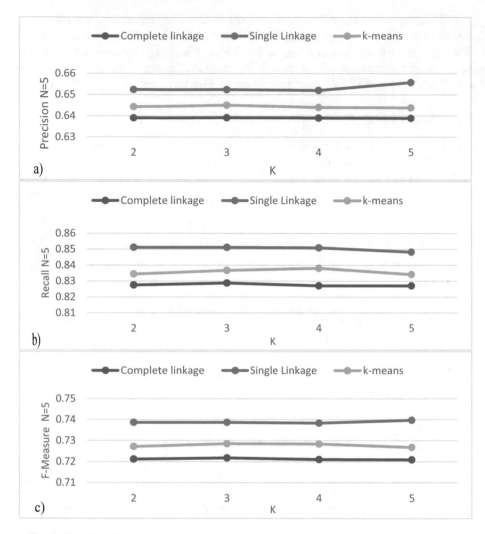

Fig. 6. Results of (a) Precision N = 5 suggestions (b) Recall N = 5 (c) F-measure N = 5.

clusters (k). The MAE metric indicates that the best result is obtained with the PC metric using the method Single Linkage. Therefore, the PC metric will be used to compare the HAC versus K-means.

Figure 5 shows the results of the metric MAE using the metric PC. It is determined that the best quality of the suggestions is obtained using the method Single Linkage. The algorithm k-means present a better performance than the Complete Linkage method.

Figure 6 shows the results of the metrics Precision, Recall and F-measure. The results obtained correspond to 5 suggestions (N = 5). The results of the metrics show that a better quality of suggestions is obtained with the Single Linkage method.

4 Conclusions

We conclude that a RS based on CF shouldn't interpret the absence of a user's vote as a zero. It should represent it as a vacuum and should not be used within the HAC process. Experimental tests showed that the proposed method generates better quality recommendations than k-means using dataset with scattered information. The cluster fusion method that best fits our dataset and gives us better results is Single Linkage. The metric that best defines the similarity between users is the PC. The metric MAE indicates that the suggestions made by the SR proposed are of high quality. The Precision metric indicates that most predictions made by the RS proposed are accepted by the user. We exclude the use of the methods of fusion Ward and Median because it is recommended to use them only with distance metrics such as Euclidean Distance, which would avoid using it with similarity metrics like JMSD or PC.

References

1. Galán, S.M.: Filtrado Colaborativo y Sistemas de Recomendación, IRC 2007, Univ. Carlos III Madrid, pp. 1–8 (2007)
2. Zinke, C., Meyer, K., Friedrich, J., Reif, L.: Digital social learning – collaboration and learning in enterprise social networks, vol. 596, pp. 3–11 (2018)
3. Covington, M.J., Carskadden, R.: Threat implications of the internet of things. In: 2013 5th International Conference Cyber Conflict, pp. 1–12 (2013)
4. Goebel, R.: Lecture Notes in Artificial Intelligence Subseries of Lecture Notes in Computer Science LNAI Series Editors (2012)
5. Adomavicius, G., Tuzhilin, A.: Toward the next generation of recommender systems: a survey of the state-of-the-art and possible extensions. IEEE Trans. Knowl. Data Eng. **17**(6), 734–749 (2005)
6. Herlocker, J.L., Konstan, J.A., Terveen, L.G., Riedl, J.T.: Evaluating collaborative filtering recommender systems. ACM Trans. Inf. Syst. **22**(1), 5–53 (2004)
7. Melville, P., Mooney, R.J., Nagarajan, R.: Content-boosted collaborative filtering for improved recommendations, no. July (2002)
8. Hartigan, J.A., Wong, M.A.: Algorithm AS 136 : A K-Means Clustering Algorithm. J. Roy. Stat. Soc. Ser. C Appl. Stat. **28**(1), 100–108 (2016). http://www.jstor.org/stable/2346830. Published by : Wiley for the Royal Statistical Society Stable
9. Ortega, J.P., del Rocio Boone Rojas, M., Somodevilla Garcia, M.J.: Research issues on K-means algorithm : an experimental trial using matlab. In: Proceedings of 2nd Working Semantic Web New Technologies, pp. 83–96 (2009)
10. Fränti, P., Virmajoki, O., Hautamäki, V.: Fast agglomerative clustering using a k-nearest neighbor graph. IEEE Trans. Pattern Anal. Mach. Intell. **28**(11), 1875–1881 (2006)
11. Menon, A.K., Chitrapura, K.-P., Garg, S., Agarwal, D., Kota, N.: Response prediction using collaborative filtering with hierarchies and side-information. In: Proceedings of 17th ACM SIGKDD International Conference Knowledge Discovery Data Mining - KDD 2011, p. 141 (2011)
12. Ekstrand, M.D., Riedl, J.T., Konstan, J.A.: Collaborative filtering recommender systems. Found. Trends® Hum.–Comput. Interact. **4**(2), 10–14 (2011)

13. Pham, M.C., Cao, Y., Klamma, R., Jarke, M.: A clustering approach for collaborative filtering recommendation using social network analysis. J. Univers. Comput. Sci. **17**(4), 1–21 (2011)
14. Müllner, D.: Modern hierarchical, agglomerative clustering algorithms, no. 1973, pp. 1–29 (2011)
15. Yang, Z., Yang, D., Dyer, C., He, X., Smola, A., Hovy, E.: Hierarchical attention networks for document classification. In: Proceedings of 2016 Conference North American Chapter Association Computational Linguistics Human Language Technologies, pp. 1480–1489 (2016)
16. Murtagh, F., Legendre, P.: Ward's Hierarchical Clustering Method: Clustering Criterion and Agglomerative Algorithm, no. June, pp. 1–20 (2011)
17. Jamain, A., Hand, D.: Mining supervised classification performance studies: a meta-analytic investigation. J. Classif. **112**, 87–112 (2008)
18. Bojorque, R., Hurtado, R.: Técnicas híbridas en Sistemas de Recomendación para optimizar el Modelo Non Negative Matrix Factorization. Universidad Politécnica de Madrid (2017)
19. Zahra, S., Ghazanfar, M.A., Khalid, A., Azam, M.A., Naeem, U., Prugel-Bennett, A.: Novel centroid selection approaches for KMeans-clustering based recommender systems. Inf. Sci. (Ny) **320**, 156–189 (2015)
20. Hernández del Olmo, F., Gaudioso, E.: Evaluation of recommender systems: a new approach. Expert Syst. Appl. **35**(3), 790–804 (2008)
21. Ricci, F.: Recommender Systems Handbook, 1003 p. Springer Science+Business Media, New York (2015). ISBN 978-1-4899-7636-9. Ricci, F., Rokach, L., Shapira, B. (eds.)

Decision Rules Mining with Rough Set

Haitao Wang[1,2], Jing Zhao[1,2], Gang Wu[1,2], Zhao Chao[1,2],
Zhang Fan[1,2], and Xinyu Cao[1,2(✉)]

[1] China National Institution of Standardization, No. 4 of Zhichun Road,
Haidian District, Beijing, China
{wanght,zhaoj,wugang,zhaochao,zhangfan,
caoxy}@cnis.gov.com
[2] AQSIQ Key Laboratory of Human Factors and Ergonomics (CNIS),
Beijing, China

Abstract. Decision rule has been widely used for its briefness, effectiveness and favorite understandability. Many methods aiming at mining decision rules have been developed. Rough set theory. Unfortunately, data is split between multiple parties in many cases. And privacy concerns may prevent these parties from directly sharing the data. This paper addresses the problem of how to securely mine decision rules over horizontally partitioned data with rough set approach. This paper integrates a general framework rather than a very specific solution for mining decision rules with rough set approach when privacy is concerned and data is horizontally partitioned.

Keywords: Rough set · Decision rule · Semi-honesty · Entropy
Commutative encryption

1 Introduction

Decision rule is an effective tool for predicting the values of specific nominal variables when the values of other variables are given. There are many practical situations where decision rule is of immense use. Examples include: a weather forecast of a special day based on a set of data, whether or not providing a loan for an applicant given his records, etc.

Though it is true for many organizations that they have gathered lots of data into their sites single handedly, large numbers of correlated data are often distributed over many sites. Here we assume those sites have the same schema, but each site has information on different entities (data is horizontally partitioned). In such case, mining on the local data often leads to inaccurate, even improper results. Thus, these parties would like to leverage their data for mutual benefit (e.g. for obtaining useful knowledge).

A key problem in this scenario is privacy concerns. The problem of securely mining decision rules in distributed circumstance is important. However, there is lack of necessary research on this problem in the context of rough set theory. In this paper, we discuss it and then present a solution. The organization of this paper is as follows: Sect. 2, we briefly introduce the related work in the area of privacy-preserving data

© Springer International Publishing AG, part of Springer Nature 2019
T. Z. Ahram (Ed.): AHFE 2018, AISC 787, pp. 357–365, 2019.
https://doi.org/10.1007/978-3-319-94229-2_35

mining. Section 3 presents the concepts in rough set theory as well as details on secure multi-party computation. Section 4 fully presents the secure multi-party mining of decision rules in the context of rough set theory. Finally, Sect. 5 concludes the paper and gives directions for future research.

2   Related Work

By far, two approaches have been advanced in privacy-preserving data mining. One is the data perturbation approach, proposed by [1, 2]. In this approach, each individual data is perturbed and the distribution of the all data is reconstructed at an aggregate level. That is to say, this technique uses probability distribution of the distorted data to generate the approximate distribution of the original data. Much work has been done by Agrawal in developing this method.

The other approach proposed by [3] is quite different from the data perturbation method. It treats privacy-preserving data mining as a special case of a more general problem, the secure multi-party computation problem. The assumption with this method is that each party is allowed to know its own data, but no one is permitted to see others' data. Whereas, the assumption with perturbation is that original data must be kept private from data mining party.

It is easy to see that the second approach deals with privacy preserving mining tasks in a distributed circumstance. It not only aims for preserving individual privacy, but also tries to preserve leakage of any information except for the final result [4]. Researches on this problem are extensive. In Lindell's work, he gives a set of protocols for securely building an ID3 decision tree where the training set is distributed between two parties. In Clifton's work, he presents a scheme for mining association rules where data is horizontally partitioned among many sites. There has also been other work (details can be found at [5, 6]). We follow this approach, but address a different problem, securely mining decision rules with rough set approach.

3   Preliminary Knowledge

In this part, we introduce concepts in rough set theory and present details on secure multi-party computation. Both of them serve as a base for our work.

Rough set approach is a highly practical method for machine learning. It always applies to mine decision rules from decision tables, which are often described as $S = \langle U, C \cup D, V, f \rangle$, where U is a nonempty set of instances, $C \cup D$ is a set of attributes, V is the domain of $C \cup D$, $f : U \times (C \cup D) \to V$, $\forall x \in U$ is described by a conjunction of attribute values. By large, C is called condition attributes and D is called decision attributes. The goal of decision rule mining is to find out how D depends on C from the given instances.

Definition 1. Let $< U, A >$ be an information system where U is a nonempty set of instances, $B \subseteq A$ be a subset of attributes, then

$$ind(B) = \{(x,y) \in U^2 : \forall a \in B, a(x) = a(y)\} \tag{1}$$

is called an indiscrimination relation of $<U,A>$.

Definition 2. Let $<U,A>$ be an information system, $B \subseteq A$, $X \subseteq U$, then the lower approximation of X under B is

$$\underline{B}X = \cup\{Y \in U/ind(B) : Y \subseteq X\}; \tag{2}$$

the upper approximation of X under B is

$$\overline{B}X = \cup\{Y \in U/ind(B) : Y \cap X \neq \phi\}. \tag{3}$$

With definition 2, we can easily calculate

$$k = \frac{|POS_C(D)|}{|U|}, \tag{4}$$

where $POS_C(D) = \cup_{X \in U/ind(D)} \underline{C}X$. Hence, the value k exactly expresses how much D depends on C.

Let $<U,A>$ be an information system, where $A = C \cup D$, C is condition attributes, D is decision attributes. For all $C' \subseteq C$, if it satisfies the following restrictions:

$$POS_{C'}(D) = POS_C(D), \tag{5}$$

there is no $C'' \subset C'$ that

$$POS_{C''}(D) = POS_{C'}(D), \tag{6}$$

However, the mining result obtained this moment is not general enough. A further mining step, named attribute value reduction, is needed.

4 Secure Decision Rule Mining

In order to securely mining decision rules with rough set approach, we need to address two issues: (1) How to complete attribute reduction securely; (2) How to go on attribute value reduction securely. The following subsections provide details on both issues.

4.1 Attribute Reduction

The purpose of attribute reduction is to find the smallest subset of attributes that satisfies (5) and (6). The main operation during this procedure is evaluation of different attribute sets with respect to the degree that decision attributes depend on them. Therefore, an index is needed to evaluate the "efficiency" of different attribute. Several schemes [7] have ever been proposed. We consider Shannon's entropy here.

Let $<U, C \cup D>$ be an information system, $U/ind(D) = \{Y_1, Y_2, \cdots, Y_m\}$, the $Entropy(D)$ is defined as follows:

$$Entropy(D) = -\sum_{j=1}^{m} P(Y_j) \log P(Y_j). \tag{7}$$

Let $R \subseteq C$, $U/ind(R) = \{X_1, X_2, \cdots, X_n\}$, the conditional entropy $Entropy(D|R)$ of R about D is given by

$$Entropy(D|R) = -\sum_{i=1}^{n} P(X_i) \sum_{j=1}^{m} P(Y_j|X_i) \log(P(Y_j|X_i)). \tag{8}$$

Based on the conditional entropy, we can compute the information gain if attribute $A \subset C$ joins $R \subset C$ to partition the instances.

$$Gain(D, R, A) = Entropy(D|R) - Entropy(D|R \cup A). \tag{9}$$

Thus, the procedure of secure attribute reduction is essentially a procedure of securely computing $Entropy(D|R)$.

Examine $Entropy(D|R)$ for an attribute R with n possible values $r_1, r_2, \cdots, r_n$ and a decision attribute D with m possible values $d_1, d_2, \cdots, d_m$.

$$Entropy(D|R) = -\sum_{i=1}^{n} P(r_i) \sum_{j=1}^{m} P(d_j|r_i) \log(P(d_j|r_i)) = \frac{1}{|T|} \sum_{j=1}^{n} |T(r_j)| \sum_{i=1}^{m} -\frac{|T(r_j, d_i)|}{|T(r_j)|} \log\left(\frac{|T(r_j, d_i)|}{|T(r_j)|}\right). \tag{10}$$

$$= \frac{1}{|T|} \left(-\sum_{j=1}^{n} \sum_{i=1}^{m} |T(r_j, d_i)| \log(|T(r_j, d_i)|) + \sum_{j=1}^{n} (|T(r_j)| \log(|T(r_j)|)) \right)$$

In (10), $|T|$ is the number of instances in the global data set, $|T(r_j, d_i)|$ is the number of all instances with decision d_i and attribute value r_j, while $|T(r_j)|$ is the global number of instances that simply take attribute value r_j. Due to horizontal partitioning of data, the global database is a union of k databases, where k is the number of different sites and each site P_i only knows its own data. The number of instances that take value r_j on attribute R can therefore be written as $|T(r_j)| = |T_1(r_j)| + |T_2(r_j)| + \cdots + |T_k(r_j)|$, where $|T_i(r_j)|$ is the number of instances, which are held by party P_i and take value r_j on attribute R. Each party $P_i(i = 1..k)$ can compute $|T_i(r_j)|$ and $|T_i(r_j, d_i)|$ independently. Therefore (10) can be written as a sum of expressions of the form

$$(v_1 + v_2 + \cdots + v_k) \log(v_1 + v_2 + \cdots + v_k). \tag{11}$$

In (11), v_i corresponds to $|T_i(r_j)|$ or $|T_i(r_j, d_i)|$. The main task is thus to compute (11) securely.

We extend his work to more parties' cases. Our strategy for securely computing (11) is to regress it to be one that only two parties are involved.

The strategy is that the first site generates a random number Rnd, uniformly chosen from a suitable range, adds this number to v_1 and send $v_{Rnd} = Rnd + v_1$ to the next site.

The random number masks the actual value of v_1, so the second site knows nothing about the first site's v_1. Note that Rnd should be a reasonable number, otherwise v_1 will be revealed. Then, the second site adds v_2 to v_{Rnd} and sends the value to the third site $\cdots$ When the last site finishes this procedure, it holds the value $v_{Rnd} = Rnd + \sum v_i$. At the moment, the first party has the value Rnd and the last party has v_{Rnd}, while $v_{Rnd} - Rnd$ is the true value of $\sum v_i$. Therefore (11) can be computed securely by Lindell's solution:

$$(v_1 + v_2 + \cdots + v_k)\log(v_1 + v_2 + \cdots + v_k) = (v_{Rnd} - Rnd)\log(v_{Rnd} - Rnd). \quad (12)$$

Protocol 1 Securely computing $(v_1 + v_2 + \cdots + v_k)\log(v_1 + v_2 + \cdots + v_k)$
{Suppose that there are k participants and everyone has a private value v_i}

1. the first site produces a random number Rnd and set $v_{Rnd} = Rnd$
2. for each site P_i do
3. participant P_i locally computes $v_{Rnd} = v_{Rnd} + v_i$
4. end for
5. call Lindell's method to compute $(v_{Rnd} - Rnd)\log(v_{Rnd} - Rnd)$.

Here is a revision of protocol 1 for an honest majority.
{Suppose that there are k participants and everyone has a private value v_i, each v_i is divided into t shares and each share is represented by $p_j(v_i), j = 1..t$}
{Rnd, v_{Rnd} are defined as in Protocol 1, $p_j(v_{Rnd}), j = 1..t$, represents each share of v_{Rnd}} (Fig. 1).

1: the first site produces a random number Rnd and set $p_j(v_{Rnd}) = Rnd/t$
2: for $j = 1$ to t do
3:   for each site P_i do
4:     participant P_i locally computes $p_j(v_{Rnd}) = p_j(v_{Rnd}) + p_j(v_i)$
5:   end for
6:   change the route from site 1 to site k (produce a new permutation with integers from 2 to $k-1$)
7: end for
8: site k get the value $v_{Rnd} = \sum p_j(v_{Rnd})$
9: call Lindell's method to compute $(v_{Rnd} - Rnd)\log(v_{Rnd} - Rnd)$

Fig. 1. Protocol 2 revision of protocol 1 for honest majority case

Though Protocol 2 is not a perfect solution, it is reasonable in real world and does work in honest majority case. With its help, attribute reduction can go on and then a subset of attributes C' that satisfies (5), (6) can be achieved. However, C' is not the final goal, there is still redundant information in attribute reduction result. A further reduction is needed.

4.2   Attribute Value Reduction

The task of attribute value reduction is to generalize the result of attribute reduction and finally present the briefest rules. Though there has been much research on this issue, none of them can be applied to our mining task because they take the assumption of one single site with all data (e.g. [8, 9]). In the cases where security is concerned and the outcome of attribute reduction is distributed over multiple sites, a new solution for attribute value reduction is needed.

Our strategy contains two steps. First a method is given for collecting the scattered result of attribute reduction while privacy of each site is preserved. Then the algorithms, which deal with the case that all data centralize on one site, will engage in our task.

Protocol 3 is an acceptable model for completing the first step of our strategy. It derives from [4]. In [4], a similar protocol is used to get the frequent item sets for finding association rules while Protocol 3 here serves for mining decision rules with rough set approach.

As commutative encryption is a key concept of Protocol 3, we give a brief description here. Firstly, we introduce the definition of quasi-commutative hash function. A hash function h is said to be quasi-commutative if for given x and $y_1, y_2, \cdots, y_m$, the value

$$z = h(h(\cdots h(h(x, y_1), y_2) \cdots, y_{m-1}), y_m) \tag{13}$$

is the same for every permutation of y_i. Commutative encryption is just the encryption method, which uses one-way and quasi-commutative hash functions as its tool. In this method, we are able to retrieve the x value when $(z, y_1, y_2, \cdots, y_m)$ is given.

Now we show the strategy for securely collecting the scattered result of attribute reduction. Let h be a one-way hash function, and every party knows h. The main idea is that each site P_i encrypts its own partial result $result_i$ with an individually decided y_i. Each site then sends $h(result_i, y_i)$ to its neighbor and encrypts the partial results from other sites. Finally every $result_i$ is encrypted by all parties. That is to say, for any $result_i$, there is a corresponding

$$z_i = h(\cdots h(h(\cdots h(h(result_i, y_i), y_{i+1}) \cdots, y_k), y_1) \cdots, y_{i-1}), \tag{14}$$

where k is the number of sites.

Now the last party P_k gathers all these z_i into a set $\{z_i\}$ and sends it to P_1. Then P_1 decrypts $\{z_i\}$ with y_1 and sends the decrypted result to site P_2. Again P_2 decrypts the result that P_1 sends to it with y_2 and sends the new decrypted result to P_3. Similarly every party $P_i (i = 3, \cdots, k)$ repeats the same operation with its own y_i as P_1 and P_2 does. At last, site P_k will get the completely decrypted $\{result_i\}$ while not knowing the derivation of each $result_i$. In addition, the one to one mapping property of h guarantees duplicates in $\{result_i\}$ will be duplicated in $\{z_i\}$ and can be deleted. Above idea is shown in protocol 3 (Fig. 2).

Protocol 3 works well when there are more than 2 parties involved. Any party who wants to disclose the attribute reduction result scattered on site P_i must collude with all parties except P_i. Otherwise it will get nothing meaningful. That is to say Protocol 3 can work in an honest majority circumstance. Unfortunately, Protocol 3 cannot prevent each site from knowing the other's information when it applies to 2 parties case.

{ k is the number of sites, $result_i$ is the attribute reduction distributed over site i }

1: for each site P_i do

2:   P_i encrypt $result_i$, $z_i = h(result_i, y_i)$

3: end for

4: P_k gathers all z_i, $\{z_i\} = \cup z_i$

5: P_k sends $\{z_i\}$ to P_1

6: for each site P_i do

7:   P_i encrypts all the z_j, where $j \neq i$

8: end for

9: for each site P_i do

10:   P_i decrypts all the z_j, where $j = 1, \cdots, k$

11: end for

Fig. 2. Protocol 3

However, this is not a problem particularly puzzling Protocol 3. Every secure computation protocol will meet it in 2 parties case, where every party will infer the opposite's information if the final result is shared between them.

In order to reduce the communication cost of protocol 3, $result_i$ is made up of patterns, of which every pattern appears only once in $result_i$ if it appears reduplicatively in the attribute reduction result of P_i, it should appear only once in $result_i$. Here's an example. Suppose that Table 1 is the attribute reduction result on P_i. We can see the records labeled 1 and 3 are actually the same pattern, so this pattern will appear only once in $result_i$. This means $result_i$ only contains the patterns labeled 1, 2, 4.

Table 1. The attribute reduction result on P_i

Label	c1	c2	c3	decision
1	1	2	3	Yes
2	2	2	3	No
3	1	2	3	Yes
4	1	1	2	Yes

4.3   Secure Computation of Support and Confidence

At last, we will address how to compute the support and confidence of a rule securely. We only talk about confidence computing here for space reason (support computing is the same). Suppose that there is a decision rule $r : condition \rightarrow decision$. The confidence of r is given by

$$P(decision|condition) = \frac{n_{cd}}{n_c}, \tag{15}$$

where n_{cd} is the number of instances which have attribute value *condition* and decision value *decision*, while n_c is the number of instances which only have attribute value *condition*. In distributed circumstance, all instances are scattered on k sites (k is the number of sites). Therefore, n_{cd} and n_c are divided into k parts, every site $i(i = 1, \cdots, k)$ has its own share n_{icd} and n_{ic}. Then

$$P(decision|condition) = \frac{\sum_{i=1}^{k} n_{icd}}{\sum_{i=1}^{k} n_{ic}}. \tag{16}$$

At this point, the secure $P(c|a)$ protocol, given in [5] for horizontally partitioned data, is employed. Using their method, it is easy to get the confidence of rule r, which is indeed the value $P(decision|condition)$.

Rough set approach is an effective tool for decision rule mining. We present a solution to this problem based on Lindell and Kantarcioglu's work. We will also investigate another case in which data is vertically partitioned among multiple parties.

Acknowledgements. This paper is supported by grants from National Key R&D Program of China (2016YFF0204205, 2017YFF0206503, 2017YFF0209004) and China National Institute of Standardization (712016Y-4941, 522016Y-4681, 522018Y-5948, 522018Y-5941, 522017Z-5853, 522017Z-5459).

References

1. Agrawal, R., Srikant, R.: Privacy-preserving data mining. In: The Proceedings of the 2000 ACM SIGMOD Conference on Management of Data, pp. 439–450 (2000)
2. Agrawal, D., Aggarwal, C.C.: On the design and quantification of privacy preserving data mining algorithms. In: Proceedings of the Twentieth ACM SIGACT-SIGMOD-SIGART Symposium on Principles of Database Systems, pp. 247–255 (2001)
3. Lindell, Y., Pinkas, B.: Privacy preserving data mining. In: Advances in Cryptology-CRYPTO 2000, pp. 36–54 (2000)
4. Kantarcioglu, M., Clifton, C.: Privacy-preserving distributed mining of association rules on horizontally partitioned data. In: The ACM SIGMOD Workshop on Research Issues on Data Mining and Knowledge Discovery (DMKD 2002), pp. 24–31 (2002)
5. Kantarcioglu, M., Vaidya, J.: Privacy preserving naïve Bayes classifier for horizontally partitioned data. In: IEEE ICDM Workshop on Privacy Preserving Data Mining, pp. 3–9, (2003)

6. Yao, A.C.: How to generate and exchange secrets. In: Proceedings of the 27th IEEE Symposium on Foundations of Computer Science, pp. 162–167 (1986)
7. Rastogi, R., Shim, K.: A decision tree classifier that integrates building and pruning. Data Min. Knowl. Discov. **4**, 315–344 (2000)
8. Hu, F., Zhang, F.J., Liu, S.H.: A rough set-based algorithm for attribute value reduction. Comput. Eng. Appl. 48–51 (2003)
9. Lin, J.Y., Peng, H., Zheng, Q.L.: A new algorithm for value reduction based on rough set. Comput. Eng. 70–71 (2003)

Social Network Modeling

Using Information Processing Strategies to Predict Contagion of Social Media Behavior: A Theoretical Model

Sara M. Levens[(⊠)], Omar Eltayeby, Bradley Aleshire, Sagar Nandu, Ryan Wesslen, Tiffany Gallicano, and Samira Shaikh

University of North Carolina at Charlotte, 9201 University City Blvd., Charlotte, NC 28223, USA
{slevens, oetayeb, baleshil, snandul, rwesslen, tgallica, sshaikh2}@UNCC.edu

Abstract. This study presents the Social Media Cognitive Processing model, which explains and predicts the depth of processing on social media based on three classic concepts from the offline literature about cognitive processing: self-generation, psychological distance, and self-reference. Together, these three dimensions have tremendous explanatory power in predicting the depth of processing a receiver will have in response to a sender's message. Moreover, the model can be used to explain and predict the direction and degree of information proliferation. This model can be used in a variety of contexts (e.g., isolating influencers to persuade others about the merits of vaccination, to dispel fake news, or to spread political messages). We developed the model in the context of Brexit tweets.

Keywords: Social media · Human behavior · Emotion contagion
Information processing strategies · Behavior transference

1 Introduction

McLuhan [1] argued that *"the medium is the message"* because the medium that people use to communicate shapes communication in profound ways. Not surprisingly, researchers have found that social media as a channel shapes how people attend to and respond to information [2–5]. Despite the ubiquity of social media use, research investigating the association between the use of social networking sites (SNS) and cognitive processing has largely been limited to applied research areas such as mental health outcomes [6–8], marketing [9–11], and education [12, 13]. Nascent research is emerging that takes knowledge from offline environments and applies it to the SNS landscape [14, 15]. However, no studies to date have applied fundamental cognitive concepts to information processing on social media, despite the explanatory potential of these concepts and how they can predict behavior in the SNS environment.

The purpose of our research is to construct a preliminary model that applies cognitive processing to the context of social media. We build upon three classic cognitive concepts in the context of the SNS milieu: the generation effect [16], psychological

© Springer International Publishing AG, part of Springer Nature 2019
T. Z. Ahram (Ed.): AHFE 2018, AISC 787, pp. 369–378, 2019.
https://doi.org/10.1007/978-3-319-94229-2_36

distance [17], and the self-reference effect [18]. This study is focused on the examination of original tweets, retweets, replies, and profile information on the social network platform, Twitter. We situate our findings in a divisive political context that had a
potentially significant impact on daily life for a vast number of people—the prospect of
the UK leaving the EU, commonly known as Brexit.

2 Cognitive Processing

Our model is informed by foundational research about information processing in an
offline context. We propose this model to explain information processing and response
behavior in the online context of social networking sites, specifically Twitter. While the
model is proposed in the context of Twitter, the concepts are generalizable to any SNS
platform (with certain adaptations).

A key concept that applies to our model is *depth of processing*, which refers to the
degree to which information that individuals attend to is encoded for current use and
storage in long-term memory [19, 20]. Irrelevant or incidental information is likely to
be processed at a relatively shallow level, making it less likely to influence behavior
and be remembered [21]. Craik and Tulving [21] found that meaningful information,
on the other hand, is typically processed at a comparatively deep level. Meaningful
information forms links with other related information, becomes integrated into an
individual's personal experiences and knowledge base, and is more likely to be
remembered and influence behavior (and changes in behavior) than cursory processed
information [22].

There are a number of established cognitive mechanisms that predict depth of
processing—and consequently behavior, learning, memory, and event recall. For our
model, we adopted the three most predominant cognitive mechanisms that predict
depth of processing: self-generation, psychological distance and self-reference [23].

Self-generation refers to the degree to which individuals construct material themselves [24–26]. For instance, in the context of Twitter, an original tweet has a higher
degree of generation than a reply, and a retweet has the lowest level of generation of
response behaviors (such as replies, mentions, and retweets) [21]. In cognition, the
generation of new material reflects high depth of processing as an individual is constructing content based on their experiences, knowledge, interests, etc. [24]. Self-
generation is also used as a bench march of learning and information proliferation—if
an individual has generated content in response to an event or message then they have
'heard' and integrated the content into their existing cognitive framework to generate
new material [24, 26].

The second concept that predicts the depth of processing is derived from construal-
level theory [17]. According to this theory, individuals process events by developing
construals, which are based on psychological distance. *Psychological distance* is the
subjective interpretation of how near or far something or someone is in terms of
temporal distance (the now), spatial distance (the here), and distance in relatedness (the
self) [17]. If an individual deems an event to have low psychological distance (i.e.,
being close to the self, here, and now), the event is expected to have a profound effect
on that individual. From the concept of psychological distance, we incorporated

temporal distance as part of our model. Temporal distance refers to the distance, in time, between an individual's response behavior (e.g., a retweet) and an instigating event (e.g., Brexit).

Degree of *self-reference* is another key concept that predicts the depth of processing; it refers to the degree to which individuals connect the material to themselves [18, 27–29].

In summary, information that is related to the self, created by the self, and close to the self is processed at a deeper level; consequently, it is encoded more deeply in memory than less relevant, distant information. When processing information in working memory, the simultaneous utilization of all three of these cognitive mechanisms (self-generation, psychological distance, and self-reference) yields the deepest level of information processing and long-term memory storage. Information that is processed utilizing only one of these three mechanisms would be processed in a comparatively cursory way. In the next section, we discuss our model that combines these cognitive mechanisms.

3 Social Media Cognitive Processing Model (SMCP Model)

We constructed the social media cognitive processing (SMCP, pronounced "SiMCaP") model to explain and predict cognitive processing in an online context based on social media behaviors (in this case, original tweets, replies, and retweets). The process of communication involves three entities: the sender, the receiver, and the message. Sender and receiver characteristics are pivotal in our model, and they are the lens through which we view content generation and response on Twitter. We use the term *dyad* to refer to the sender of a message (in this case, a tweet) and the person who replies to the message.

In Fig. 1 (left), we show the SMCP conceptual model with its component cognitive processes. Each vertex on the triangle circumscribed by the concentric circles represents the highest degree of the respective cognitive process (namely, directed generation, pairwise similarity, and temporal distance). The shaded area in the center of the triangle is where the highest degree of depth of processing would occur because it reflects the presence of all three mechanisms of cognitive processing.

In Fig. 1 (right), we show the interaction between a sender and receivers of a given message. Each message from a sender has several receivers (R_1, R_2...R_n). The shaded area (in gray) represents one dyad for which the depth of processing is calculated in the SMCP model. The model allows us to compute the depth of processing for each such dyadic interaction between senders and receivers in a given corpus.

We measured three attributes for each dyad, which are coded based on how deeply the receiver processed the sender's information. To measure the degree of self-reference for the receiver in the dyad, we defined the concept of paired similarity by calculating the likeness between a sender and a receiver, both on the level of the message and on the level of the users (explained in detail in Sect. 4.2 below). Support for this concept can also be found in the marketing and rhetorical literature. From the marketing literature, perceived similarity between sender and receiver on social media increases perceptions that the content is trustworthy, credible, and honest [30]. From

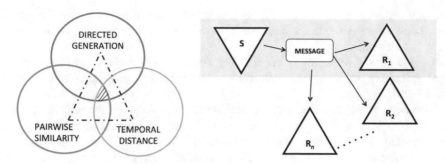

Fig. 1. (Left) SMCP model linking cognitive mechanisms of directed generation. Pairwise similarity and temporal distance. (Right) illustration of dyad relationship between sender (S) and receivers (R_1, R_2...)

the rhetorical literature, the more people identify with one another, the more consubstantial they become [31]. Moreover, a source is more likely to persuade an audience member if the audience member perceives a high level of consubstantiality with the source [31].

The second measure of depth of processing for the receiver in the dyad is called directed generation. As noted previously, it is based on the concept of self-generation (i.e., the extent to which people create their own content). We replaced "self" with "directed" because this new concept measures the degree to which the receiver generates *new* content in response to a sender's tweet. In the context of Twitter, directed generation is the tendency of a receiver to interact with the sender's content (e.g., by retweeting or replying to the sender's message). We posit that receivers who reply more frequently than retweet in response to a sender tend to process the sender's message at a deeper level than other response behaviors, such as liking or retweeting (Sect. 4.3 below).

The third measure of depth of processing in the dyad is psychological distance. We measure psychological distance in terms of the temporal distance (Sect. 4.4) in our model (other measures of psychological distance [17] like geographical distance or social distance will be included in the model as a part of our future work). Temporal distance refers to the distance, in time, between the catalyst for the sender's message (e.g., the UK vote to leave the EU), the sender's content (i.e., a tweet), and the receiver's response behavior (e.g., retweet, reply). As noted previously, temporal distance is reflective of the temporal component of psychological distance.

Having described each measure of the SMCP model, we now explain the model in the context of a specific case study and calculate the associated measures to explain depth of processing for a set of senders and receivers.

4   Method

Figure 2 outlines our overall approach, which we describe in detail in the sections below.

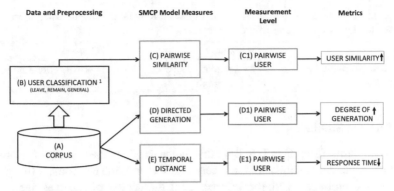

Assumptions: Users are classified by the largest number of users' Leave and Remain hashtags posted. This assumes hashtag use implies user opinion. This approach also assumes that people did not change their minds once they picked an ideological position.

Fig. 2. Outline of overall approach with measurements of each component of the SCMP model and their operationalization.

4.1 Corpus and Preprocessing Steps

We examined a corpus of approximately 2MM tweets regarding the Brexit vote in June 2016 (Step A in Fig. 2). The collected data consisted of a 20% sample of tweets that contained at least one of nine hashtags related to the Brexit event (e.g., #VoteLeave, #Brexit, #StrongerIn) collected via Twitter's GNIP Historical PowerTrack API[1]. The final dataset consisted of 2,171,135 tweets that were shared between June 16th and June 29th, 2016 from 436,474 unique Twitter IDs. From these posts, we examined the set of original tweets (tweets that are not retweets), which consisted 760,964 tweets (35%) in our corpus. We next examined cognitive engagement via the possible actions a user can take on Twitter, such as (1) posting an original tweet; (2) retweeting or sharing an original tweet; (3) replying to an original tweet; (4) mentioning other users in either an original tweet or reply; (5) favoriting an original tweet, retweet or a reply; and (6) quoting another tweet. As noted previously, our model does not account for mentions, favorites, and quotes; however, we aim to include these measures in future work. We note that these actions listed here are specific to the Twitter platform, however, similar actions can be taken on other SNS sites (e.g. Facebook) and the SMCP model can be adapted to those corresponding actions.

As part of the pre-processing steps, we assigned each user to one of three categories (1) Leave (exit the EU); (2) Remain (stay in the EU); or (3) General (ambiguous or undetermined; see Step B in Fig. 2). This classification was made on the basis of the hashtags used in the tweets for each user. If a user's tweets had a majority of hashtags advocating the leave position, the user was assumed to be in the Leave category (and similarly for the Remain category). If a user's tweets contained an equal number of leave and remain hashtags, or if the hashtags were only part of the general hashtags, the user was classified into the General category. Table 1 shows the hashtags and their categorization in our corpus.

---

[1] http://support.gnip.com/apis/.

Table 1. List of hashtags categorized as one of Leave, Remain or General categories. Hashtags were used to place users in one of these categories.

Leave	Remain	General
#strongerin, #remain, #brexitrejection	#voteleave, #leaveeu, #cleanbrexit	#brexit, #brexitnow, #brexitvote

4.2   Pairwise Similarity

Pairwise similarity (Step C in Fig. 2) is the degree to which the content of the message as well as the sender is similar to the receiver. If the message has content that is highly similar to the preferences of the receiver or if the sender of the message is highly similar to the receiver, the pairwise similarity measure would be high. We posit that this would lead to greater depth of processing, and consequently higher emotion contagion to influence offline behavior. In this article, we measure pairwise similarity as the category to which the sender and receiver belong (i.e. Leave or Remain). If the sender and receiver belong to the same category, then pairwise similarity is high (1), otherwise it similarity would be low (0).

4.3   Directed Generation Measure

The degree of self-generation in the offline context is the degree to which individuals construct original material themselves. In the online context (Fig. 2, Step D), our SMCP model assumes that a reply or a retweet by a receiver is generated material in response to (or *directed* by) a sender; however, replies have a higher degree of generation than do retweets. While retweets are more prolific, retweets represent surface contagion—the information has spread but it may not influence offline attitudes, preferences, or beliefs. Replies, on the other hand, require more effort to generate content than a retweet. We thus formulate directed generation (Fig. 2, Step D1) as the ratio of the total number of actions made by the receiver across all messages sent by the receiver, normalized by the total actions across all receivers and all messages for that sender. Directed generation can thus be characterized as the likelihood of a given receiver to engage with the sender's content, normalized by the general likelihood of which all receivers respond to the sender. We include different actions that a user can take on Twitter and use a five-point discrete scale to weight each action, with Likes being assigned a weight of 1/5, followed by Retweets (2/5), Mentions (3/5), Quotes (4/5) and Replies (5/5). As noted in Fig. 2, the higher the degree of generation, the greater the depth of processing.

4.4   Temporal Distance Measure

The third measure of depth of processing is temporal distance which is measured with respect to the time between the action a receiver takes and the event (the Brexit vote) and also the sender's message. Accordingly, if either the event or the sender's original message is closer in time, the depth of processing will be higher. This is based on the construal-level theory which states that an event closer in time to the self will lead to lower psychological distance, which in turn leads to greater effects on the depth of processing. We compute temporal distance as the average time taken by the receiver to respond to the messages sent by the sender.

5   Results and Discussion

We illustrate our approach on a subset of senders and receivers from our corpus of over 400 K users. The senders and receivers included the accounts listed in Table 2. We selected 3 senders and 4 receivers based on their classification (Leave vs. Remain), rate of activity on Twitter, and centrality in the retweet network (Fig. 3).

The 'Directed Generation' panel illustrates the tendency of the receiver to generate new content in response to a sender, normalized by the total actions across all receivers and all messages for that sender. Likelihood of the receiver generating new content in response to a sender is depicted via edge weight—thicker edges indicate greater directed generation over other response behaviors less indicative of content generation (i.e. likes). For example, the thickness of the edge between Carl #LeaveEU and LEAVE EU, is greater than the thickness of the edge between *TOM* and 'Vote Leave' indicating that there is a higher degree of depth of processing on the part of Carl #LeaveEU than *TOM*.

The 'Response Time' panel illustrates how quickly a user responds to the message of a sender. Line thickness corresponds with psychological distance such that the thinner the line the less temporal distance there is between the receiver and sender— specifically the receiver responds quickly to the content of the sender.

The SMCP model is a significant contribution because it is the first model that uses classic cognitive concepts to explain and predict responses to social media content. Grounding social media behaviors in established learning and memory theories is critical because it provides a greater understanding of how information disseminated on social media is encoded to potentially influence learning, memory, preferences, attitudes and beliefs.

Similar to other social influence diagrams, SMCP accounts for oppositional positions in mapping the relationship between a sender of a tweet and a receiver of a tweet. However, SMCP goes beyond current diagrams by recording not only the level of effort receivers expend based on their type of social media response (e.g., retweet vs. reply) but also psychological distance to the event and the extent of pairwise similarity between the sender and receiver. Together, these three dimensions have tremendous explanatory power in predicting the depth of processing a particular receiver will have in response to a particular sender's message. Moreover, the model can be used to explain and predict the direction and degree of information proliferation. This model can be used in a variety of contexts (e.g., isolating influencers to persuade others about the merits of vaccination, to dispel fake news, or to spread political messages).

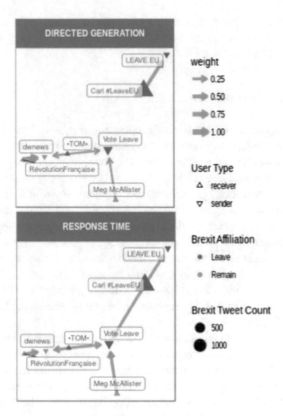

Fig. 3. Network of senders and receivers with edges showing weights of directed generation and response time.

6 Limitations and Future Work

This study represents an initial attempt to identify metrics reflecting pairwise similarity, directed generation and psychological distance on Twitter. Our model relies on several assumptions, which should be considered when interpreting this research. First, we interpreted a user's ideological position based on the hashtags the user used most often (in this case, whether each user posted primarily to the Leave category or to the Remain category). If a user's tweets had a majority of hashtags advocating the leave position, the user was assumed to be in the Leave category (similarly for Remain category). As this is our first attempt, our initial metric of pairwise similarity is simplistic and reflects the context of the sample, future work should identify more nuanced ways to measure similarity such as similarity in profile content or message content between users. We also assumed that a reply or a retweet by a receiver is generated material in response to (or *directed* by) a sender—while replied are generally directed to the sender in response to content from the sender, this is not always the case.

We were also limited to a 20% sample of the GNIP stream for the rule tags we examined. Moreover, the sample did not include favorites, quoted content or mentions

—quoted content and mentions demonstrates a new content generation and would therefore be important to include in future work that refines the directed generation metric. To illustrate the depth of processing between senders and receives in this initial concept paper we also examined only five senders with prolific tweeting and follower numbers to avoid sparsity. Future research should use a larger sample (i.e., 100% sample of GNIP rule tag use) and incorporate favorites and quotes and refine measurement of these constructs in a fuller sample.

As an emerging model, there are many opportunities to build upon this research. Future research should also explore the larger implications of this model, such as using the model to predict emotion contagion. Emotional content tends to be more deeply processed material and linguistic factors could be used to identify emotional content and perhaps create a measure of a receiver's emotional susceptibility to a sender. It will also be important to test this model in other contexts, (i.e., in response to other viral events) as well as incorporate spatial distance into the assessment of psychological distance.

References

1. McLuhan, M.: Understanding Media: The Extensions of Man. MIT Press, Cambridge (1994)
2. Alloway, T.P., Alloway, R.G.: The impact of engagement with social networking sites (SNSs) on cognitive skills. Comput. Hum. Behav. **28**(5), 1748–1754 (2012)
3. Brasel, S.A., Gips, J.: Media multitasking behavior: concurrent television and computer usage. Cyberpsychol. Behav. Soc. Netw. **14**(9), 527–534 (2011)
4. Rosen, L.D., Carrier, L.M., Cheever, N.A.: Facebook and texting made me do it: media-induced task-switching while studying. Comput. Hum. Behav. **29**(3), 948–958 (2013)
5. Booten, K.P.: A library of fragments: digital quotations, new literacies, and attention on social media. Doctoral dissertation. UC, Berkeley (2017)
6. Rosen, L.D., Whaling, K., Rab, S., Carrier, L.M., Cheever, N.A.: Is Facebook creating "iDisorders"? The link between clinical symptoms of psychiatric disorders and technology use, attitudes and anxiety. Comput. Hum. Behav. **29**(3), 1243–1254 (2013)
7. De Choudhury, M., Gamon, M., Counts, S., Horvitz, E.: Predicting depression via social media. In: ICWSM, vol. 13, pp. 1–10 (2013)
8. Frost, R.L., Rickwood, D.J.: A systematic review of the mental health outcomes associated with Facebook use. Comput. Hum. Behav. **76**, 576–600 (2017)
9. Eckler, P., Bolls, P.: Spreading the virus: emotional tone of viral advertising and its effect on forwarding intentions and attitudes. J. Interact. Advert. **11**(2), 1–11 (2011)
10. Hollebeek, L.D., Glynn, M.S., Brodie, R.J.: Consumer brand engagement in social media: conceptualization, scale development and validation. J. Interact. Mark. **28**(2), 149–165 (2014)
11. Ashley, C., Tuten, T.: Creative strategies in social media marketing: an exploratory study of branded social content and consumer engagement. Psychol. Mark. **32**(1), 15–27 (2015)
12. Traphagan, T.W., Chiang, Y.H.V., Chang, H.M., Wattanawaha, B., Lee, H., Mayrath, M.C., Resta, P.E.: Cognitive, social and teaching presence in a virtual world and a text chat. Comput. Educ. **55**(3), 923–936 (2010)
13. Junco, R., Cotten, S.R.: No A 4 U: the relationship between multitasking and academic performance. Comput. Educ. **59**(2), 505–514 (2012)

14. Valkenburg, P.M.: Understanding self-effects in social media. Hum. Commun. Res. **43**(4), 477–490 (2017)
15. Valkenburg, P.M., Peter, J.: The differential susceptibility to media effects model. J. Commun. **63**(2), 221–243 (2013)
16. Bertsch, S., Pesta, B.J., Wiscott, R., McDaniel, M.A.: The generation effect: a meta-analytic review. Mem. Cognit. **35**(2), 201–210 (2007)
17. Trope, Y., Liberman, N.: Construal-level theory of psychological distance. Psychol. Rev. **117**(2), 440 (2010)
18. Symons, C.S., Johnson, B.T.: The self-reference effect in memory: a meta-analysis. Psychol. Bull. **121**(3), 371 (1997)
19. Craik, F.I., Lockhart, R.S.: Levels of processing: a framework for memory research. J. Verbal Learn. Verbal Behav. **11**(6), 671–684 (1972)
20. Craik, F.I., Watkins, M.J.: The role of rehearsal in short-term memory. J. Verbal Learn. Verbal Behav. **12**(6), 599–607 (1973)
21. Craik, F.I., Tulving, E.: Depth of processing and the retention of words in episodic memory. J. Exp. Psychol. Gen. **104**(3), 268 (1975)
22. Ajzen, I., Sexton, J.: Depth of processing, belief congruence, and attitude-behavior correspondence. In: Dual-Process Theories in Social Psychology, pp. 117–138 (1999)
23. Radvansky, G.A., Ashcraft, M.H.: Learning and remembering. In: Cognition, 6th edn, pp. 184–188. Pearson Education, Upper Saddle River (2014)
24. DeWinstanley, P.A., Bjork, E.L.: Processing strategies and the generation effect: implications for making a better reader. Mem. Cognit. **32**(6), 945–955 (2004)
25. Slamecka, N.J., Graf, P.: The generation effect: delineation of a phenomenon. J. Exp. Psychol.: Hum. Learn. Mem. **4**(6), 592 (1978)
26. Muntinga, D.G., Moorman, M., Smit, E.G.: Introducing COBRAs. Int. J. Advert. **30**, 13–46 (2011)
27. Bellezza, F.S.: Recall of congruent information in the self-reference task. Bull. Psychonom. Soc. **30**(4), 275–278 (1992)
28. Gillihan, S.J., Farah, M.J.: Is the self special? A critical review of evidence from experimental psychology and cognitive neuroscience. Psychol. Bull. **131**, 76–97 (2005)
29. Rogers, T.B., Kuiper, N.A., Kirker, W.S.: Self-reference and the encoding of personal information. J. Pers. Soc. Psychol. **35**(9), 677 (1977)

GitHub as a Social Network

Tomek Strzalkowski[✉], Teresa Harrison, Ning Sa,
Gregorios Katsios, and Ellisa Khoja

ILS Institute, University at Albany, Albany, NY, USA
{tomek,tharrison,nsa,gkatsios,ekhoja}@albany.edu

Abstract. GitHub is a popular source code hosting and development service that supports distributed teams working on large and small software projects, particularly open-source projects. According to Wikipedia, as of April 2017 GitHub supports more than 20 million users and more than 57 million repositories. In addition to version control and code updates functionalities, GitHub supports a wide range of communication options between users, including messaging, commenting, and wikis. GitHub thus has all markings of an online social network, but how does it compare to other social media such as Twitter or Facebook? Since GitHub supports messaging between users as well as "following" it seems the answer is pretty straightforward. And yet, messaging and following do not account for the bulk of activity in GitHub, which consists largely of user initiated repository "events" related to adding, editing, and fixing the code (as well as other artifacts, such as documentation and manuals). In this sense GitHub is quite unlike Twitter, where information flows rapidly between users by being passed along to others. What information flows in GitHub, besides the actual messaging? In this paper, we discuss preliminary findings that the GitHub community displays many of the characteristics of a social network.

Keywords: Social networks · Computational sociolinguistics
Natural language processing

1 Introduction

Github, an open source software development environment based upon the git version control system, is widely recognized as providing the opportunity to engage in "social" or collaborative coding. In Github, developers take actions on their projects by using a publicly available typology of "events" related to code development and associated pursuits. Project owners can "commit" changes through a direct modification of their base code in repositories that they manage. Those without commit-rights can "fork" project code to their own repositories, creating personal copies for which they can author modifications; subsequently, they can request their changes to be incorporated into the base code through the use of a "pull" request. It is through pull requests that Github functions as a distributed model of software development. As [1] suggest, the innovation of pull requests lies in separating the development effort from decisions about whether to incorporate changes into the upstream code. Thus, anyone can initiate a pull request and potentially contribute to a project with authorship credit, but decisions about which changes to accept are managed by a core team.

© Springer International Publishing AG, part of Springer Nature 2019
T. Z. Ahram (Ed.): AHFE 2018, AISC 787, pp. 379–390, 2019.
https://doi.org/10.1007/978-3-319-94229-2_37

While Github's pull-request event provides a minimal structural basis for collaboration, scholars have explored the role of additional social actions that may enable members to collaborate productively. For example, Github members may "follow" each other and "watch" repositories, events that provide them with updates about actions taken by other users and performed within repositories. Still other actions enable free form textual communication, such as the issuance of "comments" on commits, pull requests, or issues that arise in the consideration of pull requests. As [2] report, actions taken in Github are recorded as events over time, providing users with a history of actions in repositories and related to users that is transparent and visually accessible in several different ways. Users can thus draw on this record to make inferences about projects, developers, repositories, and the appropriateness of code contributions that enable them to make useful decisions about where to invest their collaborative efforts.

Examinations of project-based and developer social networks on Github underscore the social character of these collaborations. For example, [3] observed that Github's project networks (defined by at least one common developer) were more highly interconnected than human networks and that developer networks provided for greater reach than Facebook networks. Such social connections reflect the tendency for developers to prefer to join projects where there is a prior relationship and the likelihood that they will contribute more to such projects [4].

Github's event typology and its transparency makes it possible for developers to accomplish enormous amounts of work simply by exchanging sequences of event/actions while engaging in comparatively little direct verbal communication between developers [1, 2, 5]. This has raised the question of when and how direct communication takes place on Github. In [2] it was observed that direct communication was relatively infrequent, taking place only "at the limits of transparency" when the record of actions failed to provide sufficient explanation or information to facilitate collaboration. In other work, the form, content and influence of these conversations has been more extensively addressed [5, 6].

In the work we present below, we have pursued the idea that Github functions as a social network in which repositories may be treated as interconnected information-bearing nodes. We further suggest that the actions taking place within repositories can be viewed as conversations that are composed of dialogue acts undertaken through both event/actions as well as direct verbal exchanges. Drawing on our prior work in applying sociolinguistic behaviors to online chat [7–9], we show that sociolinguistic dialogue act labels can be mapped on to Github events/actions (which are modified on the basis of textual exchanges, if any). Further we demonstrate that applying our software toolkit in these ways enables a computational analysis of certain social dynamics, such as leadership and influence roles. We expect these capabilities to lend themselves to the modeling of repository productivity and popularity.

2   The Information Flow in GitHub

One way to view GitHub as a social network is to consider repositories, which store evolving software projects, as information objects that "flow" through the network of users. This may appear counterintuitive as our first reaction is to view repositories as nodes, to which users are linked by virtue of their contributions (or attempted contributions, or simply by being observers). And yet this interpretation clearly makes sense and allows us to view GitHub at the same level as Twitter: repositories are started and "forwarded" to friends (members), which makes their followers aware of their existence, and they in turn "forward" them to their followers by getting engaged in repository activities. Further, GitHub displays trending repositories so that users can see the most popular as indicated by the number of "stars" the repositories receive from others on that day. Other accidental users may get involved by finding repositories using keyword search. So, is GitHub just like Twitter, except that the messages are much larger?

Not quite. Viewing repositories as information explains only part of the social dynamic occurring in the service. After all, repositories are not quite like Twitter messages: they are evolving objects that grow larger and more complex in time, and their lifespan is expected to be longer than a tweet. Thus, while repositories are information objects, they are also topics of interactions, and when taken together with their participants, become something akin to conversations. To be more precise, these are task-oriented conversations all centered on the main purpose of the repository, but involving a series of local topics. Each user-initiated event, such as PullRequest or Issue, etc. can be viewed as a dialogue act with which a user may introduce a new local topic into the conversation, or respond to a topic already under discussion. Social behavior is exhibited by the degree to which each participant attempts to influence the direction of conversation and how he/she relates to other participants. Each conversation has a core group of participants (members) and an evolving cast of external characters who join for a time and disappear. Figure 1 shows a schematic rendering of GitHub network of conversations.

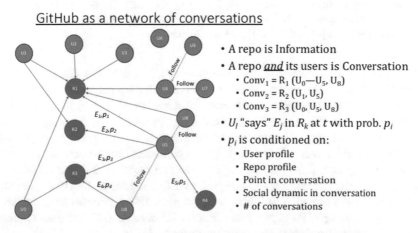

GitHub as a network of conversations

- A repo is Information
- A repo *and* its users is Conversation
  - $Conv_1 = R_1 (U_0 — U_5, U_8)$
  - $Conv_2 = R_2 (U_1, U_5)$
  - $Conv_3 = R_3 (U_0, U_5, U_8)$
- U_l "says" E_j in R_k at t with prob. p_i
- p_i is conditioned on:
  - User profile
  - Repo profile
  - Point in conversation
  - Social dynamic in conversation
  - # of conversations

Fig. 1. A schematic rendering of GitHub network of conversations. The R nodes represent repos, while the U nodes represent the users. Links from users to repos reflect involvement in a repo activities; links between users represent "following" relations.

3 GitHub Conversations

The following is a fragment of an interaction within a moderately active repository. In a fragment covering 3 months of activities, some 1600 "events" have been posted by a group of 36 participants. Each "event" is posted by a single user, and it reflects some operation performed or desired on the repository. Some events simply notify others that a change has been committed (e.g., Push Event) or that a problem has been noted (e.g., Issue Event), while other events request or notify of a future action (e.g., PullRequest Event). Yet other events are used to respond to or comment on any prior events. GitHub has a relatively small set of such approved events, which functions as a basic communication "language," albeit a lean one. When more details or clarity needs to be communicated, events are supplemented by text fields that elaborate but typically do not change the import of the event. Figure 2 shows a small fragment of in-repository discourse using GitHub commands (events), some of which include free text fields.

---

1. U1 [2:31] PullRequestEvent (opened, 123426861, Customize antd) – **Action-Directive**
   This PR will focus only on customizing Ant Design just by overriding less variables. Inconsistences between antd and Busy design will be resolved later with custom classes
2. U2 [4:34] PRCommentEvent (created, 123426861) – **Assertion-Opinion** (to U1)
   Font-size should be 16px by default; we dont use 14px font-size on the new-design.
3. U2 [4:39] PRCommentEvent (created, 123426861) – **Assertion-Opinion** (to U1)
   You can add @green too with ` #54d2a0 ` (used in topic button)
4. U1 [4:49] PRCommentEvent (created, 123426861) – **Disagree-Reject** (to U2; /3)
   This color is already there (GitHub ommited it in diff).
5. U3 5:48 PullRequestEvent (closed_merged/U1, 123426861) – **Agree-Accept** (to U1)

Fig. 2. A fragment of conversation inside a repo involving 3 users. Dialogue acts (in boldface) and communication links (in parentheses following the dialogue act label) are assigned automatically by our system. The text field, if present, is replicated under the event header. The long numerical is a unique topic id assigned by GitHub.

It is quite easy to note that the GitHub events function like dialogue acts in a conversation. Indeed, the events alone or when modified by language in the text field can be mapped onto a typology of dialogue acts automatically coded by NLP tools that we have described in [8]. A new PullRequest Event by user U1, which proposes a code update to the repository, is interpreted as an *Action-Directive* dialogue act (see e.g., [10]). This utterance compels the user U2 to comment about some aspects of the proposed code update, thus contributing an *Assertion-Opinion* dialogue act. Note that in this case we need to look into the text field to determine the correct dialogue act. U2 then suggests an alternative solution by posting another comment, to which U1 responds by *rejecting* U2's suggestion. Finally, U3 *accepts* U1's original request by closing the PullRequest Event and merging in the code.

As can be seen from this example, it is possible to assign default dialogue act labels to each event in GitHub log and thus treat these events as utterances or "turns" in a conversation. A text field, if supplied, may alter this default and also supply additional conversational elements such as communicative links (to whom is the utterance directed) as well as the sentiment. Table 1 shows a possible assignment of default dialogue acts to selected GitHub event types.

Table 1. Default dialogue act assignments to selected GitHub events currently contemplated. These defaults can be overruled when a text field is present.

GitHub event type	Description	Default dialogue act
Create	Start a new repo or a branch	Action-directive
Delete	Delete a branch or a tag	Offer-commit
Fork	Make a copy of the current repo	Action-directive
Issues	Raise an issue on any topic. Can be opened, assigned, closed, etc.	Assertion-opinion; Action-directive (if assigned)
XComment	Comment on a currently open or a recent event	Assertion-opinion
PullRequest	Request code update. Can be opened, closed, assigned etc.	Action-directive (opened); Agree-accept (closed, true); Disagree-reject (closed, false)
Push	Commit an update	Offer-commit
Watch	Bookmark a repo	Other-conventional

4 Topics and Topic Chains

A repo conversation, as any other conversation, moves through a series of topics, some of which are brief, while others may persist through many turns. A topic may be introduced by a participant posting a new event, such as a PullRequest Event, which is assigned a unique id by GitHub software. Any direct responses to this request would use the same id, thus making it easy to track the topic in the GitHub log (cf. Fig. 2). Topics may also be signaled within the text fields, and these could indicate either more specific subtopics, or alternatively, larger topics that transcend GitHub-assigned id chains, e.g., when multiple code updates relate to different aspects of the same issue. This topical continuity can be discerned from the key words used in the event text fields, as seen in the example below:

> U1 [8:59] IssuesEvent (closed, 231398889, *docker file* problem) – *Assertion-Opinion*
> Thanks for opening an issue! To help the team to understand your needs, please complete the below template to ensure we have the necessary details to assist you.
> ....
> U1 [8:59] PushEvent – *Offer-Commit, Continuation-of* (U1:1)
> *Update docker file; Merge pull request #503*

Thus, the two ways of tracking topics in GitHub logs involve event ids and key words (particularly nouns and noun-like objects) that users mention in the text fields. The long topic chains are of particular interest because they provide more opportunities for the participants to express their preferences (and occasionally sentiment), which may reveal deeper social dynamics in the group. We use the term meso-topic to refer to a topic that persists through at least N turns, not necessarily consecutive but with gaps between mentions no longer than M days. In the example repository discussed here, we used N = M = 7 which let us to identify 20 meso-topics covering jointly 10% of all turns (out of 1600), with the longest meso-topic continuing through 25 turns.

```
<turn speaker=U1 repo=R1 event_type="PullRequestEvent" action="opened unmerged"
    dialog_act="action-directive" comm_act_type="addressed-to" link_to="all-users"
    turn_no="1">id123426861 </turn>
<turn speaker=U2 repo=R1 event_type="MemberEvent" action="added" dia-
    log_act="conventional-opening" comm_act_type="addressed-to" link_to="U1"
    turn_no="2">U1 </turn>
<turn speaker=U2 repo=R1 event_type="PullRequestReviewCommentEvent" ac-
    tion="created" dialog_act="assertion-opinion" comm_act_type="response-to"
    link_to="U1:1" topic="default;size;Font" turn_no="3">id123426861 Font-size should
    be 16px by default, we dont use 14px font-size on the new-design. </turn>
<turn speaker=U2 repo=R1 event_type="PullRequestReviewCommentEvent" ac-
    tion="created" dialog_act="assertion-opinion" comm_act_type="continuation-of"
    link_to="U2:3" turn_no="4">id123426861 You can add @green too with `#54d2a0`
    (used in topic button) </turn>
<turn speaker=U1 repo=R1 event_type="PullRequestReviewCommentEvent" ac-
    tion="created" dialog_act="disagree-reject" comm_act_type="response-to"
    link_to="U2:4" topic="color" turn_no="5">id123426861 This color is already there
    (GitHub ommited it in diff). </turn>
<turn speaker=U3 repo=R1 event_type="PullRequestEvent" action="closed merged" dia-
    log_act="agree-accept" comm_act_type="response-to" link_to="U1:1"
    turn_no="6">id123426861 </turn>
```

Fig. 3. Fragment of GitHub event log partly processed by a series of automatic NLP tools, including dialogue act and communicative act tagging, and topic identification. User and repo names are replaced by anonymous symbols (U1, U2, R1, etc.)

5   Conversational Analysis of GitHub Logs

We applied standard conversation language analysis to the sequence of GitHub events in the logs of several public repositories, treating each repository as an independent conversation. (This assumption is made only for the purpose of the preliminary analysis described here as many repositories are not truly independent). Figure 3 shows an example of the output from this initial processing of conversation, which then becomes the input to the sociolinguistic analysis.

6   Social Relations and Behaviors in GitHub

We are now ready to run a computational analysis of social dynamics within a GitHub repository. We will focus on the 3 month period of the example repository mentioned earlier. This fragment covers 3 months and some 1600 events logged by 36 users. Our first question is of course what these social dynamics may be; for example, based on the use of characteristic dialogue acts, can we discern the sociolinguistic behaviors of a leader and users who are the most influential? We also want to verify that certain expected correlations of behaviors and roles in the group obtain in GitHub conversation. If these correlations hold as expected, we would have strong evidence that GitHub repositories indeed behave like any other conversations, and thus may be treated as such, for example, in predicting how repositories would evolve in the future.

The social science theory (e.g., [7, 11, 12]) postulates that a leadership role in an interaction emerges from a combination of such sociolinguistic behaviors as Topic Control, Task Control, Disagreement and Involvement. In a typical group conversation, these behaviors are expected to correlate quite tightly and jointly correlate with the degree of leadership a person has in a group. In our earlier work [9] we have verified these correlations in a series of experiments with online and off-line discourse involving small and medium groups of individuals (Fig. 4). Table 2 lists sociolinguistic behaviors covered in this article, with acronyms that appear in Figs. 4 and 5.

Table 2. Brief definitions of selected sociolinguistic behaviors mentioned in this paper. For a complete discussion see [7, 9]. All behaviors are measured as degree (in %) per speaker.

Socio-linguistic behavior	Description	Acronym
Topic control	Attempts by a speaker to impose the topic of conversation. Computed based on the rate of local topic Introduction, subsequent mentions, and average turn per speaker	TCM
Task control	An effort by a group member to define the group's goal and/or steer the group towards it. Computes the rate of directives (*Disagree-Reject* dialogue acts), process management turns by a speaker, and the ratio of successful responses by others	SCM
Involvement	Degree of engagement or participation in the discussion of a group. Computed based on proportion of turns, verbosity, and references to topics introduced by others	INVX
Disagreement	Making explicit or implicit utterances of disagreement, disapproval, or rejection. Computes the ratio of disagree-reject dialogue acts and negative sentiment per speaker	CDM
Argument diversity	Deploying a broader range of arguments in conversation. Computed based on the rate of vocabulary introduction and range used in turns	MAD
Network centrality	Degree to which others direct their comments to and/or cite topics of a speaker. Computed based on the rate of utterances addressed to the speaker and references to the topics introduced by the speaker	NCM

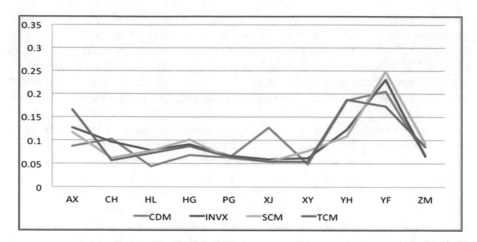

Fig. 4. Correlation between selected Leadership measures for a typical chat dialogue involving 10 individuals represented by 2-letter ids on the x-axis [9]. An average correlation observed is over 0.8 (Cronbach's alpha).

A similar phenomenon is observed for Influence, which is based on a different subset of sociolinguistic behaviors that include Topic Control, Disagreement, and also Network Centrality and Argument Diversity (Fig. 5).

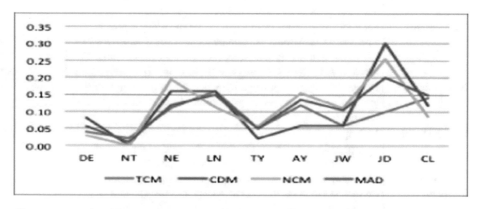

Fig. 5. Correlation between selected Influence measures for a typical chat dialogue involving 9 individuals represented by 2-letter ids on the x-axis [9]. An average correlation observed is over 0.8 (Cronbach's alpha).

Influence is of particular interest for at least two reasons. First, our definition encompasses a broader range of behaviors than what is typically used in social network analysis, where Network Centrality (in a form reduced to reflect direct links) is considered the only measure of influence. Second, Influence may cause change of behavior in the person being influenced, and thus may help us to predict the evolution of group dynamics in the future. In order to measure impact of influence, we will also need to model an additional behavior known as Positioning, which we do not discuss here (see [13]).

We have conducted sociolinguistic analysis of several GitHub repos in order to see if the selected sociolinguistic behaviors can be observed, and if they correlate as we might expect in a conversation. Specifically, we focused on the behaviors that are components of the leadership and influence social roles. The following charts, showing the distribution of sociolinguistic behaviors in our example repository, appear to confirm that the internal dynamics are indeed akin to a conversation. Figures 6 and 7 show the distribution among the 7 highest scoring participants only. Figure 8 shows the entire group of 36 participants over the 3 month span.

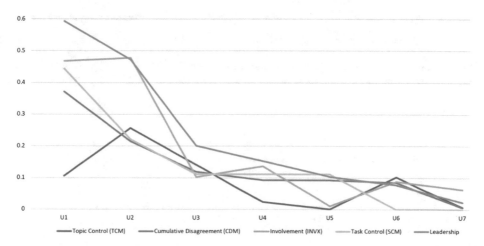

Fig. 6. Correlation between leadership measures for the top 7 highest scoring participants in the example repo. We note that U1 has the highest leadership score despite relatively lower Topic Control behavior.

These charts show an interesting dynamic within this repository, which is likely typical of any at least moderately active repository: out of all recorded users, only a handful are visibly active, and just 2 or 3 contribute most of the content and the direction of the project. A similar chart is obtained for influence, although the order of participants will be slightly different (Fig. 9).

7 Repository Profiles and Their Evolution

The sociolinguistic behaviors within a repository may be combined to form a behavioral profile of this repository. Other repositories may have different profiles and we would expect to see different dynamics in them. Another question is whether the dynamics remain stable throughout the life of a repository or do they evolve and move through different stages. We may be able to answer this question by sampling different intervals in repository lifecycle and deriving profiles from these.

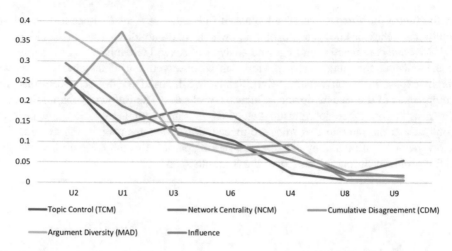

Fig. 7. Correlation between influence measures for the top 7 most influential participants in the example repo. Here U2 has the most influence with U1 a close second.

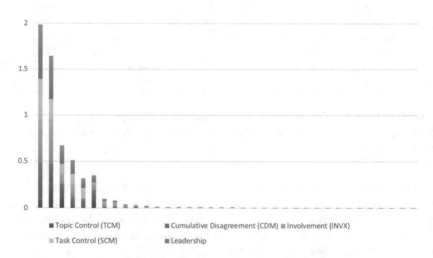

Fig. 8. Distribution of leadership scores among all 36 participants in the example repo, ordered by the degree of leadership. Each bar represents a single participant with component metrics shown as colored bands. The overall leadership score (the top band) is a linear function of the component measures, but the components are not equally weighted [9].

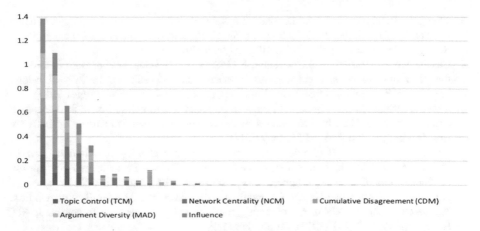

Fig. 9. Distribution of influence scores among all 36 participants in the example repo, ordered by the degree of influence. Each bar represents a single participant with component metrics shown as colored bands. The overall influence score (the top band) is a linear function of the component measures, but the components are not equally weighted [9].

8   Conclusions

In this paper, we discussed preliminary findings suggesting that GitHub indeed functions as a social network, in a manner similar to a collection of large and small "conversations". We demonstrated how the social dynamics within repositories can be extracted automatically using a software toolkit originally developed to study social dynamics in online interactions such as chatrooms, discussion forums, and more traditional social media. Understanding social dynamics within and across repositories, and over time, will help us to model the "flow" of repositories through the network of users: what makes some repositories more popular than others, what can make them go "viral", and when they die.

Acknowledgements. This work was supported by the Defense Advanced Research Projects Agency (DARPA) under Contract No. FA8650-18-C-7824. All statements of fact, opinion or conclusions contained herein are those of the authors and should not be construed as representing the official views or policies of AFRL, DARPA, or the U.S. Government.

References

1. Gousios, G., Pinzger, M., Deursen, A.V.: An exploratory study of the pull-based software development model. In: Proceedings of the 36th International Conference on Software Engineering, pp. 345–355. ACM, May 2014
2. Dabbish, L., Stuart, C., Tsay, J., Herbsleb, J.: Social coding in GitHub: transparency and collaboration in an open software repository. In: Proceedings of the ACM 2012 Conference on Computer Supported Cooperative Work, pp. 1277–1286. ACM, February 2012

3. Thung, F., Bissyande, T.F., Lo, D., Jiang, L.: Network structure of social coding in GitHub. In: Proceedings of the 2013 17th European Conference on Software Maintenance and Reengineering, CSMR, pp. 323–326. IEEE, March 2013

4. Casalnuovo, C., Vasilescu, B., Devanbu, P., Filkov, V.: Developer onboarding in GitHub: the role of prior social links and language experience. In: Proceedings of the 2015 10th Joint Meeting on Foundations of Software Engineering, pp. 817–828. ACM, August 2015

5. Tsay, J., Dabbish, L., Herbsleb, J.: Influence of social and technical factors for evaluating contribution in GitHub. In: Proceedings of the 36th International Conference on Software Engineering, pp. 356–366. ACM, May 2014a

6. Tsay, J., Dabbish, L., Herbsleb, J.: Let's talk about it: evaluating contributions through discussion in GitHub. In: Proceedings of the 22nd ACM SIGSOFT International Symposium on Foundations of Software Engineering, pp. 144–154. ACM, November 2014b

7. Broadwell, G.A., Stromer-Galley, J., Strzalkowski, T., Shaikh, S., Taylor, S., Boz, U., Elia, A., Jiao, L., Liu, T., Webb, N.: Modeling sociocultural phenomena in discourse. J. Nat. Lang. Eng. **19**, 213–257 (2012). Cambridge Press

8. Shaikh, S., et al.: DSARMD annotation guidelines, V. 2.5. ILS Technical report (2011)

9. Strzalkowski, T., Broadwell, G.A., Stromer-Galley, J., Shaikh, S., Taylor, S.: Modeling leadership and influence in online multi-party discourse. In: COLING 2012 Conference, Mumbai, India (2012)

10. Jurafsky, D., Shriberg, E., Biasca, D.: Switchboard SWBD-DAMSL Shallow Discourse Function Annotation Coders Manual (1997). http://stripe.colorado.edu/~jurafsky/manual.august1.html

11. Beebe, S.A., Masterson, J.T.: Communicating in Small Groups: Principles and Practices. Pearson/Allyn and Bacon, Boston (2006)

12. Huffaker, D.: Dimensions of leadership and social influence in online communities. Hum. Commun. Res. **36**, 596–617 (2010)

13. Lin, C.S., Shaikh, S., Stromer-Galley, J., Crowley, J., Strzalkowski, T., Ravishankar, V.: Topical positioning: a new method for predicting opinion changes in conversation. In: Proceedings of the Language Analysis in Social Media Workshop, NAACL 2013 Conference, Atlanta, GA (2013)

Applications of Fuzzy Cognitive Maps in Human Systems Integration

Nabin Sapkota[1(✉)] and Waldemar Karwowski[2]

[1] Department of Engineering Technology,
Northwestern State University of Louisiana, 175 Sam Sibley Drive,
Natchitoches, LA 71459, USA
sapkotan@nsula.edu
[2] Department of Industrial Engineering and Management Systems,
University of Central Florida, 4000 Central Florida Blvd., Orlando,
FL 32816, USA
wkar@ucf.edu

Abstract. This paper reviews applications of fuzzy cognitive maps (FCMs) in the in the field of human factors and ergonomics, with special consideration of human systems integration efforts.

Keywords: Human factors · Fuzzy cognitive maps
Human-systems integration

1 Introduction

Most of the real world problems are complex phenomena with known and unknown multivariate and higher order systems. The solutions to these problems can be ordinarily determined using conventional methods of modeling. For example, complex dynamical systems need to be analyzed by incorporating human experience and knowledge. Groumpos [1], and Papageorgiou and Salmeron [2] report that fuzzy cognitive maps (FCM) are the answer to the challenging problem of modeling and controlling such complex systems. An FCM is a directed graph with concepts as nodes and causalities as edges, where the nodes take on values from the fuzzy sets [3].

Kosko [4] first proposed FCM as an extension of cognitive maps of political elites by Axelrod [5]. Since then, several variants of original concepts have been developed including uncertainty modeling [2]. FCM approach has been used to describe and model variety of complex systems, and applied in decision sciences and operations research [1]. Some notable applications of the FCM include:

- fuzzy-neuro control [6–8],
- electrical circuits [9],
- political developments [10],
- organizational behavior and job satisfaction [11],
- economics and demographics [12],
- medicine [13, 14],
- manufacturing [15–18],

© Springer International Publishing AG, part of Springer Nature 2019
T. Z. Ahram (Ed.): AHFE 2018, AISC 787, pp. 391–399, 2019.
https://doi.org/10.1007/978-3-319-94229-2_38

- decision science [19],
- control systems [17, 20–23],
- urban design [24],
- business analysis [25, 26],
- statistical analysis [27],
- data mining of the world wide web [28, 29],
- ecological modeling [30], etc.

For a review of the above applications, readers are referred to [31].

2 The Theoretical Basis of FCM

As proposed by Kosko [4], $C_1, C_2, ..., C_n$ are the concept variables or causal variables of a complex system (see Fig. 1) that make nodes of the graph. There may exist causality among these variables. Variable that affects another variable is the causal variable, and the one that is affected is the effect variable. If causality exists between the concept variable C_i and C_j, then the degree of causality is denoted by the directed arrow with weight $W_{ij} \sim [-1, 1]$.

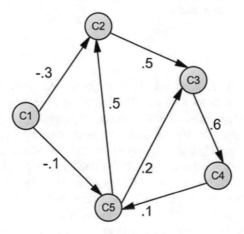

Fig. 1. A simple fuzzy cognitive map

At any moment (iteration t), each concept variable is characterized by its value A_i in the interval [0, 1], which is a fuzzy value after transformation using some function f. Generally, at the beginning of simulation (t = t_o), the values of $A_i^{t_0}$ s are 1.

$$A_j^{t+1} = f\left(\sum_{\substack{i=1 \\ i \neq j}}^{n} A_i^t W_{ij} + A_j^t\right). \tag{1}$$

where, transformation function 'f' could be bivalent, trivalent, sigmoid, or hyperbolic tangent. Generally, in most of the applications, sigmoid function as shown in (2) is applied to get the value of the concept variable.

$$f(x) = \frac{1}{1 + e^{\lambda x}}. \tag{2}$$

where, λ is the parameter of the sigmoid function that controls the slope of the function. Normally, λ equals to one has been applied in the majority of the works whenever sigmoid function was used. The slope of the function is flatter when λ is less than one (but positive) and much steeper when it is greater than one. The iterative process is continued using Eqs. 1 and 2 until, $\left| A_j^{t+1} - A_j^t \right| \forall i$ is unchanged or less than some predetermined threshold value.

3   Scenarios Modeling in FCM

Once the stationary state of the model is obtained as described in Sect. 2 above, then the values of A_j for any scenario k $(A_j^{scenario\,k})$ can be developed. A concept variable C_j is subject to a specific condition, e.g., it can take a value of 0.7 from the interval $[-1, 1]$ or it can assume any value in that interval.

Consequently, for any scenario k, the following iterative steps should be used:

If value of any C_j (for all j) is not controlled within the scenario, it is assumed to have a value of 1.0 at (t = 0). The values in $A_j^{scenario\,k}$ are not specified.

1. At the end of every iteration (after applying Eqs. 1 and 2), every A_j^{t+1} is reset to $A_j^{scenario\,k}$ if and only if, $A_j^{scenario\,k}$ is a specified quantity in step 1.
2. Step 2 is repeated until a threshold condition is reached.
3. The analysis is conducted on differences among scenarios and stationary state concerning controlled concept variables and effect variables.

4   Applications of FCM in Human Systems

FCM approach has gained wide acceptability for modeling complex systems in several different fields of study (see Sect. 1). The purpose of this paper is to review selected applications of FCM in the human factors area.

4.1   Human Decision-Making

Jones et al. [32] presented a novel approach to providing an actionable model of situational awareness (SA) using FCM that encompasses all three levels of situation awareness. The proposed model includes perception construct of SA (Level 1 SA), the comprehension (Level 2 SA) and projection (Level 3 SA) levels of situation awareness. To effectively model decision-making that reflects real-world conditions, these higher-level SA constructs should be considered. This is especially critical in military operations as sufficient data are not always available to develop a cognitive model that provides a realistic representation of the behaviors of the people involved (e.g., friendly forces, insurgents, and civilians). In the model, authors used goal driven (top-down), and data-driven (bottom-up) approaches simultaneously. In platoon leader goal-directed task analysis (GDTA), the main goal node has seven sub-goals, and each sub-goal has other related sub-tasks (or sub-sub-goals). Each sub-goal in the proposed model is modeled by sub-FCM for that sub-goal according to GDTA. Main goal influences all three levels of SA. Operator's expertise and SADemons (overload, requisite memory traps, misplaced salience, attentional narrowing, workload, fatigue and other stressors, complexity creep, errant mental models, or the out-of-the-loop syndrome) also influence all three levels of SA. Özesmi and Özesmi [30] noted that the proposed model maintains the hierarchical relationship of each SA requirement identified in the GDTA hierarchy and provides an SA score at each level.

Xue et al. [33] proposed an FCM-SA model in dynamic decision-making. First, they identified each element (concept) inside the three layers of SA decision-making model (system-interface layer, SA-Judgement layer, and operator cognition layer). Causalities among elements were then developed with the strength assigned using five degrees: highly positive influence, slightly positive influence, no influence, slightly negative influence, and highly negative influence. In the next phase, interviews were carried out with the panel consisting of ten experts and researchers from the domains of HCI, Human Factors, Teleoperation Control System, and Cognition Science in small groups of 3 or 4 or individually. The interviewees were instructed on how to draw an FCM model with examples from other areas, and were asked to draw a map, with the concepts they thought were important and should be included in the model. All the concepts were then recorded and presented to the interviewees who discussed the connections and came up with the direction of causality and its strength for their FCMs. Final weights of the connections between concepts were calculated using a score function (weighted sum). In general, this paper demonstrated how SA can be modeled using FCM approach to understand the inner dynamics of information processing of SA system.

Irani et al. [34] explored the behavior and embodied energy involved in the decision-making for information technology/information systems (IT/IS) investments. The authors described the nature of the investment through the lens of behavioral economics, causality, input-output (IO) equilibrium, and the general notion of depletion of the executive energy function. Human factors (views and needs of the relevant stakeholders and decision makers, and human resources such as management/staff time and training) and organizational factors (structure, leadership, business processes, organizational/managerial, and organizational culture) were included in the model. Data were collected through interviews, observations, and company documents. Data were coded applying behavioral economics, quantitative content analysis, and morphological analysis. Finally, data were analyzed using FCM, input-output model, and decision fatigue approach. In their FCM, authors used concepts such as anchoring effect, agency friction, confirmation bias, environment effect, feedback loops, loss aversion, reference dependence, and status quo bias.

4.2   Human Reliability

Bertolini [35] applied FCM in human factors to assess and analyze human reliability in manufacturing. The applied concept variables (number of variables) included: environment (4), work space (6), Machinery or tools (5), physical load (5), metal load (7), judgmental load (2), information and confirmation (2), and indication and communication (2). The process of FCM modeling consisted of inputs from several experts (academics, plant designer, operation manager, quality manager and safety manager of a food-processing firm) who used Delphi technique to gather information on each factor. They also provided causal connection and their signs anonymously. Later this information was given to a panel (a group of 20 or more) which then through the iterative feedback process developed a high level of agreement needed for constructing cognitive maps. Finally, the expert selected fuzzy linguistic terms (very low (VL), low (L), medium (M), high (H) and very high (VH)) to weigh the strength of all causal connections. Fuzzy causality strength (linguistic variable to fuzzy number) was determined as the sum of the product of credibility weight of individual expert and the corresponding causality strength (fuzzy number).

Gandhi and Gandhi [36] proposed an FCM model to address human factors that influence human reliability in maintenance. The concepts of emotional stability, knowledge and skill, attention and alertness, perception and memory, and motivation were used as input nodes that determine the decisive factors and human reliability as the output measure. Input factors may vary with persons and time. It was concluded that FCM could be used as a tool to understand how human factors and their interactions influence the overall human reliability in the maintenance arena.

4.3   Human Behavior

Buche et al. [37] proposed FCMs to represent internal states (internal emotions such as fear, satisfaction, love, or hate) explicitly when it comes to responses to external stimuli in the context of behavior for autonomous entities in virtual environments. Their study presents the use of FCMs as a tool to specify and control the behavior of individual agents along with a learning algorithm allowing the adaptation of FCMs through observation.

Bevilacqua et al. [38] applied a fuzzy cognitive map (FCM) approach to explore the importance of the relevant factors including workers' behavioral factor in the industrial environment. Factors that affect injury risk and the causal relationships involved are modeled to help managers to act appropriately to mitigate such risks. They included 16 concepts as causes for injuries/accidents and 12 concepts for resulting injuries/accident types. Out of 16 concepts for injuries, ten concepts partially or in totality represented workers' behavior. Through the FCM model, they calculated the maximum total effect for each type of injury and recommended, based on the results, appropriate actions to be taken in the workplace regarding policy, procedure, and training.

4.4 Usability Evaluation

Yucel et al. [39] investigated usability electronic consumer products among different age groups using FCM approach. The consumer groups were identified in three categories: younger than 40, 40–55 or older than 55 years. There were five input concepts about physical attributes of the mobile phone, smart phone, and other PDAs, five additional input concepts about cognitive attributes, and one output concept for usability in the FCM model. Causality was defined in terms of linguistic variables (very low, low, medium, high, and very high) and interviewees from different age groups were asked to rate them. These causal linguistic variables were converted to fuzzy value using triangular distribution. The results of the study on mobile phones showed that the most important dimensions of usability for each group were learnability and efficiency.

4.5 Human Health and Safety

Asadzadeh et al. [40] analyzed and assessed the integrated health, safety, environment (HSE) and ergonomics (HSEE) factors by FCM approach. They introduced six categories of concepts, namely: health, safety, environment, ergonomics, macro ergonomics, and system performance. Within the category of ergonomics, authors created six concept variables: 'pain and distress because of work', 'ergonomically unsuitable office furniture', 'difficulty to movement within workplace', 'difficulty to transport and operate machines', 'materials and tools', 'vibrations of tools and machinery on body', and 'ergonomically unsuitable machines and tools'. Within the category of macro ergonomics, seven concept variables: 'lack of instructions and education about safety and accident prevention', 'lack of respect for rules', 'work pressures', 'lack of documented instructions about works', 'unfamiliarity with organization's rules', 'lack of proper communications with managers', and 'difficulties within organization with coworkers' were used. Causal relationships and their strengths were determined using a panel of 37 experts comprising of researchers, plant managers, and people with deep experience in health and safety systems, employee relationship management, and human productivity management. Through the application of FCM, the authors were able to identify the most influential factors that contribute to the system performance.

Mei et al. [41] presented a study of disease (influenza A or H1N1) on a college/campus based on individual decision rules embedded in FCM model that incorporates both emotion-related and cognition-related components. Authors claimed that an individual's decision such as frequent washing, respirator usage, crowd contact avoidance, etc. could significantly decrease the at-peak number of infected patients, even when common policies (isolation and vaccination) are not deployed. Nine concepts representing the cognitional level of global and local epidemiological situation, panic emotion for primary emotions, the memory of secondary emotions, optimistic personality for senior emotions, and other states of infection constitute input nodes (concepts) and output concept as the desired output concept (DOC). Based on the computed value of DOC, the individual's overall assessment of the current epidemiological situation, individual decision-making rules can be set and the corresponding effect of lowering the probability of being infected to further support simulations of infectious disease propagation. The study found that influenza A would eventually die out even with no intervention.

4.6 Human Factors in Management

Dias et al. [42] proposed FCM-Quality of Interaction (FCM-QoI) model to assess the quality of interaction of learning management systems (LMS) of higher education institutions. They studied 75 professors and 1037 students for one academic year. The model used 14 input concepts and one output concept. Whenever the user interacted with LMS, 1140 metrics were acquired and then categorized into one of the fourteen concepts. The model first underwent the learning phase during which the causality weights and the concepts' parametric values were updated. In the latter testing phase, the learned model was tested and considered satisfactory when the results satisfied the predetermined criteria.

5 Conclusion

FCM is one of the viable approaches that can be used for modeling different aspects of human-system integration. Many variations of FCM approach have been developed over the years, and varieties of applications have been reported using different FCM approaches. The selected studies reviewed above demonstrate that FCM methodology has the capability to model complex systems where human behavior, experience and knowledge are important factors of system performance.

References

1. Groumpos, P.P.: Fuzzy cognitive maps: basic theories and their application to complex systems. In: Fuzzy Cognitive Maps, pp. 1–22. Springer, Heidelberg (2010)
2. Papageorgiou, E.I., Salmeron, J.L.: Methods and algorithms for fuzzy cognitive map-based modeling. In: Fuzzy cognitive Maps for Applied Sciences and Engineering, pp. 1–28. Springer, Heidelberg (2014)
3. Kandasamy, W.V., Smarandache, F.: Fuzzy cognitive maps and neutrosophic cognitive maps. Infinite Study (2003)
4. Kosko, B.: Fuzzy cognitive maps. Int. J. Man-Mach. Stud. 24(1), 65–75 (1986)
5. Axelrod, R. (ed.): Structure of Decision: The Cognitive Maps of Political Elites. Princeton University Press, Princeton (1976)
6. Sox, J.H.C., Blatt, M.A., Higgins, M.C., Marton, K.I.: Medical Decision Making. American College of Physicians, Butterworths (1988)
7. Jang, J.S.R., Sun, C.T., Mizutani, E.: Neuro-Fuzzy and Soft Computing; A Computational Approach to Learning and Machine Intelligence. Prentice-Hall, Englewood Cliffs (1997)
8. Nie, J., Linkens, D.: Fuzzy-Neural Control: Principles, Algorithms and Applications. Prentice Hall International (UK) Ltd., Upper Saddle River (1995)

9. Styblinski, M.A., Meyer, B.D.: Fuzzy cognitive maps, signal flow graphs, and qualitative circuit analysis. In: Proceedings of the 2nd IEEE International Conference on Neural Networks, vol. 2, pp. 549–556. IEEE Press, New York (1988)

10. Taber, R.: Knowledge processing with fuzzy cognitive maps. Expert Syst. Appl. 2(1), 83–87 (1991)

11. Craiger, J.P., Goodman, D.F., Weiss, R.J., Butler, A.: Modeling organizational behavior with fuzzy cognitive maps. Int. J. Comput. Intell. Organ. 1(2), 120–123 (1996)

12. Schneider, M., Shnaider, E., Kandel, A., Chew, G.: Automatic construction of FCMs. Fuzzy Sets Syst. 93(2), 161–172 (1998)

13. Papageorgiou, E.I., Spyridonos, P.P., Stylios, C.D., Ravazoula, P., Nikiforidis, G.C., Groumpos, P.P.: The challenge of soft computing techniques for tumor characterization. In: International Conference on Artificial Intelligence and Soft Computing, pp. 1031–1036. Springer, Heidelberg (2004)

14. Papageorgiou, E.I., Spyridonos, P.P., Stylios, C.D., Ravazoula, P., Groumpos, P.P., Nikiforidis, G.N.: Advanced soft computing diagnosis method for tumour grading. Artif. Intell. Med. 36(1), 59–70 (2006)

15. Stylios, C.D., Groumpos, P.P.: The challenge of modeling supervisory systems using fuzzy cognitive maps. J. Intell. Manuf. 9(4), 339–345 (1998)

16. Stylios, C.D., Groumpos, P.P.: A soft computing approach for modeling the supervisor of manufacturing systems. J. Intell. Rob. Syst. 26(3–4), 389–403 (1999)

17. Stylios, C.D., Groumpos, P.P., Georgopoulos, V.C.: Fuzzy cognitive map approach to process control systems. J. Adv. Comput. Intell. 3(5), 409–417 (1999)

18. Christova, N.G., Groumpos, P.P., Stylios, C.D.: Production planning for complex plants using fuzzy cognitive maps. In: IFAC Proceedings, vol. 36, no. 3, pp. 81–86 (2003)

19. Krishnan, R., Sivakumar, G., Bhattacharya, P.: Extracting decision trees from trained neural networks. Pattern Recogn. 32(12), 1999–2009 (1999)

20. Stylios, C.D., Georgopoulos, V.C., Groumpos, P.P.: Introducing the theory of fuzzy cognitive maps in distributed systems. In: Proceedings of the 1997 IEEE International Symposium on Intelligent Control, pp. 55–60. IEEE Press, New York (1997)

21. Stylios, C.D., Groumpos, P.P.: Fuzzy cognitive maps: a model for intelligent supervisory control systems. Comput. Ind. 39(3), 229–238 (1999)

22. Stylios, C.D., Groumpos, P.P.: Fuzzy cognitive maps in modeling supervisory control systems. J. Intell. Fuzzy Syst. 8(2), 83–98 (2000)

23. Groumpos, P.P., Stylios, C.D.: Modeling supervisory control systems using fuzzy cognitive maps. Chaos, Solitons Fractals 11(1–3), 329–336 (2000)

24. Xirogiannis, G., Stefanou, J., Glykas, M.: A fuzzy cognitive map approach to support urban design. J. Expert Syst. Appl. 26(2), 257–268 (2004)

25. Xirogiannis, G., Glykas, M.: Fuzzy cognitive maps in business analysis and performance driven change. J. IEEE Trans. Eng. Manag. 51(3), 334–351 (2004)

26. Xirogiannis, G., Chytas, P., Glykas, M., Valiris, G.: Intelligent impact assessment of HRM to the shareholder value. Expert Syst. Appl. 35(4), 2017–2031 (2008)

27. Heckerman, D., Geiger, D., Chickering, D.M.: Learning Bayesian networks: the combination of knowledge and statistical data. Mach. Learn. 20, 197–243 (1995)

28. Hong, T., Han, I.: Knowledge-based data mining of news information on the Internet using cognitive maps and neural networks. Expert Syst. Appl. 23(1), 1–8 (2002)

29. Lee, K.C., Kim, J.S., Chung, N.H., Kwon, S.J.: Fuzzy cognitive map approach to web-mining inference amplification. Expert Syst. Appl. 22(3), 197–211 (2002)

30. Özesmi, U., Özesmi, S.L.: Ecological models based on people's knowledge: a multi-step fuzzy cognitive mapping approach. Ecol. Model. 176(1–2), 43–64 (2004)

31. Papageorgiou, E.I.: Review study on fuzzy cognitive maps and their applications during the last decade. In: 2011 IEEE International Conference on Fuzzy Systems, pp. 828—835. IEEE Press, New York (2011)
32. Jones, R.E., Connors, E.S., Mossey, M.E., Hyatt, J.R., Hansen, N.J., Endsley, M.R.: Modeling situation awareness for Army infantry platoon leaders using fuzzy cognitive mapping techniques. In: Proceedings of the Behavior Representation in Modeling and Simulation Conference, pp. 216–223 (2010)
33. Xue, S., Jiang, G., Tian, Z.: Using fuzzy cognitive maps to analyze the information processing model of situation awareness. In: Sixth International Conference on Intelligent Human-Machine Systems and Cybernetics, vol. 1, pp. 245–248. IEEE Press (2014)
34. Irani, Z., Sharif, A.M., Papadopoulos, T.: Organizational energy: a behavioral analysis of human and organizational factors in manufacturing. IEEE Trans. Eng. Manag. 62(2), 193–204 (2015)
35. Bertolini, M.: Assessment of human reliability factors: a fuzzy cognitive maps approach. Int. J. Ind. Ergon. 37(5), 405–413 (2007)
36. Gandhi, M.S., Gandhi, O.P.: Identification and assessment of factors influencing human reliability in maintenance using fuzzy cognitive maps. Qual. Reliab. Eng. Int. 31(2), 169–181 (2015)
37. Buche, C., Chevaillier, P., Nédélec, A., Parenthoën, M., Tisseau, J.: Fuzzy cognitive maps for the simulation of individual adaptive behaviors. Comput. Anim. Virtual Worlds 21(6), 573–587 (2010)
38. Bevilacqua, M., Ciarapica, F.E., Mazzuto, G.: Analysis of injury events with fuzzy cognitive maps. J. Loss Prev. Process Ind. 25(4), 677–685 (2012)
39. Yucel, G., Bayraktaroglu, A.E., Unal, M.E.: A fuzzy cognitive map approach for the analysis of electronic consumer products regarding usability among different age groups. Int. J. Mob. Learn. Organ. 3(3), 322–335 (2009)
40. Asadzadeh, S.M., Azadeh, A., Negahban, A., Sotoudeh, A.: Assessment and improvement of integrated HSE and macro-ergonomics factors by fuzzy cognitive maps: the case of a large gas refinery. J. Loss Prev. Process Ind. 26(6), 1015–1026 (2013)
41. Mei, S., Zhu, Y., Qiu, X., Zhou, X., Zu, Z., Boukhanovsky, A.V., Sloot, P.M.: Individual decision making can drive epidemics: a fuzzy cognitive map study. IEEE Trans. Fuzzy Syst. 22(2), 264–273 (2014)
42. Dias, S.B., Hadjileontiadou, S.J., Hadjileontiadis, L.J., Diniz, J.A.: Fuzzy cognitive mapping of LMS users' quality of interaction within higher education blended-learning environment. Expert Syst. Appl. 42(21), 7399–7423 (2015)

Social Sensors Early Detection of Contagious Outbreaks in Social Media

Arunkumar Bagavathi$^{(\boxtimes)}$ and Siddharth Krishnan

Department of Computer Science, University of North Carolina at Charlotte,
Charlotte, NC 28223, USA
{abagavat, siddharth.krishnan}@uncc.edu

Abstract. Cascades of information in social media (like Twitter, Facebook, Reddit, etc.) have become well-established precursors to important societal events such as epidemic outbreaks, flux in stock patterns, political revolutions, and civil unrest activity. Early detection of such events is important so that the contagion can either be leveraged for applications like viral marketing and spread of ideas [4] or can be contained so as to quell negative campaigns [2] and minimize the spread of rumors. In this work, we algorithmically design social sensors, a small subset of the entire network, who can effectively foretell cascading behavior and thus detect contagious outbreaks. While several techniques (for example, the friendship paradox [3]) to design sensors exist, most of them exploit the social network topology and do not effectively capture the bursty dynamics of a social network like Twitter, since they ignore two key observations (1) Several viral phenomenal have already cascaded in the network (2) most contagious outbreaks are a combination of network flow and external influence.

In light of those two observations, we present an alternate formalism for information where we describe information diffusion as a forest (a collection of trees). Intuitively, our forest model is a more natural metaphor because most social media phenomena that go truly viral have multiple origins, thus are a combination of several trees. We show that our model serves as a solid foundation to foretell the emergence of viral information cascades. We then use the forest model in conjunction with past information cascades, to view the problem under the algorithmic lens of a hitting set and select a subset of nodes (of the social network) by prioritizing their activation time and their occurrence in the cascades.

Keywords: Social sensors · Early contagion detection
Social networks mining

1 Introduction

Social network sensors is proven to be a powerful approach to detect and characterize large outbreaks in networks [10]. A social network sensor is a set of nodes in the given large social network that forecasts the emergence of an outbreak in a sensor node to the entire network. The existing methods of designing social network sensors from a network rely on network topologies such as the connectivity of nodes, their degrees, shortest paths, and dominating paths through the network. Such methods are limited in

© Springer International Publishing AG, part of Springer Nature 2019
T. Z. Ahram (Ed.): AHFE 2018, AISC 787, pp. 400–407, 2019.
https://doi.org/10.1007/978-3-319-94229-2_39

scope since knowledge of network topology is generally difficult to obtain. And, information cascades in large scale social networks can have multiple points of entry through agents of external influence [8] or via other social media. The friendship paradox [9] one of the state-of-the-art methods to find social sensors, (which exploits the degree distribution of a network) is not effective in capturing the bursty dynamics of information contagion on a social network like Twitter.

Motivated by these observations, we develop an alternative approach by simply observing cascades propagating through the network and noting that cascades provide indirect information about the network topology without requiring costly instrumentation to observe the connectivity. However, information cascades are typically very small, and thus it is nontrivial to use them to detect large outbreaks. In designing our algorithm, we show how, by leveraging past information cascades using a hitting set formulation, we can mine a subset of nodes by prioritizing their occurrence in cascades coupled with their activation time in the cascades. The novelty of our approach lies in mining the tracking potential of a node across cascades. The rationale behind combining occurrence frequency with activation time to determine a node ranking scheme is that not only do we want nodes that participate in many cascades, but we also want to estimate their early adoption and participation rates. The key is to utilize multiple cascades and view them as a representative sampling of the network structure.

Our major contributions are:

Modeling information flow as cascade forests: We demonstrate that adopting the view point of information cascade forests, allows better characterization and detection of information flow. Our model does not just detect the outbreak, but affords the emergence of new propagation trees (which was not possible in the prior models) and thus effectively capture the bursty dynamics of information flow.

A non-network approach to designing sensors: The most popular approach uses the friendship paradox to design the sensor network, thus relying on the underlying network structure. In our work, we provide efficient algorithms that leverage observed traces of past cascades and early adopters to design social network sensors. As a result, our approach is significantly less intrusive than current methods.

Real-time outbreak detection: Since our sensor set selection relies on past information that has cascaded in the network, we demonstrate that the sensor set gleaned this way is resilient to temporal changes in cascading dynamics.

Using the standard lead time analysis approaches [3], we find that, on average we outperform the friendship paradox based method by about 42 h. We also demonstrate that we can detect real-time viral outbreaks with a lead time of approximately 19 h ahead of the state-of-the-art solutions to this problem.

2 Methodology

2.1 Problem Formulation

We present the problem similar to state-of-the-art in solving social network sensor problems [3, 7, 8]. If we consider $G = (V, E)$ as underlying social network, where V is a set of users in the social network and E is E is a set of edges representing information shared among vertices V. Network G comprise of a set of k cascades, $C = \{C_1, C_2, \ldots, C_k\}$, each occurs and fades over a period of time. Let U be the set of all nodes that have participated in at least one cascade in C, so $U \subseteq V$. Also each, $C_i \subseteq U$. A sensor design problem often includes a budget constraint of setting a number of users to be grouped into a sensor group. Let b be the budget constraint.

Then, the social sensor design problem is the following problem of maximizing lead time:

Lead-Time Maximization problem:

Given: A set of cascades $C = \{C_1, C_2, \ldots, C_k\}$, set of nodes with their corresponding activation times of the cascade forest and a budget b.

Find: A set of nodes $S \subseteq V$ such that

$$S = \mathrm{argmax}_S t_{pk}(C_i) - t_{pk}(S).$$

where the size of S is at most b and t_{pk} refers to *peak time* as defined earlier.

2.2   Sensor Design Approach

We follow a non-graph-based approach to design a sensor group. We follow two ranking schemes for nodes in the network: (1) based on occurrence frequency, (2) based on adoption time. Ranking nodes or users based on their occurrence frequency or their participation frequency across all cascades in the network is a basic approach to choose the sensor group. In addition to this ranking scheme, we also give rank to a user based on their *mean adoption time* in all cascades they have participated. By an *adoption time* t_u of a user u, we mean the time taken to infect u. Or in other words, if a user is infected by a cascade at time t_i, given the origin time of a cascade is t_0, then the adoption time of a user u for a cascade c is given by $t_u(c) = t_0 - t_i$. Thus the mean adoption time t_u of a user u, who are participated in n cascades can take a representation as

$$t_u = \frac{\sum_{k=1}^{n} t_u(c_k)}{n}$$

We order users based on their occurrence frequency and mean adoption time. A User who occurs most frequently and has the lowest mean adoption time across all cascades has highest rank and are added to the sensor group. Based on this strategy we choose first b nodes with highest ranks as a sensor group.

Since our algorithm follows an approximation algorithm to solve choosing minimum number of users in the sensor group, it has a polynomial time complexity. Our method sort the nodes based on the occurrence frequency and mean adoption time and selects nodes by iterating over the ordered set. Thus it has time complexity $O(|v|^2|c|)$, where v is the number of nodes and c is the number of cascades. The algorithm can be further refined, but that is beyond the scope of this paper.

2.3 Interpretation of Lead Time

Once we extract b sensor users from other users, we calculate peak times(p_k) of sensor $(p_k(s))$ and the other $(p_k(o))$ group of users for all cascades. The lead time I of a selected sensor user group is calculated by $(p_k(s) - p_k(o))$ for all cascades. Thus, for a single cascade c_i, following are the possible values of the lead time $I(c_i)$:

- <0; if $p_k(s) > p_k(o)$
- $=0$; if $p_k(s) = p_k(o)$
- >0; if $p_k(s) < p_k(o)$

Since we are choosing a small number of sensor users for our proposed method, in some cascades we get negative lead time. To check if our method is effective, we consider average of all lead times to be the lead time of the sensor set.

2.4 Scalability

Even though, our approach follows polynomial time complexity, it is completely parallelizable. We use the Spark framework [6] to do the distributed and parallelized ranking schemes and absorb the lead time of the sensor group. We follow distributed approach start right from data preprocessing step to determining lead time of the sensor set. Figure 1 gives a flow chart for the complete distributed process of our proposed approach. Each node in the figure is followed in a distributed fashion, whereas nodes appearing in the same level of the chart are followed in both parallel and distributed fashion.

3 Experiments and Results

We collected twitter archive data from Internet Archive (https://www.archive.org), a non-profit digital library for webpages, images, books, audio recordings and videos. The Twitter data from the Internet Archive comprises of tweets collected over the month of June 2017. Since they used the Spritzer streaming API, the data is 1% or 2% of complete tweets of June 2017. The data comprise of around 124 million tweets out of which only 16.7 million tweets have hashtags, which we use to build cascades. Figure 2 gives a timeline of tweets and tweets with hashtags over the month of July, 2017. Although the entire data from Internet Archive occupies the memory space of 45 Gigabytes, the processed data for social sensors analysis, with just hashtags (cascades), users who posted them and their corresponding posted date and time, shrinks the data size to 1.6 Gigabytes. Figure 3 gives the word cloud of to 1000 hashtags in the dataset. Based on these hashtags like 'VeronMTV2017', 'TeenChoice', 'BTS', 'KCAMexico', we believe that this data is weighted heavily on TV reality shows, music and entertainment. Our data set comprises of over 250 million tweets that span over 3.8 million cascades within total of over 1 million users.

Given the data, we rank users based on their occurrence frequency and mean

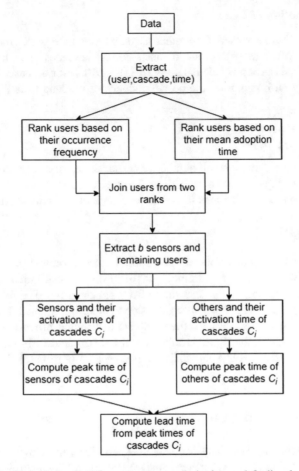

Fig. 1. Scalable method for sensor group designing and finding lead time

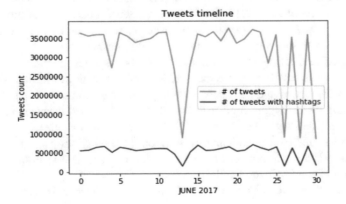

Fig. 2. Tweets frequency timeline in the data

adoption time in a distributed and parallel approach with Spark framework. From the ranked users, we take b sensor users and compute their lead time. In Figs. 4 and 5, we give lead times of sensor group designed by our approach with lead time of sensor group designed by ranking users with just occurrence frequency. In Fig. 4, we set the budget $b = 1\%$ and take top 1% users from the ranked users. Figure 4a gives the lead time of sensor group extracted from users ranked based on occurrence frequency. Figure 4b gives the lead time of sensor group extracted from users ranked based on our proposed approach. From these figures, we can note that sensor group designed by just taking 1% users by our ranking approach can forecast popularity of an event by around 20 h. Whereas, the sensor group designed just based on occurrence frequency gives only a negligible lead time.

Fig. 3. Wordcloud of top 1000 hashtags from the Twitter data used for experiment

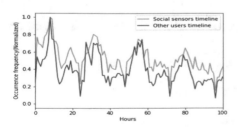
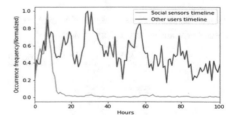

(a): Timeline of sensors and others by ranking based only on occurrence frequency

(b): Timeline of sensors and others by ranking based on occurrence frequency and mean adoption time

Fig. 4. Comparison of methods with just 1% sensor users; Our approach gets average lead time of around 20 h

Similarly, in Fig. 5 we set the budget $b = 5\%$ and give lead time of sensor group designed by our approach (Fig. 5b) and compare it with sensor set designed by occurrence frequency approach (Fig. 5a). It is notable from these figures that a sensor group of top 5% users from our ranking scheme can predict popularity 50 h ahead of other users. While, the occurrence frequency gives only a negligible lead time.

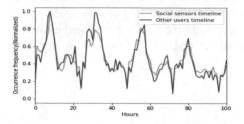 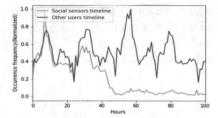

(a): *Timeline of sensors and others by ranking based only on occurrence frequency*　　　*(b)*: *Timeline of sensors and others by ranking based on occurrence frequency and mean adoption time*

Fig. 5. Comparison of methods with just 5% sensor users; Our approach gets average lead time of around 50 h

In Table 1, we give lead times of sensor group of budget values $b = 1\%, 5\%, 10\%$ and selected based on occurrence frequency and our proposed approach. In all cases, we can note that our proposed approach outperforms the occurrence frequency approach. Since all process are made in a distributed fashion, getting results of multiple budgets and comparing them with existing approach are less painstaking compared to implementing them in a serialized versions.

Table 1. Average lead time of sensor set of budgets $b = 1\%, 5\%, 10\%$ based on occurrence frequency and proposed approach

Budget b	Occurrence frequency approach	Proposed approach
1%	2 h	21 h
5%	0.8 h	50 h
10%	−0.5 h	60 h

4 Discussions and Future Works

The focal point of this work is in modelling information cascades as forests of diffusion trees. In addition to simple methods for selecting social sensors in future, we plan to add more measures like popularity of a cascade and add a weighted score for users based on their involvement in previous cascades. Substructures of a network like graphlets and motifs have been proven to be a major advantage in variety of applications like network classification [5] and network comparison [1]. With some minor improvements in our ranking schemes, we plan to add fine grained network structural properties like graphlets to improve our ranking measures. Also, we plan to test our methods on a wide number of very large scale networks.

References

1. Ahmed, N.K., Neville, J., Rossi, R.A., Duffield, N.G., Willke, T.L.: Graphlet decomposition: framework, algorithms, and applications. Knowl. Inf. Syst. **50**(3), 689–722 (2017)
2. Friggeri, A., Adamic, L.A., Eckles, D., Cheng, J.: Rumor cascades. In: ICWSM, May 2014
3. Garcia-Herranz, M., Moro, E., Cebrian, M., Christakis, N.A., Fowler, J.H.: Using friends as sensors to detect global-scale contagious outbreaks. PLoS One **9**(4), e92413 (2014)
4. Leskovec, J., McGlohon, M., Faloutsos, C., Glance, N., Hurst, M.: Patterns of cascading behavior in large blog graphs. In: Proceedings of the 2007 SIAM International Conference on Data Mining, pp. 551–556. Society for Industrial and Applied Mathematics, April 2007
5. Vishwanathan, S.V.N., Schraudolph, N.N., Kondor, R., Borgwardt, K.M.: Graph kernels. J. Mach. Learn. Res. **11**(Apr), 1201–1242 (2010)
6. Zaharia, M., Chowdhury, M., Das, T., Dave, A., Ma, J., McCauley, M., Franklin, M., Shenker, S., Stoica, I.: Resilient distributed datasets. In: A fault-tolerant abstraction for in-memory cluster computing in Proceedings of the 9th USENIX conference on Networked Systems Design and Implementation (2014)
7. Christakis, N.A., Fowler, J.H.: Social network sensors for early detection of contagious outbreaks. PLoS One **5**(9), e12948 (2010)
8. Shao, H., Hossain, K.S.M., Wu, H., Khan, M., Vullikanti, A., Prakash, B.A., Marthe, M., Ramakrishnan, N.: Forecasting the Flu: designing social network sensors for epidemics. arXiv preprint arXiv:1602.06866 (2016)
9. Eom, Y.H., Jo, H.H.: Generalized friendship paradox in complex networks: the case of scientific collaboration. Sci. Rep. **4**, 4603 (2014)
10. Sakaki, T., Okazaki, M., Matsuo, Y.: Earthquake shakes Twitter users: real-time event detection by social sensors. In: Proceedings of the 19th International Conference on World Wide Web, pp. 851–860. ACM, April 2010

Analyzing the Single-Use Plastic Bags Ban Policy in California with Social Network Model and Diffusion Model

Sekwen Kim[(✉)]

Claremont Graduate Univeristy, Claremont, CA, USA
se-kwen.kim@cgu.edu

Abstract. This study seeks to identify key social network attributes of network micro-level actors and to examine how socio-economic attributes of micro-level actors, cities, contributed to diffusion of the environmental policy, which bans the use of single-use plastic bags in grocers in California. By incorporating social network models, dynamic network model and small world phenomenon, and diffusion theory, leader and laggard model, this study seeks to answer, 'how the environmental regulatory policy which bans the use of single-use plastic bags in grocers in California has been diffused'.

Keywords: Diffusion theory · Leader and laggard model
Dynamic network model · Small world phenomenon · Environmental policy
Social network analysis

1 Introduction

It has been substantively documented by scholars that public policy is not being randomly adopted, but diffused. That is, policy is diffused either geographically or socio-economically. The objective of this study is to use social network analysis tools to examine the micro-level, city level, diffusion of the single-use plastic bags ban policy before the state referendum, which bans single-use plastic bags use in California. There are two potential major contributions from this research. The first contribution will be to further substantiate the applicability of social network analysis on micro-level policy diffusion. That is, to document that a significant pattern of micro-level policy diffusion can be comprehensively captured by using social network analysis. The second, given the scarcity of empirics to justify the diffusion of policy among micro-level actors within California, this study would further strengthen the diffusion theory in both geographically and socio-economically homogeneous settings.

The structure of this research is as follows. First, in the immediately following section, this research shall provide brief background information on the statewide ban on the single-use plastic bags in California. Second, this research includes a literature review on the key concepts utilized in this research. Third, this research shall provide core research design, data description, network generation, and measures. In this section, the explanation shall illustrate the research model of this study and provide a brief account of how this research shall collect the data as well as operationalize and

© Springer International Publishing AG, part of Springer Nature 2019
T. Z. Ahram (Ed.): AHFE 2018, AISC 787, pp. 408–418, 2019.
https://doi.org/10.1007/978-3-319-94229-2_40

measure the key variables of the research. This research shall provide result, analysis. In the fourth section, this research shall present the results and provide my interpretations of them. Lastly, in the concluding section, this research shall discuss policy implication of this study and shall suggest the future direction of the research.

2 Background

In 2014, California Governor Jerry Brown signed a state law which banned the usage of single-use plastic bags and required shoppers be charged at least 10 cents for paper bags and reusable plastic bags at grocers in California [1]. However, the plastic industry put Proposition 67, which could have repealed the ban, on the ballot in 2016 [1]. The referendum vote showed that 52% of voters voted to affirm the statewide policy banning the use of single-use plastic bags. California became the first state to ban the use of single-use plastic bags in the United States in 2016 [2].

However, the 2014 statewide ban was not the first regulation to adopt a regulatory policy on single-use plastic bags usage within California. The cities of Malibu and Fairfax both adopted policies banning single-use plastic bag in 2008 [3]. After 3 years of the inert activity of diffusion, the number of cities who adopted the similar regulatory measures on single-use plastic bags increased rapidly.

3 Literature Review

By incorporating both dynamic network theory and diffusion theory, a plausible explanation of a pattern of environmental policy diffusion can be drawn. Dynamic network theory, small world phenomenon, and diffusion theory share a similarity each emphasizing how attributes change over time. Using dynamic network theory allows this study to identify a focal subgroup as time progresses and diffusion theory allows this study to identify the socio-economic attributes of micro-level actors, cities, which strongly influence the diffusion of the environmental policy. Small world phenomenon emphasizes the accumulation of gradual incremental micro-level changes affecting radical macro-level changes.

3.1 Social Network Analysis Models

Small world phenomenon and dynamic network theory theoretically allow this study to capture the pattern and network attributes of significant subgroups. Small world phenomenon describes network development as a result of gradual incremental small changes, which cause 'phase transitions' in a network structure [4]. This implies that radical macro-level structural change is a result of aggregated micro-level structural changes [4].

Dynamic networks, as opposed to static networks, focus on changes of networks over time. By defining the human actions as "purposive, but carried out under conditions that set both opportunities for and restrictions on the achievement of these purposes", dynamic network emphasizes the importance of rate of change [4]. Dynamic network modeling is also closely related to diffusion modeling since the "rate of diffusion of ideas across a whole network may be significantly affected by relatively small local-level changes that have these macro-level effect[s]" [4]. Dynamic network recognizes the cumulation of both Dynamic network pinpoints and that there is a gradual change within a network. Cumulation of a large number of small changes can affect significant change.

3.2 Policy Diffusion Models

The diffusion theory of the policy investigates the diffusion of an innovative policy and significant attributes in its diffusion [5]. Within diffusion theory, the regional diffusion model and the leader-laggard model are useful for explaining and predicting the adoption of the single-use plastic bags ban policy.

Berry and Berry describe the regional diffusion theory as the policy diffusion being primarily influenced by proximate governments [5]. Within the regional diffusion model, the neighbor model and the fixed-regional model differ when interpreting the primary influencer of the diffusion of the policy. The neighbor model specifically emphasizes the influence of adjacent jurisdictions or states and posits that the policy adoption is significantly influenced by bordering jurisdictions [5]. While the neighbor model emphasizes the influence of bordering jurisdiction and boundary between them, the fixed regional model emphasizes the homogeneousness of influence channel within the boundary of broader jurisdictions [5]. The fixed regional model states that governments within the same jurisdiction exhibit the characteristic of emulating each other, because they share "the same channel of influence" [5]. However, Berry and Berry stated that "a more realistic regional diffusion model might assume that jurisdictions are influenced most by their neighbors but also by other jurisdictions that are nearby" and emphasized the proximity between jurisdictions as the most influential attribute influencing the diffusion of the policy [5]. Learning behavior between governments and homogeneousness of socio-economic problems are shared by proximate governments [5]. Therefore, proximity should be emphasized when explaining regional diffusion theory [5]. In addition, flexible mobility of population between jurisdictions may trigger governments to provide equivalent welfare and adopt policies, which are similar to neighboring governments [5].

The leader and laggard model of the diffusion recognizes leaders and followers in policy adoption [5]. Leaders, or pioneers, are risk-takers who are willing to test an untested policy and followers are those who are risk averse and adopt the policy after evaluating the success of once-untested policy from leaders [5]. The leader-laggard model assumes that if the government is socio-economically developed, the government is willing to take the risk and become a pioneer in adopting an untested policy [5].

4 Core Research Design

Based on the theory review and literature review on social network analysis and policy diffusion model, this research proposes the following research question:

Does the degree of policy implementation influence the diffusion of the single-use plastic bags ban policy?

This study examines the diffusion of the single-use plastic bags ban policy using a hypothesis for social network analysis and a diffusion model.

Hypothesis for social network analysis would be:

H1: There would be one distinct subgroup with certain degree of policy implementation influencing the diffusion of the environmental policy significantly.

H0: There would not be one distinct subgroup with certain degree of policy implementation influencing the diffusion of the environmental policy significantly.

5   Data Description

5.1   Data Collection

For the social network analysis to gather necessary information in regard to participating cities and degree of the single-use plastic bags ban policy implementation, this study shall construct a dataset which summarizes participating cities, policy implementing years, and degree of the single-use plastic bags ban policy implementation. Among the indices that are available online and relevant to this research, a dataset from Californians Against Waste summarizes information required for conducting this research. Using this dataset, this study shall create social network visualization, which efficiently represent the pattern of policy diffusion and shall run network analysis, which provides significant results to identify focal subgroup within the network. Therefore, the data collected and subject to analysis in this stage is quantitative in nature.

As a result of data collection, the data set includes 110 cities who adopted a single-use plastic bags ban policy in years 2008, 2011, 2012, 2013, 2014, 2015, 2016, and 2018. In addition, the degree of policy implementation was categorized into three categories: (1) a complete ban on the single-use plastic bags; (2) a 10 cents charge per reusable plastic or paper bag; and (3) a 25 cents charge per reusable plastic and paper bag.

5.2   Operationalization of Data

The key explanatory variable in the social network analysis is degree of implementation and participating year. Due to crisp categorization of both degree of implementation and participating year, using the original indices from Californian Against Waste would suffice for the operationalization of the dataset.

6   Network Generation

In order to comprehensively capture the significant patterns and attributes of networks, this study constructs a network composed of three subgroups, which are fully connected and undirected with cities as nodes and degree of policy implementation as links. The constructed network shall be composed of three subgroups:

1. A subgroup of cities with complete plastic bags ban,
2. A subgroup of cities with 10 cents charge for paper and reusable bags,
3. A subgroup of cities with 25 cents charge for paper and reusable bags.

To capture the pattern and the progress of policy diffusion, dynamic network model and small world phenomenon are utilized. As the analysis of the network of environmental regulatory policy diffusion should focus on both years of policy adoption and degree of policy implementation, dynamic network model and small world phenomenon shall be applied.

As a dynamic network model, this study presents:

1. A network of participating cities in 2008.
2. A network of participating cities in 2011.
3. A network of participating cities in 2012.
4. A network of participating cities in 2013.
5. A network of participating cities in 2014.
6. A network of participating cities in 2015.
7. A network of participating cities in 2016.
8. A network of participating cities in 2018.

Networks from 2009 and 2010 are omitted due to the absence of participating cities and the network of 2017 is not considered since state referendum was held and other cities in California were waiting for referendum result.

7 Measures

In order to draw a feasible conclusion of analysis, eigenvector centrality and density of each subgroup shall be calculated. The eigenvector centrality provides useful information on important subgroups within the "global" or "overall" structure of the network [6]. Measured eigenvector centrality would allow this study to further identify the focal subgroup, which can also be translated into important policy degree in the perspective of the single-use plastic bags ban policy in California cities. Density is another significant measure, which allows this study to capture the overall network comprehensively. Density measures the proportion of all possible links and allows this study to capture the speed of diffusion within the network among nodes [6]. By utilizing eigenvector centrality, which measures the network from a micro-level perspective and density, which measures the network from a macro-level perspective, this study can provide comprehensive analysis of the network.

8 Result

The diffusion of the policy that bans the single-use plastic bags was analyzed from three attributes of the social network analysis, which are size of the network and subgroups, eigenvector centrality of subgroups, and density of the network.

8.1   Size of the Network: Number of Participating Cities

The policy banning single-use plastic bags was first introduced in 2008 as Malibu and Fairfax first adopted the policy. This environmental regulatory policy was not introduced or adopted by other cities until 2011. In 2011, three neighboring cities of Malibu: Long Beach, Calabasas, and Santa Monica, adopted the policy. It has shown that the policy was inertly adopted until 2011. It was from 2012 until 2016 when neighboring cities have been beginning to adopt the similar environmental regulatory measure on single-use plastic bags as numbers of participating cities increased by 14, 28, 26, and 15 cities respectively.

It was shown that degree of policy implementation of 10 cents charge per reusable/plastic bags increased the most in a perspective of size, the number of participating cities increased by 9, 6, 23, 19, and 10 from the years 2012 to 2016 respectively (Figs. 1 and 2).

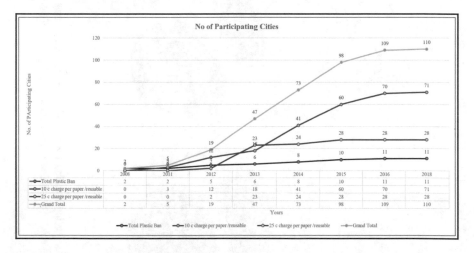

Fig. 1. As time progressed, the number of participating cities increased, however the number of participating cities with 10 cents charge per paper/reusable bags increased more rapidly than other measures of policy implementation.

8.2   Diffusion Patterns of the Policy Among Cities

2008 and 2011. Adoption of single-use plastic bags ban policy was first introduced by two cities in California, Malibu and Fairfax, in 2008. After three years of inert diffusion activity, three more cities adopted similar environmental regulatory measures in 2011. However, unlike the two cities who adopted the policy in 2008 and did not offer an alternative to the use of single-use plastic bags, the three cities who participated single-use plastic bags regulation in 2011 included a 10 cents charge per paper/reusable plastic bag.

Diffusion of Single-use Plastic Bags Ban Policy in California by Cities From 2008 to 2018

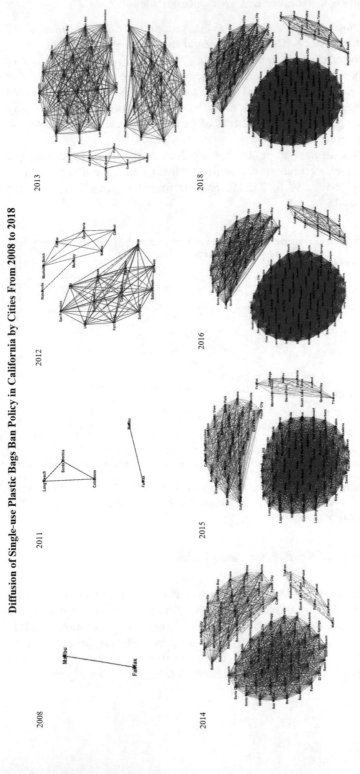

Agent with Black: Total Ban on the Single-use Plastic Bags
Agent with Blue: 10 cents charges on reusable/paper bags
Agent with Red: 25 cents charges on reusable/paper bags

Fig. 2. The environmental policy banning the use of single-use plastic bag was diffused rapidly after 2013 which is represented by the denser formation of subgroups after 2013.

2012–2018. From 2012 to 2018, the policy banning single-use plastic bags was actively diffused and the number of participating cities increased an average of 17 cities per year. Diffusion of adoption of the single-use plastic bags ban policy was most active in 2014 and 2015 by 28 and 26 cities participating respectively.

Most notable diffusion phenomena in this case is the dramatic increase in a number of participating cities of 10 cents charge per paper/reusable plastic bags. The number of cities adopting a 10 cents charge per paper/reusable plastic bags in 2014 and 2015 were 23 and 19 respectively while only two cities adopted a total ban on single-use plastic bags from 2014 to 2015. During 2014 and 2015, 1 and 5 cities respectively adopted a 25 cents charge per paper/reusable plastic bags.

Overall Diffusion. Based on social network analysis visual representation and summary statistics of participating cities, a 10 cents charge per paper/reusable plastic bag was the most actively diffused from 2014 to 2016.

8.3 Density

As the number of participating cities increased, density of the network increased as well. One noticeable change of overall density is that density changed as proportional to the change of size of the network. The structure of the network, which is composed of three fully linked subgroups, contributed to the proportional changes of density. However, since the three fully connected subgroups are not connected to each other, density cannot be measured as one, which refers to a fully connected network.

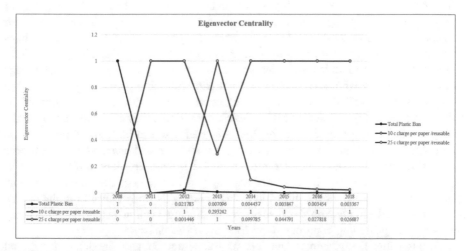

Fig. 3. With higher eigenvector centrality, which reflects the significance of each subgroup from a social network perspective, than other policy implementation, 10 cents charge per paper/reusable plastic bag is the most significant policy degree of implementation in the single-use plastic bags ban policy diffusion.

8.4 Eigenvector Centrality

Another important measurement of social network analysis is eigenvector centrality. In this study, eigenvector centrality was employed to identify important subgroups in a perspective of the diffusion of the policy.

Using eigenvector centrality analysis, it was calculated that except in 2008 when 10 cents charge per paper/reusable plastic bags was not introduced and in 2013 when number of cities who adopted 10 cents charge per paper/reusable plastic bags was outnumbered by cities who adopted 25 cents charge per paper/reusable plastic bags, a subgroup of 10 cents charge per paper/reusable plastic bags demonstrated the highest eigenvector centrality result. From the calculated eigenvector centrality measure, it can be concluded that a 10 cents charge per paper/reusable plastic bag policy was the most important policy degree of implementation in the perspective of single-use plastic bags ban policy diffusion. It is also inferred that the number of participating cities is positively correlated with eigenvector centrality of a subgroup (Fig. 3).

9 Analysis

Based on outcomes of social network analysis visual representation and eigenvector centrality analysis, it can be inferred that the 10 cents charge per paper/reusable plastic bag was the most favorable policy and the most important degree of policy implementation in the perspective of single-use plastic bag ban policy diffusion. It is also worth noting that the number of cities, who participate in the adoption of policy, influences the eigenvector centrality results and, therefore, is affecting the subgroup of being the focal subgroup. The inference could be supported by higher eigenvector centrality measure of a subgroup with 25 cents charge per paper/reusable plastic bags than that of a subgroup with 10 cents charge per paper/reusable plastic bag in 2013. In 2013, the number of cities who charged 25 cents per paper/reusable plastic bags outnumbered those who charged 10 cents per paper/reusable plastic bags.

It is also noticeable that the number of cities who participated in the adoption of the policy increased rapidly since 2013. Considering that initiating cities adopted the policy in 2008 and it took 5 years for the policy adoption to be diffused rapidly, it is congruent with the leader and laggard theory, which states risk-aversive actors observe the performance of risk-taking leading actors and imitate or adopt similar policy of leading actors. Even though the leader and laggard theory does not suggest the time span which is required for the diffusion of the policy, this study is useful to the degree that it tests and strengthens the theory by identifying leading actors and following actors and by providing congruent case to support the theory.

Through this study, general models of leader and laggard model and regional diffusion model of diffusion theory were tested and provided congruent social network analysis results, which empirically support the main argument of the theory. From diffusion model of policy perspective, the regional diffusion model provides comprehensive explanation on how geographic and socio-economic attributes of cities in California contributed to the diffusion of the single-use plastic bags ban policy in California. Malibu was the first city in southern California who adopted the plastic bag ban policy. The policy was diffused and adopted by neighboring, Santa Monica, Calabasas, Long Beach, and Manhattan Beach by following years and diffused to neighboring cities of following cities [3]. This policy diffusion in southern California is congruent with the regional diffusion model which states that governments tend to emulate the policy which was adopted by a proximate government.

The leader and laggard model is also useful to explain the policy diffusion in California. Those cities who initiated the policy adoption exhibit higher median household income than that of California as a whole and have a higher percentage of their population with Bachelor's degree or higher level of education than the average percentage of California residents [7, 8]. The leader and laggard model posits that wealthier and more educated cities are more likely to adopt an untested policy than those who are poorer and less educated [5]. As cities with higher median income and higher education level adopted an untested policy in the early stages, the diffusion of the single-use plastic bags ban policy in California explicates the leader and laggard model of policy diffusion.

10 Policy Implication

By analyzing the pattern of network development as time progresses, this study provided both quantitative and visual representation on the diffusion of the single-use plastic bags ban policy. To some degree, this study illuminates the capability of the social network analysis on related policy analysis in terms of contextual and geographical attributes. Not only environmental regulatory policy which involves both costs and benefits, but also broader field of policy area can be analyzed by utilizing the same analytical methodology.

This study also suggests that the same analytical methodology, the social network analysis, can be employed to analyze a policy diffusion of broader geographic reach. This study limited its geographical boundaries only into cities of California, due to its topic and availability of empirics. However, inter-states diffusion and even inter-countries diffusion can be analyzed and tested by using social network analysis. In this regard, the same approach and the same methodology can be applied to review and test the diffusion of either environmental or fiscal, such as tax, policies in Europe or in west coast states in the United States.

The findings of the social network analysis to portray the pattern of network development and to identify the key attributes of the network, however rough it is, are reflective of there being a need to conduct panel data regression. Furthermore, the findings of the social network analysis suggest that panel data regression would further enrich the diffusion theory by providing persuasive argument on the influence of socio-economic factors on the diffusion of the ban on single-use plastic bags policy. In order to enrich the diffusion theory model, it would be desirable to elaborate the model by conducting panel date regression with socio-economic attributes of each city in each applicable year.

11   Conclusion

This study is designed to incorporate the social network analysis to examine the diffusion of single-use plastic bags ban policy in micro-level actors, cities, in California before the state referendum.

In this study, brief background of information on the statewide ban on single-use plastic bags policy and brief description of theoretical literature review on key concepts were discussed. By utilizing, operationalizing, and calibrating data available online, this study conducted social network analysis in order to identify and measure key social network attributes which allow the diffusion of the policy to be comprehensive captured. Through the conducted social network analysis, it was identified that 10 cents per paper/reusable plastic bags was the most favorable and important degree of policy implementation during the diffusion of the policy which bans the use of single-use plastic bags and identified the correlation between the number of agents within subgroups and eigenvector centrality when the network consists of fully connected independent subgroups. This study also suggests the further capability and applicability of social network analysis, which can be incorporated when analyzing broader contextual and geographical policy topics.

It would be this study's tasks to further strengthen both theoretical and empirical argument by exploring and testing other possible socio-economic factors. By doing so, a thick description with good narrative which comprehensively explaining the mechanism of contribution of socio-economic attributes on the diffusion of the environmental policy can be derived. If socio-economic attributes influencing the diffusion of the environmental policy can be identified, this study will serve its purpose.

References

1. Rogers, P.: Voters approve plastic bag ban: what's happens next? 11 November 2016. http://www.mercurynews.com/2016/11/10/voters-approve-plastic-bag-ban-whats-happens-next/. Accessed 09 Oct 2017
2. Calefati, J.: California bag ban: voters to weigh industry's fate at the ballot box, 11 October 2011. http://www.mercurynews.com/2016/09/16/california-bag-ban-voters-to-weigh-industrys-fate-at-the-ballot-box/. Accessed 09 Oct 2017
3. Californians Against Waste: Single-Use Bags Ordinances in CA (2016). https://static1.squarespace.com/static/54d3a62be4b068e9347ca880/t/583f1f57e4fcb5d84205b330/1480531800415/LocalBagsOrdinances1Pager_072815.pdf
4. Scott, J.: Social network analysis. Sage US Census Bureau (2017). https://www.census.gov/quickfacts/fact/table/TX,NV,WA,CA,OR/PST045216. Accessed 22 Nov 2017
5. Berry, F.S., Berry, W.D.: Innovation and diffusion models in policy research. In: Theories of the Policy Process, p. 169 (1999)
6. Hanneman, R.A., Riddle, M.: Introduction to Social Network Methods (2005)
7. US Census Bureau (2017). https://www.census.gov/quickfacts/fact/table/TX,NV,WA,CA,OR/PST045216. Accessed 22 Nov 2017
8. Smith, T.F., Waterman, M.S.: Identification of common molecular subsequences. J. Mol. Biol. **147**, 195–197 (1981)

The Moderating Roles of Network Density and Redundancy in Lurking Behavior on User-Generated-Content Online Communities

Xingyu Chen[✉], Yitong Wang, Xianqi Hu, and Zhan Zhou

Department of Marketing, Shenzhen University, Shenzhen, China
celine@szu.edu.cn

Abstract. Sharing content is one of the important ways of information diffusion in online UGC (User-Generated Content), communities. Most of previous research on the sharing behavior focused on predicting the sharing behavior by the inherent characteristics of the posts. This study addressed the important role of social networking characteristics, including network structure and information density, on users' sharing behavior. Based on a social network from a large UGC platform in China, this study analyzed the panel data of 10,000 users of their daily activities. The results showed that network density and redundancy jointly influenced users' sharing behavior. This study contributes to social network theory by providing new empirical evidence on user-generated content diffusion in UGC community. In particular, it explained how network density moderating the effect of users on UGC diffusion. This study also had important management implications for platform managers to design effective product strategies to increase UGC diffusion.

Keywords: Sharing behavior · UGC community · Social network
UGC diffusion

1 Introduction

In recent years, the rapid development of the online UGC (User-Generated-Content) community has led many researchers to focus on the users' behavior in UGC communities [1, 2]. Users in UGC communities can generate their own content, and share or comment on others' content [2–4]. With the proliferation of online UGC community and online BBS, user-generated content (UGC) has become one of the most energetic forms of media on the Internet [2]. Therefore, the sharing behavior of users is an important way to diffuse content of UGC community.

Sharing content simply means retweeting or reposting others' posts. Many previous research on the sharing behavior in UGC communities focused on predicting the sharing behavior [5], and the prediction was based on the inherent characteristics of the post. By classifying frameworks based on different features related to posts or authors, those works can predict whether a post will be shared by any users, and identify the factors that are strongly associated with sharing [6, 7].

© Springer International Publishing AG, part of Springer Nature 2019
T. Z. Ahram (Ed.): AHFE 2018, AISC 787, pp. 419–427, 2019.
https://doi.org/10.1007/978-3-319-94229-2_41

In this paper, we analyze the sharing behavior through the perspective of social network. We assumed that in a dense network where every user is connected to each other, one may be exposed to the same piece of information for multiple times, especially those popular posts that currently prevail within the community. Since network structure and information density are the most important index of social network, we will explore the impact of network density and information redundancy on user sharing behavior in this study.

The main objective of this research is to study the sharing behavior on a UGC social network platform. We particularly explore the following questions:

1. Dose the information density affect sharing behavior?
2. Does the network structure affect sharing behavior?

To answer these questions, we collect a weekly data of sharing behavior by a random sample of 10,000 users over six weeks from a social network platform. We supplement our analysis with a survey of 2,622 active users in the social network. We will investigate from three perspectives:

First, we will explore the value of sharing behavior. Previous studies mostly focused on content generation behavior. However, our preliminary study found that most users are not willing to generate content. They share content more often, and such behavior can also provide a huge amount of traffic to the site.

Second, we will analyze the impact of network density on sharing behavior. When the user is in a dense network, will not be more willing to post.

Finally, by studying the impact of information redundancy on user-sharing behavior, we explore the significance of website hot posts.

2 Literature Review and Hypothesis

2.1 Sharing Behavior in UGC

The information diffusion is a topic that attracts more and more attention in many fields, including the sharing behavior in UGC community [6, 8, 9]. Most studies have focused on forecasting sharing behavior [5], and the prediction is based on the intrinsic characteristics of the article. By categorizing the frameworks based on the different characteristics associated with the article or the author, these studies could predict whether a post will be shared by any user and identify factors that are closely related to the sharing [6, 10]. Harrigan et al. [10] found that community structures, particularly reciprocal ties and certain triadic structures, substantially increased sharing behavior. Suh et al. [6] found that content features, URLs and hashtags have strong relationships with retweetability. Part of the researches also focused on the analysis of the sharing behavior, and began to analyze the motivations of sharing from the perspective of psychology [11, 12]. Starbird and Palen [11] examined the use of the retweet mechanism on Twitter, using empirical evidence of information propagation to reveal aspects of work that the crowd conducts. Lee et al. [12] found retweeting were predicted to occur when an individual was motivated for prosocial motivations.

2.2   Social Networks

Social network theory reveals the phenomenon of association in society (related nodes can be people, things, events, businesses, groups, cities and even countries), and provides a method of analyzing relational data (not attribute data) Social science concerns [13, 14]. In the field of management science, the fruits of extensive use of social networks to study organizational behavior are published in top journals such as the Academy of Management Journal. They studied the impact of social network variables such as joint, centrality, structural holes and network density on outcome variables such as individual performance, team performance, research and innovation [15–18].

In recent years, some scholars have gradually started their research in combination with social networks. Most scholars regard social networks as research content and analytical methods to study various relationships in marketing. For example, Iacobucci and Hopkins proposed the use of social network analysis to analyze two-way relationships and social networks in marketing [19]; Rindfleisch and Moorman studied the problem of access to and use of information in new product alliances from the perspective of coordination [20]; Antia and Frazier used centrality and network density as the reason for contract implementation among inter-company channel relationships [21]; Janssen and Jager used social network structure and small-world theory to study the dynamics of the market and so on [22].

It was not until 2010 that some scholars applied social networks to online community research. For example, Kozinets et al. studied the process of network communication among consumers, and then proposed that the essence of group word-of-mouth marketing is the systematic change of marketing information and connotation in its embedded social network [23].

Network Density. Network density describes the overall situation of the interaction between all members of the self-center network, belonging to the concept of organizational level, rather than individual level. As shown in Fig. 1, the graph on the left has a network density (0.8) greater than the density on the right (0.3) because there is more association between users in the left structure. Users interaction can affect [24] network density (the more interactions, the greater the density). Research shows that the similarity of personal information drives a sense of empathy among users and thus resonates [25, 26]. It can be inferred that similarities or commonalities (i.e., commonalities of communities) of online UGC community users in terms of values, hobbies, etc., will also narrow the distance between network members and increase the network density. Among them, if the overall quality of the online interaction of the UGC community is good, the online UGC community atmosphere and community commonality presents a greater opportunity to present the network thus strengthening the network density. Therefore, online interactive quality may serve as a regulatory variable.

It is worth mentioning that the network density in the network structure, which is used to measure the degree to which users and friends communicate with each other, may have some impact on user sharing behavior. For example, the network density may potentially cause information redundancy, that is, in dense networks where users are closely connected to each other, the same user may publish or share the same information multiple times, especially those that are prevalent in the community [27].

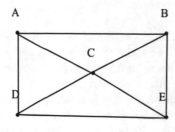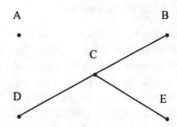

Fig. 1. Network density

Sohn finds a moderating role of network density on the relationship between valence of information and transmission of that information [28]. More specifically, participants in the study are more likely to transmit positive information only in a sparse network, as they perceive the information to be more valuable. In the process of idea generation, Stephen et al. [27] find that local network with low density is more valuable because people in such a network can generate more different ideas than in another with high density where they see similar information airing in the network and their creativity is impaired. Sun et al. (2016) document a decreasing tendency to adopt new products if it is popular among friends. Given the various social network factors that affect diffusion, it is attractive to develop rationalized seeding strategies and start a viral marketing campaign from a specific type of consumers.

Information Redundancy. Our conceptual model examines the association between the popularity of content and users sharing behavior. We use an observed metric, the popularity of contents, as a summary statistic of all motives. Besides, we investigate the moderating role of network density on the differential responsiveness and detect a condition information redundancy, namely the reaction when users are situated in a dense network and facing popular contents. In this regard, we extend the seminal idea in Stephen et al. (2016) that uses network density to measure information redundancy, and use it to understand the pattern of users in content diffusion.

Users sharing in any UGC platform is a key element to make the content viral and improve the activeness of the platform. Many factors may affect users sharing intention in the UGC platform. A UGC user is more likely to share interesting stories [30], useful information [29, 31, 32], posters by famous people [3, 33, 34], or because the user has many followers [10, 29, 35]. From the perspective of platform's manager, however, most of the motives documented in the literature are difficult to measure or manipulate. We study a readily available and easy-to-implement metric, the popularity of the posts, which is usually a flag attached to the post indicating its popularity [36]. Exposure to popular posts may affect users sharing intention.

Based on the researches and suggests above, we assumed that:

H1: Users share more contents when they are exposed to more popular posts.

Information redundancy occurs when users are in a dense network and face popular content. In this case, his friends relate to each other, and most popular posts shared by a friend are likely to be shared by other friends. Note that in a sparse network, the less contact between friends, even the hot posts shared by friends will not produce the problem of information redundancy. In general, users will not be exposed multiple times in the same article. Hence, we propose the following hypothesis:

H2: In a dense network, users are more responsive when exposed to popular posts.

In a dense network, users have more opportunity to interact with their friends. And sharing some contents, especially the widely viewed or shared post, will create common topics and enhance their emotion.

3 Method

3.1 Participants

We collect our data from an online UGC platform in China, Qiushi-Baike (QB hereafter). Established in 2005, QB has attracted over 20 million registered users who together have accumulated more than 100 million posts. Users on QB disclose their own funny or embarrassing stories or repost stories by others. Like Facebook, users have friends network on QB, and they can only see posts from their friends. We track a random sample of 2,622 users in the final dataset. We keep users that were active in both of the two periods: (1) three-month period from April 26, 2016, to July 26, 2016, and (2) six weeks from July 26, 2016, to September 5, 2016.

3.2 Measures

Summary statistics of all measures, listed below, are presented in Table 1.

Table 1. Descriptive statistics & Pearson correlations

	1	2	3	4	5	6
1 Sharing	-					
2 Posts seen	0.14	-				
3 Hot (percentage of hot post seen)	0.01	0.00	-			
4 Density (clustering coefficient)	0.03	−0.05	−0.04	-		
5 Age	0.10	0.07	−0.05	0.03	-	
6 Tenure	−0.01	0.04	−0.01	−0.04	0.07	-
Mean	0.30	229.45	0.28	0.50	28.97	2.08
Standard dev.	6.42	1006.78	0.30	0.50	15.83	1.04

Sharing and Posts Seen: We calculate the (flow) number of shares and (flow) number of posts in the circle of friends for each user in each week.

Density: Network density measures the degree of contact between a focused user's friends. We use clustering coefficients to measure the density of the network. The clustering factor for a focused user is the ratio of the actual connection between his friends and the possible connection between his friends. When the clustering coefficient is close to 1, we call it a dense network; when the clustering coefficient is close to 0, we call it a sparse network.

Age and Tenure: We have captured the users' age and tenure data. In the sample, the average user age was 29 years, with a standard deviation of 16 years. Tenure is defined as the year since registration, and users have been registered for an average of more than two years.

4 Results

After a two-step of data collection, we analyzed the data initially by SPSS v.22, and got some preliminary results. We examine the sharing behavior of users primarily using field data from QB. Table 2 reports the main results. Column 1 estimates the model on all users without controlling for network density. Column 2 estimates the users in a sparse network, with clustering coefficient smaller than the median; Column 3 estimates for users in a dense network, with clustering coefficient above the median. Column 4 uses the dummy for clustering coefficient above the median, and column 5 uses the actual value of clustering coefficient, as a measure of network density. From the value of Pseudo-R2, our model fits the data well and explains 85% to 90% of the variation of the sharing behavior.

Table 2. Preliminary results

		Median-split		Density metrics	
		Below	Above	Dummy	Raw
Posts seen	0.11	0.16**	0.10	0.12	0.12
Hot post	0.08**	0.09***	0.06	0.08***	0.09
Hot*density				−0.04	0.04
Pseudo-R2	0.88	0.84	0.90	0.88	0.88

Notes. *Standard errors are clustered at the level of users.*
***, **, * indicate a significance level of .01, .05, and .10, respectively.

Overall, we find that more posts seen from friends promote users share more contents. 1% increase in the number of posts seen lead to an increase of sharing by 10% (column 3, p = 0.45) to 16% (column 2, p = 0.05). This is strictly less than one, as the marginal propensity to share is decreasing in the number of posts seen. The results supported our hypotheses above.

Effect of Hot Contents. We assumed that these hot posts have a positive effect on sharing behavior, and users share more contents when they were exposed to more popular posts (H1). This hypothesis is supported. We found that 1 SD increase in the hot posts (0.3, see in Table 1) increases the sharing by 8% (p = 0.02, see in Table 2). Such an increase may come from the fact that contents with better quality promote more sharing. Namely, users share more posts when facing more popular contents. The labels for popularity works effectively in triggering sharing.

Effects of Network Density. Although insignificant, our analysis above hints that users may be more responsive than nonpopular contents (0.27, p < 0.001). We thus examine users sharing behavior under information redundancy, i.e., when they are situated in a dense network facing popular contents. When located in a sparse network, where there are very few connections between friends, users are more likely to share when seeing popular contents. One standard deviation increase in the proportion of popular posts leads to an increase of 9% (p < 0.01) in sharing. However, the opposite is true in dense networks. Most of users prefer to share.

5   Conclusions

Our preliminary study explores the impact of network density and redundancy of information to the user's sharing intention. We found that redundant information will make users more willing to share some content. When users seeing more popular posts in a well-connected network, will be better informed about the aggregate behavior of their peers, and their interest to conform will be triggered, leading to increased probability of sharing. Users can use this to get in touch with other friends in order to have more common topics. This enhances user interaction. In the meantime, we explored the impact of network density on users' willingness to share. We found that while hot posts have a positive effect on users, the intensity of the network can have a stronger impact on this role. That is, the denser the network, the more willing the users are to share content.

As a preliminary analysis of the panel data, we obtained the above suggestions. But the data is much more than that, and they may allow us to get even more surprising results. Use more complex models or ways to research them and give them value.

References

1. Li, H., Liu, Y.: Understanding post-adoption behaviors of e-service users in the context of online travel services. Inf. Manag. **51**(8), 1043–1052 (2014)
2. Zhang, K., Evgeniou, T., Padmanabhan, V., Richard, E.: Content contributor management and network effects in a ugc environment. Mark. Sci. **31**(3), 433–447 (2012)
3. Liu-Thompkins, Y., Rogerson, M.: Rising to stardom: an empirical investigation of the diffusion of user-generated content. J. Interact. Mark. **26**(2), 71–82 (2012)
4. Presi, C., Saridakis, C., Hartmans, S.: User-generated content behaviour of the dissatisfied service customer. Eur. J. Mark. **48**(9/10), 1600–1625 (2014)

5. Luo, Z., Osborne, M., Tang, J., Wang, T.: Who will retweet me?: finding retweeters in Twitter, pp. 869–872 (2013)
6. Suh, B., Hong, L., Pirolli, P., Chi, E.H.: Want to be retweeted? Large scale analytics on factors impacting retweet in Twitter network. In: IEEE Second International Conference on Social Computing, pp. 177–184. IEEE (2010)
7. Yang, Z., Guo, J., Cai, K., Tang, J., Li, J., Zhang, L., et al.: Understanding retweeting behaviors in social networks. In: ACM International Conference on Information and Knowledge Management, pp. 1633–1636. ACM (2010)
8. Gomez-Rodriguez, M., Leskovec, J., Krause, A.: Inferring networks of diffusion and influence. ACM Knowl. Discov. Data Min. **5**, 1019–1028 (2011)
9. Lerman, K., Ghosh, R.: Information contagion: an empirical study of the spread of news on Digg and Twitter social networks. Comput. Sci. **52**, 166–176 (2010)
10. Harrigan, N., Achananuparp, P., Lim, E.P.: Influentials, novelty, and social contagion: the viral power of average friends, close communities, and old news. Soc. Netw. **34**(4), 470–480 (2012)
11. Starbird, K., Palen, L.: (How) Will the revolution be retweeted?: information diffusion and the 2011 egyptian uprising, pp. 7–16. DBLP (2012)
12. Lee, M., Kim, H., Kim, O.: Why do people retweet a Tweet?: altruistic, egoistic, and reciprocity motivations for retweeting. Psychologia **58**(4), 189–201 (2017)
13. Kwak, H., Lee, C., Park, H., Moon, S.B.: What is Twitter, a social network or news media?. In: International Conference on World Wide Web, pp. 591–600 (2010)
14. Borgatti, S.P., Mehra, A., Brass, D.J., Labianca, G.: Network analysis in the social sciences. Sciences **323**(5916), 892 (2009)
15. Cross, R., Cummings, J.N.: Tie and network correlates of individual performance in knowledge-intensive work. Acad. Manag. J. **47**(6), 928–937 (2004)
16. Soda, G., Usai, A., Zaheer, A.: Network memory: the influence of past and current networks on performance. Acad. Manag. J. **47**(6), 893–906 (2004)
17. Nerkar, A., Paruchuri, S.: Evolution of R & D capabilities: the role of knowledge networks within a firm. Manag. Sci. **51**(5), 771–785 (2005)
18. Sparrowe, R.T., Liden, R.C., Wayne, S.J.: Social networks and the performance of individuals and groups. Acad. Manag. J. **44**(2), 316–325 (2001)
19. Iacobucci, D., Hopkins, N.: Modeling dyadic interactions and networks in marketing. J. Mark. Res. **29**(1), 5–17 (1992)
20. Rindfleisch, A., Moorman, C.: The acquisition and utilization of information in new product alliances: a strength-of-ties perspective. J. Mark. **65**(2), 1–18 (2001)
21. Antia, K.D., Frazier, G.L.: The severity of contract enforcement in interfirm channel relationships. J. Mark. **65**(4), 67–81 (2001)
22. Janssen, M.A., Jager, W.: Simulating market dynamics: interactions between consumer psychology and social networks. Artif. Life. **9**(4), 343–356 (2003)
23. Kozinets, R.V., De Valck, K., Wojnicki, A.C., Wilner, S.J.: Networked narratives: understanding word-of-mouth marketing in online communities. J. Mark. **74**(2), 71–89 (2010)
24. Stringer, C.: Modern human origins: progress and prospects. Philos. Trans. R. Soc. Lond. B Biol. Sci. **357**(1420), 563–579 (2002)
25. McKenna, K.Y., Bargh, J.A.: Causes and consequences of social interaction on the internet: a conceptual framework. Media Psychol. **1**(3), 249–269 (1999)
26. Wellman, B., Gulia, M.: Virtual communities as communities. Commun. Cybersp. 167–194 (1999)

27. Stephen, A.T., Zubcsek, P.P., Goldenberg, J.: Lower connectivity is better: the effects of network structure on customer innovativeness in interdependent ideation tasks. Soc. Sci. Electron. Publ. **53**(2), 263–279 (2015). 150619065151001
28. Sohn, D.: Disentangling the effects of social network density on Electronic Word-of-Mouth (eWOM) intention. J. Comput.-Med. Commun. **14**(2), 352–367 (2009)
29. Venkatesh, V., Morris, M.G., Davis, G.B., Davis, F.D.: User acceptance of information technology: toward a unified view. MIS Q. **27**(3), 425–478 (2003)
30. Berger, J., Milkman, K.L.: What makes online content viral? J. Mark. Res. **49**(8), 192–205 (2009)
31. Davis, F.D., Bagozzi, R.P., Warshaw, P.R.: User acceptance of computer technology: a comparison of two theoretical models. Manag. Sci. **35**(8), 982–1003 (1989)
32. Workman, M.: New media and the changing face of information technology use: the importance of task pursuit, social influence, and experience. Comput. Hum. Behav. **31**(1), 111–117 (2014)
33. Ahn, D.Y., Duan, J.A., Mela, C.F.: Managing user-generated content: a dynamic rational expectations equilibrium approach. Mark. Sci. **35**(2), 284–303 (2016)
34. Marett, K., Joshi, K.D.: The decision to share information and rumors: examining the role of motivation in an online discussion forum. Nucl. Phys. A **24**(1), 47–68 (2009)
35. Cha, M., Haddadi, H., Benevenuto, F., Gummadi, P.K.: Measuring user influence in Twitter: the million follower fallacy. In: International Conference on Weblogs and Social Media, ICWSM 2010, Washington, DC, USA, May, vol. 14. DBLP (2010)
36. Wu, F., Huberman, B.A.: Novelty and collective attention. Proc. Natl. Acad. Sci. USA **104** (45), 17599–17601 (2007)

Online Social Media Addictive Behavior: Case Study of Thai Military Officers

Siriporn Poolsuwan[✉]

Information Sciences Department, Suan Sunandha Rajabhat University,
Bangkok, Thailand
siriporn.po@ssru.ac.th

Abstract. This research aims to examine the objectives in social media usage and social media usage behavior of Thai military officers. Data was gathered from 95 commissioned officers and 190 non-commissioned officers who are working at 21st Military Circle, by using questionnaire. Percentage, Mean, standard deviation and T-Test were applied for data analysis. The findings showed that most of the respondents access social media at home. The social media mostly used were Facebook, YouTube, Line, and Instagram. The frequency of usage social media's day were 3–4 days/week, while working between 08.01 A.M.–11.00 A.M., and spending more than 3 h a day was the most usage time. The devices which they used to access the social media were smartphones, PC/Notebook, tablets, and Ipad. Their main purpose of social media usage were for knowledge and self improvement. The comparison of social media usage between commissioned officers and non-commissioned officers was not statistically significant different at .05.

Keywords: Internet addiction · Social media usage
Online media usage behavior

1 Introduction

With the development of computer and communication technology, the new innovation as an internet was emerged a large, which useful knowledge source. Internet is not only in forms of text but also sound, image, news, movie and as information sources include an entertainment. At the present time, it plays an important role in education, routine life, working of people in all over the world. In December 2017, the reports of United Nation [1] revealed that there were 7,550,262,000 people, among these people, 4,050,247,583 people were the internet users. While Asian people were 4,504,428,000 people or 55% of the world's population, 2,023,630,194 people or 48.7% were the internet users [2]. Furthermore, there were 66,188,503 people in Thailand and 57,000,000 of them were the internet users [3]. Mobile phone or smartphone was a tool to access internet. The new digital in 2017 global overview report from We Are Social and Hootsuite [4] revealing that more than half of the world's population now uses the internet and more than half the world now uses a smartphone. Furthermore, they were using mobile phone or smart phone as a tool to access the new media.

© Springer International Publishing AG, part of Springer Nature 2019
T. Z. Ahram (Ed.): AHFE 2018, AISC 787, pp. 428–438, 2019.
https://doi.org/10.1007/978-3-319-94229-2_42

The new emerging media which play the crucial role today and influence a society to change rapidly are the important communication channel, a part of communication process, news media, news notification, opinion exchange, and presentation is social media. Combined with a computer technology advancement and internet network, the social network has emerged as a new innovation. Consequently, a person can utilize a social network via Social Networking Sites (SNSs). SNSs are virtual communities where users can create individual public profiles, interact with real-life friends, and meet other people based on shared interests [5]. SNSs enables multiple responses in social media, creates an interaction between people via internet by communications and reactions, etc. Types of social media in social networking sites are i.e. web blog, data/knowledge (i.e. Wikipedia, Google Earth), online games, community (such as Facebook), photo management (such as Instagram, Pinterest), media (such as You-Tube), business/commerce (such as Amazon, eBay, Lazada). Consumption behavior in media of 67 million [6] of Thai people have been changed by the expanded growth of internet and 4G communication network from mainly consuming television and radio media to increasingly consuming other media via other devices. Owing to one of an important aspect of online social media, people can use it anytime and anywhere resulting in news perception or in real time [7] and it can be important channels for performing in-depth news broadcasting. People can use it to communicate with each other without a medium, the mass media. Contents in the online social media consist of knowledge, consciousness and social opinions or sharing of happiness and entertainments, thus, supporting the media to be more widely acknowledged and have a clear tendency such online society or social media will be the people's mainstreamed media in the future having a great influence toward ideas, attitudes and way of life. Therefore, this study, an online social media addictive behavior attached to 21st Military Circle, should be conducted.

2 Literature Reviewed

Social media is forms of electronic communication (Social Networking Sites (SNSs) such as websites for social networking and microblogging) through which users create online communities to share information, ideas, personal messages, and other content (such as videos) [8]. The social media is not always an online distraction or procrastination platform. While some may be addicted to their social media networks, it is one of the best ways to stay informed. Major news outlets, corporations and persons of interest use social media to deliver messages to the masses. With items posting immediately, the public stays informed. Some issues cause controversy, but social media does more good than harm in retrospect. [9] and Social Networking Sites (SNSs) as web-based services.

SNSs are virtual communities where users can create individual public profiles, interact with real-life friends, and meet other people based on shared interests [10]. However, on the other hand it has also affected the society in the negative way. The addictive part of the social media is very bad and can disturb personal lives as well.

A lot of people are the most affected by the addiction of the social media. They get involved very extensively and are eventually cut off from the society. It can also waste individual time that could have been utilized by productive tasks and activities. These is some of the advantages and disadvantages of social media. However, these is the enough advantages and disadvantages to decide which way to go on the social media. As for popularity of using social networking sites in the report of Top 15 Most Popular Social Networking Sites and Apps in February 2018 consist of Facebook, YouTube, Instagram, Twitter, Reddit, Pinterest, Vine (In January 2017, The Vine became the Vine Camera), Ask.fm. Tumblr, Flickr, Google+, LinkedIn, VK, Classmates, Meetup [11].

A survey concerning internet user's behavior in Thailand has been conducted through several websites and online social media from middle of June to end of July, 2017. Subsequently, 25,101 have answered the survey questionnaire and it showed that in regard to internet users, they spent their times with internet in holidays averagely more than working day/school day in minor level: 6 h and 48 min per day for holidays and 6 h and 30 min for working day/school day, if compared, they were higher than 2015: 6 h 24 h per day. Places for using internet were mostly residences (85.6%), offices (52.4%) and using internet while traveling i.e. on a train, at bus-stop, etc. (24.1%) in opposition to the third rank in 2016: educational institutes. Activities in internet usage according to a 5-year survey showed that the top 5 ranks of activities were i.e. receiving-sending email, data search, social media, reading e-book and watching online TV/listening to music; however, they have been shifted from time to time. Nevertheless, with respect to the types of current Thai social media, it was found that Youtube was in the 1st rank (97.1%), 2nd one was Facebook (96.6%) and Line was come as 3rd rank (95.8%); nevertheless, the rankings have been changed. Usages of social media whether Line, Facebook, Instagram and Youtube, etc. were for communicating, watching online movies, lives and calling via application (86.9%), secondarily, data search via internet (86.5), receiving-sending email (70.5%), watching online TV/listening to online music (60.7%) and buying goods/services via online (50.8), respectively. Difficulties in internet usages that users encountered in activities were i.e. annoying advertisements in great numbers in activities such as listening to music, watching clip video or watching recorded dramas, etc. (66.6%), secondarily a delayed connection/slow internet speed (63.1%), difficulties in connecting to the internet/frequent disconnection (43.7%), facing problems when using internet for activities but being unaware which personnel can provide assistance (39.6%) and annoying junk e-mail (34.2%), respectively [12]. SNS use is also driven by a number of other motivations. From a uses and gratifications perspective, these include information seeking (i.e., searching for specific information using SNS), identity formation (i.e., as a means of presenting oneself online, often more favorably than offline) and entertainment (i.e., for the purpose of experiencing fun and pleasure) In addition to this, there are the motivations such as voyeurism and cyberstalking that could have potentially detrimental impacts on individuals' health and wellbeing as well as their relationships [13].

3 Methodology

This survey research is a case study of Thai military officers in 21st Military Circle, Thailand. The research aims to examine objectives in online media usage and online media usage behavior of Thai military officers. Data was gathered by using questionnaire from 95 commissioned officers and 190 non-commissioned officers in 21st Military Circle, Thailand. The questionnaire utilized in the study consisted of checklist, Likert scales and comprised four main parts. The first part of questionnaire included the questions of demographic concerning gender, age, education, rank, and income. The second part included the respondents' social media usage behavior, such as the objectives in online social media usage, types of social media, content of social media, and the access sources. The third part concerned with the problems in social media addiction. And the recommendation of the respondents is the last part. Data collection was done by Multi-Stage Sampling. The duration of the study was from November 1, 2017 to Dec 1, 2017. Percentage, Mean, standard deviation and T-Test were applied for data analysis. The descriptive statistics was used to assess online social media addictive behavior. It is shown as below:

4 Results and Conclusion

Table 1. Profile of the Thai military officers in 21st Military Circle

Description	n (person)	Percentage
1. Age		
Below 20 years	23	8.07
20–25	105	36.84
26–30	38	13.33
31–35	32	11.23
36–40	15	5.26
41–45	14	4.91
46–50	36	12.63
Over 56 years	22	7.72
Total	285	100
2. Marital status		
Single	156	54.74
Divorced/widowed	24	8.42
Married and cohabited	101	35.44
Married-separated	4	1.4
Total	285	100

(*continued*)

Table 1. (*continued*)

Description	n (person)	Percentage
3. Education degree		
Below upper-secondary education	20	7.02
Upper-secondary education	98	34.39
Below Bachelor's degree	35	12.28
Bachelor's degree	129	45.26
Master's degree	3	1.05
Doctor's degree	0	0
Total	285	100
4. Current position(s)		
Command	65	22.81
Directorate	101	35.44
Conscription	40	14.04
Reserved officers' training	20	7.02
Musical band	7	2.46
Military police	45	15.79
Detention	7	2.46
Total	285	100
5. Experience		
Below 1 year	61	21.4
1–10 year(s)	137	48.07
11–20 years	39	13.68
21–30 years	22	7.72
Over 31 years	26	9.12
Total	285	100
6. Income(s)/month		
Below 15,000 Baht	130	45.61
15,000–35,000 Baht	135	47.37
35,001–50,000 Baht	15	5.26
50,001–65,000 Baht	3	1.05
Over 65,000 Baht	2	0.7
Total	285	100

4.1   Profile of the Respondents

All respondents were male (85.26%) and female (14.74%). Most of them were single
(54.74%), 20–25 years old (36.84%), Bachelor's degree graduated (45.26%), work
experience 1–10 year (48.07%) and their rank were Lance Corporal (25.96%). They
have a monthly income of 15,000–35,000 Thai baht (Table 1).

Table 2. Experience of respondents on social media usage

Type(s) of social media usage	n (person)	Percentage
1. Experience in online social media usage		
Used to	259	90.9
Not used to	26	9.1
Total	285	100.0
2. Current experience in online social media usage		
Below 1 year	27	9.47
1–2 year(s)	34	11.93
3–4 years	42	14.74
5–6 years	56	19.65
Over 6 years	100	35.09
Total	259	100.00
3. Frequent usage time(s)		
Morning (04.01–08.00 h)	48	18.53
Late in the morning (08.01–11.00 h)	54	20.85
Noon (11.01–13.00 h)	30	11.58
Afternoon (13.01–16.00 h)	10	3.86
Evening (16.01–19.00 h)	47	18.15
Night time (19.01–21.00 h)	42	16.22
Late at night time (21.01–24.00 h)	27	10.42
After midnight (00.01–04.00 h)	1	0.39
Total	259	100.00
4. Usage duration(s) of online social media		
Over 3 h	95	36.7
2.30–3 h	18	6.9
2 h–2.30 h	2	0.8
1.30 h–2 h	27	10.4
1 h–1.30 h	24	9.3
30 min–1 h	65	25.1
Below 30 min	28	10.8
Totals	259	100.0
5. Place(s) for online social media usages		
Residence	177	68.34
Office	58	22.39
Learning center	2	0.77
Department store	4	1.54
Internet café	2	0.77
Bistro/restaurant	4	1.54
On vehicle	12	4.63
Total	259	100.00

(*continued*)

Table 2. (*continued*)

Type(s) of social media usage	n (person)	Percentage
6. Device for accessing online social media		
Smartphone	210	81.08
PC/notebook	45	17.37
IPad	1	0.39
Tablet	3	1.16
Total	259	100.00

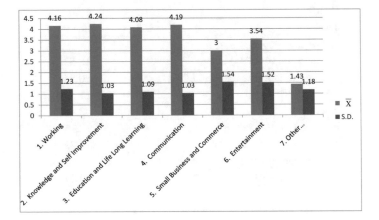

Fig. 1. Objectives of social media usage of Thai military officers

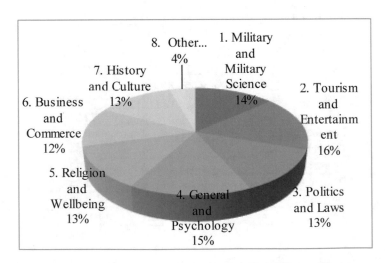

Fig. 2. Content of social media usage of Thai military officers

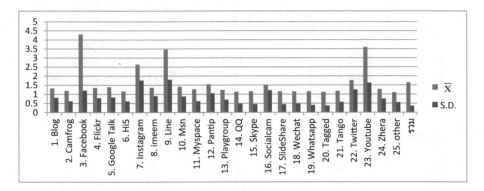

Fig. 3. Accessible social network sites of Thai military officers usage

4.2 Social Media Addictive Behavior of the Respondents

According to the study, the results of social media addictive behavior of the respondents were the devices they used to access the social media were smartphones (73.68%), PC/Notebook (15.79%), tablets (1.05%) and Ipad (0.35%). The frequency of usage online media's day were 3–4 days/week (90.24%) with a using time while working between 08.01 A.M.–11.00 A.M. (20.85%) and spending more than 3 h a day (33.33%) was the most usage time. Most of the respondents (62.10%) access online media at home (Table 2). From Fig. 1, the purpose of online media usage of respondents in overview was in moderate level ($\bar{x} = 2.80$, S.D. $= 0.73$), as in each aspect found that the purpose of online media usages were for knowledge and self-improvement ($\bar{x} = 4.24$, S.D. $= 1.03$), communication ($\bar{x} = 4.19$, S.D. $= 1.03$), working ($\bar{x} = 4.16$, S.D. $= 1.23$), and education ($\bar{x} = 4.08$, S.D. $= 1.09$), respectively. The online media mostly used by the respondents (as shown in Fig. 3) were Facebook ($\bar{x} = 4.30$, S $= 1.20$), YouTube ($\bar{x} = 3.59$, S $= 1.63$), Line ($\bar{x} = 3.46$, S $= 1.80$), and Instagram ($\bar{x} = 2.64$, S.D. $= 1.76$). From Fig. 2, most of the content of social media usage of Thai military officers were tourism and entertainments (16%), general psychologies (15%) and military science (14%).

Table 3. The comparison of social media addictive behavior between commissioned officers and non-commissioned officers, Thai military officers, 21st Military Circle

Rank	N	Mean	S.D.	t	Sig.
Commissioned officers	80	2.75	0.56	−0.88	0.39[*]
Non-commissioned officers	179	2.82	0.62		
Total	259	5.57	1.18		

[*]$p < .05$

4.3   The Comparative Study of Social Media Addictive Behavior of Thai Military Officers

The result of the comparison of social media addictive behavior between commissioned officers and non-commissioned officers, Thai military officers, 21st Military Circle, was found that no statistically significant different at .05.

5   Discussion

Online social media addictive behavior: case study of Thai military officers attached to 21st military circle signified that the most accessible devices in the sample group (n) were smartphone, PC/Notebook, tablet and Ipad in similarity with "Hootsuite & We are social" [4] having reported internet user behavior around the world in 2018, February, that smartphone was used as a tool for accessing social media for 91% compared to 71% of Thai people, 25% with laptop and 12% using desktop computer or PC with a daily frequency and in conformity with the research conducted by EIC [15] with the results: Thai people spent their leisure times with the internet or consuming online social media as it was easier for current consumers to be able to choose what they want to consume on social media via smartphone and tablet; furthermore, a high competitive service in mobile phones combined with various promotions leaded to an important change in the mobile business sector and towards consumers in Thailand in combination with a high speed internet technology and high diversity of business in media from local to international service providers. With respect to usage durations, it was found that the sample groups frequently used online social media at late in the morning (08.01–11.00 h) and mostly used at residences in similarity with the survey of Hootsuite & We are social [4] as they found that the most frequently used place for accessing internet for consuming online social media were residences (85.6%) while the secondarily used places were offices (52.4%) and 24% for using while travelling. Taking in to consideration, it concludes that the military officers attaching to 21st Military Circle have been provided with government's accommodation welfare and such accommodation was not far from their units combined with their duty shifts starting from 8.30 A.M.–16.30 P.M. [14]. Therefore, they can consume online social media at their residence before reporting for the duty shift for over 3 h in each time in relation to the report by Electronic Transactions Development Agency (Public Organization) or ETDA [12] with survey details: the internet users spent their time in holidays averagely more than working days/school days in minor degree: 6 h and 48 min per day for holidays and 6 h and 30 min for working day/school day, if compared, they were higher than 2015: 6 h 24 h per day. Places for using internet were mostly residences (85.6%).

The purposes of military officers mostly use online social media for gaining more knowledge and self-development, communication and working or for performing on their duty. However, according to the study conducted by ETDA [12], it was concluded that most weekly activities for consuming online social media of Thai people were Line, Facebook, Instagram, Youtube, etc. for various activities i.e. communicating, watching online movie, lives and calling via applications, secondarily data search via internet (86.5%), receiving-sending email (70.5%), watching TV/listening to online music (60.7%) and buying online goods/services (50.8%).

As for information contents frequently consumed via online social media, it was found that tourisms, entertainments, general psychologies, military, histories and cultures; politics and laws were the most frequently used ones in similarity with the research conducted by Economic Intelligence Center or EIC [15]. Most Thai people (more than 70%) were mainly interest in a tourism and tended to increase along with their ages; moreover, compared to the past 5 years, 55% of consumers have traveled more frequently, signifying the lifestyles emphasizing travel and gaining new experience.

In the aspect of sources for accessing online social media, This study was found that Facebook, Youtube, Line and Instagram were the most frequently used ones in conformity with Hootsuite & We are Social [4] which found that the most popular online social media were Facebook (75%), Youtube (72%), Line (68%) and Instagram (50%) in opposition with ETDA [12] as they suggested that the most frequently used online social media are Youtube (97.7%), Facebook (96.6%) and Line (95.8%).

Acknowledgments. This study was successful by financial support from Suan Sunandha Rajabhat University and many helps from my colleagues.

References

1. United Nations: Department of Economic and Social Affairs, Population Division: World Population (2017). https://esa.un.org/unpd/wpp/Publications/Files/WPP2017_Wallchart.pdf. Accessed 21 Feb 2018
2. Internet World Stats Usage and Population Statistics: the Internet Big Picture (2018). https://www.internetworldstats.com/stats.htm. Accessed 21 Feb 2018
3. Internet World Stats Usage and Population Statistics: Internet Usage in Asia Internet Users, Facebook Subscribers & Population Statistics for 35 countries and regions in Asia (2018). https://www.internetworldstats.com/stats3.htm. Accessed 21 Feb 2018
4. Kemp, S.: Digital in 2018: World's Internet Users Pass the 4 Billion Mark, 30 January 2018. https://wearesocial.com/blog/2018/01/global-digital-report-2018. Accessed 21 Feb 2018
5. Griffiths, M.D., Kuss, D.J., Demetrovics, Z.: Social networking addiction: an overview of preliminary findings. In: Behavioral Addictions: Criteria, Evidence, and Treatment, pp. 119–141. Elsevier Inc. (2014). https://doi.org/10.1016/b978-0-12-407724-9.00006-9
6. Mahidol University, Institute for Population and Social Research: Mahidol Population Gazette, vol. 27, pp. 1–2 (2017)
7. Ansongkhram, A.: The Effect of Social Media on Working Citizens in Bangkok and the Surrounding Areas. School of Communication Arts, Sripatum University, Khonkaen (2015)
8. Definition of social media. In: Merriam-Webster Dictionaries (2018). https://www.merriam-webster.com/dictionary/social%20media. Accessed 21 Feb 2018
9. Agrawal, A.J.: It's Not All Bad: The Social Good Of Social Media (2018). https://www.forbes.com/sites/ajagrawal/2016/03/18/its-not-all-bad-the-social-good-of-social-media/#447e087756fb. Accessed 21 Feb 2018
10. Kuss, D.J., Griffiths, M.D.: Online social networking and addiction: a review of the psychological literature. Int. J. Environ. Res. Public Health **8**, 3528–3552 (2011). https://doi.org/10.3390/ijerph8093528

11. Kallas, P.: Top 15 Most Popular Social Networking Sites and Apps, February 2018. https://www.dreamgrow.com/top-15-most-popular-social-networking-sites/. Accessed 21 Feb 2018
12. Electronic Transactions Development Agency (Public Organization): Thailand Internet User Profile 2017. Office of Strategy, Bangkok (2017)
13. Kuss, D.J., Griffiths, M.D.: Social networking sites and addiction: ten lessons learned. Int. J. Environ. Res. Public Health **14**(3), 311 (2017). https://doi.org/10.3390/ijerph14030311
14. The Prime Minister's Office: Announcement of the Office of the Prime Minister on Working Time and Public Holiday, No. 12 (1959)
15. Economic Intelligence Center: Thai Lifestyle in Thailand 4.0 (2017). https://www.smartsme.co.th/content/70344. Accessed 5 Jan 2018. Volume Editor should be left blank

Human Factors in Energy Systems: Nuclear Industry

A Guide for Selecting Appropriate Human Factors Methods and Measures in Control Room Modernization Efforts in Nuclear Power Plants

Casey Kovesdi[✉], Jeffrey Joe, and Ronald Boring

Idaho National Laboratory, Idaho Falls, ID, USA
{Casey.Kovesdi,Jeffrey.Joe,Ronald.Boring}@inl.gov

Abstract. Many of the U.S. nuclear power plants are approaching the end of their 60-year licensing period. The U.S. Department of Energy Light Water Reactor Sustainability Program is conducting targeted research to extend the lives and ensure long-term reliability, productivity, safety, and security of these plants through targeted research, such as integrating advanced digital instrumentation and control technologies in the main control room. There are many challenges to this, one being the integration of human factors engineering in the design and evaluation of these upgrades. This paper builds upon recent efforts in developing utility-specific guidance for integrating human factors engineering in the control room modernization process by providing commonly used data collection methods that are applicable at various phases of the upgrade process. Advantages and disadvantages of each method are provided for consideration of an optimal human factors evaluation plan to be used throughout the lifespan of the upgrade process.

Keywords: Control room modernization · Human-systems interface evaluation
Human factors engineering · Nuclear power plants

1 Introduction

Nuclear power annually accounts for approximately 20% of the electrical generation in the United States (U.S.), and has done so over the past couple decades. However, many of these U.S. nuclear power plants (NPPs) are now approaching the end of their 60-year licensing period. In order to extend the lives of these NPPs beyond their existing licensing period and to ensure their long-term reliability, productivity, safety, and security, one area that is being explored under the U.S. Department of Energy (DOE) Light Water Reactor Sustainability (LWRS) Program concerns main control room (MCR) modernization. Specifically, the integration of advanced digital instrumentation and control (I&C) technologies is being researched and implemented to ensure safety and enhance productivity and situation awareness of the plant.

These upgrades can vary in scope and level of effort; nonetheless, the modernization process often entails a stepwise approach where individual systems of the plant are upgraded and scheduled around planned outages. The result of such differences in

© Springer International Publishing AG, part of Springer Nature (outside the USA) 2019
T. Z. Ahram (Ed.): AHFE 2018, AISC 787, pp. 441–452, 2019.
https://doi.org/10.1007/978-3-319-94229-2_43

the upgrade process requires grading the level of the human factors engineering (HFE) design and evaluation effort [1]. For utilities, this graded approach is no easy feat, often requiring HFE expertise on hand to sift through the required regulatory documentation and to select the appropriate HFE methods and measures to apply.

Recent work to support the industry has focused on developing a user-centered design process for utilities to use as part of ensuring that HFE is properly integrated into their planned upgrade process. Namely, this process is described as the Guideline for Operational Nuclear Usability and Knowledge Elicitation (GONUKE) [2]. While GONUKE provides a user-centered design process to follow, no specific HFE methods or measures are explicitly suggested to utilities. Hence, there is opportunity to elaborate on the specific HFE methods and measures that are commonly used within each of the phases and types of evaluation described in GONUKE.

The next section describes the existing HFE documentation to support control room modernization: the U.S. Nuclear Regulatory Commission (NRC) HFE Review Model, the Electric Power Research Institute (EPRI) HFE Guidance for Control Room and Digital Human-System Interface (HSI) Design and Modification, the Institute of Electrical and Electronics Engineers (IEEE) Guide for the Evaluation of Human-System Performance in Nuclear Power Generating Stations, and the GONUKE process. This paper concludes with providing and exemplifying commonly used HFE data collection methods and measures used for HFE activities specific to control room modernization.

2   Existing HFE Guidance for Control Room Modernization

2.1   NUREG-0711: Human Factors Engineering Program Review Model

The U.S. NRC NUREG-0711 Rev. 3, *Human Factors Engineering Program Review Model,* supports NRC staff in HFE reviews for new NPP builds, as well as for upgrades of existing NPPs [3]. NUREG-0711 essentially serves as a detailed resource for the NRC staff to verify that state-of-the-art HFE principles have been incorporated throughout each phase of development. These phases are described as *Planning and Analysis, Design, Verification and Validation (V&V),* and *Implementation and Operation.* There are twelve review elements arranged within each of the four phases.

A comprehensive description of these elements is beyond the scope of this paper; the reader should refer to NUREG-0711 for further details. However, it is important to note that each element is interlinked. For example, a modification is assumed to at least in part be driven by lessons learned from operating experience of past events, which drives what functions should and should not be automated. Human actions are then expected to be analyzed in detail through analytical approaches such as task analysis to identify important human actions that have implications to plant safety. This information is then used to drive the design of the HSI(s), procedures, and training. The integration of the HSI(s), procedures, and training are then evaluated during V&V. The validated design is then implemented into the plant, where at this point, human performance is continuously done.

NUREG-0711 emphasizes performance measurement in the selection of appropriate HFE methods and measures, especially in integrated system validation (ISV) during V&V. NUREG-0711 discusses a need for a hierarchical set of performance measures including aspects of plant performance, personnel task performance, situation awareness, cognitive workload, and anthropometric/physiological factors. *Plant performance* concerns measures of performance for various plant functions, systems, or components. *Task performance* concerns measures of time, accuracy, amount accomplished, subjective reports, and behavioral categorization by observers. *Situation awareness* concerns the degree to which personnel's perception of plant parameters and understanding of the plant's condition correspond to its actual condition at any given time and influences of predicting future states. *Workload* comprises the physical, cognitive, and other demands that tasks place on the plant personnel. Finally, *anthropometric/physiological factors* concern the visibility of displays, accessibility of control devices, and ease of manipulating control devices.

2.2 EPRI Human Factors Guidance for Control Room and Digital HSI Design and Modification

EPRI 3002004310, *Human Factors Guidance for Control Room and Digital Human-System Interface Design and Modification*, provides comprehensive guidance for utilities, their suppliers, and contractors to integrate HFE into the overall design efforts [5]. One important piece to EPRI 3002004310 is the emphasis on tailoring and grading the HFE approach based on the scope of the modernization effort. For instance, an HFE program should be appropriately based on the impact the modification may have to the safety of the plant. Modifications with direct impact to plant safety should take greater priority where greater rigor is applied to the design and evaluation process compared to lower-risk modifications. For selecting appropriate HFE methods and measures, it may not be necessary to use costlier data collection for a finer grained analysis when the proposed modification is less impactful on plant safety. However, modifications that impact multiple systems or have direct safety consequences should use more rigorous data collection methods that build higher confidence that no new human error modes were introduced and that state-of-the-art HFE principles are reflected in the design.

2.3 IEEE Guide for the Evaluation of Human-System Performance in Nuclear Power Generating Stations

The IEEE *Guide for the Evaluation of Human-System Performance in Nuclear Power Generating Stations,* IEEE Std 845-1999(R2011) outlines specific evaluation characteristics that should be considered when evaluating human-system performance related to systems, equipment, and facilities in NPPs [6]. Application of IEEE Std 845-1999 (R2011) in NPP MCR modernization activities can range from gathering informal design input such as user opinions to tightly controlled experimental techniques used for answering specific design questions (i.e., hypotheses). IEEE Std 845-1999(R2011) emphasizes the use of a diverse set of measures to the extent that it is cost justifiable, as a single measure may not provide sufficient validity. Lastly, a distinction is made between subjective and objective measures. *Subjective* measures are comprised from data collected from judgement and opinions of users or experts. *Objective* measures are comprised from data collected from observable human behavior. IEEE Std 845-1999 (R2011) describes how objective measures are less apt at being biased from opinion like subjective measures. However, objective measures can be difficult to collect from unobservable human behavior like cognitive processes.

2.4   The GONUKE Process for Control Room Modernization

The intent of the GONUKE process is to provide direct support for utilities in NPP MCR modernization activities that ensures HFE is adequately and consistently integrated throughout the upgrade process [2]. Hence, GONUKE fits within the review model of NUREG-0711 by covering key HFE activities for each of the four phases covered. However, one important distinction with GONUKE is that there is greater emphasis on HFE involvement in the earlier phases to ensure success at later phases.

GONUKE describes three types of evaluation: expert review (i.e., herein described as verification), user testing (i.e., herein described as validation), and knowledge elicitation. The latter two involve data collection methods with plant personnel while the former involves human factors subject matter experts using human factors standards like NUREG-0700 [4] in analyses.

There are four fundamental HFE evaluation phases as described by GONUKE, which correspond to the phases of NUREG-0711 [3]. The first phase, Planning and Analysis, includes baseline evaluation (i.e., for validation) and cognitive walkthrough (i.e., for knowledge elicitation) as data collection activities. Baseline evaluation entails evaluating the as-built system so that it can serve as a benchmark for the new system. Cognitive walkthrough entails collecting plant personnel input to capture needs and expectations of the new system. Next, Design (i.e., described as *formative* evaluation) is done through usability testing and feedback about the HSI design to support refining the HSI design through this collection of early feedback; identification of acceptance criteria can also be considered. These formative activities during Design are most successful when done iteratively [7]. Third, a *summative* evaluation (i.e., V&V) is done through ISV and feedback about operator performance after the design process is complete to confirm the usability and performance of a design. Finally, operator training and operator experience reviews are collected with the built design as part of Implementation and Operation for continuous monitoring of human-system performance.

3   Common HFE Data Collection Methods and Measures

Over the past several years, human factors researchers from the LWRS Program have conducted numerous workshops with various utilities. This work used a range of HFE data collection methods for collecting design input, as well as evaluating human-system performance through operator-in-the-loop studies. These methods were applied in various phases described in NUREG-0711 [3]. There was a range of testing environments, spanning from less controlled platforms such as informal discussions and field visits to controlled studies using full-scope testbeds. A collection of common HFE methods previously used from this past work is presented in Fig. 1, which is an adaptation of Rohrer's *Landscape of User Research Methods* [8].

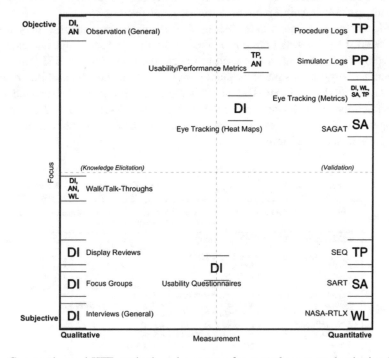

Fig. 1. Commonly used HFE methods and measures for control room modernization in NPP.

The y-axis denotes the focus of evaluation, being the degree to which the method is subjective or objective. As discussed in IEEE Std 845-1999(R2011), objectivity is the degree to which the method comprises data from observed human behavior whereas subjectivity is the degree to which the method comprises data from judgement and opinions of users or experts [6]. IEEE Std 845-1999(R2011) suggests using a diverse set of these measures to strengthen validity of findings.

On the x-axis, the methods lie on the degree of measurement, denoted as being qualitative or quantitative. *Qualitative* measurement provides descriptive information regarding the qualities of a topic [9]. For example, qualitative data may take the form of the description and categorization of various higher-level themes extracted from open-ended comments. This information is important for identifying and correcting potential HSI design issues [10]. Contrarily, *quantitative* measurement provides information regarding the quantity of a topic [9]. Quantitative analyses deal with both discrete and continuous data to make numerical measurement for both descriptive and comparative purposes. One may reasonably assume that quantitative measures are most important in later-staged efforts like ISV; however, quantitative measures still can be useful earlier in the development cycle [7]. To that end, qualitative measures can carry important insights in later phases such as collecting feedback about operator performance.

A third dimension of Fig. 1 pertains to the human performance dimension(s) that can be collected from the HFE method. These dimensions include: *DI* – Design Input, *PP* – Plant Performance, *TP* – Task Performance, *SA* – Situation Awareness, *WL* – Workload, and *AN* – Anthropometric/Physiological Factors.

The latter five correspond to the hierarchical set of performance measures described in NUREG-0711 [3]. *Design input* is defined as the thoughts, feelings, and experiences expressed by the users that can be used to inform the design of the HSI. For example, comments and suggestions made by plant personnel to enhance the usability of the HSI would classify as design input. The methods presented in Fig. 1 are discussed in detail in the subsequent sub-sections, provided with a short description and a list of their advantages and disadvantages. Their applications to GONUKE [2] and to NUREG-0711 [3] are summarized in Fig. 2.

	Planning and Analysis	Design	Verification and Validation	Implementation and Operation
Verification	*Verification incorporates use of various human factors standards and guidelines as opposed to data collection methods from plant personnel.*			
Validation	**Baseline Evaluation** • *Eye Tracking (Heat Maps)* • *Usability Questionnaires* • *Usability/performance Metrics* • *Procedure Logs* • *Simulator Logs* • *Eye Tracking (Metrics)* • *SAGAT* • *SEQ* • *SART* • *NASA-RTLX*	**Usability Testing** • *Eye Tracking (Heat Maps)* • *Usability Questionnaires* • *Usability/performance Metrics* • *Procedure Logs* • *Simulator Logs* • *Eye Tracking (Metrics)* • *SAGAT* • *SEQ* • *SART* • *NASA-RTLX*	**Integrated System Validation (i.e., ISV)** • *Usability/performance Metrics* • *Procedure Logs* • *Simulator Logs* • *Eye Tracking (Metrics)* • *SAGAT* • *SEQ* • *SART* • *NASA-RTLX*	**Operator Training** • *Procedure Logs* • *Simulator Logs* • *SART* • *NASA-RTLX*
Knowledge Elicitation	**Cognitive Walkthrough** • *Observation (General)* • *Walk/talk-throughs* • *Interviews (General)* • *Usability Questionnaires*	**Operator Feedback on Design** • *Walk/talk-throughs* • *Focus Groups* • *Display Reviews* • *Interviews (General)* • *Usability Questionnaires*	**Operator Feedback on Performance** • *Focus Groups* • *Interviews (General)*	**Operator Experience Reviews** • *Observation (General)* • *Focus Groups* • *Interviews (General)*

Fig. 2. HFE methods and measures mapped to GONUKE (*rows*) and NUREG-0711 (*columns*).

3.1   Description of Common HFE Data Collection Methods and Measures

Observation (General). Observation is a general HFE technique used to collect data about the physical and verbal aspects of a task or scenario [11]. It can be used to inform design or to inform task analysis. In this context, observation here is described as passive (i.e., without intervening or interfering with what operators are doing). Various tools of ranging complexity can support in observation such as audio/video recordings, spreadsheets, as well as pen and paper [12]. *Advantages:* Provides rich contextual information, minimally intrusive, does not necessarily require specialized equipment.

Disadvantages: Cannot formally evaluate performance, does not reveal plant personnel's thoughts and rationale for making decision observed or why an observed error occurred.

Walk/Talk-Throughs. A walkthrough is an HFE technique where an expert user 'walks through' or demonstrates a set of tasks or scenario to describe a task, highlight potential issues with the system, or highlight important actions that may be influenced by the upgrade. A talkthrough is a verbal demonstration of a walkthrough [5]. *Advantages:* Provides accurate descriptions of a task, observers can query particular topics in 'real time,' collects behavioral and attitudinal aspects of human interactions. *Disadvantages:* Provides only descriptive data (i.e., cannot be used for validation), requires access to expert users such as plant personnel, requires a degree of effort synthesizing data collected from verbal reports and observations made.

Display Reviews. In a display review, static HSI displays are systematically evaluated with expert users (e.g., plant personnel) to uncover potential design issues [5]. This activity can be completed remotely, at the plant, or at a simulator. Likewise, this activity can be done one-on-one or as a group (i.e., see Interviews versus Focus Groups). *Advantages:* Provides readily actionable design recommendations, incorporates identified issues and suggestions directly from actual users of the system, very flexible to administer. *Disadvantages:* Identified issues and suggestions are not directly tied to observable human-system performance data, issues and suggestions may be biased by past experience with an existing system.

Focus Groups. A focus group is a general HFE technique used to collect qualitative (i.e., attitudinal) data about a specific topic. This activity is done with a group of users (ideally around 5 individuals), where verbal notes are collected from semi-structured questions administered by a facilitator [5]. Data can be collected in digital (e.g., spreadsheet) or pen and paper format. Raw notes are ultimately synthesized to make meaningful insights to the research question at hand. *Advantages:* Data collection is more efficient than with traditional interviews. *Disadvantages:* Collects only attitudinal data (not behavioral), can be strongly susceptible to response bias between individuals compared to interviews.

Interviews (General). Like focus groups, interviewing is a general HFE technique used to collect qualitative (i.e., attitudinal) data about a specific topic. However, these activities are done one-on-one, where verbal notes are recorded from the interviewee [11]. *Advantages:* Less prone to response bias compared to focus groups, can provide rich data when done with observation. *Disadvantages:* Cannot formally evaluate performance, data collection is more labor intensive than focus groups.

Usability/Performance Metrics. Usability/performance metrics, as described in ISO 9241-11 [14], incorporate measurement to the extent of effectiveness, efficiency, and satisfaction with which a specified user achieves specified goals in a particular environment. Typical measures include task success, error frequency, completion time, and usability ratings. It should be noted that there are different formal integrated platforms to which usability/performance metrics are collected. These platforms include the Operator Performance Assessment System (OPAS) [15] and the Supervisory Control and Resilience Evaluation (SCORE) framework [16]. These methodologies incorporate both structured evaluation of operator actions as well as expert evaluation. The reader should refer to the provided references for an in-depth discussion of these methodologies. *Advantages:* Can formally evaluate objective performance, low cost, does not require specialized equipment, data collection and post-processing is not labor intensive. *Disadvantages:* Sensitivity of measures are controversial for HFE studies in NPP MCRs, methods like OPAS and SCORE require considerable upfront effort identifying important human actions.

Eye Tracking (Heat Maps). A heat map is a visualization technique that uses different colors to show the amount of fixations or dwell durations over an area (i.e., such as a control board) [13]. These visualizations illustrate where users looked and for how long. *Advantages:* Illustrates through objective data where users looked and for how long. *Disadvantages:* Cannot formally evaluate performance, requires specialized eye tracking equipment, data collection and post-processing is labor intensive, does not explain why users looked at a particular place, does not provide information about visual scanning patterns or sequence of eye movements.

Usability Questionnaires. A usability questionnaire provides a set of structured questions that address different design elements of usability through allowing users to either rate or comment about each design element. Information collected from the questionnaire can be used to support identifying potential design issues and areas for improvement with the HSI. Questions can be created from content in NUREG-0700 [4], EPRI 3002004310 [5], or other relevant resources (e.g., operating experience). *Advantages:* Provides both qualitative and quantitative data, information collected can be adaptable depending on project needs, can be used to compare different designs, easy to develop and administer, low cost. *Disadvantages:* Provides only subjective data from user opinions, custom questions need to be pilot tested to ensure they address the usability topic at hand, no normative benchmark scores to compare and validate are available since questions are customized each time.

Procedure Logs. Procedure logs are a form of observation done in a controlled setting/lab, where the observer collects step-by-step actions from plant personnel. Each step is then time-stamped. For design and evaluation, the data collected here can be used to identify correct path as well as collect completion times per step. *Advantages:* Provides objective performance data of task performance, low cost, does not require specialized equipment. *Disadvantages:* Requires formal acceptance criteria for evaluation or an alternative design to compare to, data collection can be labor intensive.

Simulator Logs. Simulator logs capture the values of key parameters through the lifetime of a scenario. The data captured here is fully applicable to observing plant performance. *Advantages:* Provides objective data of plant performance. *Disadvantages:* Requires a testbed that is capable of recording and readily exporting key process values.

Eye Tracking (Metrics). Eye tracking is a general methodology used in HFE to capture measures of visual attention, mental workload, and situation awareness [17]. As such, physiological data can be collected to make inferences about a certain HSI design. *Advantages:* Captures a rich set of performance measures which are continuously tracked throughout the course of a scenario. *Disadvantages:* Requires specialized eye tracking equipment, data collection and post-processing is labor intensive, requires additional level of inference when associating measures to HFE constructs (e.g., workload), not all testing environments or individuals are capable of using the equipment (e.g., issues range from lighting considerations to individual differences [17]).

Situation Awareness Global Assessment Technique (SAGAT). SAGAT is a freeze-probe/recall HFE method used to assess the three levels of situation awareness. SAGAT provides an objective means to evaluate situation awareness where participants' responses to key queries are compared to what actually happened during the scenario. Questions are crafted to address Level 1, Level 2, and Level 3 situation awareness [18]. *Advantages:* Provides objective data of situation awareness. *Disadvantages:* Freeze-probes used can be highly intrusive, requires testing environment capability to freeze and blank out indication status, requires considerable upfront effort developing meaningful queries to administer during each probe.

Single Ease Question (SEQ). The SEQ is a standardized single post-scenario usability question that asks users to rate (i.e., typically using a seven-point scale; 1 = very difficult; 7 = very easy) their overall ease of completing a task [7]. The SEQ can be administered digitally, via pen and paper, or verbally after completion of a task. *Advantages:* Very easy to administer, low cost, provides quantitative data of perceived task difficulty. *Disadvantages:* Not diagnostic regarding identification of usability issues when used alone, may not necessarily correlate with observed task performance.

Situation Awareness Rating Technique (SART). The SART is a post-scenario rating method used to derive a measure of perceived situation awareness [11]. The questionnaire comprises 10 dimensions (i.e., questions) using a seven-point rating scale (1 = low; 7 = high). A composite situation awareness score is derived from SART where a greater value denotes greater situation awareness. A simpler version (i.e., the 3D-SART) is also available, comprising of only 3 questions. *Advantages:* Easy to administer, low cost, not intrusive. *Disadvantages:* Provides subjective (perceived) data of situation awareness, may be more representative of workload than situation awareness.

National Aeronautics and Space Administration Raw Task Load Index (NASA-RTLX). The NASA-RTLX is a post-scenario rating method to assess workload, comprising six different dimensions [11]. Each dimension (i.e., question) typically uses a 20-point scale (1 = low; 20 = high) where higher values denote greater workload. The NASA-RTLX is a shortened version of the NASA-TLX where the set of 15 pairwise comparison is omitted. Workload can be evaluated by each dimension and holistically from aggregating the individual scales. *Advantages:* Easy to administer, low cost, not intrusive, there is a rich database of benchmark values across various industries to compare to [19]. *Disadvantages:* Provides a subjective (perceived) assessment of workload, responses can be confounded with task performance.

3.2   Application of HFE Data Collection Methods and Measures

Researchers from the LWRS Program partnered with a large utility as part of integrating HFE into the control room upgrades for several non-safety related systems. The upgrade significantly affected I&C behind the board. Moreover, there was also a number of indications and controls from control boards of the MCR that were affected, including several new digital HSIs and single loop interface module (SLIM) replacements for manual-auto stations. In this partnership, LWRS human factors researchers were able to successfully integrate common HFE methods within the utility's major engineering project milestones such as their software-in-the-loop (SWIL) testing, factory acceptance testing (FAT), and licensing operator requalification training (LORT). For instance, researchers iteratively incorporated lightweight methods like usability questionnaires, interviews, and display reviews to collect design feedback and knowledge elicitation with licensed operators between planned activities for each milestone.

Data collected from each milestone was used to inform the HSI design to which updated versions were used in a full-scale HFE workshop to evaluate the upgrades in an operational context. This workshop used formal objective methods such as eye tracking metrics and usability/performance metrics, as well as subjective methods such as usability questionnaires, SEQ, SART, and NASA-RTLX. Data collected was evaluated against industry accepted criteria or plant expected criteria for usability testing. Results from the study showed significant improvements to the design of the HSIs. No safety-critical issues were identified.

4   Conclusions

This paper summarizes commonly used HFE methods and measures for MCR modernization in NPPs. As addressed in the paper, there are certain advantages and disadvantages of each method that should be accounted for when planning HFE integration into control room modernization efforts. A recommended approach is to use a diverse set of methods to collect human factors data across all quadrants described in Fig. 1. By including a balanced and diverse set of these methods and measures, this paper offers a complete, but not exhaustive, set of methods that can be used for design input and validation of human-system performance as described in NUREG-0711 [3]. Nonetheless, future work should investigate alternative HFE methods that may serve useful.

Acknowledgments. INL is a multi-program laboratory operated by Battelle Energy Alliance LLC, for the United States Department of Energy under Contract DE-AC07-05ID14517. This work of authorship was prepared as an account of work sponsored by an agency of the United States Government. Neither the United States Government, nor any agency thereof, nor any of their employees makes any warranty, express or implied, or assumes any legal liability or responsibility for the accuracy, completeness, or usefulness of any information, apparatus, product, or process disclosed, or represents that its use would not infringe privately-owned rights. The United States Government retains, and the publisher, by accepting the article for publication, acknowledges that the United States Government retains a nonexclusive, paid-up, irrevocable, world-wide license to publish or reproduce the published form of this manuscript, or allow others to do so, for United States Government purposes. The views and opinions of authors expressed herein do not necessarily state or reflect those of the United States government or any agency thereof. The INL issued document number for this paper is: INL/CON-17-44009.

References

1. Joe, J.C., Boring, R.L., Persensky, J.J.: Commercial utility perspectives on nuclear power plant control room modernization. In: Proceedings of the 2012 American Nuclear Society Nuclear Power Instrumentation and Control and Human Machine Interface Technology Conference, pp. 2039–2046 (2012, in press)
2. Boring, R.L., Ulrich, T.A., Joe, J.C., Lew, R.T.: Guideline for operational nuclear usability and knowledge elicitation (GONUKE). Procedia Manuf. **3**, 1327–1334 (2015)
3. U.S. Nuclear Regulatory Commission. Human Factors Engineering Program Review Model, NUREG-0711, Rev. 3 (2012)
4. U.S. Nuclear Regulatory Commission. Human-System Interface Design Review Guidelines, NUREG-0700, Rev. 2 (2002)
5. Electric Power Research Institute. Human Factors Guidance for Control Room and Digital Human-System Interface Design and Modification: Guidelines for Planning, Specification, Design, Licensing, Implementation, Training, Operation, and Maintenance for Operating Plants and New Builds. EPRI, Palo Alto, CA. 3002004310 (2015)
6. Institute of Electrical and Electronics Engineers. IEEE Guide to the Evaluation of Human-System Performance in Nuclear Power Generating Stations (IEEE Std. 845-1999). Institute of Electrical and Electronics Engineers, New York (1999)
7. Sauro, J., Lewis, J.R.: Quantifying the User Experience: Practical Statistics for User Research. Morgan Kaufmann, Burlington (2016)
8. Rohrer, C.: When to use which user experience research methods. Jakob Nielsen's Alertbox (2014). https://www.nngroup.com/articles/which-ux-research-methods/. Accessed 24 Jan 2018
9. Kuniavsky, M.: Observing the User Experience: A Practitioner's Guide to User Research. Morgan Kaufmann, Burlington (2003)
10. Kovesdi, C.R., Joe, J.C.: A novel tool for improving the data collection process during control room modernization human-system interface testing and evaluation activities. In: 10th International Topical Meeting on Nuclear Plant Instrumentation, Control, and Human Machine Interface Technologies, pp. 1261–1271 (2017)
11. Stanton, N., Salmon, P.M., Rafferty, L.A.: Human Factors Methods: A Practical Guide for Engineering and Design. Ashgate Publishing Ltd., Farnham (2013)
12. Kirwan, B., Ainsworth, L.K. (eds.): A Guide to Task Analysis: The Task Analysis Working Group. CRC Press, Boca Raton (1992)
13. Bergstrom, J.R., Schall, A. (eds.): Eye Tracking in User Experience Design. Elsevier, New York (2014)
14. ISO/IEC. 9241-11 Ergonomic Requirements for Office Work with Visual Display Terminals (VDT)s- Part II Guidance on Usability. ISO/IEC 9241-11,1998 (E) (2014)
15. Skraaning, G.: Experimental Control Versus Realism: Methodological Solutions for Simulator Studies in Complex Operating Environments, HWR-361, Halden Reactor Project, Halden, Norway (2004)

16. Braarud, P.Ø., Eitrheim, M.H.R., Fernandes, A.: SCORE – an integrated performance measure for control room validation. In: 9th International Topical Meeting on Nuclear Plant Instrumentation, Control, and Human Machine Interface Technologies, pp. 2217–2228 (2015)
17. Kovesdi, C., Rice, B., Bower, G., Spielman, Z., Hill, R., LeBlanc, K.: Measuring Human Performance in Simulated Nuclear Power Plant Control Rooms Using Eye Tracking, INL/EXT-15- 37311. Revision 0 (2015)
18. Endsley, M.R.: Designing for Situation Awareness: An Approach to User-Centered Design. CRC Press, Boca Raton (2016)
19. Grier, R.A.: How high is high? A meta-analysis of NASA-TLX global workload scores. In: Proceedings of the Human Factors and Ergonomics Society Annual Meeting, vol. 59, no. 1, pp. 1727–1731. SAGE Publications, Los Angeles, September 2015

Quantifying the Contribution of Individual Display Features on Fixation Duration to Support Human-System Interface Design in Nuclear Power Plants

Casey Kovesdi[✉], Katya Le Blanc, Zachary Spielman, Rachael Hill, and Johanna Oxstrand

Idaho National Laboratory, Idaho Falls, ID, USA
{Casey.Kovesdi, Katya.LeBlanc, Zachary.Spielman, Rachael.Hill, Johanna.Oxstrand}@inl.gov

Abstract. The integration of new digital instrumentation and control (I&C) technologies like advanced human-system interfaces in U.S. nuclear power plant main control rooms is important for addressing long-term aging and obsolescence of existing I&C. Nonetheless, attention should be made to ensure these technologies reflect state-of-the-art human factors engineering (HFE) principles. Often, there is conflicting guidance from one guideline to another, requiring the analyst to make a judgment call on addressing these 'tradeoffs.' The objective of this research was to inform the analyst of these tradeoffs through an empirical investigation of how certain display features that characterize common HFE guidelines concerning visual clutter and saliency influence information processing in a naturalistic context. By understanding the unique contribution of each display feature using a multilevel model, the HFE analyst should have an understanding of the interrelations of each feature with its impact on cognitive processes. Results and implications are discussed in this paper.

Keywords: Human factors engineering · Human-system interface design
Nuclear power plants · Control room modernization · Cognitive processes

1 Introduction

Domestic electricity demands in the United States (U.S.) are expected to increase at an average rate of one percent each year. Meanwhile as most of the existing U.S. nuclear power plant (NPP) fleet begin to reach the end of their 60-year operating licenses, the U.S. Department of Energy (DOE) Light Water Reactor Sustainability (LWRS) Program continues to develop a scientific basis to extend their operating life through targeted research and development (R&D) pathways. Particularly, the LWRS Advanced Instrumentation, Information, and Control (II&C) Systems Technologies pathway conducts targeted R&D to address long-term aging and obsolescence of existing instrumentation and control (I&C) technologies, where one focus area consists of testing and implementing new digital I&C technologies such as integrating advanced

© Springer International Publishing AG, part of Springer Nature (outside the USA) 2019
T. Z. Ahram (Ed.): AHFE 2018, AISC 787, pp. 453–464, 2019.
https://doi.org/10.1007/978-3-319-94229-2_44

human-system interfaces (HSIs) in NPP main control rooms. There are many advantages to modernizing with these digital I&C technologies, including [1]:

- Improved safety through reduced frequency of challenges to the plant,
- Improved capacity factor of the plant,
- Improved computational processing power and access to information,
- Prepared main control room (MCR) I&C systems for future needs,
- Addressed past human engineering discrepancies (HEDs) in the MCR,
- Reduced operations and maintenance costs through the reduction or elimination of specialized maintenance on analog systems that are nearing their end of life or are obsolete, and
- Increased productivity levels in plant staff where staffing levels, especially outside of the MCR during normal operations, could be further reduced.

Nonetheless, careful attention must be paid to ensuring that these technologies reflect sound human factors engineering (HFE) design principles so that no new human error modes are introduced and that human-system performance is optimal. The U.S. Nuclear Regulatory Commission (NRC) currently provides state-of-the-art HFE design principles in *Human-System Interface Design Review Guidelines* (NUREG-0700, Rev. 2 [2]), which lists nearly 2000 guidelines for HSI design guidance. While these HSI design guidelines are comprehensive, the guidelines were written 'uni-dimensionally' such that one guideline does not account for any interactions with other relevant guidelines [3]. There may be conflicting guidance from one guideline to another, requiring the analyst to make a judgment call on how to address these 'tradeoffs.' For instance, Kovesdi and Joe [4] investigated common NUREG-0700 guidelines violated in various HSI displays, combining several HFE evaluations undergone across various utilities over the past three years. One higher prioritized guideline that was often violated concerned the use of higher contrast colors to ensure adequate visual salience of information for readability (i.e., see NUREG-0700 1.6.1-2). However, Kovesdi and Joe [4] also identified that the overuse of salient colors resulting in visual clutter, described in NUREG-0700 1.3.8-1, as well as the assignment of the same color (e.g., red) across multiple codes (e.g., alarms and valve/pump status), described in NUREG-0700 1.3.8-8, can decrease human-system performance.

Indeed, the application of these state-of-the-art HFE guidelines require an understanding of the tradeoffs between each of their application to HSI design and their impact on human-system performance. To date, the relationship of these variables on human-system performance in an integrated process control context is not clearly understood. Specifically, these tradeoffs encourage further investigation of how each guideline individually contributes to human-system performance when integrated into various HSI concepts and investigated in a naturalistic context. Such findings should help clarify when an HSI designer should consider adding a design feature based on specific HFE human-system performance criteria versus operator preference. While design input elicited from operator preferences is important to HSI design, such feedback integrated into the HSI should not introduce any new human error modes or HEDs that are important to plant safety.

The objective of this research is to support in informing the analyst of these tradeoffs through an early empirical investigation of how certain display features that characterize common NUREG-0700 guidelines concerning visual clutter and display color saliency influence human-system performance when integrated in a naturalistic context. To this end, a description of relevant human-system performance considerations for cognitive processes is described in the following sub-section. Further, the following sub-sections describe measures of visual clutter, color saliency, and color conflict, as well as other considerations that may affect information processing. Specific hypotheses are described to close this section.

1.1 Human-System Performance: Measuring Cognitive Processes

The Institute of Electrical and Electronics Engineers (IEEE) *Guide for the Evaluation of Human-System Performance in Nuclear Power Generating Stations*, IEEE Std 845-1999(R2011) [5], provides guidance for evaluating human-system performance related to systems, equipment, and facilities in NPPs. In this document, several considerations associated with evaluating human-system performance are discussed that have direct relevance to this study. First, IEEE-Std 845-1999(R2011) discusses the need to consider *cognitive processes* as part of diversely measuring human-system performance. Second, this document discusses the importance of using *experimental techniques* to evaluate human-system performance as a way of providing objective results concerning system performance. Lastly, the selection of a testbed is discussed where varying levels of fidelity can be used, which should be driven by the research question at hand. For instance, using a *lower fidelity* testbed such as static prototypes of HSI display concepts can be appropriate in earlier phases of the development process, especially with making more generalized inferences of candidate design concepts [6, 7].

Eye movements are a useful measure of cognitive processing. Fixations, or the pauses in eye movements, have been known to trace to where one's focused attention is being visually allocated to in space. Focal vision, representative of focused attention, refers to the narrowed region in space (i.e., $\sim 2°$) where the fovea is positioned [8]. The duration of a fixation is a known measure for evaluating information processing in the human factors literature [e.g., 9], described as a temporal length of the pause in a given fixation [10]. Longer durations can infer that more time is spent processing, or interpreting, the information being presented on a certain region of an interface [11]. From a neuroergonomic perspective, multiple neural pathways are used to guide eye movements depending on whether the eye movement was invoked due to a response to salient stimuli or deliberately being guided for visual search [12]. Both cases have implications to HSI display design; deliberate search should only be disrupted in cases of important events that require immediate resolution.

1.2 Display Feature Considerations

Display feature considerations included measures of *visual clutter, color saliency, color conflict*, and *label frequency*. These measures are discussed next.

Visual Clutter. Visual clutter can be defined as the state in which there are an excess of visual items to the point that visual search performance can be degraded [13]. For instance, Rosenholtz and colleagues [13] discuss several measures of visual clutter that have been studied in previous studies, spanning from simply the number of items in a scene to sophisticated algorithms such as Feature Congestion and Subband Entropy. *Edge Density* was of particular interest since it was shown to be an exclusive measure of the spatial distribution of visual clutter, excluding color attributes.

Color Saliency. Color saliency, or the extent to which a color provides attention-gaining properties, is important to efficient extraction of visual information from a display [7]. Within HSI display design for NPPs, a recent design approach, Information Rich Design (IRD), reserves the use of highly salient colors (i.e., high contrast) for only emergent information like alarms that signify an abnormal situation [14]. The notion of reserving salient colors for only important information in IRD follows the belief that there are diminishing returns in attention gaining qualities in highly contrast colors when overused; hence, only the highest priority information should include higher contrast colors such as alarms. IRD thus will portray most information in a "dull screen" appearance such as through a low contrast gray color. On the other hand, recent HFE studies found that some operators prefer higher contrast colors such as red and green to convey valve and pump status to support viewing from across the control room [15]. Additionally, NUREG-0700 1.3.8-9 specifies that symbols should be 'readily discriminable against background colors under all expected ambient lighting conditions' [2]. Indeed, based on the literature, it can be inferred that the overuse of color can have diminishing returns on mental workload; however, having too little color may also have a negative impact. The right level of color is still an open question, often requiring the HSI designer to make a judgement call. For measurement of saliency, NUREG-0700 suggests using the International Commission of Illumination (CIE) color distance metric, Delta E, to evaluate the differences between foreground to background colors. Hence, Delta E was used to measure color salience in this study.

Color Conflict. NUREG-0700 1.3.8-8 specifies that any given color should only be assigned to a single meaning [2]. For example, the use of red is commonly reserved for alarm notification. The assignment of multiple meanings to a single may hence result in 'hindering proper assimilation of information' presented by the HSI display, as described in NUREG-0700. Interestingly, Kovesdi and Joe [4] identified a common practice in HSI display design to use red/green color combinations to convey valve and pump statuses, which violates NUREG-0700 1.3.8-8. Based on possible disparity between NUREG-0700 guidance and the application of IRD, a measure of color conflict (i.e., the density of red) was explored in this study.

Label Frequency. The number of items to be searched has shown to impact fixation duration; the presence of more items results in increased fixation duration [16]. In the case with HSI displays that use a mimic layout, the number of components within view may vary depending on where the user's attention is directed. Thus, a measure of item frequency was of interest. This study used label frequency to quantify the number of items since industry guidance [2] suggests that all components contain labels (i.e., NUREG-0700 1.2.8-2).

1.3   Additional Considerations

Other considerations extraneous to display design were considered in this study, including *fixation sequence* and *inter-trial differences*, which are discussed next.

Fixation Sequence. Prior research has shown that fixation duration increases in succession with the sequence of fixation [17]. One possible explanation for this phenomenon is that scene perception is cumulative where each fixation supports 'building' a representation of the scene over time [18].

Inter-trial Differences. A final consideration in this study was possible inter-trial differences. For instance, certain information on the display might be simply more familiar to operators based on from prior experience.

1.4   Hypotheses

As mentioned, this research aims at informing the HFE analyst of tradeoffs commonly found in NUREG-0700 guidelines concerning the selection of colors to convey information in HSI displays for NPP main control rooms, through quantification of the contribution of specific display features. Hence, it was expected that:

(1) Edge density, Delta E, red density (i.e., for color conflict), and label frequency each contribute uniquely to explained variance in fixation durations, even with extraneous variables (i.e., fixation sequence and inter-trial differences) being accounted for.
(2) Increasing edge density, red density, and label frequency increases fixation duration. Decreasing Delta E increases fixation duration.

By understanding the unique contribution of each display feature, the HFE analyst should have an understanding of the interrelations of each feature regarding its impact to fixation duration, a measure of information processing, when put into a naturalistic context. Further, results from this study should support future HFE reviews using guidelines from NUREG-0700 through providing an empirical basis in prioritizing and classifying potential risk of certain design issues related to use of color. Finally, this study should offer potential measures for evaluating HSI display characteristics concerning display clutter, display color saliency, and color conflict.

2   Method

2.1   Participants

A total of twelve auxiliary operators from a commercial U.S. NPP ($N = 12$) participated this study, consisting of the entire auxiliary operator population for the plant. Two participants left midway through data collection. Additionally, the eye tracking system used failed to accurately calibrate two other participants. Hence, a sample of eight participants ($n = 8$) were used for analysis. Their ages ranged from 24 to 63 years ($M = 49$, $SD = 13$) and their experience in the nuclear industry ranged from 2 to 31 years ($M = 18$, $SD = 10$).

2.2   Apparatus

This experiment used a series of nine questions. Seven questions were related to whether either opening or closing a particular valve from within the liquid radiological waste evaporator system would change the state (e.g., increase or decrease) of a certain evaporator component (e.g., vapor body). Two trials were embedded alarm detection trials (i.e., as indicated by a red colored trend somewhere on the display) where participants were instructed to ignore the trial question and respond to the alarm; these trials were treated as a secondary task. There were four different HSI design concepts tested, equating to a total of 36 experimental trials. The experimental program would show a static stimulus image of a concept HSI display and received responses from participants through keypresses where 'z' denoted an increased state and '/' denoted a decreased state. Participants pressed 'spacebar' if an alarm was present rather than pressing 'z' or '/' for the primary task. The stimulus and alarm trials were randomly assigned to control for carryover effects. The experimental program was installed on a Windows laptop with a 17″ screen, set at a 1440 × 2160 resolution. The program captured accuracy, response times, and eye tracking measures through syncing with a SensoMotoric Instruments (SMI) 60 Hz remote eye tracking system. Participants all sat in the same relative position and distance from the eye tracking system (i.e., ∼24 in. away).

2.3   Procedure

Participants were initially provided a brief training session to become familiar with: (1) the liquid radiological waste evaporator system, (2) the different HSI design concepts used under this study, and (3) the task instructions for this study. Participants were asked to respond as quickly and accurately as possible. Next, participants were calibrated to the eye tracking system through a nine-point calibration protocol. Participants were asked to view each individual fixation dot at nine different locations on the screen. Calibration would be repeated until average x- and y-coordinates were within $0.2°$ accuracy. Participants completed a brief set of practice trials that would ensure they understood what each keypress denoted (e.g., decrease or increase). For data collection, participants were presented a question (e.g., 'With the system in the current configuration what would be the result on level of E 01 if we close LV 208B?') located in the upper region of the screen. Once understood, the participant would press 'enter' where the program would show a fixation display with a crosshair located in the center of the screen for 1000 ms; this was done to ensure all participant's initial gaze was in the same relative location. Finally, the stimulus image of a concept HSI display, with the question available, was presented for as long as the participant needed.

2.4   Experimental Design

Participants completed the nine trials from each HSI design concept, in a randomized order, before moving on to the next HSI design concept. The order of HSI design concept was randomly assigned to control for carryover effects. However, due to the loss of data for four participants, the order of HSI design concept was not balanced.

The four different HSI design concepts included different combinations of valve and pump colors being white/gray or red/green. Further, HSI design concepts either contained colored flow streams to differentiate the different product steams (e.g., nitrogen versus distillate), or uniformly colored flow streams. Each two-level variable here essentially created four different HSI design concepts.

2.5   Image Processing: Quantification of Display Features

Image processing for quantification of display features was completed in R Version 3.3.3 [19]. Extraction of display features involved primarily use of the EBImage [20], imager [21], colorscience [22], and tesseract [23] packages. A summary of the process involved for fixation and display feature extraction is discussed next.

Fixation Extraction. For each fixation, the x- and y-coordinates were extracted and a circular window of roughly 2° was rendered for subsequent image processing. The window represented a rough estimate of foveal vision.

Edge Density. Fixation images were read into R using the EBImage [20] package. A high-pass Laplacian filter was then applied to the fixation image for edge detection. Absolute pixel values over the circular window were then summed to measure density. Higher values denoted greater edge density.

Delta E 2000. Color saliency was calculated by using the colorscience [22] package in R. Fixation images were read into R. Next, the image was compressed to a 50×50 pixel version to reduce computational load. Next, Delta E 2000 was calculated using a reference color from the HSI display background for each pixel within the circular window. The average Delta E 2000 from each fixation image was used as a measure of display color saliency. Greater Delta E 2000 values denoted greater color differences, suggesting that the image contained more salient content.

Red Density. Using EBImage [20], each fixation image was parsed into red, green, and blue color channels. Green and blue channels were subtracted to include only the red channel. Red density was then summed across pixels. Greater red density indicated more color conflict (i.e., displaying more red pumps or valves).

Label Frequency. Label frequency was calculated using the tesseract [23] package for optical character recognition (OCR) processing across each fixation image. Extracted labels through OCR were summed to create a label frequency measure.

2.6   Data Analysis

Prior to applying statistical analyses, outliers were identified and removed using Tukey Fences based on values greater than 1.5 times the interquartile range across both dependent and independent variables. This study used a multilevel model (MLM) to test the significance of each display feature and extraneous factors on fixation durations throughout the course of each trial. A MLM was used to account for the nested nature, or 'non-independence' [24, 25], of the data; fixations were nested within trials and trials were nested within participant. MLM analyses were built in R using the nlme [26] package. Model parameter estimation used the maximum likelihood criterion. Fixed effects included question type (inter-trial differences), fixation sequence, label frequency, edge density, Delta E 2000, and red density. Participants were included in as a random effect. Edge density, Delta E 2000, and red density were centered (using the sample mean) and scaled into z scores to support interpretation of the model [25]. Each scaled variable had a mean of 0 and standard deviation of 1.

3   Results

Results from the MLM are outlined in Table 1. In general, fixation sequence, edge density, Delta E 2000, and red density significantly predicted fixation durations. There was no significant relationship observed for question type and label frequency.

Table 1. MLM summary table.

	b	SE	t	p
Intercept	300.73	5.23	57.47	**0.00***
Question 2	9.16	4.94	1.85	0.07
Question 3	−0.33	5.33	−0.06	0.95
Question 4	−4.18	5.05	−0.83	0.41
Question 5	−1.53	5.94	−0.26	0.80
Question 6	2.65	5.19	0.51	0.61
Question 7	2.35	4.92	0.48	0.63
Alarm trial 1	4.01	6.87	0.58	0.56
Alarm trial 2	−5.82	7.09	−0.82	0.41
Fixation sequence	0.28	0.08	3.45	**0.00***
Label frequency	−1.19	1.94	−0.61	0.54
Edge density	5.51	2.38	2.31	**0.02***
Delta E 2000	−8.77	1.76	−4.99	**0.00***
Red density	4.81	2.08	2.31	**0.02***

b denotes estimates of regression coefficients, SE denotes the standard errors, t denotes t statistics, p denotes p values. Significant p values are **bolded** and labeled with * ($p < .05$).

Table 2 reports the statistics commonly used to report the explained variance with adding each predictor into the MLM [27]: the marginal R^2 for fixed effects and conditional R^2 for fixed and random effects.

Table 2. Marginal R^2 and conditional R^2.

	$R^2_{MLM(m)}$	Δ in $R^2_{MLM(m)}$	$R^2_{MLM(c)}$	Δ in $R^2_{MLM(c)}$
+Question type	0.26%	–	99.44%	–
+Fixation sequence	0.43%	+0.16%	99.55%	+0.11%
+Label frequency	0.43%	+0.00%	99.55%	+0.00%
+Edge density	0.57%	+0.14%	99.49%	−0.06%
+Delta E 2000	0.86%	+0.30%	99.53&	−0.05%
+Red density	0.94%	+0.08%	99.54%	+0.01%

Figure 1 illustrates the relationship, using the coefficient estimates (*b*), of the observed significant display features predicted on fixation duration.

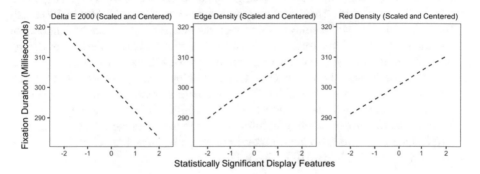

Fig. 1. Predicted fixation duration values by significant display features from MLM.

4 Discussion

4.1 Research Question 1: Did Each Display Feature Uniquely Contribute to the Explained Variance for Fixation Duration?

Edge density, Delta E 2000, and red density uniquely contributed to the explained variance for fixation duration, as shown in Table 1. These findings generally suggest that the selected measures used to evaluate impact of visual clutter (i.e., edge density), color saliency (Delta E 2000), and color conflict (red density) were sensitive and uniquely explained variance for fixation duration. It is worth emphasizing, however, that the total contribution of explained variance that these variables had on fixation duration was relatively small, as shown by the marginal R^2 statistics in Table 2. These findings should not suggest that the effects observed here are not practically significant. Rather, these findings suggest that cognitive processes are complex, likely involving several factors that were not formally accounted for in this study (e.g., operational experience or other measures of clutter or saliency).

Label frequency did not uniquely predict fixation duration. It is unclear why label frequency was not a significant predictor for fixation duration. It is possible that counting the number of labels simply was not sensitive enough for this study. For example, label frequency ranged from 0–17, with a mean of 4.76, and standard deviation of 3.83. Another possibility could have been that label frequency was not a valid measure for visual clutter; for instance, it is possible that the frequency of labels does not adequately capture the intricacies of visual clutter created from increasing display complexity. Future research should explore alternative ways of measuring clutter.

4.2   Research Question 2: Was the Relationship with Display Features on Fixation Duration in the Direction as Expected?

The directionality between edge density, Delta E, and red density with fixation duration followed what was hypothesized (see Fig. 1). Edge density and red density both had a positive relationship on fixation duration. Delta E had a negative relationship on fixation duration. From a cognitive processing standpoint, increasing the amount of graphical detail in a display (i.e., increasing edge density) may have a negative impact on information extraction. This finding supports the application of NUREG-0700 1.2.8-1 [2] when evaluating HSI displays, which suggests maintaining the minimal amount of information needed to depict plant components on a mimic display. Further, assigning multiple meanings to a single color (e.g., use of red for alarms and valve/pump state) may also have a negative impact on cognitive processing (i.e., see NUREG-0700 1.3.8-8 [2]). Though, the measure of red density should be further explored in future research to validate it as an applicable measure of color conflict. Finally, the application of different colors to convey meaningful information to the task at hand (e.g., in this case to differentiate each product stream) may have had a positive impact on information extraction (i.e., see NUREG-0700 1.2.8-6 [2]).

4.3   The Connection Between Fixation Duration and Cognitive Processing

An important point worth addressing is that the measure of fixation duration is not a direct measure of cognitive processing [7]. Rather, it is inferred that increased fixation duration is indicative of greater information processing requirements [11]. While there is a strong body of literature to support this interpretation, there may be alternative explanations. It is possible that there was another mechanism that can explain the relationships observed between the display features and fixation durations in this study. For instance, information rich display regions may have simply helped with refocusing visual attention to those regions independent of information processing demands (e.g., these regions were more interesting to look at). It may be worth examining the effects these display features on other measures of cognitive processing such as pupil diameter, other eye movement metrics (e.g., saccade amplitude), subjective reports, or other physiological measures such as heart rate variability in future research.

5   Conclusions

This research aimed at supporting in informing the HFE analyst of tradeoffs commonly found in NUREG-0700 guidelines concerning the selection of colors to convey information in HSI displays for NPP main control rooms, through quantification of the contribution of specific display features. Findings from this study suggest that (1) increased graphical display features, (2) the application of more than one meaning to a given single color, and (3) the overuse of the "dull screen" IRD principle each contributed to increases in fixation duration, a measure of cognitive processing. Hence when designing HSI displays, the findings from this paper support the need to ensure that only the minimal amount of visual detail necessary for a meaningful representation is used, color codes convey a single meaning (unless other performance-based evidence strongly suggests otherwise), and that information important for operations is salient. Future research should examine the role of these display features in a larger context, such as inclusion of different tasks, demographic characteristics (e.g., years' experience), and other display characteristics such as display complexity [28] for a more holistic understanding of contributing factors that may influence fixation duration and cognitive processes.

Acknowledgments. INL is a multi-program laboratory operated by Battelle Energy Alliance LLC, for the United States Department of Energy under Contract DE-AC07-05ID14517. This work of authorship was prepared as an account of work sponsored by an agency of the United States Government. Neither the United States Government, nor any agency thereof, nor any of their employees makes any warranty, express or implied, or assumes any legal liability or responsibility for the accuracy, completeness, or usefulness of any information, apparatus, product, or process disclosed, or represents that its use would not infringe privately-owned rights. The United States Government retains, and the publisher, by accepting the article for publication, acknowledges that the United States Government retains a nonexclusive, paid-up, irrevocable, world-wide license to publish or reproduce the published form of this manuscript, or allow others to do so, for United States Government purposes. The views and opinions of authors expressed herein do not necessarily state or reflect those of the United States government or any agency thereof. The INL issued document number for this paper is: INL/CON-17-43991.

References

1. Kovesdi, C.R., Joe, J.C., Boring, R.L.: A human factors engineering process to support human-system interface design in control room modernization. In: 10th International Topical Meeting on Nuclear Plant Instrumentation, Control, and Human Machine Interface Technologies, pp. 1843–1855 (2017)
2. U.S. Nuclear Regulatory Commission: Human-System Interface Design Review Guidelines, NUREG-0700, Rev. 2 (2002)
3. Kirwan, B., Ainsworth, L.K. (eds.): A Guide to Task Analysis: The Task Analysis Working Group. CRC Press, Boca Raton (1992)
4. Kovesdi, C.R., Joe, J.C.: A review of human-system interface design issues observed during analog-to-digital and digital-to-digital migrations in U.S. nuclear power plants. In: 10th International Topical Meeting on Nuclear Plant Instrumentation, Control, and Human Machine Interface Technologies, pp. 1568–1580 (2017)
5. Institute of Electrical and Electronics Engineers: IEEE Guide to the Evaluation of Human-System Performance in Nuclear Power Generating Stations (IEEE Std. 845-1999). Institute of Electrical and Electronics Engineers, New York (1999)

6. Kantowitz, B.H.: Selecting measures for human factors research. Hum. Factors **34**(4), 387–398 (1992)
7. Wickens, C.D., Hollands, J.G., Banbury, S., Parasuraman, R.: Engineering Psychology & Human Performance. Psychology Press (2000)
8. Jacob, R.J., Karn, K.S.: Eye tracking in human-computer interaction and usability research: Ready to deliver the promises. Mind **2**(3), 4 (2003)
9. Just, M.A., Carpenter, P.A.: Eye fixations and cognitive processes. Cogn. Psychol. **8**, 441–480 (1976)
10. Kovesdi, C., Rice, B., Bower, G., Spielman, Z., Hill, R., LeBlanc, K.: Measuring Human Performance in Simulated Nuclear Power Plant Control Rooms Using Eye Tracking, INL/EXT-15- 37311. Revision 0 (2015)
11. Goldberg, J.H., Kotval, X.P.: Computer interface evaluation using eye movements: methods and constructs. Int. J. Ind. Ergon. **24**(6), 631–645 (1999)
12. Parasuraman, R., Rizzo, M. (eds.): Neuroergonomics: The Brain at Work. Oxford University Press, Oxford (2008)
13. Rosenholtz, R., Li, Y., Nakano, L.: Measuring visual clutter. J. Vis. **7**(2), 17 (2007)
14. Braseth, A.O., Nurmilaukas, V., Laarni, J.: Realizing the information rich design for the loviisa nuclear power plant. In: American Nuclear Society International Topical Meeting on Nuclear Plant Instrumentation, Control, and Human-Machine Interface Technologies (NPIC&HMIT), vol. 6, May 2009
15. Lew, R., Ulrich, T.A., Boring, R.L.: Nuclear reactor crew evaluation of a computerized operator support system HMI for chemical and volume control system. In: International Conference on Augmented Cognition, pp. 501–513 (2017)
16. Moffitt, K.: Evaluation of the fixation duration in visual search. Atten. Percept. Psychophys. **27**(4), 370–372 (1980)
17. Irwin, D.E., Zelinsky, G.J.: Eye movements and scene perception: Memory for things observed. Atten. Percept. Psychophys. **64**(6), 882–895 (2002)
18. Unema, P.J., Pannasch, S., Joos, M., Velichkovsky, B.M.: Time course of information processing during scene perception: the relationship between saccade amplitude and fixation duration. Vis. Cogn. **12**(3), 473–494 (2005)
19. R Core Team: R: A language and environment for statistical computing. R Foundation for Statistical Computing, Vienna, Austria (2015). https://www.R-project.org/
20. Pau, G., Fuchs, F., Sklyar, O., Boutros, M., Huber, W.: EBImage - an R package for image processing with applications to cellular phenotypes. Bioinformatics **26**(7), 979–981 (2010)
21. Barthelme, S.: imager: Image Processing Library Based on 'CImg'. R package version 0.40.2 (2017)
22. Gama, J.: colorscience: Color Science Methods and Data. R package version 1.0.4
23. Ooms, J.: tesseract: Open Source OCR Engine for R. R package version 1.6 (2017)
24. Barr, D.J.: Analyzing 'visual world' eye tracking data using multilevel logistic regression. J. Mem. Lang. **59**(4), 457–474 (2008)
25. Nuthmann, A.: Fixation durations in scene viewing: modeling the effects of local image features, oculomotor parameters, and task. Psychon. Bull. Rev. **24**(2), 370–392 (2017)
26. Pinheiro, J., Bates, D., DebRoy, S., Sarkar, D., R Core Team: nlme: Linear and Nonlinear Mixed Effects Models. R package version 3.1-131 (2017)
27. Nakagawa, S., Schielzeth, H.: A general and simple method for obtaining R2 from generalized linear mixed-effects models. Methods Ecol. Evol. **4**(2), 133–142 (2013)
28. Hugo, J., Gertman, D.: A qualitative method to estimate HSI display complexity. Nucl. Eng. Technol. **45**(2), 141–150 (2013)

Autonomous Algorithm for Start-Up Operation of Nuclear Power Plants by Using LSTM

Deail Lee and Jonghyun Kim[✉]

Department of Nuclear Engineering, Chosun University, 309 Pilmun-daero,
Dong-gu, Gwangju 501-709, Republic of Korea
dleodlfl004@chosun.kr, jonghyun.kim@chosun.ac.kr

Abstract. Autonomous operation is one of the technologies of the forth-industrial revolution that is attracting attention in the world due to the development of new computer algorithms and the hardware performance. Its main core technology is based on artificial intelligent (AI). Autonomous control, which is a high level of automation, is having the power or ability of self-governance in the overall system without human intervention. This study aims to develop an autonomous algorithm to control the NPPs during start-up operation by using Long-Short Term Memory (LSTM) that is one of the recurrent neural network (RNN) methods. RNN, which is a type of AI method, is suitable for application to the NPP system because it can help to calculate the interaction of non-linear parameters as well as to capture the pattern of time series parameters. This study suggests a conceptual design for autonomous operation during start-up operation from 2% power to 100% power in nuclear power plants.

Keywords: Nuclear power plant · Long-short term memory
Autonomous algorithm

1 Introduction

In recent years, the core technologies of the forth-industrial revolution are attracting attention in the world due to the development of new computer algorithms and the hardware performance. Its main core technology in autonomous operation based on artificial intelligent (AI).

Autonomous control, which is a high level of automation, is having the power or ability of self-governance in the overall system without human intervention [1]. It can have many advantages such as safety improvement through the highly automatic system, optimal control in the system, adaptability even with a changing environment, improved reliability of the operating system, and cost saving. Due to the advantages of autonomous operation, it has been applied in many industrial areas, especially in those areas that require high-level of precision and can reduce human burden.

Nuclear power plants (NPPs) are operating with manual control by operators as well as automatic control by an automated algorithm in the system. The digital technology provides advantages such as processing of numerous data, improvement of

© Springer International Publishing AG, part of Springer Nature 2019
T. Z. Ahram (Ed.): AHFE 2018, AISC 787, pp. 465–475, 2019.
https://doi.org/10.1007/978-3-319-94229-2_45

system reliability, the flexibility of adding new functions, automation of periodic tests, self-diagnostics, and improved operation [2]. NPPs have been able to increase safety and efficiency as well as to reduce operator's burden by applying digital automation system. However, the level of automation for controlling NPPs during start-up operation is not at par with the autonomous control, because the intervention by operators is still required in a large portion of the control.

Start-up operation is to raise the reactor power to 100% and thereafter generate the electricity. It is normally performed by operators using operating procedures. It is generally known that the burden of operators in this operation is higher than during normal operations, because the operator has to determine the control strategy of components by monitoring many physical parameters in the NPP. Therefore, to decrease the burden of operators an increased automation level of NPPs to an autonomous control is expected to reduce the burden of operators.

This study aims to develop the autonomous power startup/shutdown control algorithm by using AI method. In order to apply an AI method, the requirements have been developed based on the general operating procedure as well as the operating strategy of current NPPs. This study suggests an operator operation-based control module framework, based on procedure and operating strategy, and the Long-Short Term Memory (LSTM) method that is one of the recurrent neural network (RNN) methods. The RNN, which is a type of AI method, is regarded as a suitable approach for the NPP system because it can help to calculate the interaction of non-linear parameters as well as to capture the pattern of time series parameters. For designing the algorithm of the power start-up/shutdown control function, Sect. 2 introduces the suggested operation-based control module framework as well as the LSTM. Section 3 explains the power start-up/shutdown control algorithm in detail.

2   Approach

2.1   Autonomous Operation System During Start-Up/Shutdown Operation

This study proposes to alleviate the heavy workload of operators by leveraging the advances of artificial intelligence. The purpose of this system is to operate NPPs by one operator during start-up/shutdown operation. An autonomous system can perform functions such as control, diagnosis, self-validation, decision making, and adaptation, to achieve its purpose. Figure 1 shows the suggested framework of the autonomous operation system. It consists of functions such as power start-up/shutdown control function, accident diagnosis/protection control function, performance monitoring function, strategy selection function, and operator interaction function.

Power start-up/shutdown control function that is the target function of this study can control and monitor NPPs based on the operator action by using the operating procedure. Accident diagnosis/protection control function can diagnose the accident through monitoring NPPs, and control the NPPs to mitigate the accident. Performance monitoring function can monitor the performance of power start-up/shutdown control function as well as plant parameters. Strategy selection function can select the

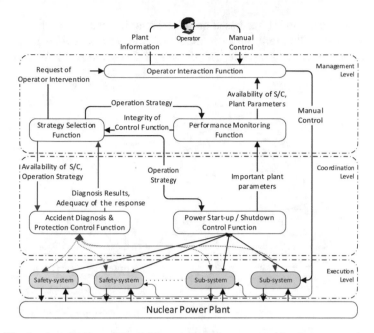

Fig. 1. Autonomous control framework during start-up/shutdown operation

operation strategy through the output results that are generated at the accident diagnosis/protection function and the performance monitoring function. Operation interaction function not only can exchange information between the operator and an autonomous operation system but also can transfer manual control signal to an autonomous control system.

2.2 Operator Module Framework

This study suggests the operator module framework for designing the power start-up/shutdown control function. The power start-up/shutdown control should be considered by existing NPP operators to be performed by four operators. Operator module framework uses modular operators that reflect current operator's role. In NPPs, operator's role is a controller role to perform the situation assessment, system monitoring, detection, planning response, and action. The cognitive ability of operators is applied to design the automation system as well as operation support system [3]. Therefore, each operator module can perform the function considered current operator's duties.

In order to design power start-up/shutdown control function, the operator module framework includes four modular operators. Currently, operators in NPPs consist of a senior reactor operator (SRO), a reactor operator (RO), a turbine operator (TO), and an electrical operator (EO). They control components of NPPs in the main control room (MCR) through decision-making of operators based on operating procedures [4]. Figure 2 shows simplified information exchange process between operators and control panel in MCR. To reflect current operation strategy, the process of exchanging information between modules is based on the existing information exchange process of operators in NPPs. Therefore, operator module framework is developed as four operator modules such as SRO module, RO module, TO module and EO module.

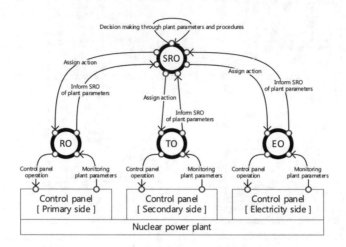

Fig. 2. The simplified interaction between operators

In addition, operator module framework can control and monitor sequentially or simultaneously because of current operation strategy. The current operators control components sequentially or simultaneously based on operating procedure. Procedures are essential to plant safety because they support and guide personnel interactions with plant systems and personal responses to plant-related events [5]. The structure of operating procedure is divided into an operation mode definition, an initial condition check, a monitoring limitation/precaution, and performing operation steps. The operation steps are written on the operating procedure in order. However, the operator can perform simultaneously or sequentially because of actions that require controlling and monitoring simultaneously. Therefore, in order to consider sequential or simultaneous control and monitoring, each operator module has their procedure steps that are based on current operating procedure. Figure 3 shows the process in which the existing procedural steps are assigned to the SRO/RO/TO/EO module.

Each module has functions considering the operation process of the existing operator. For designing functions, Table 1 shows the function design process in a structured way. Function design process can help for reflecting all the duties of the operator during power start-up/shutdown operation in the module.

2.3 Long Short-Term Memory (LSTM)

To decide the control signal in each procedure, procedure steps in the operator module used LSTM network as a calculation tool. LSTM network is one of the recurrent neural network (RNN) that is an AI method. Generally, LSTM network can calculate output

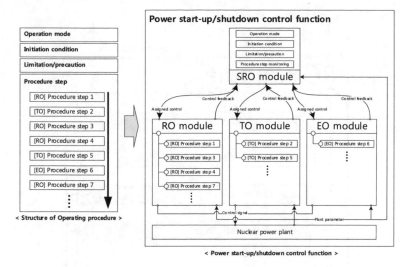

Fig. 3. The design process of SRO/RO/TO/EO module based on operating procedure

Table 1. The function design process

No.	Process
1.	Duty analysis of existing operators
2.	Design of operation algorithm through analyzed duty
3.	Define individually designed operation algorithms as functions and define input/output of algorithms
4.	Combination of defined functions according to input/output of algorithms
5.	Define the input/output of the module

data from multiple input data. Besides, it can process long temporal sequences of data because of structural characteristics. In each procedure, the process for calculating control signal is similar to figure out the output data from multiple input data in LSTM network. Each procedure step is the process for calculating control signal through parameter trend and state monitoring of NPPs. Each operator module can calculate control signal in each procedure step by using LSTM network. Figure 4 shows an example of RO module applied LSTM network.

LSTM network can also generate control signal included the experiential operation of operators using operation history data. The learning process is required to use the LSTM network. LSTM network, which uses supervised learning, is trained through training data. Trained LSTM network using the operation history data can have operation ability based on operating procedure as well as experiential operation.

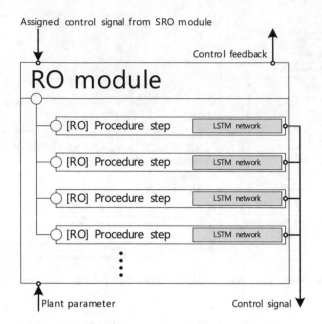

Fig. 4. Example of RO module applied LSTM network

3   Design of Power Start-Up/Shutdown Control Function

To design the power start-up/shutdown control function in the autonomous operation system, this study suggested an operator module framework for an LSTM network. LSTM network focuses on applying the operator module framework. LSTM network modeling, training, and validation are not considered in this paper.

This section explains the deigned power start-up/shutdown control function in detail by applying the operator module framework and LSTM network, which are proposed in the previous section. Figure 5 shows the overall structure of the power start-up/shutdown control function. The purpose of this function is to control and monitor NPPs based on the operating procedure as well as the current operation strategy during start-up/shutdown operation. The input/output of the function consider the input/output of other functions. Therefore, the function has systemic interaction with other functions. The input consists of plant parameter, which includes physical value and component states, and operation strategy from the strategy selection function. The output includes important plant parameter and control signal.

To reflect current operation strategy, the function has SRO/RO/TO/EO module that are modeled module through current operator's duties. SRO module is able to monitor plant parameter and instruct RO/TO/EO module to control NPPs. RO/TO/EO module can control NPPs as well as report the results of the control to SRO module.

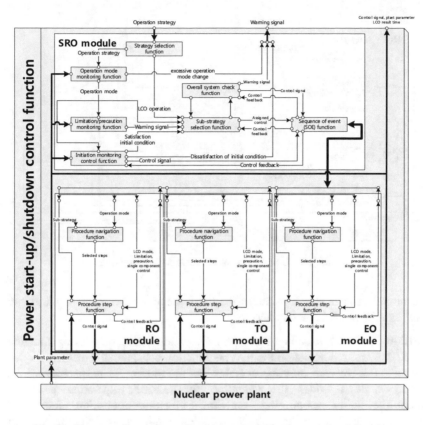

Fig. 5. The overall structure of power start-up/shutdown control function

3.1 Senior Reactor Operator (SRO) Module

The purpose of the SRO module is to provide RO/TO/EO module operation signals and monitor operation based on operating procedures. Functions of SRO module are defined as the existing operation process of SRO. Current SRO has duties that are operating strategy decision, determining operation mode, checking limitation/precaution condition, checking initiation condition, monitoring plant parameter as well as the performance of operators, and instructing operation strategy to RO/TO/EO based on operating procedure during start-up/shutdown operation. Figure 6 shows the connection of modeled function in the SRO module through the SRO actions.

Strategy Selection Function. To receive operation strategy of the upper level, strategy selection function is used to an input gate in SRO module. Strategy selection function can transmit the operation strategy, which includes limiting condition for operation (LCO), general power start-up/shutdown operation, accident operation and the manual control by an operator, to other functions in SRO module. In addition, this function is

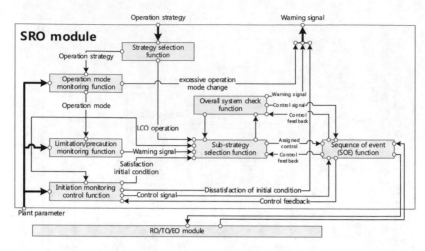

Fig. 6. Connection of function in SRO module

able to stop the power start-up/shutdown control function when receiving the signal for manual control or accident operation.

Operation Mode Monitoring Function. Operation mode monitoring function can calculate current/goal operation mode. This function considers the process for defining operation mode by the existing operator by monitoring plant parameter. Current operation mode provides the guideline for selecting operating procedure in limitation/prevention monitoring function, initiation-monitoring function, and sub-strategy selection function. Goal operation mode is used in the sub-strategy selection function to identify the satisfaction of operation goal. In addition, operation mode monitoring function requests a modified operation strategy to the strategy selection function. It needs to deal with the wrong operation strategy when there is excessive operation mode change.

Initiation Monitoring Control Function. Initiation monitoring control function can monitor initial conditions according to current operation mode and operate the components to meet the required initial conditions. To start operating procedure steps, operators check initial condition suggested in procedure. Moreover, operators control components to satisfy initial condition. Therefore, initiation monitoring control function is possible by instructing the component control through RO/TO/EO module.

Limitation/Precaution Monitoring Function. The operator controls component according to limitation/precaution condition based on operating procedure. To reflect limitation/precaution condition in SRO module, this function monitors the plant parameter about limitation/precaution condition. It can also send the warning signal to the sub-strategy function. The warning signal means that the RO/TO/EO module operate out of limit and precaution condition.

Sub-strategy Selection Function. Sub-strategy selection function allows the RO/TO/EO module to assign the control signal to RO/TO/EO module by using the output parameter of other functions. Sub-strategy function has been added to consider operator duty that is instructing operation strategy to RO/TO/EO based on operating procedure. Sub-strategy function can perform actions like (1) requisition for checking overall system state (i.e., when a problem occurs with RO/TO/EO module a control feedback is sent to SRO module), (2) setting up sub-strategy, (3) system monitoring, (4) assigning the control signal to RO/TO/EO module through operation mode, sub-strategy and limitation/precaution condition. Table 2 shows an example parameter of the sub-strategy selection function.

Table 2. Example parameter of the sub-strategy selection function

Type	Parameter	Data
Input	Operation strategy	Increase reactor power from 2% to 50% within 50 min
Input	Operation mode	Mode 2 (current) $\rightarrow$ Mode 2 (goal)
Input	LCO mode	None
Input	Initiation condition	Satisfaction
Output	Sub-strategy selection	Increase turbine set-point to 1800 RPM with 8% reactor output within 5 min
Output	RO module	Maintain 8% $\pm$ 1% reactor power
Output	TO module	Increase turbine set-point to 1800 RPM within 5 min
Output	EO module	None
Input	Control feedback	Satisfaction
Output	Self-checking request	None
Input	Self-checking result	None
Input	limitation/precaution	Excess reactor power increasing rate

The Sequence of Event (SOE) Function. In order to operate some components during power start-up/shutdown operation, sequential operation is required. SOE function allows RO/TO/EO module designed to enable the sequential operation of the component. In addition, it monitors the performance of RO/TO/EO module.

Overall System Check Function. Overall system check function is added for checking the overall system or component state according to the request of the sub-strategy selection function. Overall system check function can try assignment of the control signal to RO/TO/EO module. If the assigned control signal cannot operate the component, it sends problems to the upper-level function. This warning signal helps to modify the operation strategy in upper-level function.

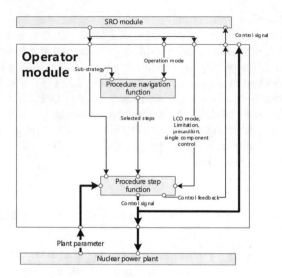

Fig. 7. Connection of function in operator module

3.2 Reactor Operator (RO), Turbine Operator (TO), Electrical Operator (EO) Module

RO/TO/EO module allow for controlling components through assigned control signal of SRO module as well as plant parameter. In order to control the NPPs, functions consist of procedure navigation function and procedure-step control function. Figure 7 shows the connection of functions in RO/TO/EO module.

Procedure Navigation Function. The procedure navigation function has been added to provide operation steps to the operator module. This function can find operation steps through output data, which is the sub-strategy as well as the operation mode. Each of RO/TO/EO modules have their operation steps.

Procedure Step Function. The procedure step function is to allow the generation of the control signal of the component via plant parameter monitoring, procedure steps, LCO mode condition and limitation/precaution condition. Moreover, it can control each component according to the initiation monitoring control function and the overall system check function in SRO module. The procedure step function receives defined procedure steps from the procedure navigation function. LSTM network is applied in the procedure step function to calculate control signal. LSTM network can calculate control signal through the plant parameter as well as input data of the procedure step function. Figure 8 shows an example of the LSTM network and the parameter of the procedure step function for a specific operating step.

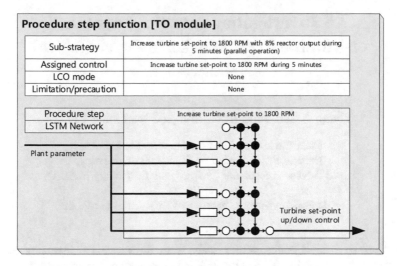

Fig. 8. The LSTM network and the parameter of the procedure step function

4   Conclusion

This study attempted to design power start-up/shutdown control function in the autonomous operation framework. In order to design the function, this study suggested the operator operation-based control module framework and LSTM method, which is one of the AI methods. This approach enables the systematic analysis of the power plant system and the implementation of the LSTM network in the designed function. This study shows a detailed design of power start-up/shutdown control functions, their relationships, and their sub-functions. In order to develop the autonomous control system, this designed function will be used for applying LSTM network.

References

1. Antsaklis, P.J., Passino, K.M., Wang, S.: An introduction to autonomous control systems. IEEE Control Syst. **11**(4), 5–13 (1991)
2. Jung, C., Kim, J., Kwon, K.: An integrated approach for integrated intelligent instrumentation and control system (I 3 CS) (1997)
3. Korea Institute of Nuclear Safety: Analysis of the operation performance in advanced information display (KINS/RR-485) (2007)
4. Kim, S., et al.: Development of extended speech act coding scheme to observe communication characteristics of human operators of nuclear power plants under abnormal conditions. J. Loss Prev. Process Ind. **23**(4), 539–548 (2010)
5. University of Nevada, Reno Commission: Human Factors Engineering Program Review Model (NUREG-0711, Rev. 2). Washington, DC, February 2004

An Investigation into the Feasibility of Monitoring a Worker's Psychological Distress

Young A Suh, Jung Hwan Kim, and Man-Sung Yim[✉]

Nuclear Environment and Nuclear Security Laboratory,
Department of Nuclear and Quantum Engineering,
Korea Advanced Institute of Science and Technology (KAIST),
Daejeon, Republic of Korea
{dreameryounga, poxc, msyim}@kaist.ac.kr

Abstract. The objective of this study is to investigate the feasibility of developing a worker psychological distress monitoring system using Electroencephalogram (EEG). Psychological impairment has emerged as a key security (insider threat) and safety (human error) issue at Nuclear Power Plants (NPPs) as well as other industries. Although the U.S. Nuclear Regulatory Commission (NRC) highlighted the importance of NPP workers' Fitness-For-Duty (FFD) to ensure personnel reliability, current FFD programs only consider drug and alcohol testing and fatigue management. However, today's bio-signals technology makes it possible to monitor the physical and mental state of workers. Thus, this study examines the feasibility of using EEG indicators to identify potentially-at-risk workers, especially those with acute psychological distress. We reviewed historical cases of insider threat and human error at nuclear facilities, and analyzed these cases from the perspective of a suspect's mental health. Based on bio-signal literature, a variety of EEG indicators identified at risk workers with a psychological impairment. As such, we selected the following: (1) Frontal EEG asymmetry; (2) EEG coherence; and (3) the variations of frequency domain EEG indicators (Theta, Alpha, Beta and Gamma) at certain brain area. To verify the appropriateness of these EEG indicators in realistic situations, this study performed a pilot experiment. The resting states of EEG (Eye Closed and Eye Open) were recorded on 56 student subjects (36 healthy and 20 with a high score for depression and anxiety symptoms). The resting states of EEG results showed a statistically significant difference between at-risk students and healthy students. This means specific EEG indicators can be used to classify the mental status of workers. These results can be applied directly to the mental health monitoring system of nuclear power plants as well as the industries requiring high reliability (aerospace, military and transportation).

Keywords: Electroencephalogram (EEG) · Fitness-For-Duty (FFD)
Nuclear industries · Psychological distress check-up

© Springer International Publishing AG, part of Springer Nature 2019
T. Z. Ahram (Ed.): AHFE 2018, AISC 787, pp. 476–487, 2019.
https://doi.org/10.1007/978-3-319-94229-2_46

1 Introduction

Fitness-For-Duty (FFD) refers to the physical and/or mental ability of an employee to perform safely the essential functions of his or her job [1]. The Nuclear Regulatory Commission's (NRC's) regulation 10 CFR part 26, highlights the importance of FFD programs. Specifically, nuclear facilities should have reasonable assurance that workers are (1) not under the influence of any substance, legal or illegal, and (2) not mentally or physically impaired from any cause [2]. To comply with this regulation, nuclear industries implemented a drug & alcohol test and fatigue management (through reporting an employee's working hours) [3]. However, current FFD programs are limited in evaluating psychological distress.

According to the literature, mental health problems are caused by work stress, job dissatisfaction, occupational exposure, marital conflict, martial satisfaction, human relationship and other personal reasons [4]. Psychosocial distress is one of the major health hazards for NPP workers [5, 6]. There are situations where NPP workers are required to work with multiple layers of personal protection equipment under stressful conditions. This was true for Chernobyl clean-up workers, who ultimately experienced depression, anxiety disorders, post-traumatic stress disorder, headaches and suicidal thoughts, at higher rates than the general population [7]. According to the research, the cleanup workers' psychosocial stresses continued for 18 years after the Chernobyl nuclear accident. Likewise, a growing number of Japanese workers, who risked their health to shut down the crippled Fukushima Daiichi nuclear power plant, are suffering from depression, anxiety about the future and a loss of motivation [8]. Thus, monitoring a worker's mental health is important.

Nuclear industries understand the importance of mental health services, but hesitate to implement an assessment program due to concerns associated with the difficulties of implementing regular check-up and a shortage of experts. Currently, TEPCO is considering hiring a full-time psychiatrist to provide counseling services for the TEPCO workers on a monthly basis [9]. However, this action has limitations due to its subjective and potentially biased results; this is mostly a reactive approach; and the services are infrequent. Fortunately, today's healthcare bio-signals technology makes it possible to monitor the physical and mental state of individuals. Thus, this study examines the feasibility of using EEG signal indicators to identify Potentially-at-risk workers, especially those under psychological distress.

2 Historical Cases Related to Mental Health Issues

Actual case studies [10] allow us to understand the importance of mental health problems in nuclear facilities. This section reviews historical cases where a psychologically impaired person caused an incident in a nuclear or non-nuclear, high-reliability facility. The first case addresses the theft of UO2 (two 5 gal cans) at a GE Low Enrichment Uranium Plant, Wilmington, United States, in 1979 [11, 12]. David Learned Dale was a temporary employee of a GE subcontractor who was suffering from depression. The financial impact of the incident was about a million-dollar loss to the GE Company.

The second case involved the murder of 13 people at the Fort Hood Army base. Major Nidal Malik Hasan was a 39-year-old U.S. Army psychiatrist, of Palestinian descent. For the last six years, supervisors gave him poor evaluations and warned him that he was doing substandard work. Unfortunately, no action taken, even though there may have been cues to his potential for committing a bad action. Peers described him as "disconnected," "aloof," "paranoid," "belligerent," and "schizoid". Ultimately it was determined that his motivation for the killings was ideological combined with a mental problem.

The third case was the March 2015 German airplane crash. The cause of the accident was not a mechanical defect but rather a mental problem with the flight's first officer. He was suffering from depression, and he decided to commit suicide killing all passengers and crew onboard.

There have been numerous cases where a tragic or costly incident resulted from a worker's psychological impairment, especially depression. In addition, both depression and anxiety are reported to be serious modern diseases caused by job stressor [13–15]. Thus, this study selected depression and anxiety symptom for investigating mental health problems using Electroencephalogram (EEG).

3   Electroencephalogram (EEG) Indicators

Before performing a pilot experiment investigating the feasibility of monitoring a worker's mental status using EEG, this study reviewed significant EEG indicators from previous studies. Based on this literature review [16–24], a variety of EEG indicators can be used to identify an at risk worker having a psychological impairment. The EEG indicators selected were: (1) Frontal EEG asymmetry; (2) EEG coherence; and (3) the variations of frequency domain EEG indicators (Theta, Alpha, Beta and Gamma) at certain brain area. Table 1 describes the specific features associated with depression and anxiety, and possible EEG indicators to measure these symptoms.

Most studies focused on two EEG indicators, asymmetry and coherence, for interpreting these psychological symptoms. To determine the adequacy of this approach, the third indicator was analyzed to clarify its value in detecting a worker's mental health.

Table 1. EEG Indicators related to depression and anxiety

Specific features	Possible EEG indicators
Depression anxiety	- **Alpha frontal asymmetry:** Vulnerability to depression and anxiety. Greater right than left activity in case of depression and anxiety [16–21] - **Alpha parietal asymmetry:** Leftward shift in parietal asymmetry in case of depression [22–24] - **Alpha coherence in high right parietal activity** in case of anxiety [23]

4   Pilot Experiment

Participants. Atotal of 56 subjects (36 healthy and 20 high score of depression and anxiety symptoms students) participated in this experiment. The subjects were engineering undergraduate or graduate students at the Korea Advanced Institute of Science and Technology (KAIST). The mean age of the sample population was 23.8 years with an age range of 19–29 years. None of the subjects reported a history of brain or heart disease, including their family members. The participants were not smokers, and they avoided alcohol and caffeine consumption the day before the experiment. In addition, none of the subjects took drugs commonly prescribed for minor ailments (indigestion, headache, cold). Prior to conducting the experiment, the KAIST Ethics Committee for Research on Human Subjects approved this study. Fully informed about the purpose, methods, and possible risks associated with the study, the students provided written consent before participating.

To group the participants into categories that signified a recognized range of depression and anxiety, we prepared a survey based on the short-form version of the Depression Anxiety Stress Scales (DASS-21) [25]. The DASS-21 consists of 21 negative emotional symptoms. Participants rated the extent to which they experienced each symptom over the past week, on a 4-point severity/frequency scale. Participant scores for the depression, anxiety and stress were determined by summing the scores for the seven questions in each of the three categories. Multiplying each summed score by two, allowed us to compare our results with the normal DASS (42 items). Scores can range from 0 to 56. The scores represent the severity of the subject's symptoms: Depression (above 21), Anxiety (above 15), and Stress (above 26). Scores above 26 indicate the subjects are suffering severe clinical symptoms. However, there were no severe depression and anxiety students based on the collected data. Thus, the ten students suffering from moderate level of depression and anxiety were categorized into the mental problem group.

The EEG Recording. An EEG system with 19 channels (BrainMaster Discovery 24ETM (Brain Master Technologies Inc.)) recorded EEG data with Linked ears reference (LE). The 19 channels are labeled Fp1, Fp2, F3, F4, F7, F8, Fz, C3, C4, Cz, T3, T4, T5, T6, P3, P4, Pz, O1, and O2. The electrodes are arranged in the international 10–20 system. Participants wore an electro cap with matching channel positions. The EEG data was recorded in the resting status (eye closed, eye open). During the EEG recording, the impedance was kept below 5kohm. All channels of EEG are acquired with the sampling rate of 256 Hz, 24 bits resolution.

EEG Analysis. A Quantitative EEG (QEEG) analysis [26] is a computer analysis of the EEG signal using 19 or more channels of a simultaneous EEG recording. First, raw digital EEG data are recorded, then analyzed and compared against a reference database of "normal subjects" found in the literature (NeuroGuide normative database). To evaluate the digital EEG Frequency-based analysis, the Fast Fourier Transform (FFT) algorithm was used. EEG frequency bands were categorized into the fourteen bands: (1) Delta: 1–4 Hz; (2) Theta: 4–8 Hz; (3) Alpha: 8–12 Hz; (4) Beta: 12–25 Hz; (5) High Beta: 25–30 Hz; (6) Gamma: 30–40 Hz; (7) High Gamma: 40–50 Hz,

(8) Alpha1: 8–10 Hz; (9) Alpha2: 10–12 Hz; (10) Beta1: 12–15 Hz; (11) Beta2: 15–18 Hz; (12) Beta3: 18–25 Hz; (13) Gamma1: 30–35 Hz; and (14) Gamma2: 35–40 Hz [27, 28].

EEG raw data were subjected to a Fast Fourier Transform (FFT) algorithm to calculate the absolute (μV^2) power and relative (%) power and the FFT Power Ratio (Arb). Absolute Power is the actual power (voltage) in a subject's EEG database (Power is microvolts squared). Relative Power is the relative power of each given band/sum of power from 1 to 50 Hz. FFT Power Ratio is calculated by one given band/other given band.

***Multivariate Analysis of Variance (MANOVA) Analysis* [29].** The MANOVA uses SPSS 24 software. It compares the variations between groups, with the variations within groups. MANOVA is used when there are two or more dependent variables. To perform the MANOVA analysis, requires meeting some assumptions. We assumed normal distribution of dependent variables, linearity, and homogeneity of variances and covariance [30, 31]. The results from the MANOVA analysis can be evaluated using p-value. The p-value or probability value is the probability for a given model that, when the null hypothesis is true, the statistical summary (such as the sample mean difference between two compared groups) will be the same as or of greater magnitude than the actual observed results [32]. Statistical significance (p-value) is something that can be measured to a given confidence level: P* < 0.05 and P** < 0.01.

5   Experiment Results

Whether a subject suffered from psychological distress or not, we grouped them into a fitness status: (A) Normal, (B) Moderate depression and (C) Moderate anxiety. To diagnose their symptoms clearly, they answered DASS-21 questionnaires. After grouping, we examined the fourteen values (i.e., Delta, Alpha, Theta, Beta, High Beta, Gamma, High Gamma, Alpha1, Alpha2, Beta1, Beta2, Beta3, Gamma1 and Gamma2) of absolute power and relative power using an EEG during the resting states (Eye closed and Eye open), respectively. In addition, the variation of 10 FFT power ratios (Delta/Theta, Delta/Alpha, Delta/Beta, Delta/High-Beta, Theta/Alpha, Theta/Beta, Theta/High-Beta, Alpha/Beta, Alpha/High Beta and Beta/High Beta) were compared depending on three different groupings above.

To categorize the differences in the three groups, this study examined the multi-variate test of significance, using Wilkys' lambda (λ), which relates directly to the F-distribution. Wilks' lambda is a test statistic used in MANOVA to test whether there are differences between the means of the identified groups of subjects, for a combination of dependent variables [33]. Accordingly, we will reject the null hypothesis if Wilk's lambda is small (close to zero). This test considers the dependent values of 28 absolute EEG power, 28 relative EEG power and 20 FFT Power ratio (called Band-to-Band Power ratio). These values were affected by the independent variable, mental status ($\lambda = 0.455$; F = 7.659; p < .0001). Based on the mental status factor, the MANOVAs showed significant effects 60 possible EEG indicators from 76 (total) EEG indicators.

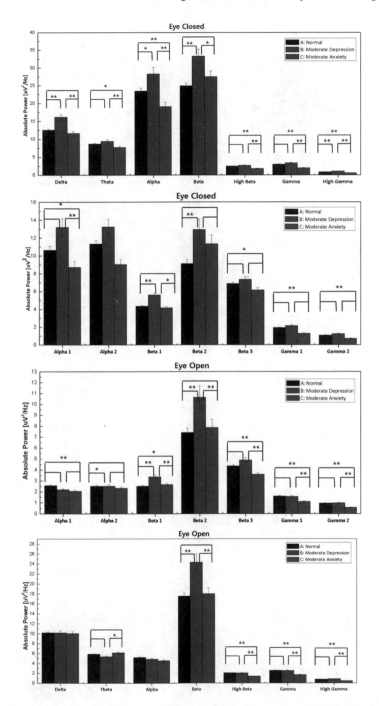

Fig. 1. The Absolute Power differences of three groups: Group mean (±*S.E.*) Absolute Power difference (y-axis) between three groups for fourteen indicators (x-axis) during Eye Closed and Eye Open. The significant indicator (P < 0.01 and P < 0.05) differences between the two groups (A vs B, A vs C, B vs C) were marked as ** and * in the Figure.

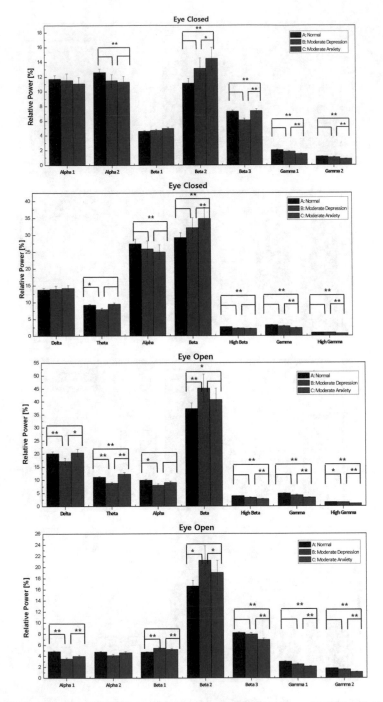

Fig. 2. The Relative Power differences of three groups: Group mean ($\pm S.E.$) Relative Power difference (y-axis) between three groups for fourteen indicators (x-axis) during Eye Closed and Eye Open. The significant indicator ($P < 0.01$ and $P < 0.05$) differences between the two groups (A vs B, A vs C, B vs C) were marked as ** and * in the Figure.

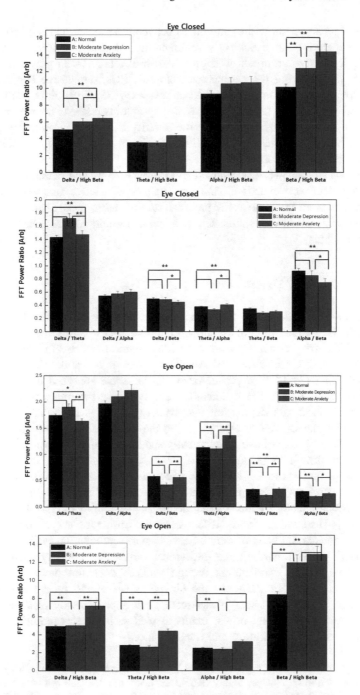

Fig. 3. The FFT Power Ratio differences of three groups: Group mean ($\pm S.E.$) Power Ratio difference (y-axis) between three groups for ten indicators (x-axis) during Eye Closed and Eye Open. The significant indicator ($P < 0.01$ and $P < 0.05$) differences between the two groups (A vs B, A vs C, B vs C) were marked as ** and * in the Figure.

Figures 1, 2 and 3 show the differences between the mean values of possible EEG indicators depending on a subject's different mental status. Compared the normal group, the depressive group showed greater overall relative beta power and absolute beta power [34]. During the resting eyes closed EEGs, depressive subjects showed elevated alpha and beta compared to the control group [35]. In addition, delta and theta band activity is increased [36, 37]. Depressive symptoms relate to decreases in SDNN, RMSSD and HF [38, 39]. The greater the severity of the symptoms, the greater the reduction in heart rate variability. In addition, the Heart Rate Variability test may reflect the severity of the symptoms [38, 40].

In previous studies [41, 42], reduced EEG alpha activity and increased beta activity appeared to reflect cortical activation. These changes were associated with anxiety. Our research results were the same as the previous studies. From these results, 60 indicators can define a subject's mental status. These differences could detect an NPP worker with an abnormal status.

6   Summary and Discussion

This paper investigated the use of power spectrum analysis on EEG data to identify a worker's mental state. A total of 56 subjects' resting states of EEG were recorded. For mental problem (depression and anxiety) identification, we examined fourteen frequency bands of EEG indicators: (1) Absolute power of 14 bands; (2) Relative power of 14 bands; and (3) 10 FFT power ratios. These were analyzed depending on the independent variables (76 EEG indicators) and dependent variable (subject's mental status: normal, moderate depression and anxiety). Using MANOVA, the pilot test results were validated. These results showed the resting states (Eye closed and Eye Open) for EEG indicators have a statistically significant difference for at-risk students compared to healthy college students.

There were limitations of this study. First, given the small sample size, we urge readers to be cautious in interpreting our data. Second, there were limitations in the existing self-report scales for anxiety and depression by Gotlib and Cane [43] and Clark and Watson [44] research. Third, EEG signals are different for each person, so individual difference should be considered. Steven Luck's publication [45] mentioned that a participant's EEG recording has individual differences. Finally, depression and anxiety related to EEG indicators, though studied, does not have an established, commonly accepted theory. Known and Unknown issues will be examined in future work. Despite unresolved issues, these results can apply directly to a mental health monitoring system at nuclear power plants as well as industries requiring high reliability such as aerospace, military and transportation.

Acknowledgments. This work was supported by the Nuclear Safety Research Program through the Korea Foundation Of Nuclear Safety (KoFONS) using the financial resource granted by the Nuclear Safety and Security Commission (NSSC) of the Republic of Korea. (No. 1703009) This research was also supported by the KUSTAR-KAIST Institute, KAIST. This work was in part supported by the BK21 plus program through the National Research Foundation (NRF) funded by the Ministry of Education of Korea. This research was also supported by Basic Science

Research Program through the National Research Foundation of Korea(NRF) funded by the Ministry of Science, ICT & Future Planning (NRF-2016R1A5A1013919).

References

1. Americans with Disabilities Act of 1990, as amended by the ADA Amendments Act of 2008, Pub. L. No. 110-325. http://www.ada.gov/pubs/adastatute08.htm. Accessed 9 Feb 2011
2. Barnes, V., et al.: Fitness for duty in the nuclear power industry: a review of technical issues. No. NUREG/CR-5227; PNL-6652; BHARC-700/88/018. Nuclear Regulatory Commission, Washington, DC (USA). Div. of Reactor Inspection and Safeguards; Battelle Human Affairs Research Center, Seattle, WA (USA); Pacific Northwest Lab., Richland, WA (USA) (1988)
3. United States Nuclear Regulatory Commission. NRC regulations 10 CFR Part 26 Fitness 25 for Duty Programs (2008). https://www.nrc.gov/reading-rm/doc-collections/cfr/part026. Accessed 28 Feb 2018
4. Parkinson, D.K., Bromet, E.J.: Correlates of mental health in nuclear and coal-fired power plant workers. Scand. J. Work Environ. Health 9(4), 341–345 (1983)
5. Mori, K., Tateishi, S., Hiraoka, K.: Health issues of workers engaged in operations related to the accident at the Fukushima Daiichi Nuclear Power Plant. In: Psychosocial Factors at Work in the Asia Pacific, pp. 307–324. Springer International Publishing (2016)
6. Hiraoka, K., Tateishi, S., Mori, K.: Review of health issues of workers engaged in operations related to the accident at the Fukushima Daiichi Nuclear Power Plant. J. Occup. Health 57 (6), 497–512 (2015)
7. Loganovsky, K., Havenaar, J.M., Tintle, N.L., Guey, L.T., Kotov, R., Bromet, E.J.: The mental health of clean-up workers 18 years after the Chernobyl accident. Psychol. Med. 38 (4), 481–488 (2008)
8. Shigemura, J., Tanigawa, T., Saito, I., Nomura, S.: Psychological distress in workers at the Fukushima nuclear power plants. JAMA 308(7), 667–669 (2012)
9. Sano, S.Y., Tanigawa, T., Shigemura, J., Satoh, Y., Yoshino, A., Fujii, C., Tatsuzawa, Y., Kuwahara, T., Tachibana, S., Nomura, S.: Complexities of the stress experienced by employees of the Fukushima nuclear plants. Seishin Shinkeigaku Zasshi 114, 1274–1283 (2012). (in Japanese with English abstract)
10. Pope, N.G., Hobbs, C.: Insider Threat Case Studies at Radiological and Nuclear Facilities, USA (2015). https://doi.org/10.2172/1177991
11. Howard, E.M.: 'Attempted Extortion – Low Enriched Uranium', NRC Information Notice No. 79-02 (1979). http://www.nrc.gov/reading-rm/doc-collections/gencomm/info-notices/1979/in79002.html. Accessed 2 Feb 1979
12. Howard, E.M.: 'Attempted Extortion – Low Enriched Uranium', NRC IE Circular No. 79-08 (1979). http://www.nrc.gov/reading-rm/doc-collections/gencomm/circulars/1979/cr79008.html. Accessed 2 Feb 1979
13. Sincero, S.M.: Three Different Kinds of Stress (2012). Explorable.com: https://explorable.com/three-different-kinds-of-stress. Accessed Feb 2017
14. Glazer, S., Beehr, T.A.: Consistency of implications of three role stressors across four countries. J. Organ. Behav. 26(5), 467–487 (2005)
15. Whitfield, M., Cachia, M.: How does workplace stress affect job performance? (2018)
16. Coan, J.A., Allen, J.J.B.: Frontal EEG asymmetry as a moderator and mediator of emotion. Biol. Psychol. 67(1–2), 7–49 (2004). https://doi.org/10.1016/j.biopsycho.2004.03.002

17. Thibodeau, R., Jorgensen, R.S., Kim, S.: Depression, anxiety, and resting frontal EEG asymmetry: a meta-analytic review. J. Abnorm. Psychol. **115**, 715–729 (2006). https://doi.org/10.1037/0021-843x.115.4.715

18. Gotlib, I.H.: EEG alpha asymmetry, depression, and cognitive functioning. Cogn. Emot. **12**, 449–478 (1998). https://doi.org/10.1080/026999398379673

19. Henriques, J.B., Davidson, R.J.: Regional brain electrical asymmetries discriminate between previously depressed and healthy control subjects. J. Abnorm. Psychol. **99**, 22–31 (1990). https://doi.org/10.1037/0021-843x.99.1.22

20. Tomarken, A.J., Davidson, R.J.: Frontal brain activation in repressors and nonrepressors. J. Abnorm. Psychol. **103**, 339–349 (1994). https://doi.org/10.1037/0021-843x.103.2.339

21. Sun, L., Peräkylä, J., Hartikainen, K.M.: Frontal alpha asymmetry, a potential biomarker for the effect of neuromodulation on brain's affective circuitry. Front. Hum. Neurosci. **11**, 584 (2017)

22. Grünewald, B.D., Greimel, E., Trinkl, M., Bartling, J., Großheinrich, N., Schulte-Körne, G.: Resting frontal EEG asymmetry patterns in adolescents with and without major depression. Biol. Psychol. **132**, 212–216 (2018)

23. Bruder, G.E., Fong, R., Tenke, C.E., Leite, P., Towey, J.P., Stewart, J.E., Quitkin, F.M.: Regional brain asymmetries in major depression with or without an anxiety disorder: a quantitative electroencephalographic study. Biol. Psychiat. **41**(9), 939–948 (1997)

24. Stewart, J.L., Bismark, A.W., Towers, D.N., Coan, J.A., Allen, J.J.: Resting frontal EEG asymmetry as an endophenotype for depression risk: sex-specific patterns of frontal brain asymmetry. J. Abnorm. Psychol. **119**(3), 502 (2010)

25. Lovibond, S.H., Lovibond, P.F.: Manual for the Depression Anxiety & Stress Scales, 2nd edn. Psychology Foundation, Sydney (1995)

26. Nuwer, M.: Assessment of digital EEG, quantitative EEG, and EEG brain mapping: report of the American Academy of Neurology and the American Clinical Neurophysiology Society. Neurology **49**(1), 277–292 (1997)

27. Bendat, J.S., Piersol, A.G.: Engineering Applications of Correlation and Spectral Analysis, 315 p. Wiley-Interscience, New York (1980)

28. Otnes, R.K., Enochson, L.: Digital Time Series Analysis. Wiley, Hoboken (1972)

29. Dunteman, G.H.: Introduction to Multivariate Analysis. Sage Publications, Thousand Oaks (1984)

30. Morrison, D.F.: Multivariate Statistical Methods. McGraw-Hill Book Company, New York (1967). XI11 + 338 S

31. Tabachnick, B.G., Fidell, L.S.: Using Multivariate Statistics. Allyn & Bacon/Pearson Education (2007)

32. Wasserstein, R.L., Lazar, N.A.: The ASA's statement on p-values: context, process, and purpose (2016)

33. Everitt, B.S., Dunn, G.: Applied Multivariate Data Analysis, vol. 2. Arnold, London (2001)

34. Knott, V., Mahoney, C., Kennedy, S., Evans, K.: EEG power, frequency, asymmetry and coherence in male depression. Psychiat. Res.: Neuroimaging **106**(2), 123–140 (2001)

35. Pollock, V., Schneider, L.: Quantitative, waking EEG research on depression. Biol. Psychiat. **27**, 757–780 (1990)

36. Knott, V., Lapierre, Y.D.: Computerized EEG correlates of depression and antidepressant treatment. Prog. Neuropsychopharmacol. Biol. Psychiat. **11**, 213–221 (1987)

37. Kwon, J., Youn, T., Jung, H.: Right hemisphere abnormalities in major depression: quantitative electroencephalographic findings before and after treatment. J. Affect. Disord. **40**, 169–173 (1996)

38. Kemp, A.H., Quintana, D.S., Gray, M.A., Felmingham, K.L., Brown, K., Gatt, J.M.: Impact of depression and antidepressant treatment on heart rate variability: a review and meta-analysis. Biol. Psychiat. **67**(11), 1067–1074 (2010)
39. Licht, C.M., de Geus, E.J., van Dyck, R., Penninx, B.W.: Association between anxiety disorders and heart rate variability in The Netherlands Study of Depression and Anxiety (NESDA). Psychosom. Med. **71**(5), 508–518 (2009)
40. Agelink, M.W., Boz, C., Ullrich, H., Andrich, J.: Relationship between major depression and heart rate variability: clinical consequences and implications for antidepressive treatment. Psychiat. Res. **113**(1), 139–149 (2002)
41. Kiloh, L.G., Osselton, J.W.: Clinical Electroencephalography. Butterworths, London (1961)
42. Enoch, M.-A., Rohrhaugh, J.W., Davis, E.Z., Harris, C.R., Ellingson, R.J., Andreason, P., Moorc, V., Varner, J.L., Brown, G.L., Eckardt, M.J., Goldman, D.: Relationship of genetically transmitted alpha EEG traits to anxiety disorders and alcoholism. Am. J. Med. Genet. **60**, 400–408 (1995)
43. Gotlib, I.H., Cane, D.B.: Self-report assessment of depression and anxiety (1989)
44. Clark, L.A., Watson, D.: The General Temperament Survey. Southern Methodist University, Dallas, TX (1990, unpublished manuscript)
45. Luck, S.J.: An Introduction to the Event-Related Potential Technique. MIT Press, Cambridge (2014)

Accident Diagnosis and Autonomous Control of Safety Functions During the Startup Operation of Nuclear Power Plants Using LSTM

Jaemin Yang, Daeil Lee, and Jonghyun Kim[✉]

Department of Nuclear Engineering, Chosun University, 309 Pilmun-daero, Dong-gu, Gwangju 501-709, Republic of Korea
jonghyun.kim@chosun.ac.kr

Abstract. Accident diagnosis is regarded as one of the complex tasks for nuclear power plant (NPP) operators. In addition, if the accident occurs during the startup operation, it is hard to cope with the situation appropriately because the initial conditions are different from the normal operation mode. Although operating procedures are provided to operators, accident diagnosis and control for recovery are difficult tasks under extremely stressful conditions. In order to achieve safe operation during the startup operation, this study proposes algorithms not only for accident diagnosis but also for protection control using long short-term memory (LSTM), which is an advanced version of recurrent neural networks, and functional requirement analysis (FRA). Using the LSTM, the network structures of algorithms are built. In addition, FRA is performed to define the goal, functions, processes, systems, and components for protection control. This approach was trained and validated with a compact nuclear simulator for several accidents to demonstrate the feasibility of diagnosis and correct response under startup operation.

Keywords: LSTM · FRA · Accident diagnosis · Protection control

1 Introduction

The goals of nuclear power plants (NPPs) are safe operation and power generation. To ensure safety, specific procedures and training programs are served to operators in preparation for not only the general operations but also transients or anomalies. The operator should diagnose the NPP state correctly and respond appropriately by using these procedures.

However, both the diagnostic activity and control to deal with the situation are known to be difficult and complex tasks. To diagnose the status of an NPP, operators should recognize plant states and alarms as well as check numerical instruments such as indicators. Due to the high quantities of variables generated by these measurements, it can increase the difficulty of diagnosis. In addition, the continuous monitoring under abnormal situations can be a mentally burdensome task for operators because they have to find success paths and respond to anomaly at the same time [1].

© Springer International Publishing AG, part of Springer Nature 2019
T. Z. Ahram (Ed.): AHFE 2018, AISC 787, pp. 488–499, 2019.
https://doi.org/10.1007/978-3-319-94229-2_47

Moreover, in case of startup operation, it has different features that can disrupt awareness of the NPP state. In terms of the diagnostic activities under full power operation, generally, the status of NPP variables is stable. Thus, the detection of anomaly or abnormal condition is easier than startup operation. Also, the procedures are well prepared depending on the type of anomalies or accidents. However, accident diagnosis in startup operation is more complex than full power because several operation modes can exist. Each operation mode has different and various plant states (e.g., pressure condition, reactor coolant temperature, and water level) so that operators cannot interpret the situation easily, even it can cause the wrong diagnosis. Even though there are procedures for responding to specific events, the situation may be different. This is due to the unavailability of components and systems for various reasons (e.g., maintenance activity, bypassing, and unsatisfactory condition to use), so that they cannot be operable at a necessary time. Therefore, the safety of NPP can be decreased by the weakening of the defense in depth concept and lack of risk management [2].

In this light, this study aims to develop an accident diagnosis algorithm and autonomous control algorithm to protect the NPP under abnormal situations in the startup operation. Functional requirement analysis (FRA) is performed to figure out the structure of NPP safety functions for control. In accordance with the time series feature of NPP datasets, an improved recurrent neural networks (RNN), long short-term memory (LSTM), which is a kind of the artificial neural networks (ANN), is applied to develop the algorithms. The network is trained through a compact nuclear simulator (CNS) that is the implementation model of a Westinghouse three-loop, 930MWe pressurized water reactor (PWR). The demonstration of its applicability is validated with some test datasets.

2 Methodology

This study proposes two methodologies (i.e., LSTM and FRA) to develop the algorithms. The LSTM is applied to developing the algorithm structure to diagnose the accident and control the functions under abnormal situation or emergency. FRA is performed for decomposing safety functions to control considering the goal of safety and to draw inputs and outputs of those functions.

2.1 Long Short-Term Memory (LSTM)

Varieties of diagnostic algorithms and control algorithms have been suggested to reduce the burden of operators and help respond correctly to anomalies in NPPs. For instance, ANN, fuzzy logic, the hidden Markov model (HMM), and the support vector machine (SVM) are representative examples and these kinds of algorithms applied approaches based on artificial intelligence (AI) techniques generally. The ANN is regarded as one of proper approaches to cope with datasets in the pattern recognition to find features within the given datasets that are non-linear and dynamic. It is especially a promising method for the accident diagnosis and control because a characteristic of accident diagnosis and control of NPPs is feature extraction and response in context [3–7].

Among the artificial intelligence approaches, especially the RNN shows good performance at analyzing time series data (i.e., data which include sequential information). Unlike the conventional ANN-based methods, RNN assumes that the input and output are not independent from each other, that is, sequential data can be used as input for an algorithm. In addition, by every element of a sequence, the same calculation is performed and the output result is affected by the previous calculation result. Therefore, the structure of arrangement of units looks like a circulation shape, thus, all units share the same parameters in contrast with general ANNs with different parameters. Due to the circular structure, it can reflect the features of sequential data so that it can naturally represent dynamic systems and can capture the dynamic behavior of a system.

However, even though RNN can learn past values through time (i.e., back propagation through time), too much back propagation for a long time causes either the blowing up or vanishing gradient problem. Because the RNN cell is composed of multiplication operations, regardless of whether the value is small or not, a repetitive multiplying operation can result in wrong consequences. In the case of blowing up, it may cause the oscillation of weights, whereas vanishing gradient problem may lead weights to be almost zero.

The LSTM was proposed to resolve the above mentioned issues of RNN. It has been developed based on RNN architecture for processing long sequential data. Although the LSTM is based on RNN, it can deal with long sequential data because the cell unit structure is different. In case of LSTM, it uses a specific cell unit, which is called a memory cell in place of RNN neuron. Each LSTM cell unit consists of three gates (i.e., input gate, forgetting gate, and output gate), and through these gates, it adjusts the output values. In case of the input gate, it determines how much to reflect the input value, whereas the forgetting gate determines the degree to forget the value of previous cell state. In addition, the output gate determines the degree of output. Figure 1 shows the LSTM cell architecture applied in this study [8]. The Eqs. (1) to (4) stand for each gate denoted by 'i', 'o', 'f', and input node denoted by 'g' which has *tanh* activation function denoted by 'ϕ'.

$$g_l^{(t)} = \phi\left(W_l^{gx}h_{l-1}^{(t)} + W_l^{gh}h_l^{(t-1)} + b_l^g\right) \tag{1}$$

$$i_l^{(t)} = \phi\left(W_l^{ix}h_{l-1}^{(t)} + W_l^{ih}h_l^{(t-1)} + b_l^i\right) \tag{2}$$

$$f_l^{(t)} = \phi\left(W_l^{fx}h_{l-1}^{(t)} + W_l^{fh}h_l^{(t-1)} + b_l^f\right) \tag{3}$$

$$o_l^{(t)} = \phi\left(W_l^{ox}h_{l-1}^{(t)} + W_l^{oh}h_l^{(t-1)} + b_l^o\right) \tag{4}$$

These equations update a layer of memory cells $h_l^{(t)}$ where $h_{l-1}^{(t)}$ represents the previous layer at the same sequence step and $h_l^{(t-1)}$ represents the same layer at the previous sequence step. Because of these advantages, the LSTM has been applied for a

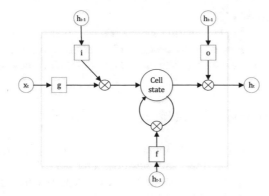

Fig. 1. The architecture of the LSTM cell

variety of tasks for varying-length sequential data, such as natural language processing, sentiment capture from writings, video classification on the frame level, automatic speech recognition, genomic analysis and so on [9–11].

2.2 Functional Requirement Analysis (FRA)

To verify the functions that must be carried out for responding to accidents or anomalies, an FRA has been performed for the goal of plant safety. This approach is a top-down comprehensive analysis of the requirements for developing control algorithm. Through the FRA, it is possible to provide the framework with functions necessary for ensuring safety and to identify the system and the device that are responsible for satisfying the functions [12].

There are four steps to perform FRA as follows:

Step 1. Determining the scope of FRA
Before performing the FRA, the scope of FRA to design the model should be determined considering the goal of the project. The scope can be identified in consideration of goal-related documents such as design descriptions of the plant, operation procedures, etc. Then, high-level functions that are related to the goal can be identified as outputs (e.g., functions which are critical for safety).

Step 2. Decomposing the high-level functions
The high-level functions consist of low-level functions, which are processes and systems necessary for achieving the goal. From decomposing the high-level functions, processes and systems to achieve the goal can be extracted as outputs. To decompose the high-level functions, the research and investigation of operation methods for the relevant function should be performed.

Step 3. Characterizing the functions
The combination of systems and components used to achieve a function should be characterized. Identification of how systems and components perform the functions are performed. The outputs from this step are the fundamental components and subsystems related to functions.

Step 4. Verification of FRA

The final step of the implementation of the FRA is verification of the results. In accordance with the review of NRC criteria [12], verification should be performed. In addition, it may change existing safety functions or introduce new functions.

3 Accident Diagnosis Algorithm

This chapter introduces accident diagnosis algorithm considering the plant dynamics and operational mode. To reflect the time-series characteristics, the LSTM network structure is used. The algorithm is trained with the data obtained from the compact nuclear simulator. The validation of network is performed with the test datasets.

To model the algorithm, a desktop computer with the following hardware configurations is used: NVIDIA GeForce GTX 1080 8 GB GPU, Intel 4.00 GHz CPU, Samsung 850 PRO 512 GB MZ-7KE512B SSD, and 24 GB memory. Python 3.6.3 is used for coding language that is one of the most popular computer languages for machine learning and deep learning. The libraries developed to model the algorithm for machine and deep learning (e.g., Keras, Tensorflow) were used for implementation.

3.1 Classification of Operational States

This study identifies different operational states during the startup operation. The framework of accident diagnosis algorithm was developed considering both the complex NPP characteristics and dynamic states under startup operation. In case of accident diagnosis under startup operation, operation modes are different so that initial conditions under accident or anomaly are different by mode. In addition, even if the operation mode is same, the availability of components or systems can be different depending on the time of occurrence. Thus, the accident diagnosis algorithm should be built in consideration of modes and important steps that can change the availability or plant state.

3.2 Network Modeling for Accident Diagnosis

The network model based on the LSTM for accident diagnosis can be regarded as a multi-classification problem [13]. Therefore, the model is designed as a many-to-one structure. Figure 2 shows the network structure of the model applied in this study. The network is composed of three LSTM layers and one output layer, which batch sizes are 32 and 8, respectively. The structure of network can be customized to optimize the network considering its performance. In addition, the softmax function, which can assign the probabilities to diagnosis results, is included in the output layer to decide the ranking. It is commonly used for the deep learning model to classify several classes (i.e., more than three classes). In addition, 10 time steps (i.e., 10 sets of NPP datasets, which means NPP trend) are considered in this model.

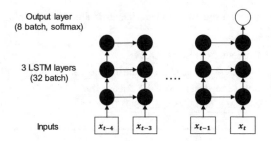

Fig. 2. The network structure of the model for multi-label classification

3.3 Pre- and Post-processing

Pre-processing of input data and Post-processing of output data are performed to be used in the input layer and output layer, respectively. Because the normalization can prevent the reduction of learning speed and getting stuck in local minima (i.e., not global minima but one of the minima among the several minimum points), which can cause performance degradation of a network.

In case of pre-processing, the min-max scaling method is applied via the following Eq. (5).

$$X_{norm} = (X - X_{min})/(X_{max} - X_{min}) \tag{5}$$

The input data are scaled from zero to one considering minimum and maximum of the collected data [14], whereas the output data are scaled within zero to one by the softmax function in output layer through the Eq. (6). Despite the transformation, the magnitude relation among the results does not change [15].

$$S(y_i) = e^{y_i} \Big/ \sum e^{y_i} \tag{6}$$

3.4 Training

The training of the network is performed using the CNS that was originally developed by the Korea Atomic Energy Research Institute (KAERI). The reference plant of CNS is the Westinghouse 3-loop, 930 MWe PWR. To select the inputs for training, a total of 51 parameters were selected based on procedures and importance, which can affect the NPP states and component availability. Then, 65 scenarios (2% power) with 11,571 datasets (i.e., 11,571 s of data including 51 parameter values in each time step) are used for training, as shown in Table 1.

3.5 Test Results

The validation of network is performed with 17 scenarios with 3,395 datasets, as shown in Table 2. Figure 3 shows the results of validation based on accuracy and loss. The X-axis and Y-axis represent the epochs and accuracy (or loss). The accuracy value is

Table 1. Scenarios used for accident diagnosis algorithm training (2% power)

Initiating events	Numbers
Loss of Coolant Accident (LOCA)	32
Main Steam Line Break (MSLB) inside the containment	12
Main Steam Line Break (MSLB) outside the containment	12
Steam Generator Tube Rupture (SGTR)	9
Total	65

Table 2. Scenarios used for accident diagnosis algorithm test (2% power)

Initiating events	Numbers
Loss of Coolant Accident (LOCA)	8
Main Steam Line Break (MSLB) inside the containment	3
Main Steam Line Break (MSLB) outside the containment	3
Steam Generator Tube Rupture (SGTR)	3
Total	17

almost 0.94 and the loss is almost 0.13, also, the validation accuracy and loss are almost converged to training accuracy and loss. In other words, it means the training of algorithm is carried out well without overfitting or underfitting.

The accident diagnosis algorithm has been tested with test scenarios. LOCA in loop2 cold-leg with the 40 cm^2 break size and SGTR in loop1 with the 10 cm^2 break size are used. The malfunction of each scenario is injected at 10 s. Figure 4 shows the diagnosis results. The solid line and dotted line mean the actual value of test data and the diagnosis results of the algorithm, respectively. The X-axis means the time, and Y-axis stands for the diagnosis result. In case of LOCA, after 20 s, diagnosis result is constantly converged to almost one. In case of SGTR, after 35 s, diagnosis result is constantly converged to almost one.

4   Protection Control Algorithm

In order to respond correctly in abnormal situations, this study suggests an algorithm to control systems and components related to safety functions (i.e., protection control algorithm). As well as the accident diagnosis algorithm, protection control algorithm is based on LSTM. To model the algorithm, identification of components and systems for the safety functions is performed using FRA. In accordance with FRA results, inputs and outputs for the algorithm are determined. The protection control algorithm consists of four networks by the mode (i.e., cold shutdown, hot shutdown, hot standby, startup), as like the accident diagnosis algorithm. The pre-processing of inputs is performed in the same manner as accident diagnosis algorithm. In addition, the post-processing of outputs is performed to control the systems and components accurately.

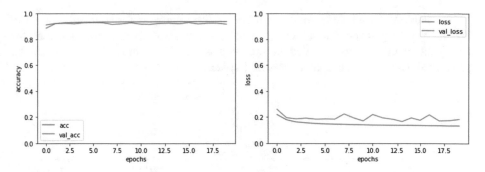

Fig. 3. Validation results of the trained algorithm with accuracy and loss

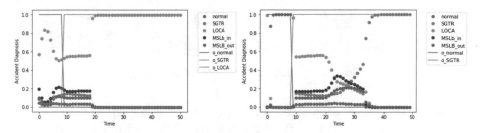

Fig. 4. Diagnosis results (left: 40 cm^2 LOCA in loop 2 cold-leg, right: 10 cm^2 SGTR in loop1)

4.1 Safety System Modeling Using FRA

This study models NPP safety systems by using FRA. The reference plant for FRA is a Westinghouse 3-loop PWR same as accident diagnosis algorithm. To determine the scope of FRA, the goal is defined as the safe operation to respond the anomalies. In accordance with the goal of safety, seven critical safety functions (CSFs) are extracted. Table 3 shows seven CSFs and their purposes.

Table 3. Seven CSFs

CSFs	Purposes
Subcriticality	• Shut reactor down to reduce heat production
Core cooling	• Transfer heat from the core to cool it down
Heat sink	• Ensure the integrity of heat sink
RCS integrity	• Transfer heat out of the coolant system • Maintain pressure and temperature of reactor coolant system
Containment integrity	• Close valves penetrating containment • Control pressure and temperature
RCS inventory	• Maintain volume or mass of reactor coolant system
Maintenance of vital auxiliaries	• Maintain operability of systems needed to support safety systems

In addition, the whole framework of NPP safety functions for control is shown in Fig. 5. Not only seven critical safety functions, but also the relevant processes for satisfying functions are modeled in the framework. Then, those processes can be performed using relevant components and systems.

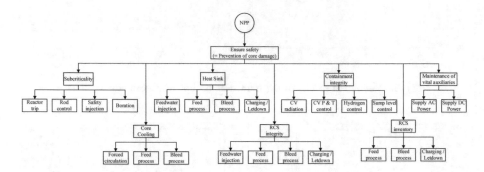

Fig. 5. The framework of NPP safety functions

Table 4 shows an example of processes, systems, and components to satisfy the function of reactivity control. Plant protection system (PPS), digital control rods system (DCRS), safety injection system (SIS) and chemical and volume control system (CVCS) are modeled under subcriticality function. In case of SIS and CVCS, they include relevant pumps, tanks, and valves, respectively. Then, the input/output can be identified as parameters indicating related system states and components states for control.

Table 4. Processes, systems, and components to satisfy the subcriticality function

Processes	Systems	Components
Reactor trip	Plant protection system	Control element drive mechanism
Rod control	Digital control rods system	Control element drive mechanism
Safety injection	Safety injection system (SIS)	Safety injection pump
		Safety injection tank
		Safety injection valves
Boration	Chemical and volume control system (CVCS)	Boric acid storage tank
		Safety injection tank
		CVCS valves

4.2   Network Modeling for Safety Function Control

Based on the FRA results, the network is modeled using LSTM for control to satisfy the CSFs. This study applies many-to-one structure to design the prediction model.

Figure 6 shows the network structure of the model for prediction. The network is composed of an input layer, two hidden layers, and one output layer. The hidden layers have 60 hidden nodes. A total of 168 input parameters and component states from the NPP are used as inputs to the LSTM network. They consist of 74 physical parameters and 94 states of components and systems. The outputs of the network will be the values within 0 to 1 indicating the states of components and systems.

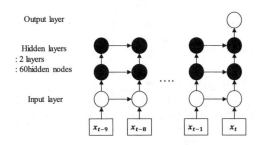

Fig. 6. The network structure of the model for prediction

4.3 Data Processing

Input pre-processing is performed with min-max normalization method to prevent getting stuck in local optima and reduce the time for training network. More details are written in Sect. 3.3.

In addition, post-processing of output is performed to convert the discrete control values (e.g., open or close, start or stop). In case of the output of discrete control components or systems, only 1 and 0 can be inputs to control, whereas the output of LSTM is the printed values that range from 1 to 0. Thus, to convert outputs to 1 and 0, the standard value was given according to the average error that is the difference between training data and predicted data of On/Off parameter. As a result of the calculation, the standard value is decided to 0.05, so when the LSTM output reaches 0.95 in the downward direction or 0.05 in the upward direction, the state of the component is changed. Figure 7 shows an example for post-processing of output for the discrete control.

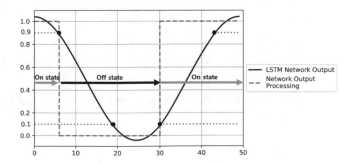

Fig. 7. Post-processing of output for the discrete control

4.4 Training

To train the network, the data have been collected from the CNS. Then, 106 scenarios (2% power) with 102,237 datasets (i.e., 102,237 s of data including 168 parameter values in each time step) are used for training, as shown in Table 5.

Table 5. Scenarios used for protection control algorithm training

Initiating events	Number
LOCA	20
LOCA + safety injection failure	20
SGTR	18
SGTR + safety injection failure	18
Main steam line break (MSLB) inside the containment	15
Main steam line break (MSLB) outside the containment	15
Total	106

4.5 Test Results

To validate the algorithm, untrained LOCA scenario with different break size has been used. The validation is performed to compare the performance of the algorithm and the automation + human control for safety functions in LOCA as shown in Fig. 8. The left figure shows the steam generator PORV position, whereas the right figure shows the steam generator water level. In case of PORV, the protection control algorithm starts control almost 250 s faster than automation + human control. Not only the PORV but also the other components, which are not shown in this paper show the similar results about control (e.g., starts faster, open faster, close faster). The right figure shows that the steam generator water level has recovered faster, and even maintains a higher level.

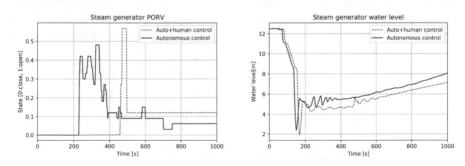

Fig. 8. Comparison of autonomous control and automation + human control in LOCA.

5 Conclusion

This study proposes accident diagnosis algorithm and protection control algorithm for safety functions of NPPs by using the LSTM network as well as the FRA. Both networks that are applied in algorithms were trained using a CNS and also validated to demonstrate the effectiveness of the algorithm. In case of accident diagnosis algorithm, it is expected that the safety of NPP during startup operation can be improved by application of the algorithm that can unload task of operators in abnormal situations. In addition, the protection control algorithm can improve the feasibility to respond to accident appropriately or even better.

References

1. Maruo, O., et al.: HSI for monitoring the critical safety functions status tree of a NPP. In: International Nuclear Atlantic Conference, Recife, PE, Brazil, November, pp. 24– 29 (2013)
2. Jang, S.C., et al.: Development of Risk Assessment Technology for Low Power, Shutdown and Digital I and C System. No. KAERI/RR–2794/2006. Korea Atomic Energy Research Institute (2007)
3. Santosh, T.V., et al.: Application of artificial neural networks to nuclear power plant transient diagnosis. Reliab. Eng. Syst. Saf. 92(10), 1468–1472 (2007)
4. Uhrig, R.E., Hines, J.: Computational intelligence in nuclear engineering. Nucl. Eng. Technol. 37(2), 127–138 (2005)
5. Fantoni, P.F., Mazzola, A.: A pattern recognition-artificial neural networks based model for signal validation in nuclear power plants. Ann. Nucl. Energy 23(13), 1069–1076 (1996)
6. Lee, S.J., Seong, P.H.: A dynamic neural network based accident diagnosis advisory system for nuclear power plants. Prog. Nucl. Energy 46(3–4), 268–281 (2005)
7. Mo, K., Lee, S.J., Seong, P.H.: A dynamic neural network aggregation model for transient diagnosis in nuclear power plants. Prog. Nucl. Energy 49(3), 262–272 (2007)
8. Hochreiter, S., Schmidhuber, J.: Long short-term memory. Neural Comput. 9(8), 1735–1780 (1997)
9. Auli, M., Galley, M., Quirk, C., Zweig, G.: Joint language and translation modeling with recurrent neural networks. In: Empirical Methods in Natural Language Processing (EMNPL), vol. 3 (2013)
10. Sutskever, I., Vinyals, O., Le, Q.V.: Sequence to sequence learning with neural networks. In: Advances in Neural Information Processing Systems (NIPS), vol. 27, pp. 3104–3112 (2014)
11. Pollastri, G., Przybylski, D., Rost, B., Baldi, P.: Improving the prediction of protein secondary structure in three and eight classes using recurrent neural networks and profiles. Proteins: Struct. Funct. Bioinf. 47, 228–235 (2002)
12. O'Hara, J.M., et al.: NUREG-0711. Human Factors Engineering Program Review Model. Rev 3 (2012)
13. Lipton, Z.C., et al.: Learning to diagnose with LSTM recurrent neural networks. arXiv preprint arXiv:1511.03677 (2015)
14. Jain, Y.K., Bhandare, S.K.: Min max normalization based data perturbation method for privacy protection. Int. J. Comput. Commun. Technol. 2, 45–50 (2011)
15. Lei, Y., et al.: An intelligent fault diagnosis method using unsupervised feature learning towards mechanical big data. IEEE Trans. Ind. Electron. 63(5), 3137–3147 (2016)

Applications in Energy Systems

Exploring the Potential of Home Energy Monitors for Transactive Energy Supply Arrangements

Andrea Taylor[1(✉)], Bruce Stephen[2], Craig Whittet[1],
and Stuart Galloway[2]

[1] The Glasgow School of Art, Glasgow, UK
{A.Taylor, C.Whittet}@gsa.ac.uk
[2] University of Strathclyde, Glasgow, UK
{Bruce.Stephen, Stuart.Galloway}@strath.ac.uk

Abstract. There has been considerable investment in micro energy generation from both domestic consumers and small-scale providers. However, current metering arrangements and home energy monitoring products are too basic to enable real-time billing and remuneration, limiting the effectiveness of this investment. This paper describes the exploration of home energy monitors as a technical enabler to unlock the local trading potential of the investment in micro energy generation, and the human factors involved in interacting with these products that might pose obstacles to successful uptake. First, a human factors analysis of eight home energy monitors was conducted, which identified a number of usability issues. Next, a range of design concepts were developed to address the key usability problems identified, incorporate the forward-looking facility for alternative energy supply models, and stimulate further investment in energy prosumption. This study contributes an understanding of the potential of home energy monitors for transactive energy supply arrangements.

Keywords: Human factors · Energy transactions · Interface design
Product Design Engineering

1 Introduction

The UK Government has legislated ambitious policies to reduce greenhouse gases, with the Climate Change Act (2008) mandating emissions reductions from 1990 levels of 34% by 2020 and 80% by 2050 [1]. The resulting penetrations of small scale low carbon technologies including renewable energy sources and distributed storage have led to resilience driven aspirations of decentralized power system control and operation at local and neighborhood level. Although not currently realized, coordination of local distributed generation sources and a greater degree of demand flexibility offers the potential to lower the cost of energy at source and to enable remuneration for consumer participation, addressing rising costs of energy supply which impacts strongly on all consumers, in particular the fuel poor.

Realizing consumer participation in the delivery of energy services in the UK currently has a number of barriers in the form of limited metering resolution at

© Springer International Publishing AG, part of Springer Nature 2019

T. Z. Ahram (Ed.): AHFE 2018, AISC 787, pp. 503–513, 2019.
https://doi.org/10.1007/978-3-319-94229-2_48

settlement points, regulation governing retail and demand response services being limited to only the largest energy users (e.g. industrial customers). Power demand and supply from residential buildings is a domain of growing complexity as on-site renewables, heat and electricity storage and electric vehicles begin to gain popularity at distribution network level. Understanding how highly variable household power demand and supply, driven by stochastic occupant behaviors and influenced by new technologies, can be integrated with energy trading technology is a key challenge. By far the greatest barrier though to end user participation in power system operation is the overhead associated with dealing with large numbers of small players: aggregate contributions would provide a valuable operational response but the size of cohort required to provide it would generate transaction cost overheads far greater than the value. Automation provides a solution to this, but for the householder, how this can interface in a transparent and intuitive manner is an additional obstacle to implementation [2].

Technological enablers are required though that will address the technical and social issues surrounding the use of automated technology for energy supply and services. Existing models use a very conservative point-measurement/central-billing method that is supported by usage estimates to determine customer bills. This often leads to a breakdown of trust in the sector. Furthermore, the quarterly or monthly resolution at which bills are produced has compounded this trust through lack of transparency. Moving to a digitized based settlement process combined with automation[1] makes multiple small transactions possible, supported by higher resolution metering. Suppliers and consumers authenticating such systems instill trust in the business arrangement. Finally, secure digitized solutions can support much more representative billing algorithms, allowing for dynamic pricing throughout the day and thus short term economic support for demand side response (the use of demand flexibility to address shortfalls or surpluses of generation as might be expected from intermittent renewable sources) [3].

1.1   The Study

The EPSRC (Engineering and Physical Sciences Research Council) project TESA (Transactive Energy Supply Arrangements) aimed to landscape the social, technical, regulatory, and design pathways to enable the future supply of energy to customers that will further stimulate investment in electricity generation both from small-scale providers and customers themselves, reducing the demand for large-scale infrastructure investments [4]. A key objective of the TESA project was to inform on the enablers and obstacles for future energy supply arrangements at the neighbourhood level as a means of supporting the 'people' side of digital transactive models. This study explored the potential of home energy monitors, formally known as In Home Displays (IHDs), as a technical enabler to unlock the local trading potential of the investment in micro energy

---

[1] According to a study reported in IEEE Spectrum (Oct 2017), many major companies are working on integrating Blockchain Technology into their products. This is seen as a disruptive development as we move towards a more 'digital future' and at the macro level it is seen as a better way, a more cost effective way, of managing data and trust interfaces.

generation (small-scale power generation) from both domestic consumers and small-scale providers, and the human factors currently involved in interacting with these products that might pose obstacles to successful uptake. The information that will enable engagement in future energy supply arrangements is multifaceted and has to be presented carefully, and with consideration to the householder. The aim is to avoid placing an excessive burden on householders through their interaction with energy products. The study builds on and extends our previous work, which involved a practical review of energy saving technology for ageing populations [5].

Home energy monitors are designed to increase householders' awareness and understanding of energy usage, connecting routine behavior to consumption in order to motivate conservation behavior and reduce energy bills. Home energy monitors are commercially available or provided by energy suppliers when installing a smart meter. Commercially available energy monitors are typically made up of three parts: an in home display, a sensor, and a transmitter. The householder is required to clamp the sensor on to a power cable connected to the electricity meter, which measures the current passing through it. The transmitter sends the data wirelessly to the display unit. Energy monitors provided by energy suppliers also incorporate wireless transmission to connect to an in-home display [6], however householders are not required to install the technology. Typically, electricity usage is displayed in units of energy used (kWh), cost (£) or carbon emissions (CO_2). In the UK, in home displays are required by standard to provide tariff information, consumption, time of day and in some cases information pertaining to remaining credit. Some displays offer additional features such as alerts when a set amount of electricity has been used.

2 Method

The research was conducted by 20 Product Design Engineering students from The Glasgow School of Art and University of Glasgow in the UK, in 2017, supported by academic staff working on the TESA project. Each student selected a user group to focus their research on, primarily based on who they deemed likely to practice and benefit from 'energy prosumption' (production and consumption of energy) in the future. The students then identified and recruited a small group of householders to take part in the study, mostly through their social networks, who matched the profile of their chosen user group. The user groups included: students/young people living in shared accommodation who value the environment and live on a budget; family home owners (all ages: young to elderly people) who produce or plan to produce renewable energy; and young professionals who value the environment and have a disposable income to invest in renewable energy devices and technologies in their home. Approximately 110 householders living in residential areas of South West Scotland participated in the study. Actively engaging with householders at all stages of the study helped to ensure the likelihood that the resulting design concepts and future scenarios are both fit for purpose and desirable for the people who will use them.

2.1   Human Factors Analysis

Working individually or in small teams, the students conducted a human factors analysis of eight home energy monitors: Amphiro B1 Connect by Amphiro [7], Elite Classic by Efergy [8], EM02 Power Meter by TACKLife [9], In Home Display by Scottish Power [10], Minim+ by geo [11], OWL +USB by OWL [12], OWL Micro+ by OWL [13], and Smart Energy Tracker by Scottish and Southern Energy (SSE) [14]. The Amphiro B1 Connect is somewhat different to the other products reviewed in that it monitors energy consumption from a particular device: a shower. The product fits directly to the shower and displays the current water consumption and temperature. The other products reviewed monitor electricity consumption in the home. Additionally, the In Home Display by Scottish Power and the Smart Energy Tracker by SSE monitor gas consumption. None of the existing product base currently utilizes blockchain technology, and none of the products reviewed monitor energy generated by renewable sources such as solar panels. The monitors were chosen to provide an overview of the market sector: none were supplied by manufacturers or sales agents.

The analysis itself was exploratory. First, a high-level task analysis was conducted based on the 'out of the box' product experience. This included everything from the process of unpacking and revealing the product for the first time to installing (if required), configuring and operating it (Fig. 1). Procedures and processes were documented, and a high-level sequential task description was produced. Next, the high-level task analyses were used to define representative scenarios that covered all aspects of energy monitoring. The students then took each product/scenario, and working with participants, performed a verbalized/observed product walkthrough using the procedure laid down in [15]. The full spectrum of scenarios, from unpacking to operating the product were covered, and the scenarios were frozen at key points to allow for in-depth questioning. Key insights were documented, recorded and/or photographed as appropriate. Finally, procedures and processes identified during the product walkthrough as being particularly problematic were subject to further research. The students selected from a range of 12 established usability methods (Table 1) to develop a more tailored analysis, primarily based on the issues discovered and the insights that each method is designed to provide.

At the conclusion of the data collection phase, the students and principle investigators convened and undertook a presentation workshop in order to synthesise, cross check and group the findings.

2.2   Design Concepts and Future Scenarios

Based on the results of the human factors analysis, the students explored and developed a set of design concepts and future scenarios to: address the key usability problems identified during the human factors analysis, incorporate the forward-looking facility for true transactive energy supply arrangements, and stimulate further investment in energy prosumption (production and consumption of energy). An iterative design methodology was adopted, involving the continual improvement of the design concepts. First, each student brainstormed ideas. They then selected a design concept for further development and generated low-fidelity prototypes e.g. sketches and foam

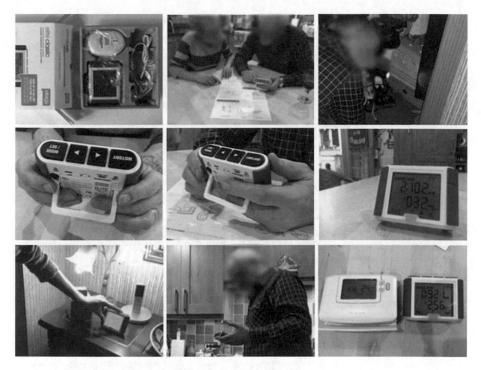

Fig. 1. Observations of an elderly couple setting up a home energy monitor.

models. Next, the concept was presented to several participants as test users, any problems and opportunities were noted, and the concept was evaluated and refined accordingly. As required (time permitting), the students repeated the preceding step until the design concept was resolved to a level required for final presentation.

3 Results

3.1 Human Factors Analysis: Key Problems Identified

The human factors analysis identified a number of usability problems with the home energy monitors reviewed, some of which were discovered in our earlier practical review of energy saving technology for ageing populations [5]. This is perhaps not too surprising as three of the products reviewed were the same in both studies: Elite Classic, OWL Micro+ and OWL +USB. However, our earlier review was conducted in 2012, five years prior to the current study, and it is disappointing to discover that little progress has been made on these issues. The key problems identified were:

- Accessing the electricity meter if it is located in a hard-to-reach/dark location
- Identifying the main power cable from the other electrical cables
- Attaching the sensor to the power cable e.g. due to the stiffness of the clip
- Following the set-up instructions e.g. due to a large volume of text

Table 1. The analysis of the energy monitors involved selecting from 12 methods.

Method	Insights provided
Heuristics (e.g. [16])	A flexible subjective approach in which observations during product usage are recorded
Immersion analysis (e.g. [17])	Immersion in the research process from the point of view of the user for design empathy and insight e.g. simulation of physical disabilities
Focus groups (e.g. [18])	Group interviews where participants are selected on the basis they would have something to say on the topic and would be comfortable talking to the interviewer and to each other
Semi-structured interviews (e.g. [15])	Pre-ordered questions and themes presented to participants by the interviewer, but with scope to branch off into other relevant areas as required
Function flow diagrams (e.g. [19])	Graphical representation of device function and events that occur during the performance of a task
Abstract hierarchy (e.g. [20])	A model of the system in terms of a hierarchy of functions, from the most abstract of functions to the most local of processes, and their relationships to one another
Microsoft product reaction cards (e.g. [21])	A set of cards with a word (adjective) written on each card that participants are asked to select from to describe their response to a product
Personas (e.g. [17])	A synthesis of the full participant pool into a subset of fictional characters that embody the dominant traits in the sample
Cambridge impairment simulator (e.g. [22])	Filters are applied to image and sound files to simulate some of the main effects of common visual and hearing impairments
Co-creation/video prototypes (e.g. [17])	Visual prototypes that are taken to participants in order to gather initial design feedback
Photo diary (e.g. [17])	A technique whereby participants themselves can take photos of device issues as they are encountered over a longer time period and without an analyst being present
Design with intent toolkit (e.g. [23])	Cards and worksheets that act as different 'lenses' through which common problems can be viewed and new perspectives gained

- Finding/understanding tariff information (needed for an accurate display of cost)
- Pairing the transmitter and the display
- Operating awkward to reach controls e.g. buttons located at the rear of the display
- Reading fixed LCD screens with narrow viewing angles and no backlighting
- Understanding electricity usage when expressed as kWh or CO_2
- Understanding visually busy information displays
- Understanding energy consumption as normal/high/low (compared to others)
- Knowing how to save energy based on the information provided
- Losing interest with the monitor e.g. due to lack of noticeable feedback
- Finding the monitor aesthetically unattractive
- Setting up profiles (accounts) for individual members of a household.

A range of design concepts were developed to address the key problems identified with the energy monitors reviewed. For example, one concept proposed a 'tips' information screen that provides personalized suggestions on how to save energy based on appliance usage (Fig. 2 *left*); and a second concept proposed a display with illuminated edges that change color along a spectrum of blue to red, to provide noticeable information 'at a glance' on current energy usage e.g. a change to red provides a visual cue that usage levels are high (Fig. 2 *right*).

3.2   Design Concepts and Future Scenarios

A wide range of design concepts were explored and developed. In particular, concepts focused on provision for neighbors to trade energy directly, more efficiently, with each other; new interfaces for energy forecasting (demand and pricing) to incentivize householders to use energy during periods of high production and low demand; and new interfaces to incentivize householders to generate renewable energy, fitting in with the alternative transactive models of TESA. Generally, the scenarios assume that energy suppliers will have installed smart meters in all households, and tariffs will be based on the time of day where energy suppliers charge more at times of low energy production and high energy demand, and vice versa, in order to help reduce strain on the national power grid. Time of use tariffs are already available in some countries (most notably France) and will become more flexible as technology permits, although there is a risk of 'information overload' for the householder [24].

Provision for Neighbors to Trade Energy Directly. Several of the design concepts explored local market trading between domestic prosumers (households who produce and consume energy) at the neighbourhood level. For example, the *OWL solar+* concept is designed for a potential future where Government has, through subsidies, funded installation of solar panels on the rooftops of blocks of flats [25]. In this scenario, energy generated by the solar panels is split equally between the flats as a network. The *OWL solar+* encourages householders to share (and request) surplus energy with neighbours, in the spirit of community, as any unused energy is fed into the conventional power grid. The *Hub* concept similarly enables neighbours to trade energy (Fig. 3). The interface includes options for householders to select how much electricity or water they want to trade and to negotiate a price with a neighbour. The interface also includes a weather forecast (e.g. amount of rain expected to fall in the next few days) to aid decision-making.

New Interfaces for Energy Forecasting. Residential energy use is very diverse and follows the routines of dwelling occupants rather than localized trends making it difficult to predict and accommodate [26]. To persuade householders to follow neighborhood group patterns, a common directive could be imparted through an in home display. Consequently, other design concepts explored new interfaces for energy forecasting to incentivize householders to use energy during periods of high production and low demand, and to highlight the opportunity to use affordable and clean energy. For example, the *EnergyApp* concept displays the current price of energy, color-coded to communicate 'at a glance' when prices are high or low, and the predicted price of energy over the upcoming period (Fig. 4). In addition, the *EnergyApp* displays a

Fig. 2. Design concepts were developed to address the key usability problem.

Fig. 3. The *Hub* concept allows neighbors to trade electricity or water.

percentage breakdown of how energy is currently being generated (% renewable energy, fossil fuels, nuclear) to encourage householders to consume energy, e.g. charge their electric car, when levels of renewable energy are high. The *CLU* concept is designed for a potential future where there is an increase in communal living (co-living) as a solution to the problem of population growth, and in particular, the large percentage of the population living in cities. The monitor, which is designed for communal use, displays predicted peaks in energy generated by renewable sources, giving householders the opportunity to choose clean energy.

New Interfaces to Incentivize Householders to Produce Renewable Energy. Several of the design concepts explored how householders might be incentivized to generate renewable energy. For example, the *Future of Energy Monitoring and Exchange* concept is designed for a potential future where householders require to be convinced of the cost-benefit of producing energy. In this scenario, Government will fund solar panels to be installed on local government buildings (e.g. community center, library), and households in the vicinity will be given an energy monitor to buy (and trade) energy with these buildings, thereby supporting local services, or from (and with)

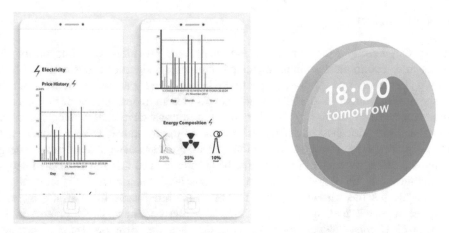

Fig. 4. The *EnergyApp* concept (left) displays energy forecast and energy composition diagrams. The *CLU* concept (right) displays predicted production peaks in renewable energy.

neighbors who produce energy. A motivational feature of the interface is notification of how much energy could be produced if the householder installed solar panels. The *Social Energy Network* concept is designed for a potential future where communities use energy generated by the community rather than a conventional energy provider (Fig. 5). The interface incentivizes householders to produce renewable energy by making it easy to trade with neighbours and by giving more control. A key layout provides a graphical overview of those neighbours who are on the network and how much energy they wish to trade; selecting a neighbour opens a dialog box with more details on, and options to, conduct the transaction.

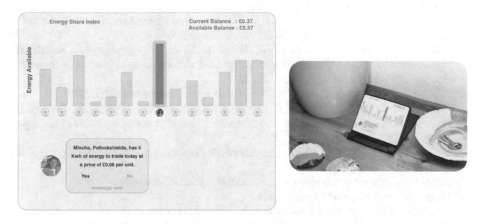

Fig. 5. The *Social Energy Network* is designed for a community of domestic prosumers.

4 Conclusion

Transacting directly with consumers and small-scale providers deep into utility networks in an economically viable manner is currently a significant business and technological challenge with not all barriers well understood, particularly those from the perspective of human factors. This study contributes an understanding of the potential enhancement of home energy monitors to support transactive energy supply arrangements and the obstacles posed by the failure to apply human factors principles to current products. Low carbon generation technologies are inevitable in future energy systems meaning that intermittency must be accommodated through associated supply arrangements; in the domestic context, the householder can have the opportunity to provide this but this can only happen reliably if the interface to these arrangements is designed to make participation effortless. With the continued developments in the Internet of Things, engineering systems design and the growing sector of artificial intelligence and smart devices technologies, there is a very strong possibility that consumer choice will be further restricted by the pace of change due to the scale of the large utility providers. Through understanding consumer behavior, applying sound human factors methods, working with energy providers (all scales), and developing responsible, practical, reliable and intuitive physical and digital products, the current and future consumers of energy will have access to products that are able to meet the growing demands of the energy sector.

Acknowledgements. The study is funded by EPSRC (EP/R002312/1). We thank all those who were involved for their time and support.

References

1. UK Houses of Parliament: Climate Change Act 2008, Chap. 27
2. Stephen, B., Galloway, S., Burt, G.: Self-learning load characteristic models for smart appliances. IEEE Trans. Smart Grid 5(5), 2432–2439 (2014)
3. Allison, J., Cowie, A., Galloway, S., Hand, J., Kelly, N., Stephen, B.: Simulation, implementation and monitoring of heat pump load shifting using a predictive controller. Energy Convers. Manag. **150**, 890–903 (2017)
4. EPSRC. http://gow.epsrc.ac.uk/NGBOViewGrant.aspx?GrantRef=EP/R002312/1
5. Walker, G., Taylor, A., Whittet, C., Lynn, C., Docherty, C., Stephen, B., Owens, E., Galloway, S.: A practical review of energy saving technology for ageing populations. Appl. Ergon. **62**, 247–258 (2017)
6. GOV.UK. https://www.gov.uk/government/consultations/smart-metering-equipment-technical-specifications-second-version
7. Amphiro. https://www.amphiro.com/en/produkt/amphiro-b1-connect/
8. Efergy. http://efergy.com/uk/elite-classic
9. TACKLife. https://www.tacklifetools.com/product/product/index/id/84
10. Scottish Power. https://www.scottishpower.co.uk/energy-efficiency/smart-meters/
11. Geo. https://www.geotogether.com/consumer/product/minim-electricity-monitor-led-sensor/
12. OWL. http://www.theowl.com/index.php/energy-monitors/standalone-monitors/owl-usb/
13. OWL. http://www.theowl.com/index.php?cID=185

14. SSE. https://www.sse.co.uk/DXP/assets/files/SSE%20IHD%20Full%20Guide%20AUG16%20cropped.pdf
15. Stanton, N.A., Salmon, P.M., Rafferty, L., Walker, G.H., Baber, C., Jenkins, D.P.: Human Factors Methods: A Practical Guide for Engineering and Design. Ashgate, Farnham (2013)
16. Stanton, N.A., Young, M.S.: What price ergonomics? Nature **399**, 197–198 (1999)
17. Hanington, B., Martin, B.: Universal Methods of Design: 100 Ways to Research Complex Problems, Develop Innovative Ideas, and Design Effective Solutions. Rockport Publishers, Beverly (2012)
18. Langford, J., McDonagh, D.: What can focus groups offer us? In: McCabe, P.T., Hanson, M. A., Robertson, S.A. (eds.) Contemporary Ergonomics 2002, pp. 502–506. CRC Press, Boca Raton (2002)
19. Kirwan, B., Ainsworth, L.K.: A Guide to Task Analysis: The Task Analysis Working Group. CRC Press, Boca Raton (1992)
20. Naikar, N.: Work Domain Analysis: Concepts, Guidelines, and Cases. CRC Press, Boca Raton (2013)
21. Benedek, J., Miner, T.: Measuring desirability: new methods for evaluating desirability in a usability lab setting. In: Proceedings Usability Professionals Association Conference, pp. 8–12 (2002)
22. Cambridge Engineering Design Centre. https://www-edc.eng.cam.ac.uk/
23. Lockton, D., Harrison, D., Stanton, N.A.: The design with intent method: a design tool for influencing user behaviour. Appl. Ergon. **41**, 382–392 (2010)
24. Fischer, D., Stephen, B., Flunk, A., Kreifels, N., Byskov Lindberg, K., Wille-Haussmann, B., Owens, E.H.: Modelling the Effects of variable tariffs on domestic electric load profiles by use of occupant behavior submodels. IEEE Trans. Smart Grid **8**(6), 1949–3053 (2017)
25. Ofgem Feed-in Tariffs. https://www.ofgem.gov.uk/environmental-programmes/fit
26. Stephen, B., Tang, X., Harvey, P.R., Galloway, S., Jennett, K.I.: Incorporating practice theory in sub-profile models for short term aggregated residential load forecasting. IEEE Trans. Smart Grid **8**(4), 1591–1598 (2015)

The Operation of Crude Oil Pipeline: Examination of Wax Thickness

Fadi Alnaimat[✉], Bobby Mathew, and Mohammed Ziauddin

Department of Mechanical Engineering, The United Arab Emirate University,
P.O. Box 15551, Al Ain, United Arab Emirates
falnaimat@uaeu.ac.ae

Abstract. Solid and wax precipitation during production, transportation, and storage of petroleum fluids is a common problem faced by the oil industry throughout the world. Continuous wax deposition is a critical issue for offshore transport pipeline and for the oil and gas industry. Significant amount of effort and capital are typically spent annually by the Oil & Gas companies for the prevention and removal of wax in production and transportation lines. This study examines a new technology to monitor wax deposition thickness in sub-sea pipelines. The wax thickness predictions from the wax thermal sensing method is expected to be promising compared with conventional measurement techniques. It is expected that this sensing method can support the operators of Oil & Gas companies to predict accurately wax deposition which enables operators to adjust operation quickly to prevent a complete clogging of pipelines.

Keywords: Wax deposition · Modeling · Simulation · Oil pipeline

1 Introduction

Flow assurance has recently become very important in the oil and gas industry. Flow assurance refers to ensuring successful flow of hydrocarbon stream from oil reservoir to the point of sale. Flow Assurance developed because traditional approaches are inappropriate for deep under water production due to extreme distances, depths, and temperatures. Flow assurance in deep underwater pipelines is extremely diverse, it includes many specialized subjects and many forms of engineering field. Flow assurance includes network modeling and transient multiphase simulation. In addition, it involves handling many solid deposits, such as, gas hydrates, asphaltene, wax, scale, and naphthenates. These solid deposits can interact with each other and can cause blockage formation in pipelines and result in pipeline failure. Flow assurance is considered to be the most critical task during deep water energy production because of the high pressures and low temperature involved under these conditions. Pipe damage and disturbance in production due to the flow assurance failure can cause huge financial losses.

Crude oil in a reservoir flows to the surface through the pipe by the reservoir pressure. If the pressure drop between reservoir and the receiving facilities become significant, the wells stop producing and the flow in the line will stop due to the large

© Springer International Publishing AG, part of Springer Nature 2019
T. Z. Ahram (Ed.): AHFE 2018, AISC 787, pp. 514–522, 2019.
https://doi.org/10.1007/978-3-319-94229-2_49

pressure drop in the system. The operational life time of a well is a transient process because the flow of crude oil decreases over time due to the increase in pressure drop in pipe. As a result, the wells shut down. In multiphase flow the fluid phases differ in various parts of the pipeline system and at different production life time due to temperature, pressure and flow rates. Density can have significant effect on pressure drop.

Hydrates formation in pipe is also considered to be very important in flow assurance. Hydrates are formed by natural gas molecules getting into hydrogen-bonded water cages at temperatures well above normal water freezing. Hydrate prevention is one of the key issues in flow assurance. A possible solution to alleviate this issue is to use of inhibitor Methanol or MEG. Another prevention method that is typically used in addition to inhibition is insulation of pipelines. Pipeline insulation is used to keep crude oil and natural gas temperature in pipelines above hydrate formation temperature. The calculation of thermal insulation requires thermal calculations which can be extensive.

As crude oil flow through the pipeline, its thermal energy is gradually lost by heat transfer to the surroundings. Wax can deposit at inner walls if the temperature is below WAT. Wax deposit in pipeline can be significant as shown in Fig. 1. Wax precipitation is a highly temperature dependent process. Therefore, thermal method can be very effective for detection, and preventing of wax precipitation. Direct heating is considered be an effective method to solve issues related to wax deposition by keeping the temperature higher than solidification temperature. The basic principle of direct heating involves passing electric current through the pipeline wall to generate heat. It is the most reliable option for deep water field operation of crude oil pipe.

Fig. 1. Pipeline with solid deposits, the left is a wax deposit and right is a scale deposit [1].

The pipeline material is conductive to electric current, and through it the electric heating occurs. Electric power is supplied from the platform through two cables. One of the two cables are connected to the near end of the pipe, and the other cable connected to the end of the pipe [2]. In this arrangement, each pipeline's end is connected to the two cables making a closed circuit. Direct Heating method typically ensures that the temperature of the produced fluid is kept above WAT, approximately 20 °C, during shutdowns, and also increasing fluid temperature above ambient temperature which is around 6 °C.

This study examines a new technology to monitor wax deposition thickness in sub-sea pipelines. It presents heat transfer calculation in pipe carried out using CFD software Ansys Fluent. It is expected that this sensing method can support the operators of Oil & Gas companies to accurately predict wax deposition which in turn would enable operators to adjust operation quickly to prevent a complete clogging of pipelines.

Different methods for the detection of wax deposition thickness have been developed such as the use of electric resistance [3], pressure pulses [4], heat transfer techniques [5], ultrasound and strain gauges [6], and radiography [7, 8]. Thermal wave processing [9] and thermal methods used in relation to nondestructive testing [8] have potential for the detection of waxy deposits in pipes. Temperature oscillation methods have also been developed in relation to heat transfer in heat exchangers [11–13] to detect and monitor fouling of the heat transfer surfaces. Although some researchers attempted to use a pressure pulse technique to predict the deposition thickness, the downside of this technique is that there is a high degree of uncertainty associated with the results and that carrying out measurements is a complex and expensive procedure. Currently there is no accurate online wax thickness measurement technique available. There are various deposition models that that can be used to predict the influence of temperature on deposition rate [14–17]. The important parameter that is missing and needed to test these models is accurate prediction of the wax thickness.

2 Mathematical Modeling

The variation of fluid temperature in the axial direction will be used to examine the wax thickness. It is expected that as wax thickness increases, the temperature increases. Thus, based on the surface temperature and average velocity detected, one can detect wax buildup thickness. If the temperature of crude oil decreases below the wax appearance temperature (WAT), the precipitation of the wax molecules occurs. WAT is the temperature where the first wax crystal begins to precipitate out from the crude oil. The crystallization of the wax particles can develop and advance to a strong solid structure near pipe wall. This traps the crude oil causing it to experience a gel consistent with low flow ability.

A cylindrical coordinate system (r, z) is appropriate for the boundary value problem detailed above. The axial velocity for the flow in the pipe in cylindrical coordinate is given by the Poiseuille profile:

$$u(r) = -\frac{1}{4\mu}\frac{\partial p}{\partial z}\left(r_o^2 - r^2\right) \tag{1}$$

The energy equation for the pipe is approximated by:

$$u\frac{\partial T}{\partial z} = k\,\rho\,cp\frac{1}{r}\frac{\partial}{\partial r}\left(r\frac{\partial T}{\partial r}\right) \tag{2}$$

In this study, the variation of crude oil viscosity is modeled to account for the effect of temperature near the wall which simulates the behavior of wax formation of near the wall. The oil becomes very viscous once the temperature becomes close to WAT forming a wax layer on the inner wall of pipeline as shown in Fig. 2.

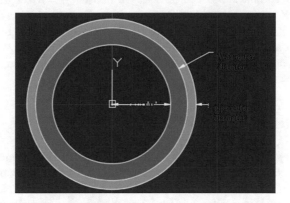

Fig. 2. Wax layer in pipeline inner surface

The thermal sensing method is based on thermal resistance of wax. Thermal resistance in the inner pipe due to the flow crude oil inside the pipe, $R_{conv, inner}$, and thermal resistance due the movement of underwater flow on the outside pf pipeline, $R_{conv, outer}$, are given as:

$$R_{conv,inner} = \frac{1}{(h1)A_{inner}}, \quad R_{conv,outer} = \frac{1}{(h2)A_{outer}} \tag{3}$$

Also thermal resistance in the pipe wall is calculated as,

$$R_{wall} = \frac{\ln\left|\frac{r3}{r2}\right|}{2\pi(k_{wall})L} \tag{4}$$

Also thermal resistance in the wax layer formed is calculated similarly as,

$$R_{wax} = \frac{\ln\left|\frac{r2}{r1}\right|}{2\pi(k_{wax})L} \tag{5}$$

The total thermal resistance and the total heat loss can be calculated as,

$$R_{Total} = R_{conv,inner} + R_{wax} + R_{wall} + R_{conv,outer} \tag{6}$$

$$\dot{Q} = \frac{T_{oil} - T_{outer\ water}}{R_{tot}} \tag{7}$$

The oil viscosity is a function of temperature and it is given as:

$$\mu(T) = p_1 T^3 + p_2 T^2 + p_3 T + p_4 \tag{8}$$

where p_1, p_2, p_3, p_4 are constants obtained for by correlating properties data reported in [18, 19]. In the current study, the moving wax is set to deposit on the surface if the temperature reaches below 301 K [18, 19]. Oil density ρ = 815 kg/m^3, specific heat C_p = 1950 J/kg.K, thermal conductivity of oil k_{oil} = 0.1344 W/m.K; thermal conductivity of wax k_{wax} = 0.15 W/m.K. The pipe is made of steel and its dimension used in the simulation are L = 1 m, and inner D_{in} = 10 cm, and pipeline thickness is 1 cm. The operating parameters that are set in the simulation study are V_{in} = 0.02 m/s, T_{in} = 320, $T_{w,o}$ = 290.

3   Computational Solution Method

It is aimed in this study to examine the significance of thermal resistance of wax in dirty pipeline and to compare with a clean pipeline. Thus the simulation study is carried out on a flow domain that is modeled such that it includes an electric heater at the outer surface in middle of pipe as shown in Fig. 3. In this arrangement, the heat is applied by the electric heat which passes through the steal (pipe wall) and then to the oil in the pipeline. If the heat is sufficiently high, this will prevent formation of wax particles in pipe at that location and this enables a useful comparison. The flow domain and simulation is carried using a 2D- axisymmetric model as shown in Fig. 2a. The flow domain discretized mesh is shown in Fig. 2b. The simulation is carried out using a CFD software Fluent.

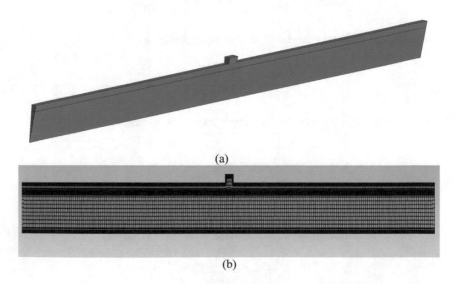

(a)

(b)

Fig. 3. (a) axis-symmetry flow domain (b) flow domain mesh

4   Results and Discussion

Figure 4 shows a contour of velocity distribution in the pipeline along the axial direction for fixed outer surface temperature. It is shown clearly that velocity near the wall is very small compared to that in the inner pipeline. It is also shown that the velocity near the wall decreases in the axial direction. This can be attributed to the fact that oil loses its thermal energy reduces in temperature which becomes very close to wax deposition neat the wall. Figure 5 shows a contour of temperature distribution in the pipe along the axial direction for fixed outer surface temperature. It is shown that the heat is applied via the electric heater and passed to the wall surface and the moving crude oil.

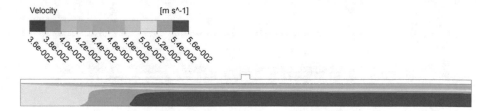

Fig. 4.   Contour of velocity distribution in a segment of pipeline.

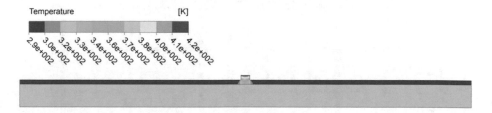

Fig. 5.   Contour of temperature distribution in a segment of pipeline.

Figure 6 shows temperature profile in radial direction at middle section and outlet section. It is shown that the temperature decreases slightly in the radial direction until near the wall. It is shown that the temperature increases significantly near the middle region which is near the electric heater. Also shown the radial temperature decreases at the outlet section due to the decrease in wall temperature. The temperature variation in the radial direction allows to determine the temperature difference and calculate thermal resistance of the pipe wall, wax layer, and the moving fluid. This leads to easily determining the wax thickness based on the temperature difference.

Figure 7 shows temperature profile in the axial direction at inner-wall, outer-wall, and near-wall. It is clearly shown that temperature decreases gradually in the axial direction near the wall. It is noted that the temperature is increases sharply around the middle of pipe which is close to the electric heater. Also noted that the outer wall

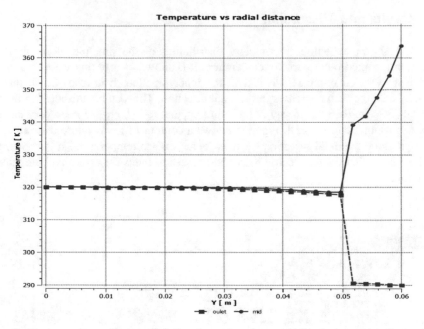

Fig. 6. Temperature profile in radial direction at locations: middle section and outlet section

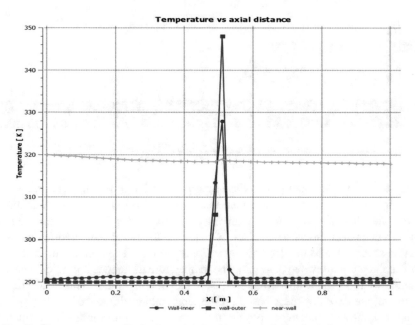

Fig. 7. Temperature profile in the axial direction at wall-inner, wall-outer, near-wall.

surface temperature is fixed at 290 K, and near the electric heater increases sharply as heat passes through the middle section. The inner wall surface temperature is slightly higher than the outer surface temperature along the pipe except near the electric heater increases. In this middle section, the inner surface temperature increases sharply but with less magnitude compared to the outer surface temperature.

5 Conclusions

Wax deposition in pipe causes blockages of oil transportation pipelines, and this leads to production losses and damaging of oil flow. In this study, thermal sensing method is examined to predict wax deposition thickness. Predictions of wax thickness using thermal sensing is expected to be promising technique as it is considered to be non-intrusive method. Accurate prediction of wax layer thickness is very critical component in flow assurance of pipeline. The proposed detection method of wax thickness can be promising with the main advantage of being a non-intrusive accurate measurement tool.

Acknowledgments. The authors acknowledge financial support received from the UAEU research fund, Grant no. 31N158, UAE.

References

1. Offshore magazine: Deepwater challenges paraffins. http://www.offshore-mag.com/
2. Nysveen, A., Kulbotten, H., Lervik, J.K., Børnes, A.H., Høyer-Hansen, M., Bremnes, J.J.: Direct electrical heating of subsea pipelines—technology development and operating experience. IEEE Trans. Ind. Appl. **43**(1), 118–129 (2007)
3. Chen, X.D., Li, D.X.Y., Lin, S.X.Q., Ozkan, N.: On-line fouling/cleaning detection by measuring electric resistance—equipment development and application to milk fouling detection and chemical cleaning monitoring. J. Food Eng. **61**, 181–189 (2004)
4. Gudmundsson, J.S., Durgut, I., Celius, H.K., Korsan, K.: Detection and monitoring of deposits in multiphase flow pipelines using pressure pulse technology. In: 12th International Oil Field Chemistry Symposium, Geilo, Norway (2001)
5. Cordoba, A.J., Schall, C.A.: Application of a heat transfer method to determine wax deposition in a hydrocarbon binary mixture. Fuel **80**, 1285–1291 (2001)
6. Zaman, M., Bjørndalen, N., Islam, M.R.: Detection of precipitation in pipelines. Petrol. Sci. Technol. **22**, 1119–1141 (2004)
7. Harara, W.: Deposit thickness measurement in pipes by tangential radiography using gamma ray sources Russ. J. Nondestruct. Test. **44**, 796–802 (2008)
8. Edalati, K., Rastkhah, N., Karmani, A., Seiedi, M., Movafeghi, A.: The use of radiography for thickness measurement and corrosion monitoring in pipes. Int. J. Press. Vessels Pip. **83**, 736–741 (2006)
9. Gleiter, A., Mayr, G.: Thermal wave interference. Infrared Phys. Technol. **53**, 288–291 (2010)
10. Liu, J., Yang, W., Dai, J.: Research on thermal wave processing of lock-in thermography based on analyzing image sequences for NDT. Infrared Phys. Technol. **53**, 348–357 (2010)

11. Freund, S., Kabelac, S.: Investigation of local heat transfer coefficients in plate heat exchangers with temperature oscillation IR thermography and CFD. Int. J. Heat Mass Transfer **53**, 3764–3781 (2010)
12. Roetzel, W., Das, S.K., Luo, X.: Measurement of the heat-transfer coefficient in plate heat-exchangers using a temperature oscillation technique. Int. J. Heat Mass Transfer **37**, 325–331 (1994)
13. Roetzel, W., Luo, X.: Extended temperature oscillation measurement technique for heat transfer and axial dispersion coefficients. Rev. Gen. Therm. **37**, 277–283 (1998)
14. Bagatin, R., Busto, C., Correra, S., Margarone, M., Carniani, C.: Wax modeling: there is need for alternatives. In: SPE Russian Oil & Gas Technical Conference and Exhibition (2008)
15. Hoffmann, R., Amundsen, L.: Single-phase wax deposition experiments. Energy Fuels **24**, 1069–1080 (2010)
16. Singh, P., Venkatesan, R., Fogler, H.S., Nagarajan, N.R.: Aging and morphological evolution of wax-oil gels during externally cooled flow through pipes. In: 2nd International Conference in Petroleum Phase Behaviour and Fouling, Copenhagen, Denmark (2000)
17. Handal, A.D.: Analysis of some wax deposition experiments in a crude oil carrying pipe. Master thesis (2008)
18. Ukrainczyk, N., Kurajica, S., Šipušić, J.: Thermophysical comparison of five commercial paraffin waxes as latent heat storage materials. Chem. Biochem. Eng. Q. **24**(2), 129–137 (2010)

A Generalized Ergonomic Trade-off Model for Modularized Battery Systems Particularly for ICT Equipment

Victor K. Y. Chan[✉]

Macao Polytechnic Institute, School of Business, Rua de Luis Gonzaga Gomes,
Macao SAR, China
vkychan@ipm.edu.mo

Abstract. The author and a co-inventor earlier patented a method and a sub-system to be incorporated into battery management systems for practically optimizing the ergonomics of battery system charging and discharging in a bid to tackle batteries' long-standing conundrums of slow charging and costly charging infrastructures. The key idea is to modularize battery systems and prioritize charging and discharging of their battery modules so as to minimize periodic human effort to unload, load, and/or (re)charge the battery modules. The author also proposed earlier a mathematical trade-off model to address the ensuing issue of optimizing the extent of modularization, assuming Poisson probability distribution of depleted battery modules in each discharge cycle and zero incremental overhead mass of additional packaging materials and electronics associated with further modularization of a battery system. This article attempts to generalize the model by relaxing these two restrictive assumptions.

Keywords: Battery ergonomics · Battery modularization · Human effort
Trade-off · ICT equipment

1 Introduction

It is commonly known and detailed in an earlier article by the author [1] that for decades, batteries have been deployed to electrically power electrical and electronic equipment, inclusive of instruments, devices, and machineries. Batteries' role in powering mobile phones, tablet computers, other mobile information and communication technology (ICT) devices, and electrical vehicles is especially in the limelight these days when e-commerce and environmentalism are such overarching issues around the globe. Any such equipment comprises and/or is electrically connected to housings where batteries are usually installed and is thus powered by the batteries' electrical energy released from their chemically stored energy during their electrical discharge (i.e., during the conversion of their chemically stored energy into electrical energy as their output). Conversely, the chemically stored energy in the batteries is topped up or replenished through electrically (re)charging them (i.e., through conversion of the inputted electrical energy into chemical energy to be stored in them).

© Springer International Publishing AG, part of Springer Nature 2019
T. Z. Ahram (Ed.): AHFE 2018, AISC 787, pp. 523–534, 2019.
https://doi.org/10.1007/978-3-319-94229-2_50

Thereby, throughout the life of a typical battery, it repetitively undergoes electrical discharge and charge cycles respectively to provide electrical energy for powering equipment and to have its stored chemical energy topped up, in essence, serving as a transient storage of electrical energy for equipment, particularly mobile equipment. In view of their limited pollutant emission and falling operational cost [2], batteries are considered more environment-friendly [3–5] and economical [5], than many other energy sources. In particular, their portability and mobility qualify them for being the most widely adopted category of mobile energy sources for powering mobile equipment.

The author's earlier article [1] also delves into the fact that research effort from academia and industry have focused on material science, battery microstructures, and battery manufacturing technologies to revolutionize battery chemistry, anode/cathode materials, battery manufacture optimization [6] in order to optimize the trade-off between the major well-known properties of batteries, namely:

(a) the cycle life, i.e., the number of charge/discharge cycles for a battery to undergo before its energy capacity degrades significantly [7],
(b) the power density, i.e., the maximum amount of power (or energy per unit time) that a battery can supply per unit of its volume or mass [8],
(c) the energy density, i.e., the energy capacity of a battery per unit of its volume or mass [9],
(d) the capital cost of a battery per unit of its energy capacity [2],
(e) the charging speed, i.e., the time to (re)charge a battery from a certain lower state of charge to a certain higher state of charge) [10], and
(f) the capital cost of the charging infrastructures to charge batteries of certain type [11].

In particular, battery applications are especially beset by the limitations on the last two properties (e) and (f) above [12].

2   A Method and Subsystem for Battery Management Systems

The author together with a co-inventor earlier patented a method and a subsystem to be incorporated into battery management systems for practically optimizing the ergonomics of battery system charging and discharging in a bid to address the two aforementioned limitations without recourse to any revolutionary technological breakthroughs in material science, battery microstructures, or battery manufacturing technologies [13, 14]. Though having already been delineated in the author's earlier publications [1, 13, 14], the key philosophy behind the method and the subsystem is again expounded in this section for the readers' reference before proceeding with the rest of this article.

Leaving aside various auxiliary features, the method and the subsystem chiefly adopt modularization of each battery system, prioritization of the discharging (and charging) of the individually unloadable and loadable battery modules in the battery system according to users' human effort to unload the battery modules from and/or to (re)load the battery modules into the battery system's housing, and/or to load replacement battery modules into the housing such that during the discharge of the battery modules to power equipment, discharge priority is accorded to the battery modules in order of their ease of being unloaded and/or (re)loaded and/or the ease of their replacement battery modules being loaded.

Thereby, by the time the battery system needs (re)charging, only such preferentially discharged battery modules as a subset of all those in the battery system, as opposed to the battery system as a whole, are depleted or low and in need of (re)charging. Thus, instead of simply (re)charging the whole battery system in situ, users may opt to individually unload from the housing only such depleted or low battery modules, convey them to and (re)charging them with, for example, the domestic mains supply, and (re)load them individually back into the housing for continuing to power the equipment.

Alternatively, instead of unloading the battery system as a whole and loading a fully charged replacement battery system as a whole into the housing, the users may opt to individually unload from the housing only such depleted or low battery modules and load fully charged replacement battery modules individually into the housing for continuing to power the equipment. As the charging speed is rather irrelevant to the users' experience with domestic charging, for example, at the users' homes overnight or at the users' workplaces during their working hours, low-power, ordinary chargers suffice to substantially charge such depleted or low battery modules within the allowable charging hours in daily practice, rendering specialized, high-power, high-speed charging [15] infrastructures superfluous.

Therefore, the most vexing limitations on (e) and (f) above can no longer be concerns. The essence of the philosophy lies in the modularization of the battery system, as in Fig. 1 for example, which in effect breaks the otherwise large battery system into smaller battery modules, ending up with these relatively smaller battery modules amenable to being individually unloaded and loaded probably without specialized, heavy-duty loading/unloading facilities.

$[1, 1]$	$[1, 2]$	...	$[1, q - 1]$	$[1, q]$
$[2, 1]$	$[2, 2]$	...	$[2, q - 1]$	$[2, q]$
...	...	...	...	...
$[p - 1, 1]$	$[p - 1, 2]$	...	$[p - 1, q - 1]$	$[p - 1, q]$
$[p, 1]$	$[p, 2]$	...	$[p, q - 1]$	$[p, q]$

Fig. 1. This is an exemplary layout of the battery modules in a modularized battery system, which comprises $p \times q$ battery modules. Here, each rectangle denotes a battery module whereas the outer thick lines signify the housing. Without loss of generality, assume that the housing's opening is at the top, as indicated by the missing of a thick line. Also, despite the columns appearing vertical, they can be alternatively oriented, for example, horizontally.

In contrast, specialized, heavy-duty facilities are mandatory in the case of unloading and loading the otherwise much larger battery system as a whole. Likewise, thanks to modularization, domestic (re)charging of relatively small battery modules with low-power, ordinary chargers is made practically realizable whereas charging the otherwise much larger battery system as a whole conceivably necessitates specialized charging facilities of a much higher power and thus much more sophisticated electrical safety protection.

In addition, prioritization of the discharge of battery modules likely results in the depleted and low battery modules (in need of (re)charging) being those easily unloadable and/or loadable and/or those of which replacement battery modules are easily loadable, further reducing the users' human effort to unload and load such battery modules for (re)charging or to load replacement battery modules and further eradicating the need for specialized, heavy-duty loading/unloading facilities.

Despite all its advantages above, the method and the subsystem are nevertheless confronted with an ensuing dilemma as to the extent of modularization or how many battery modules a battery system should be broken into or how small the battery modules should be so as to minimize users' human effort to unload, convey, and load the battery modules during the daily electrical charge and discharge cycles. In essence, this is a trade-off of users' human effort between unloading/conveying/loading a large number of battery modules of small masses/sizes and unloading/conveying/loading a small number of battery modules of large masses/sizes. For the sake of putting the aforementioned method and subsystem in real-world practice, the author earlier devised a mathematical model (hereinafter, the "first trade-off model") [1] of such a trade-off by assuming Poisson probability distribution of the number of depleted or low battery modules at the end of each discharge cycle of a battery system, an modified arithmetic progression relationship between the number of such depleted or low battery modules and the human effort to unload and load them, and a power relationship between the mass of a battery module and the human effort to unload, convey, and load it.

3   The First Trade-off Model: Why not Absolutely Realistic?

First, in the first trade-off model [1], the number of depleted or low battery modules in need of (re)charging at the end of each discharge cycle of a battery system (or equivalently at the beginning of each charge cycle of a battery system) is assumed to follow a Poisson probability distribution [16, 17], which typically models the number of occurrences of an event (e.g., arrivals or appearances of a specific thing) within a fixed time interval or space and assumes each occurrence being independent of the time since the last occurrence or independent of the temporal or spatial distance between the occurrence and any neighboring occurrences.

For a battery system adopting the method and the subsystem of the last section, assuming a Poisson probability distribution of the number of depleted or low battery modules at the end of each discharge cycle connotes assuming that the depletion or lowness of a battery module occurs at a random time after another battery module has been depleted or run low. This in turn assumes users' usage behavior or habit being that after a battery module has just been depleted (though probably unknown to the users), whether the users keep on using the battery system to power equipment without first (re)charging the battery system until the depletion of another battery module (though probably also unknown to the users) is a random choice of the users. In other words, the users' usage behavior or habit is that the number of depleted or low battery modules at the end of each discharge cycle is highly random except that the users in the long term maintain a constant mean number of depleted or low battery modules at the end of each discharge cycle, which is another assumption of the Poisson probability distribution [16, 17].

Nonetheless, such an assumed usage behavior or habit is empirically revealed to be unrealistic especially when it comes to users of ICT equipment, such as mobile phones. Rahmati et al. [18] categorizes mobile phone users into two types regarding their human-battery interaction, namely:

A. Those who regularly charge their mobile phones regardless of the associated batteries' charge levels (i.e., states of charge), for example, every 1 or 2 days, or whenever convenient.
B. Those who charge their mobile phones based on charge level feedback from the mobile phones' battery interface.

For Type A users, the state of charge and thus the number of depleted or low battery modules at the end of each discharge cycle are rather random, and the latter can thus be approximated by Poisson probability distribution quite well. In contrast, Each Type B user appears to target a preferred state of charge of his/her own choice at the end of each discharge cycle before he/she chooses to (re)charged his/her mobile phone's battery. It is these Type B users that the first trade-off model [1] with the underlying Poisson probability distribution assumption fails to describe realistically, and it is partly due to these Type B users that the derivation of the generalized model (hereinafter, the second trade-off model) of this article is justified. Second, the first trade-off model [1] assumes zero incremental overhead mass (or weight) of additional packaging materials and electronics associated with further modularization of a battery system (i.e., with breaking the battery system into more battery modules). However, Plett [19] indicates that this assumption oversimplified the structural designs of battery systems in that each additional battery module involves its own packaging materials and controller electronics, so a battery system of a large number of small battery modules comprises substantially more packaging materials and electronics than a battery system of a small number of large battery modules even if both systems have the same energy capacity. This provides another rationale for developing the second trade-off model by revising the first trade-off model [1].

4 The Second Trade-off Model

The study of Rahmati et al. [18] uncovered that each battery system user can have his/her own usage behavior or habit and thus his/her own probability distribution of the number of depleted or low battery modules in need of (re)charging at the end of each discharge cycle of his/her battery systems (or equivalently at the beginning of each charge cycle of his/her battery systems). In general, the probability distribution function of the number of depleted or low battery modules at the end of each discharge cycle among the population of all the discharge cycles of a particular brand and model of battery systems used by a cohort of users to power some equipment is denoted by

$$P(K = k; \lambda) \tag{1}$$

where

$P(K = k; \lambda) =$ the probability of the event that the number of depleted or low battery modules at the end of each discharge cycle of that brand and model of battery systems used by that cohort of users is equal to k,

$K =$ the random variable for the number of depleted or low battery modules at the end of each aforesaid discharge cycle, and

$\lambda =$ the mean number of depleted or low battery modules at the end of each aforesaid discharge cycle.

Dependent on users' usage behavior or habit, the parameter λ characterizes the probability distribution function in (1) and can be empirically estimated by averaging K among a sample of the discharge cycles of that particular brand and model of battery systems used by that cohort of users whereas $P(K = k; \lambda)$ is empirically determinable by collecting the frequencies distribution of K among that sample like what Rahmati et al. [18] did.

As with the first trade-off model [1], take the example of the layered layout in Fig. 1 being the way the battery modules are installed in the battery system's housing. The battery modules in the layer least deep inside the housing (i.e., those closest to the housing's opening at the top or [1, 1] to [1, q]) are easiest to unloaded from and loaded into the housing, those in the layer second least deep inside the housing (i.e., [2, 1] to [2, q]) are second easiest to unloaded and loaded, and so on. Needless to say, the battery modules in the layer deepest inside the housing (i.e., those farthest from the housing's opening at the top or [p, 1] to [p, q]) are least easy to unloaded and load. Hence, discharge priority is also accorded in the above order with [1, 1] to [1, q] having highest priority, [2, 1] to [2, q] second highest, and [p, 1] to [p, q] lowest, and thus the battery modules upon powering equipment become depleted or low also in the above order with [1, 1] to [1, q] becoming so first, [2, 1] to [2, q] next, ..., and [p, 1] to [p, q] last. Also as with the first trade-off model [1], it is reasonably assumed that the difference in the ease of unloading and loading battery modules between those in any two consecutive layers is a constant. Then, users' human effort to unload and load k depleted or low battery modules at the end of a discharge cycle of a battery system is given by

$$E_L = E_{L1} + (E_{L1} + E_{LD}) + (E_{L1} + 2E_{LD}) + (E_{L1} + 3E_{LD}) + \cdots$$
$$+ [E_{L1} + (l - 1)E_{LD}] + \frac{r}{L}(E_{L1} + lE_{LD}) \tag{2}$$

$$= \frac{l}{2}[2E_{L1} + (l - 1)E_{LD}] + \frac{r}{L}(E_{L1} + lE_{LD}) \tag{3}$$

$$= \frac{\left\lfloor \frac{k}{L} \right\rfloor}{2}[2E_{L1} + (\left\lfloor \frac{k}{L} \right\rfloor - 1)E_{LD}] + \frac{\left\lceil \frac{k}{L} \right\rceil}{L}(E_{L1} + \left\lfloor \frac{k}{L} \right\rfloor E_{LD}) \qquad (4)$$

where

E_L =   users' human effort to unload and load $k = lL + r$ battery modules,

l =   the (integral) number of layers fully occupied by the k battery modules,

L =   the number of battery modules in a layer,

r =   the (integral) number of battery modules in the layer partially occupied by the k battery modules such that $0 \leq r < L$,

E_{L1} =   users' human effort to unload and load all the battery modules in the layer with most easily unloadable and loadable battery modules, i.e., in the layer least deep inside the housing or the layer nearest to the housing's opening or the layer of [1, 1] to [1, q] in Fig. 1,

E_{LD} =   the (incremental) difference in users' human effort to unload and load all the battery modules in a layer between any two consecutive layers,

$\left\lfloor \frac{k}{L} \right\rfloor$ =   the quotient of k divided by $L = l$, and

$\left\lceil \frac{k}{L} \right\rceil$ =   the remainder of k divided by $L = r$.

The equality between (2) and (3) is due to the fact that all the terms in (2) other than the last one belong to an arithmetic progression. L is known to the users given the known layout of the battery modules in the battery system. E_{L1} and E_{LD} can be empirically estimated respectively by averaging a sample of users' ratings of their human effort to unload and load all the battery modules in the layer of [1, 1] to [1, q] in Fig. 1 and averaging a sample of their ratings of the (incremental) difference in their human effort to unload and load all the battery modules in a layer between any two consecutive layers.

It is common experience that human effort to convey anything increases only marginally even if its mass more than doubles when its mass remains on the low side but more than doubles even if its mass increases only marginally when its mass is approaching the physical limit of the person concerned. In other words, such human effort increases with the mass as per a convex (downward) curve if the former is plotted on the vertical axis against the latter on the horizontal axis. The same is true of the users' human effort to unload a battery module from the housing, carry it between the housing and the charging facility, and load it back into the housing afterwards. As such, as with the first trade-off model [1], the typical relationship between users' human effort to convey a battery module (i.e., to unload it from the housing, carry it between the housing and the charging facility, and load it back into the housing) and its mass can reasonably be modeled by a power function with a power greater than one as follows:

$$E_C = m^s \qquad (5)$$

where

E_C =   users' human effort to convey the battery module,

m =   the mass of the battery module, and

s =   the power of the power function, which is greater than 1.

It is noteworthy that the terminology "power" in this paragraph should be interpreted mathematically and by no means connotes electrical power of batteries of any kind. In practice, the parameter m in (5) is measurable by simply weighing the battery module. The parameter s in (5) can be estimated by regression given that (5) can be transformed, by taking logarithm on both sides, into

$$\ln E_C = \ln m^s$$
$$\text{or equivalently, } \ln E_C = s \ln m$$

which in turn forms the basis of the regression equation

$$\ln E_{C,i} = s \ln m_i \tag{6}$$

where $E_{C,i}$ and m_i are respectively the values of E_C and m for the i-th battery module, which are measurable respectively by a sample of users' average rating of the human effort to convey the i-th battery module and by weighing the i-th battery module. By empirically collecting a sufficiently large sample of value pairs $(E_{C,i}, m_i)$, or equivalently $(\ln E_{C,i}, \ln m_i)$, for $i = 1, 2, \dots$ and substituting them into (6), the parameter s can be estimated as the regression coefficient in (6).

In consideration of the maximization of the economies of scale in the manufacturing of the battery modules, this article takes it for granted that identical battery modules (probably, except for their levels of aging) are adopted by the battery systems that (1) concerns. All things being equal, if the net mass m_0 of each battery module (solely inclusive of the mass of the energy storage structures in each battery module but exclusive of the overhead mass of the packaging materials and electronics) in the aforementioned battery systems now doubles, then the energy capacity of each battery module doubles, half the number of battery modules are needed to be discharged till depletion or lowness in order to power equipment consuming the same energy as before, and thus the mean number λ of depleted or low battery modules at the end of each discharge cycle of the aforementioned battery systems halves. As such, m_0 is inversely proportional to λ, all things being equal, or mathematically,

$$m_0 \lambda = C \tag{7}$$

where C is a constant, which can be determined by empirically estimating the value of λ as in the discussion on (1) particularly for the aforementioned battery systems with battery modules of a given value of mass m_0 as weighed and multiplying this given value of m_0 by the value of λ. Alternatively, and more precisely, C can be determined by repeating the above multiplication, each time for a different value of m_0 and thus a different value of λ, and averaging the values of C from the multiple multiplications.

For the energy storage structures in battery modules of different sizes, they are simply the proportional miniaturizations or enlargements of each other. Thus, the volume V_0 and the mass m_0 of the energy storage structures in a particular battery module are related proportionally such that

$$V_0 = k_0 m_0 \tag{8}$$

where k_0 is a constant to be trivially estimated.

Reasonably assuming a roughly cubic shape of each battery module, the surface area A of the packaging materials covering each battery module is given by:

$$A = 6 \times \left(\sqrt[3]{V_0} \right)^2 \tag{9}$$

$$= 6 \times \left(\sqrt[3]{k_0 m_0} \right)^2 \tag{10}$$

$$= 6 \times (k_0 m_0)^{\frac{2}{3}} \tag{11}$$

where (10) is arrived at by substituting (8) into (9). If the battery modules are not of a cubic shape, (9) to (11) may require some refinement but they as what they are here can still serve as a good approximation if the shape does not differ much from a cube.

The mass m of a battery module is composed of three components, namely,

- the mass m_0 of the energy storage structures in the battery module,
- the mass of the packing materials covering the battery module, which is proportional to the surface area A of the packaging materials, and
- the mass of the electronics associated with the battery module, which is a percentage of m_0 [19].

In other words,

$$m = m_0 + k'A + k''m_0 \tag{12}$$

$$= (1 + k'')m_0 + 6k' \times (k_0 m_0)^{\frac{2}{3}} \tag{13}$$

upon substituting (11) into (12) where both k' and k'' are constants to be trivially estimated.

Substituting (13) into (5) gives

$$E_C = [(1 + k'')m_0 + 6k' \times (k_0 m_0)^{\frac{2}{3}}]^s \tag{14}$$

Incorporating (7) into (1) gives

$$P(K = k; \frac{C}{m_0}) \tag{15}$$

Therefore, the probabilistic expected value of users' total human effort to (re)charge a battery system at the end of each of its discharge cycles, upon substituting (4) and (14), is given by

$$E = E(E_L + kE_C)$$

$$= \sum_{k=0}^{n} \left\{ \begin{array}{l} \frac{\lfloor \frac{k}{L} \rfloor}{2}[2E_{L1} + (\lfloor \frac{k}{L} \rfloor - 1)E_{LD}] \\ + \frac{\lceil \frac{k}{L} \rceil}{L}(E_{L1} + \lfloor \frac{k}{L} \rfloor E_{LD}) + k[(1+k'')m_0 + 6k' \times (k_0 m_0)^{\frac{2}{3}}]^s \end{array} \right\} P(K = k; \frac{C}{m_0})$$

$$(16)$$

where

E = the expected value of users' total human effort to (re)charge a battery system at the end of each of its discharge cycles,

$E(\cdot)$ = the probabilistic expected value function, and

n = the total number of battery modules in each battery system.

In practice, n is known to the users given the known layout of the battery modules in each battery system. Equation (16) forms the second trade-off model.

5   Ultimate Objective: Optimizing m_0 for Minimizing Human Effort

As with the first trade-off model, the ultimate objective of formulating this second trade-off model is of course to optimize, or more precisely minimize, E in (16) upon substituting into it the values of n, L, E_{L1}, E_{LD}, k_0, k', k'', s, C, and $P(K = k; \frac{C}{m_0})$ as known, estimated, or determined in ways delineated in the last section for battery systems that the last section concerns. Mathematically, the minimization is

$$\min E = \min_{m_0 \in \{M_L, \frac{M_L}{2}, \frac{M_L}{3}, \frac{M_L}{4}, \cdots\}} \sum_{k=0}^{n} \left\{ \begin{array}{l} \frac{\lfloor \frac{k}{L} \rfloor}{2}[2E_{L1} + (\lfloor \frac{k}{L} \rfloor - 1)E_{LD}] \\ + \frac{\lceil \frac{k}{L} \rceil}{L}(E_{L1} + \lfloor \frac{k}{L} \rfloor E_{LD}) \\ + k[(1+k'')m_0 + 6k' \times (k_0 m_0)^{\frac{2}{3}}]^s \end{array} \right\} P(K = k; \frac{C}{m_0})$$

$$(17)$$

where M_L = the total mass of the energy storage structures in each layer of battery modules in each battery system, and is accomplished by searching for the optimal m_0 out of the feasible values M_L, $\frac{M_L}{2}$, $\frac{M_L}{3}$, $\frac{M_L}{4}$, ..., which correspond to splitting each layer into different integral numbers of battery modules. In practice, this search needs not to be indefinite but can stop at the possible minimum mass of the energy storage structures in each battery module as limited by the manufacture and practical usage.

As with the search for the optimal m in the first trade-off mode, the search for the optimal m_0 here can be performed simply by substituting M_L, $\frac{M_L}{2}$, $\frac{M_L}{3}$, $\frac{M_L}{4}$, ... into m_0 to find the minimum E value that results or by means of genetic algorithms.

6 Discussion

This article aims to propose the second trade-off model (16) as a mathematical model of human effort trade-off concerning the method and the subsystem for battery design that were published earlier by the author and a co-inventor and are now covered by several patents and patents pending around the globe. It also suggests (17) for the optimization, or more precisely, minimization of users' human effort to (re)charge the battery systems in question in the context of the aforesaid trade-off. The minimization is to be accomplished through the search for the optimal mass m_0 of the energy storage structures in each battery module of the battery systems. This second trade-off model improves the first trade-off model proposed earlier by the author in that the former and the human effort minimization based on it relax the restrictive and unrealistic assumptions underlying the latter and its corresponding human effort minimization. The improved and more realistic human effort minimization can better position the method's and the subsystem's related real-life manufacture and commercialization in the highly competitive market through better optimizing m_0 for the sake of users' best ergonomic benefit of minimum human effort to (re)charge the battery systems at the end of each of their discharge cycles during their daily usage. Whilst the method's and the subsystem's key philosophy is to modularize battery systems, such a mass m_0 is a key parameter as it dictates the extent of the modularization, i.e., the number of battery modules that each battery system is to be broken into.

Acknowledgments. This article was supported by Grant (004/2013/A) of the Science and Technology Development Fund, the Government of Macao Special Administrative Region.

References

1. Chan, V.K.Y.: The modeling of technological trade-off in battery system design based on an ergonomic and low-cost alternative battery technology. In: Ahram, T., Karwowski, W. (eds.) Advances in Human Factors, Software, and Systems Engineering, Advances in Intelligent Systems and Computing, vol. 598, pp. 122–130. Springer, Cham (2018)
2. Wood III, D.L., Li, J., Daniel, C.: Prospects for reducing the processing cost of lithium ion batteries. J. Power Sour. **275**, 234–242 (2015)
3. Hülsmann, M., Fornahl, D. (eds.): Evolutionary Paths Towards the Mobility Patterns of the Future. Lecture Notes in Mobility. Springer, Heidelberg (2014)
4. Barbarossa, C., Beckmann, S.C., De Pelsmacker, P., Moons, I., Gwozdz, W.: A self-identity based model of electric car adoption intention: a cross-cultural comparative study. J. Environ. Psychol. **42**, 149–160 (2015)
5. Shiau, C.-S.N., Samaras, C., Hauffe, R., Michalek, J.J.: Impact of battery weight and charging patterns on the economic and environmental benefits of plug-in hybrid vehicles. Energy Policy 37, 2653–2663 (2009)
6. Lin, M.-C., Gong, M., Lu, B., Wu, Y., Wang, D.-Y., Guan, M., Angell, M., Chen, C., Yang, J., Hwang, B.-J., Dai, H.: An ultrafast rechargeable aluminium-ion battery. Nature **520**, 324–328 (2015)

7. Omar, N., Monem, M.A., Firouz, Y., Salminen, J., Smekens, J., Hegazy, O., Gaulous, H., Mulder, G., Van den Bossche, P., Coosemans, T.: Lithium iron phosphate based battery – assessment of the aging parameters and development of cycle life model. Appl. Energy **113**, 1575–1585 (2014)

8. Shousha, M., McRae, T., Prodić, A., Marten, V., Milios, J.: Design and implementation of high power density assisting step-up converter with integrated battery balancing feature. IEEE J. Emerg. Sel. Top. Power Electron. **5**, 1068–1077 (2017)

9. Liu, Q.-C., Liu, T., Liu, D.-P., Li, Z.-J., Zhang, X.-B., Zhang, Y.: A flexible and wearable lithium-oxygen battery with record energy density achieved by the interlaced architecture inspired by bamboo slips. Adv. Mater. **28**, 8413–8418 (2016)

10. Hsieh, G.-C., Chen, L.-R., Huang, K.-S.: Fuzzy-controlled li-ion battery charge system with active state-of-charge controller. IEEE Trans. Ind. Electron. **48**, 585–593 (2001)

11. Yilmaz, M., Krein, P.T.: Review of battery charger topologies, charging power levels, and infrastructure for plug-in electric and hybrid vehicles. IEEE Trans. Power Electron. **28**, 2151–2169 (2013)

12. Ralston, M., Nigro, N.: Plug-in electric vehicles: literature review. Technical report, Center for Climate and Energy Solutions (2011)

13. Chan, K.Y.V., Leong, S.L.: A control method to charge and discharge battery modules and the related system. Chinese Patent ZL 2015 1 0646435.8, 11 July 2017

14. Chan, K.Y.V., Leong, S.L.: A control method to charge and discharge battery modules and the related system. Japanese Patent Pending 2017-539044 (2017)

15. Engel, B., Wussow, J., Mummel, J.: Grid integration of conductive and inductive high-power charging systems. In: Liebl, J. (ed.) Netzintegration der Elektromobilität 2017, p. 73. Springer, Wiesbaden (2017)

16. Poisson, S.D.: Recherches sur la Probabilité des Jugements en Matière Criminelle et en Matière Civile, Précédées des Règles Générales du Calcul des Probabilitiés. Bachelier, Paris (1837)

17. Render, B., Stair Jr., R.M., Hanna, M.E., Hale, T.S.: Quantitative Analysis for Management, 13th edn. Pearson, Essex (2018)

18. Rahmati, A., Qian, A., Zhong, L.: Understanding human-battery interaction on mobile phones. In: 9th International Conference on Human Computer Interaction with Mobile Devices and Services, pp. 265–272. ACM, New York (2007)

19. Plett, G.L.: Battery Management Systems, Volume II: Equivalent-Circuit Methods. Artech House, Norwood (2016)

Analysis of the Dangers of Professional Situations for Oil and Gas Workers of Various Professional Groups in the Arctic

Yana Korneeva[✉], Natalia Simonova, and Tamara Tyulyubaeva

Northern (Arctic) Federal University named after M.V. Lomonosov,
Arkhangelsk, Russia
ya.korneeva@narfu.ru

Abstract. The reported study was funded by RFBR according to the research project № 18-013-00623. The activities of oil and gas specialists are carried out by the shift method, a number of dangerous situations arise due to extreme climatic, geographical, industrial and social factors. Employees assess these situations differently. To some of them, workers are adapted, they know how to act, which reduces their subjective assessment of the danger, and other part still causes certain difficulties and requires more attention from the management of enterprises. With the help of the questionnaire, a subjective assessment of the dangers of occupational situations that might occur during the shift was conducted. Employees assessed the following 18 situations. The most dangerous for employees are situations where relatives have problems at home, and you cannot help, when it is necessary to perform a work hazardous to health. This study showed the differences in the assessment of the danger of occupational situations by shift workers of various occupational groups.

Keywords: Safety · Dangerous of professional situations · Oil production
Professional groups · Shift work · Arctic

1 Introduction

The reported study was funded by RFBR according to the research project № 18-013-00623. The Russian Federation has a unique resource potential, due to which it occupies one of the leading places in the world fuel and energy system. The growth of gas transportation volumes largely depends on the development of new gas condensate fields in the Arctic of Russia. Significant removal of these objects from settlements with developed infrastructure promotes the use of shift work method in companies.

Activities in extreme conditions make other requirements for the professional adaptation of workers, which is due to the unpredictability of the occurrence of stressful or emergency situations that require rapid response and resolution [1–3]. The safety of the worker's behavior and the success of his professional activity in extreme conditions depends to a large extent on the characteristics of a person's personality, his subjective sensation, self-regulation, and protection from danger [4, 5]. Therefore, the study of the worker's safety behavior is necessary not only taking into account the organization and

© Springer International Publishing AG, part of Springer Nature 2019
T. Z. Ahram (Ed.): AHFE 2018, AISC 787, pp. 535–540, 2019.
https://doi.org/10.1007/978-3-319-94229-2_51

working conditions, but also the features of his psychological safety, as a potential opportunity to avoid making mistakes in the workplace [6].

Psychological safety, like any professional characteristic, depends on:

(1) the stable individual qualities of a person;
(2) the current status of a person;
(3) attitudes towards the activities performed and its safety; and
(4) the professionally important qualities and skills.

Due to the fact that the activities of specialists in oil and gas companies are carried out mainly by the shift method, a number of dangerous situations arise due to group isolation conditions and extreme climatic and geographic conditions of the Arctic. They are: absence of the possibility to leave the shift camp in adverse weather conditions, difficulties with transportation, limited means of communication, limited availability of medical assistance, etc. [7, 8]. These situations are assessed differently by the employees: to some of them the workers are adapted, they know how to behave and act, and therefore their subjective assessment of its danger is reduced. Other situations still causes certain difficulties and requires more attention from the management of companies. The present study is devoted to the study of such situations. *The research aim* is to study the subjective assessment of the danger of occupational situations by oil and gas workers of different professional groups by the shift method in the Arctic.

2   The Research Materials and Methods

The study involved 70 oil workers in the territory of the Nenets Autonomous Okrug (the duration of the shift 30 days) between the ages of 24 and 60 (mean age 38.7 ± 9.7). The work experience of the shifted method varies from 0.5 to 31 years (9.53 ± 7.6). Employees participated in the study with their personal consent; the selection by other parameters was not required.

With the help of the questionnaire, a subjective assessment of the dangers of occupational situations that may occur during the shift period was conducted. The employees assessed the following 18 situations:

(1) the relatives have problems at home, but you cannot help;
(2) hazardous work;
(3) an error due to which you or your colleagues may suffer;
(4) the situation when you are sick, and a doctor's consultation is required;
(5) prevent risks associated with testing new equipment;
(6) a colleague violates safety techniques;
(7) uncorrected equipment;
(8) work without personal protective equipment;
(9) the situation of weather changes, as a result of which there is no exit from shift camp;
(10) smoke or fire;
(11) rendering of the first medical aid;
(12) absence of colleagues in the workplace when assistance is needed;

(13) work without prior instruction in safety precautions;
(14) a situation of power failure;
(15) a deprivation of communication facilities;
(16) a situation of water supply shutdown;
(17) there are chronic diseases, but you have forgotten the necessary medications;
(18) an injury.

The situation was determined on the basis of the results of previous expedition studies and analysis of documentation on labor protection of oil and gas companies. The assessment was made on a 7-point scale, where the 1 situation is estimated by the employee as minimally dangerous; 7 as the most dangerous situation. The statistical methods are: descriptive statistics, multi-dimensional variance analysis MANOVA. The processing was carried out using the SPSS 22.00 software package (licensing agreement No. Z125-3301-14, NArFU).

3 Research Results and Discussion

In order to determine the most dangerous situations for the workers, the analysis of descriptive statistics was used (Table 1).

Table 1. Results of descriptive statistics on the subjective assessment of the dangers by workers in oil and gas companies

Situations	Mean value	Standard error of mean
1. The relatives have problems at home, but you cannot help	3.18	0.307
2. A hazardous work	3.15	0.308
3. An error due to which you or your colleagues may suffer	2.89	0.324
4. The situation when you are sick, and a doctor's consultation is required	2.80	0.288
5. Prevent risks associated with testing new equipment	2.80	0.276
6. A colleague violates safety techniques	2.76	0.281
7. An uncorrected equipment	2.71	0.268
8. A work without personal protective equipment	2.69	0.262
9. The situation of weather changes, as a result of which there is no exit from shift camp	2.47	0.240
10. Smoke or fire	2.44	0.299
11. Rendering of the first medical aid	2.40	0.284
12. Absence of colleagues in the workplace when assistance is needed	2.35	0.201
13. Work without prior instruction in safety precautions	2.29	0.242
14. A situation of power failure	2.27	0.251
15. A deprivation of communication facilities	2.15	0.233
16. A situation of water supply shutdown	2.04	0.235
17. There are chronic diseases, but you have forgotten the necessary medications	2.02	0.264
18. An injury	2.02	0.250

According to Table 1, the most dangerous for employees are situations where relatives have problems at home, and you cannot help them (3.18 ± 0.307) and when it is necessary to perform work hazardous to health (3.15 ± 0.308), (Fr2 = 2.42, with p < 0.001).

Due to the fact that according to working conditions the professions that are in demand in the oil and gas industry have significant differences, we assumed that workers will also assess the danger of occupational situations in different ways. Therefore, we conducted a professional analysis of all the professions of specialists who took part in the study and divided them into four professional groups: (1) oil and gas operators, (2) operators of cleaning facilities and a boiler house, (3) drivers and (4) engineering and technical personnel.

The first group includes oil and gas operators, which perform work to service wells and ensure their uninterrupted operation under the supervision of persons of technical supervision. They regulate the operation of the equipment in accordance with the specified regime, serve the ground equipment of the wells, participate in the installation, dismantling and repair of field equipment. Their professional activity is carried out mainly in the open air.

In the second group, operators of sewage treatment plants and a boiler house are singled out. Whose work goals are broadly consistent with the goals of oil and gas operators' work: monitoring, assessment, diagnostics, and maintenance of equipment. However, operators for maintenance of treatment facilities, boiler house most often perform their professional tasks using automated equipment, monitoring the operation of automatic devices, checking the correctness of the instrument readings, eliminating minor malfunctions. At the same time, professional activity is carried out in the open air, but more often in closed premises.

The drivers of various kinds of vehicles were united in the third group. Workers of this category carry out transport management, maximally while ensuring the safety of passengers, also ensure the technically sound condition of the vehicle itself. Their work is often in the open air.

The fourth group consisted of engineers and technicians (an oil preparation engineer, a mechanic, a master, etc.). The main function is to monitor the operation of oil and gas equipment, maintain documentation, introduce new technologies in the preparation and transportation of oil. Their professional tasks also include the organization and maintenance of maintenance and repair of mechanical equipment, quality control of oil preparation, ensuring control over the quality of the work performed by oil and gas production for team members. Their professional activities are carried out both in closed premises and in the open air. Thus, the goals of their work are multifunctional, since they cover not only control, assessment and diagnostics, but also consist in transformation and maintenance. At the same time, the means of labor include manual and simple adaptations, mechanical, automatic, portable and stationary.

In order to study the differences in the assessment of the dangers of occupational situations by shift workers of various professional groups, a multidimensional analysis of variance was used. Multivariate tests show a statistically significant effect of the factor professional group on the results of the study (p = 0.001). The results of one-dimensional tests indicate that this effect is valid for the following situations:

(1) the situation of weather changes, which is why there is no exit (p = 0.003);
(2) a situation where you are sick and require a doctor's consultation (p = 0.03);
(3) there are chronic diseases, but you forgot the necessary medications (p = 0.054);
(4) deprivation of communication (p = 0.019); and
(5) an injury (p = 0.043).

The greatest danger is the four situations out of five (the situation of changing weather, which means there is no exit; the situation when you are sick and you need a doctor's consultation; there are chronic diseases, but you forgot the necessary medicines) are for drivers (Fig. 1). This is directly related to the characteristics of their professional activity. On the one hand, drivers work in the open air, and therefore the possibility of its implementation and efficiency directly depends on the weather conditions, on the other hand, while fulfilling their duties, they are significantly removed from the fishery where the medical station is located, these situations are most at risk.

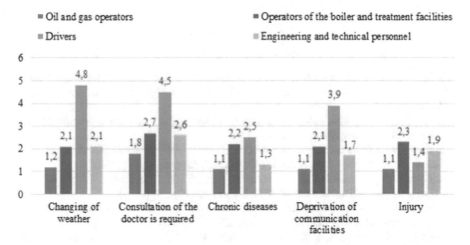

Fig. 1. Subjective assessment of occupational hazards by workers of various occupational groups

The injuries are assessed as the most dangerous for operators and maintenance technicians. Maintenance specialists in the maintenance and repair of equipment often face minor injuries; these are occupations with the highest risk of injury, so they really assess the potential threats of such situations and realize the possible consequences. Operators' activities have minimal risks of traumatism. Therefore, they can highly appreciate the danger of such situations arising from the lack of experience of behavior in them, the complexity of predicting the consequences, their uncertainty about the correctness of actions in them.

According to Fig. 1, as the most dangerous, workers of all groups assessment the situation when the doctor's consultation is required. Which may be due to the fact, that during the shift period there are often non-standard health-related situations that are not always successfully resolved. One of the possible reasons for this assessment may be the fact that in these situations the employee is sent to hospitals and polyclinics for a nearby city or a city of permanent residence.

Most specialists are wary of the occurrence of such situations, as this leads to the inability to continue working and reduces their pay. In this connection, they try to hide such diseases and continue to carry out their professional activities, understanding all the health risks.

Thus, as a result of our research, it is established that there are differences in the assessment of the dangers of professional situations by the shift workers of four groups, such as:

(1) weather changes, where there is no exit;
(2) the situation when you are sick and you need medical advice;
(3) there are chronic diseases, but you forgot the necessary medications;
(4) deprivation of communications and
(5) the injury.

References

1. Kotik, M.A.: Psychology of the safety of activities: from the first publications in Dorpat to modern studies at the University of Tartu. Cogn. Reg. Act. Hist. Ontogenet. Appl. Prob. **894**, 135 (1990)
2. Kotik, M.A., Emel'janov, A.M.: The nature of human operator errors: the example of vehicle management. Moscow Transport, p. 251 (1993)
3. Shlykova, N.L.: Formation of a new direction in psychology - psychological safety of subjects of professional activity. Hum. Fact. Probl. Psychol. Ergon. **3**, 6–8 (2006)
4. Zaharova, R.R., Kalimullina, G.N., Romanov, V.S.: Working conditions and health status of employees of oil refineries. Lab. Med. Ind. Ecol. **4**, 120–122 (2015)
5. Korneeva, Y., Simonova, N.: Psychological safety of oil and gas workers in the conditions of group isolation of the arctic. Adv. Sci. Lett. **23**(11), 10511 (2017)
6. Degteva, G.N., Korneeva, Y.A., Simonova, N.N.: Personal resources of oil and gas workers for the purposes of adaptation to the negative arctic climate and geographical conditions. Hum. Ecol. **9**, 15 (2017)
7. Belykh, S.L., Simonova, N.N., Korneeva, Y.A., Voitekhovich, T.S.: Shift mental representations as a factor of personnel professional adaptation. Psikhologicheskii Zhurnal **37**(5), 32 (2016)
8. Korneeva, Y.A., Simonova, N.N., Degteva, G.N., Dubinina, N.I.: Shift workers adaptation strategies in the far North. Hum. Ecol. **9**, 9 (2013)

Safety Management Principles in Electric Power Industry Based on Human Factors

Yang Song[✉]

China Southern Power Grid Extra High Voltage Transmission Company Liuzhou
Bureau, #22 Haiguan Road, Liuzhou, China
colaittis@163.com, songyang@ehv.csg.cn

Abstract. Safety is the most crucial factor in energy industry, meanwhile the
safety topic is also not an integrity discipline by now. The paper introduced the
reasons why safety discipline is incomplete and presented 15 safety principles in
human factors, devices factors and safety system aspects. Safety is a subject
interdisciplinary and the researchers were hardly proved correct in certain
questions, however they can put forward some principles to make our industry
safe enough. The paper presented kinds of principles actually applied in the
power industry.

Keywords: Human factors · Safety management principles · Safety systems
Electric power industry

1 Introduction

Electric power industry is one of the most complex system in the world. This system
offers interrupt power supply, in nationwide its too crucial to tolerance even power off
for a little while. The fatal factor in this field is the safety and stability of the electric
power grid. The industry includes design, plan, construction, maintenance, repairing,
upgrading and so on, which placed to strict desire to numerous engineers in the
industry. Not only they should assure the grid and the devices are under control, but
also they should keep safe when working in the field.

However, the safety problems didn't seem to be well organized as a discipline with
complex reasons. By now in the academic these is debate about the safety theory. Some
experts recognized the safety is a subjective concept while some others hold a controversy viewpoint. This paper agreed with the former opinion, but also thought it can
be judged by objective standards. So safety in electric power industry actually means
some suitable methods rather than correct answers.

Before mentioned the concrete safety principles, the paper would answer a primary
question: what is safety and how to describe it? In Marvin Rausand's works [1], Safety
has two identification. The paper concluded that safety is the state and confidence
without unexpected losses, not mentioned the cases with intent to do harm. And safety
has more important figuration than identification.

(1) For individuals, safety mind has been form when childhood. Usually people think
    about the safety problems by common sense eventually.

© Springer International Publishing AG, part of Springer Nature 2019
T. Z. Ahram (Ed.): AHFE 2018, AISC 787, pp. 541–550, 2019.
https://doi.org/10.1007/978-3-319-94229-2_52

(2) The action for safety can hardly been proved correct but easily been proved wrong.
(3) Safety investment are necessary but benefits always are invisible.
(4) For a company or an organization, safety system and concept take shape after its carrier has formed. So it unlikely to design a brand new safety system, as the safety system is not independent.

Most research about safety focus on some specific area by now, such as how to prevent electric shock or gas hazard, there is rare principles of safety guides about how to organize and improve the overall safety extend. The paper recognized that safety is too complex to summarize in a single way, electric power enterprises and organizations run their own methodical safety or risk system. The paper summarized 15 principles in three aspects as below.

2　Safety Management Principles

2.1　Safety Principles of Devices Design Based on Human Factors

In former research [11], it has been proved that human factors should be considered to reduce common human errors. The paper presented principles about how to realize it.

2.1.1　Reliable and Distinguishable Indicator and Interlock Relationship for Power Grids and Devices

The same with other industry such as subway system [12], the power system rely on SCADA, protection devices and interlock relationship to keep out of human errors. The operators monitor the power grid and its substation or plants mainly by SCADA (Supervisory Control And Data Acquisition) system. And each main devices has projection relay to cut off the fault devices immediately. Furthermore, the entire grid are protected by some more complex system such as interlock relationship devices, safety and stability control devices. Generally speaking the Protection relay, control system and SCADA, interlock relationship devices are irreplaceable and its signal or criterion should show the state of the grid and devices precisely. Power system is a continuous system even if only for a while, instantaneous, its energy transmission and fault will release huge energy and need timely removal and recovery in order to prevent the failure affect the larger range, so controlling and protection devices has been placed along with the transmission grid. For example, both the state of switch and the current can be showed or judged for whether a transmission line in use, but the later one is more reliable than the former one [2]. So more reliable and distinguishable indicator or criterion should be chosen when design to differ the similar but different cases. More than one indicators can show the state of devices but the most appropriate one or their logic combination should be chosen.

2.1.2　Redundancy Principles

The same with other complex system, electric power system consists of masses series and parallel connections [3]. The system should maintain the frequency and voltage, cut off the fault part, it is far more than a single system in fact.

Redundancy design should be based on the keeping system security and reliability, and usually focus on some key components or functions. When a system failure occurs, such as a equipment damage occurs, the redundant configuration of components can be put into operation or remain in normal to maintain the working condition. Most important devices have redundancy figuration, how to choose and how much redundancy is a problem. The paper recognized that the redundancy should be consider as following conditions:

(i)   Failure rate is beyond the average component failure rate;
(ii) Failure cannot be detected easily;
(iii) The failure influence the whole system.

2.1.3 Unique Named

From a substation to a variety in the software program, everything should have its unique name in some scope.

The name of the device should be significant distinguished. After a long period of development, many devices from manufacture have its own nomenclature rules, which largely avoid duplication and confusion, and also avoid many security risks. However, due to the high integration of the power grid, the naming of physical quantities in equipment and software is not uniform, and it is difficult to accurately predict the large number of carriers that need to be named. But the false name or rename itself is a threat to security, and the rules of name need to be regulated according to the actual situation. Especially in the software, the name of the physical quantity or the unit is always the only way for checking. The false name or rename hardly be detected but would cause potential safety hazard.

2.1.4 Unify Configuration and Composition Principles

The unify configuration and composition of equipment is an important guarantee of safety. Although it does not cause the fault directly, it is important considering the influence of human factors. The unify configuration and composition of equipment means less probability of mistakes in maintaining obviously. The author or had analyzed 100 incidents and attempted accident, found that 82% of them include "unsafe and non-artificial factors", about half of which existed unified equipment or identification.

2.1.5 Power Grid and Devices Visible and Manageable Principles

The monitoring of the power grid and its equipment can be classified by monitoring of the power grid and monitoring of transmission equipment. At present, the maximum monitoring system reveals the power grid level, and the minimum monitoring reaches the flash or CPU of Industrial PC (IPC). Accidents due to the state of key component invisible is popular [4]. It is hard to define what is visible level is best considering the economic cost. As the visibility is aim to being manageable, the devices which visible level rely on should be more reliable than the devices they are supervising, and the indicators visible level show to human can make a difference. Except visibility, being manageable requires Human-Machine Interactive(HMI) function as well. In Prof.

Nancy Leveson's research, the HMI function should include incremental control and error tolerance [13].

The paper took the temperature relay of electric reactor for example. The standard <IEC 60255-8 Electric Relays: Part 8 Thermal Electric Relays>, presented temperature curves of electric reactor depending on different circumstance. However the temperature sensor is not reliable enough for protection relay and supervise. So the designer chose electric current of the electric reactor to calculate the temperature, as the electric current is the only crucial factor for temperature. When the incremental temperature archived the setting the projection relay will cut off the reactor. When the temperature rises but not high enough, the SCADA will remind the operator to take appropriate action. And SCADA and 2 sets of protection relay have independent sensor and circuit so any single failure wouldn't lead to accident in theory.

2.1.6   Oriented Numerous Devices Management: Comprehensive Management and Professional Management Crossing Principle

Most electric power enterprises and institutions are holding comprehensive management and professional management. Comprehensive management orients all kind of devices in an area or power plants, while professional management focus on a certain kind of devices in spread area or wider range. Comprehensive management is in charge with recover and eliminate the defect of the devices. Professional management observe all the information about certain kind of devices and detect the hidden risk. For example, if this kind of devices from some company breakdown in some special environment, it may be a hidden risk to other companies in this kind of environment too. Professional management would recover and eliminate the hidden risk [5] before it happens in their own company. Professional management actually play a external supervise role. Most devices are included in both comprehensive management and professional management.

2.2   Safety Principles Human Factors

2.2.1   Human Factors Date Analysis Attribution Principle

As the paper introduction indicated, safety is a subjective concept in essence. Staffs often contribute to different factors when analyzing the same accident. Furthermore, accidents are something negative and rare people can see all of the details. As Haddon presented in his research, countermeasures to accidents should focus on reducing future loses rather than casual factors [14]. The accidents and their countermeasures are asymmetric. Introduction of data analysis is an objective method for accidents attribution. Any accident has many causes but people are always attribute to the factors they knew and thought can be changed. So an objective method of analysis is collect more accidents or near misses to conclude. Any accident is due to some of the conclusions, with unpredictable time and address.

The paper analyzed full picture of 100 accidents or near miss belong to a major. The analysis included two steps: primary analysis oriented accident factors, further analysis oriented the reasons based on the primary analysis. The primary analysis select 12 possible factors (top 9 listed in Fig. 1). The further analysis is attributed to lack of risk control based on the primary analysis. The paper classified the risk control by 4 types: institution control, organization control, technical skills control and safety culture control. Institution control is setting standards of procedures, rules and measures to control working procedure or equipment specification rule-based, making the work more measurable too. Organizational control is to minimize the safety risk through appropriate deployment of resources including human resources, tools, time and schedule. The technical control is reducing the risk by experienced workers or supervisors with higher safety skill. Safety culture control is that individuals have a strong sense of safety awareness, the department or team has rich safety culture overall.

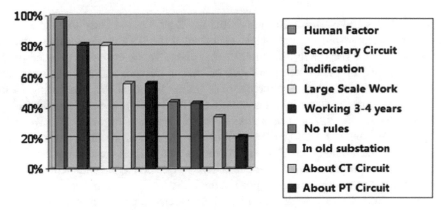

Fig. 1. Primary analysis: factors statistics of 100 accidents

Any accidents can be attributed to the lack of some aspects. The reasons statics are listed in Fig. 2 and the accident countermeasures can be easily made up. And the accident countermeasures probably effect more than one aspects, ones effected higher accumulation aspects should have priority. After countermeasures has been put in use, the number of accidents and near misses of this major decreased to less than half.

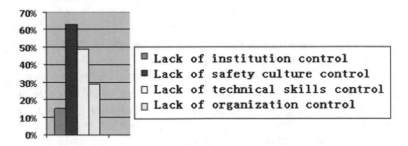

Fig. 2. Further analysis: reasons statistics of 100 accidents

This classification is according to the aspects which the company or organization can make a difference. The reasons mentioned above might not be fit for other companies or circumstance.

2.2.2 Minimum Number of Individuals Principle

Supervise institution are proceeding in field work of electric power industry. It means any operation or field work requiring at least 2 workers, expect qualified employees making an inspection tour for equipment. Most work or operation need at least 2 field workers and 2 others to permit. So it usually takes four people to start a field work, the number is going to be more if the wok is complex. This has been proven to be reasonable and written into China's electricity safety institution [6]. In former research the human error are classified into slips and mistakes [15]. At least 2 field workers and 2 others to permit institution decrease the probability of both.

2.2.3 Definite and Suitable Responsibility for Each Post of Duty Principle

Each safety responsibility corresponds a post of duty. The paper assumed that no person or post can see the full picture of safety risks. The assumption is based on the that for the issue of safety, everyone's experience is limited, and it is not even clear what the key factors are in maintaining the safety for someone. According to the author's experience, rare ones even knows that he needs to know more than 50% of the total amount of safety information and knowledge. Everyone has his or her own safety sense based on their knowledge from his or her childhood. Professional knowledge aims to make the technical problems as "common sense". It is the common sense that really dominates people's behavior about safety. If you need to keep the work safe, you shouldn't make your job look more professional. Instead, you should make your work look more ordinary and logic, so more people can understand and their safety knowledge and viewpoint can be put in use. Making the work "seemed ordinary" should be the responsibility of grassroots managers. Top manages do not necessarily have an experience of each professions, they need the grassroots level managers express safety risk visualization to make risk pre-control decisions [7]. Prof. Nancy Leveson presented that bottom-up decentralized decision may be right but cannot prevent accident [16]. The paper agreed with that but should point out bottom-up information is necessary. Generally speaking grassroots managers or middle level in a company or organization are in charge with the bottom-up and up-bottom safety information dealing and transfer. Also they are important connection for unified safety culture. Their duty and role should be definite and suitable.

2.2.4 Rules and Standards Principle: Rules and Standards Should Offer the Workers a Good Choice in Any Circumstance

The power industry, like all other industries, has a well-established regulatory system. Although the specific procedures adopted by each country varied, the role of rules is same. In equipment management, should have the manufacture procedure, technical regulations, design procedure, transport regulations, construction installation procedures, commissioning acceptance procedures, operation maintenance procedure and equipment life cycle management regulations, etc.; In terms of personnel management, it should be equipped with scheduling operation procedures, electric power safety operation regulations, technical specifications for electric work tickets and operation instructions for specific work tasks. In addition, the investigation in some specific areas such as accident, operation maintenance strategy based on the power grid operation need to issue the specification.

The rules and standards normally provided a safe way to complete the work. However some of them are hardly put in use in some circumstance. When this happens, the staff have to disobey the rules and standard. The worse effect is rules and standards lose some authority in the safety culture and cause safety culture degradation further more [8].

2.3   Principles of Safety Management System

Safety system is widely spread system consists of human and non-human factors, the concept of safety system the paper mentioned is oriented to an organization or a department at least. As Prof. Nancy Leveson indicated in her book [10], the safety system cannot rely on the reliability of the components. The paper presented principles about how the safety system works.

2.3.1   "No Dead Ends and More Crossing Points" Principle
Draw the flow diagram of the safety system and it should have no dead ends and consist more "crossing points":

If the whole process of safety management in the form of flow charts, so there should be no dead end in this diagram, there should be "circle" situation. More "crossing points" means more check procedure and more safer, "dead ends" means not being check easily.

The following chart showed the common safety procedure. The author analyzed 100 accidents and near miss which he knew well and found errors in every step. The sum number that errors occurred showed in the chart. The red steps means more error occurs, and they are all "dead ends" step. The green steps has seldom errors and most of them are "crossing points" as Fig. 3 showed.

2.3.2   Multi-majors Assemble Principle
Approved working plan by multiple levels institution, assemble the engineers from all relative aspects to talk about certain problem.

Any works or on the operation of the equipment need to be permitted by the relevant levels and professional, which is the on behalf of the company risk control. While the assessment of risk is unlikely to be complete, it is easy to judge when it comes to majors. Multi-disciplinary review can help risk control in the work.

In terms of safety, people can imagine the problems faced by the others, but can not completely stand in the position of the other roles to consider a problem. It's much easier to find others' problem than to find his or her own. So examination of work plan or project design by multi-majors background professionals is more likely to be a safe way.

2.3.3   Safety Management System Orientation Principle
Safety management system is neither an independent system nor a single system. Individuals have different viewpoints and ideas from different positions, everyone has his or her own safety system thinking. So different level or positions should play different roles. Each role is oriented and define the unacceptable losses [17]. Operators'

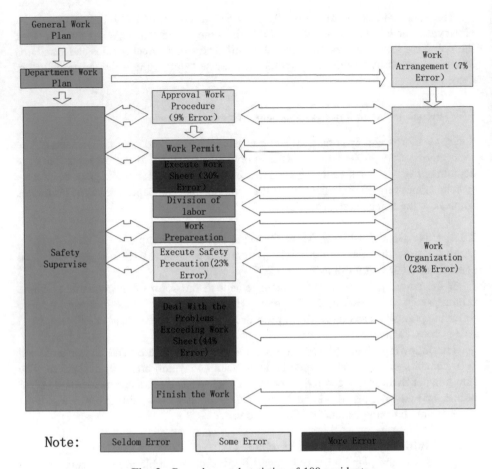

Fig. 3. Procedure and statistics of 100 accidents

error may be a direct cause for accidents but not the only reason. The paper presented one of the appropriate way as follows.

Safety management include different levels, different levels should have targeted orientation. The top level operate the whole system, the middle level system manege the human resources and devices, the base level focus on the concrete risk of work. For the base level the safety risk comes from the concrete work; for the middle level the safety risk are based on human resources and devices; for the top level, making sure the whole system operating normally is the most crucial charge [9].

2.3.4   Tolerance Safety Management System Principle

Tolerance the unknown factors existed and different opinions about the risk is feature of advanced safety management system.

No perfect safety management system exited but the system should be clearly visible for controlling. The problem is easily found when something goes wrong. Acknowledgment the unknown factor existed in the system is necessarily.

When different individuals or minority have different opinions about the risk in the work, they should not be thought wrong. In the system view, any thoughts are root in the system and culture. Different opinions maybe mean less important but do not mean wrong. The advantage of system thinking is assembling the resources in a certain direction but the disadvantage is lost some viewpoint from other sides. As the paper mention above, no one can see the whole view about safety system so the tolerance is necessary.

2.3.5   The Compatibility Principle of Technology Development

Technology never stopped developing so the safety system should have the compatibility for both new and dated technology, especially in communication field. In electric industry unify communication statute is applied and tested to make sure the compatibility the different devices but factual condition is more complex. When new tech are already been use, the compatibility is not easily fixed any more. So new devices should be consider about the compatibility of the whole system in the safety view.

3   Conclusion

The paper introduced the reasons which safety discipline is incomplete and presented 15 safety principles in human, devices and system aspects. Safety is a subject theory so the researchers hardly were proved correct in certain question, however they can put forward some principles to make our industry safe enough.

The author has proceed 10 year's research about the safety theory and found it is hard to conclude a unified way to keep our system safe. However, the principles and objection in safety system have a lot in common. So the paper presented the principles about safety and hope they can be consider in more area.

References

1. Rausand, M.: Risk Assessment Theory, Methods and Applications, p. 61. Wiley, Hoboken (2011)
2. Minhui, J.: Electric Power System Projection Relay Case Analysis, pp. 362–372. China Electric Power Press (2012)
3. JIhong, Y.: Reliability and intelligent maintenance, p. 51 (2011)
4. Leveson, N.G.: Engineer a Safer World, System Thinking Applied to Safety, pp. 66–67. MIT Press, Cambridge (2011)
5. Shouxin, S.: Electricity Safety Management Serious: Human Factors, pp. 44–45 (2008)
6. Chaoyin, X., et al.: Working Regulations of Power Safety, p. 17 (2015)
7. Leveson, N.G.: Engineer a Safer World, System Thinking Applied to Safety, pp. 441–442. MIT Press, Cambridge (2011)
8. Liming, C., et al.: Electricity Safety Management Serious: Safety Culture, pp. 51–53 (2009)
9. Yun, L.: Safety Discipline, pp. 189–191 (2015)
10. Leveson, N.G.: Engineer a Safer World, System Thinking Applied to Safety, pp. 7–9. MIT Press, Cambridge (2011)

11. Leveson, N.G.: Engineer a Safer World, System Thinking Applied to Safety, pp. 284–285. MIT Press, Cambridge (2011)
12. Leveson, N.G.: Engineer a Safer World, System Thinking Applied to Safety, p. 190. MIT Press, Cambridge (2011)
13. Leveson, N.G.: Engineer a Safer World, System Thinking Applied to Safety, pp. 282–283. MIT Press, Cambridge (2011)
14. Haddon Jr., W.: The prevention of accidents. In: Preventive Medicine, pp. 592–613. Little, Brown, Boston (1967)
15. Norman, D.: Human Factors in Hazardous Situations, pp. 139–143. Clarendon Press, Wotton-under-Edge (1990)
16. Leveson, N.G.: Engineer a Safer World, System Thinking Applied to Safety, p. 14. MIT Press, Cambridge (2011)
17. Leveson, N.G.: Engineer a Safer World, System Thinking Applied to Safety, p. 181. MIT Press, Cambridge (2011)

A Theoretical Assessment of the Challenges Facing Power Infrastructure Development in Low-Income Countries in Sub-Sahara Africa

Emmanuel Ayorinde[1,2(✉)], Clinton Aigbavboa[1,2],
and Ngcobo Ntebo[1,2]

[1] Department of Civil Engineering, University of Johannesburg,
Johannesburg, South Africa
engrkulz@gmail.com, {caigbavboa,ntebon}@uj.ac.za
[2] Department of Construction Management and Quantity Surveying,
University of Johannesburg, Johannesburg, South Africa

Abstract. Power infrastructure development is the pillar for every nation's economic development. The constraints facing the development of energy in the Low-Income Countries (LICs) of Sub-Sahara Africa (SSA) have greatly complicated their economic growth. The purpose of the paper is to identify the challenges facing power development in LICs. A confirmatory literature relating to power infrastructure development was undertaken to identify the challenges affecting power development in the LICs of SSA. The factors affecting the investment in power development sector of the LICs were found to be lack of funding, unfavorable policy framework, lack of technological knowledge, low electrification tariffs due to low incomes and lack of preparedness from the government were identified as the major cause of underdevelopment of the power sector in LICs. In order to improve the development of power infrastructure in LICs, favorable policies must be adopted that will bring about active participation of private investment in the energy sector. Likewise, the adoption of the Green-House-Gas (GHG) emission charters must be implemented to attract finance through the Clean Development Mechanisms (CDM) to develop the power infrastructure in these regions. The study contributes to the improvement of power sustainability in the LICs, which will directly improve the economic development, eradicate poverty, and contribute to power development in Africa.

Keywords: Economic growth · Infrastructure
Low income countries power development

1 Introduction

Power infrastructure development is the basis for every nation's economic and industrial development, also for low-income countries (LICs) in Sub-Sahara Africa (SSA). Secured access to energy infrastructure will improve lives and support the sustainable development goals. Most of the 1.3 billion people in the world without

© Springer International Publishing AG, part of Springer Nature 2019
T. Z. Ahram (Ed.): AHFE 2018, AISC 787, pp. 551–563, 2019.
https://doi.org/10.1007/978-3-319-94229-2_53

access to electricity are from SSA countries and South Asia [1]. The lack of adequate power infrastructure in the region has about 81% of the people relying greatly on the traditional biomass for cooking. Since most people in the LICs of SSA lives in the rural areas, these traditional biomass usage systems have had negative influence on air pollution, soil decay, deforestation etc. [2]. This trend of biomass usage will increase by 10% from 2009 to 2030. It will continue in this way because the access to electricity in the region cannot meet up to the size of the growing population due to lack of power infrastructure development in the region [3].

The high rate of power under-development in the region, led the United Nations secretary general Banki Moon in enacting a platform called "sustainable energy for all initiative" this will comprise of civil society, private and public collaborations, with the aim of meeting the universal access to global energy. These will also increase power infrastructure sufficiency and it will increase the global renewal energy market [4]. Anthropogenic greenhouse gas (GHG) emitting from the LICs of SSA is about 68% of the power-related services, and this trend will rise if there is improvement in the power infrastructure in the region [5]. The improvement in the power infrastructure development in the LICs of SSA will improve the lives and economy in the region but it should be in a sustainable way. The adoption of the following renewal energy trends such as; biomass power, wind power, wind energy, solar power, hydroelectricity, thermal energy will definitely reduce the absorption of GHG to promote sustainable development for all, secured supply of energy, improve economic growth and also supports the millennium development goals initiatives (MDGs) [6].

In general terms every nation in the world have a scale in measuring per capita earnings, energy resources, economic goals, population size, investments and technologies [7]. Therefore, the region requires a viable power infrastructure institution to enable the region achieve the global energy for all initiatives which were enacted by the United Nations. For the countries in Africa or the LICs in Africa to achieve this initiative by the year 2030, the power generation capacity of the region must rise to about 13% growth on a yearly basis. Since the past two decades ago, the power generation of the region has increased by just 2%, leaving a gap of 11% [8]. The impact of the rate of under-development in the power infrastructure sector of the LICs of SSA is evident in the area of high level of poverty and the least less industrialized regions in the world [5]. This region is blessed with excess renewal energy resources, but the problem of funding power infrastructure has limited the development electricity in the region. There are studies that showed for LICs in SSA to achieve the general access to energy for all initiative, the rate of electrification I the region must improve by 60%. Anything less than this will result in the realization of just 80% of the united nation general access to energy for all initiatives by 2030 [9]. Expressly said, it can be seen that for the LICs countries to achieve the global access energy for all initiatives there must be exploration of effective policies and mechanisms to mitigate the challenges facing power infrastructure in the region [4].

2   Literature Review on Power Development in the LIC's

The impact of power infrastructure development in a country cannot be over exaggerated. Various collaborators in the sector have recognized the amount of economic growth that power infrastructure development can bring to a particular region [10]. For instance, the Forum of Energy Ministers of Africa agreed that an improvement in the power infrastructure development in the LICs of SSA will improve the quality of lives of the citizens, promotes sustainable development in the region, create wealth and employment, improved industrialization and also in realizing the MDGs objectives. And also, the studies of Duvenage et al. [11], Suberu et al. [5] and Wood et al. [12] agrees that realizing a secured power supply system and also, eliminating the power sector aid to climate change, both are the challenges facing the realization of the sustainable energy agenda in the region. The rate of anthropogenic GHG emission is relative to the rate of energy output in the LICs of SSA. It has been assumed that by the year 2030, while the demand for energy supply will rise to 53% in the region also, the amount of carbon dioxide emission will increase by 55% in the region. As result of these anthropogenic GHG emission that is relative to energy output, will cause about 1.5 billion in the world will lack access to power supply [13].

However, with the great abundance of renewable energy resources in the LICs of SSA, these enormous potential is an advantage when can be utilized in realizing of a safe and secure environment and promotes sustainable energy development in the region. An example of such abundance is the 19-fold usage of hydroelectric power in Africa [4]. The study of Jumbe [14] and Haliu [15], agrees that the result of low power under-development in the LICs, can be linked to the following; lack of finance, fuel price, technological rates, subsidy, policy framework and economic level. The application of the United Nations initiatives by the LICs of SSA will result to the improvement of power accessibility of the region from the present 30.5%.

The motive of making energy supply accessible with the reduction in the prices of tariffs, these tend to complications in the policy framework because of the rate of poverty in the society. Therefore, the act of subsidizing the power infrastructure sector to make it cheap for the citizens, makes realizing a stable supply of electricity in the region unachievable [16]. The studies of Singh et al. [17] and Roy et al. [18], agrees that the absence of mechanisms such as; taxes, bonds, market instruments, tax exclusion and subsidy will complicate the promotion of renewable energy development. Countries like Malawi, Botswana and South Africa have an enormous potential of renewable energy but the adoption of fossil fuel in the generating electricity will increase the emission of the GHG, thereby making the environment not safe for the lives in the region [4]. The influence of policy framework in the development of power infrastructure development in the region is part of the setbacks the region has experienced. Therefore, there is need for new and effective ways of funding power infrastructure development in the LICs of SSA, to enable the region achieve the general access to energy for all initiatives [19].

2.1   An Overview of the Factors Affecting Sustainable Energy in LICs

According to United Nations reports in 2012, the following are the factors affecting the development of sustainable energy in the LICs of SSA.

Less Financial Strength: Due to the enormous resources needed to finance power infrastructure development in the region and due to lack of private sector participation to aid government spending. Since government is the traditional financier of power projects in the region.

Policy Framework: There must be a well laid down favorable policies by the central government, through legislations that will attract the participation of private investors. When there are favorable policies, it will encourage private investors in the power infrastructure development in the LICs of SSA to advance their development in the sector.

Lack Technological Knowledge: This factor has affected the development of power infrastructure in the LICs of SSA. The rate of lack of awareness of the technological advancement in the power infrastructure sector in the LICs has hindered the region from realizing an efficient power supply and the United Nations energy for all initiatives.

Tight Power Generating Plan: This factor also come up as a limitation in the development of power infrastructure in the LICs and the United Nations energy for all initiatives.

Low Rate for Electrification: The amount for electrification also, comes up as a limitation in the development of power infrastructure in the LICs. The less the rates for tariffs for electricity in the region, the less the private investors are willing to invest in the region. This is because private businesses are profit oriented and not interested in rendering services without making profit off the investment.

2.2   Ensuring the Development of Power Infrastructure in the LICs

All know the importance of power infrastructure development. Different shareholders in the power industries and policy institutions have noted the significance power infrastructure development means to the economic development of country [10]. The report of The Forum of Energy Ministers of Africa in 2006 said that it is accepted that a boost to the energy sector can influence sustainable development, create employment, industrial growth, empowers citizens and promotes the achievements of the MDGs. The study of Suberu et al. [5], Duvenage et al. [11] and Wood et al. [12] agrees that making power supply stable and restraining the power sector grants to climate change; both are the problems faced in achieving the sustainable agenda. The rate of anthropogenic green gas emission (GHG) which is related to energy output in the SSA. It has been framed that by 2030, the demand for energy will increase by 53% and the carbon dioxide emission will increase by 55%, making a massive population in the world of about 1.5 billion to lack adequate access to power supply [13]. With the enormous potentials Africa has in renewable energy, it can be utilized in achieving a safe environment to meet the demand of sustainable energy in the region; an example is the potential of hydropower in Africa with "19-fold" usage. In agreement to this Haliu [15] and Jumbe [14] stated that the nature of low power supply in the region could be attributed relationships between the following; finance, "grid connectivity", comparative fuel price, price of technology, subsidy agenda, power policies and earnings level. Countries in the SSA has the perspective to holistically boost the power access rate from the latest 30.5% by the applying the united nations initiates in curbing these challenges.

The drive to make power supply accessible in line with reduction in the rates makes it a complicated policy because majority of the citizens are poor. Therefore, by subsidizing the electricity sector makes power cheap for the citizens and in turn affects the supply of electricity in the region [16]. Singh et al. [17] and Roy et al. [18] agrees that the absence of important instruments like taxes, market mechanisms, grants, tax exclusion and subsidies will contrive the renewable energy deployment. There is abundance of renewable sources in countries like Malawi, Botswana and South Africa, but utilizing the fossils energy increases the emission of carbon dioxide in the region thereby having negative influences on the environment, making the environment unsafe for the citizens [4]. Most cases the resources for the development of the power infrastructure is made available but the negative influence of recognized institution, policies framework has hindered its development as shown in Fig. 5 below. Therefore, there is a call for immediate news ways of financing power infrastructure to enable the region to have access to a total supply of electricity in a sustainable manner [19].

2.3 Power Investment in SSA Low-Income Countries (LIC)

Energy investment planning comprises of effective agendas, objectives, programs and well-structured regulations that can be used to attain social-economic platform in the power sector to boost the approach to dependable, economical suitable, general acceptance, reasonable and climate-safe energy for sustainable growth and empowerment of the citizens [20]. The absence of political drive, lack of preparedness and unfavorable policies in fixing the energy sector challenges has hindered the progress of the sector [21]. There are great needs of private sector participation in the power sector of (LIC) in SSA, to bring about general energy access to all initiatives by the year 2030 [22]. The more private sector investment in the sector and effective policies will greatly improve the power sector, because private sector participation attracts additional finance that can bridge the gap financially, technically and managerial aspect of the power industry [23]. With the participation of private investment in the power sector, this will allow government to shift attention to other sector in need of development example; health care facilities. The challenges facing private participation in the SSA countries in well-known and the way other nations has been success in streamlining these challenges in promoting power investment differs greatly, mainly because the interwoven relationship with the problems [24]. The population structure of the SSA is about 910.4 million people, 37% of the people stay in the urban centers. The electrification ratio is unequal with 59.5% for urban centers while 14.2% for rural centers that inhabits most of the population, these makes them on a pitfall in the access of power supply, socio-economic development and employment [25].

There are rise in social problems in SSA due to lack of jobs and inadequate lack of sufficient basic social facilities for the growing populations [26]. Predominantly power sector has a significant role to play, in equally promoting the access to electricity in the rural centers that can create jobs and developments for the citizens and improve their lives positively. The main aim of private investment is to make profit i.e. value for money, not to render social services to the citizens, therefore motivating them has its challenges [27]. Despite its obvious disadvantages in motivating private investors in the power infrastructure projects, there have been some success cases of private investment in the area of renewable energy in different nations. One example was private has improved the power sector is the Addax Bioenergy Ltd services in Sierra Leone, where more than $258 million was invested in the Makeni Power Project. The power project comprises of "sugar plantation and bio-ethanol facility" with strength to distil 385,000l of ethanol in a day, and can produce up to 32 MW energy via the generating plant. Addax Bioenergy was able to get more funding through the clean carbon credits by achieving the Clean Development Mechanisms (CDM) and got the government support energy purchase agreement with the national electricity authority to purchase energy generated by the project. These cases did not only portray that private investment can promote renewable energy development in Africa, it also shows that effective government policies can attract private investment for economic growth.

2.4 Financing LIC in SSA Through Climate Finance

Africa requires about US$100 billion yearly, in order to stay in tune with the safe environment power generation agenda. The inability to provide safe environment energy to a certain level will bring about unsafe environment for citizens and can lead to about 100 million citizens living in abject poverty by 2030. Therefore, climate finance for safe environment energy was created by multinational relationships to deliver a safe environmental energy for populace in SSA. Climate finance is projected to rise to US$100 billion annually by 2020 to complement infrastructure finance, for the access to power and assist safe environment energy planning services and adoption in SSA [3]. Policy makers in the SSA region must adapt to favorable policies that can influence climate finance in the region, to reduce the recent statistics of other regions benefiting more than the SSA on climate finance, as shown in the Fig. 1 below. Climate finance has the potential the complement the United Nations sustainable access to energy for all initiative by 2030 in diverse ways [5]. The sustainable energy agenda comprises of policies and framework aimed at attracting finances for renewable energy development in the SSA. Climate finance mechanisms are independent of the sustainable energy goals, but can play a significant role in facilitating the circulations of technologies for renewable energy and influencing efficient energy, by making sure energy is generally accessed in a sustainable manner and pattern thereby promoting the sustainable energy for all initiatives [4].

2.5 Malawi Experience Climate Finance

Malawi is a country in Africa, located in the Southern part of Africa, has boundaries with Zambia in the northwest, Tanzania to the northeast and Mozambique on the west,

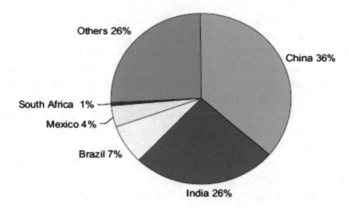

Fig. 1. Climate finance distribution

south and east border respectively. With a population of 15 million people, whereby 85% lives in the rural area and a growth rate of 2.8% [28]. The percent of poverty is high in Malawi because of environmental pollution, lack food security for the citizen, gender issues, and unsafe climatic factors influencing the poverty ratio to a very high proportion. Insecure power generation has affected the lives of the people of Malawi. The non-availability of stable source of power has hindered the development of the production industry, tourism and mining sectors of the economy, which ought to diversify the economy to boost the economic development, the failure of the energy sector to deliver adequate stable and safe source of energy, has resulted to increase in poverty and development of the country [17]. The economic impact of power unavailability is estimated to be 6.5% GDP when compared to 5.5% GDP in Uganda and 4% GDP in Tanzania, therefore, it is very vital for the development of sustainable power infrastructure in the region to aid economic development, improves lives, and alleviates poverty [1].

Most developing countries contributes little to the greenhouse gas (GHG) emission in the world, thereby making them not qualified for climate finance through the CDM programs [29]. While which contributes greatly to the greenhouse emission gas (GHG), due to the size of their development tend to attract more climate finance to their region. This is ironical because countries with low rate of emission need the climate finance to develop the low-carbon economy, to influence development [30]. Lack of adequate finance is said to have hindered to development of sustainable power infrastructural development in Malawi [31]. Below is the rural electrification program in Malawi aimed at financing renewable energy projects.

2.6 Rural Electrification in Malawi

In a bid to improve the access of electricity in Malawi, in 2003 the government of Malawi created a body called Malawi Rural Electrification programs (MAREP) responsible for improving electricity access in both rural and urban centers, with favorable legal and political for inclusive development for all aimed at creating employment and reduce poverty by a guaranteed power security (Government of

Malawi, 2013). In the frame of implementing the Energy Policy in Malawi, several policies were enacted like the power supply Act, Rural Electrification policy, power regulation policy, electricity policy. For the purpose of financing, managing, aiding and regulating, led to the enactment of the Energy Policy Act in 2004. The vital part of the Act created the Malawi Rural Fund that consists of funding disseminated by legislators with the motives of the finances and rural sales of electricity dues. The finances gotten from the rural electrification fund act, it has been significant to the growth sustainability energy goals of the MAREP the project is in the seventh stage 2003. For the purpose of efficiency with flow of finances in the MAREP Act, apart from the traditional means of financing infrastructure, the fund gotten from the price of petrol and diesel is injected to make funds available and nation's law guides this act. In the same vain budgets, allocation and rates from fossils fuel are directed to finance the renewable energy development agenda in Malawi.

2.7   Lake Turkana Wind Power Project (LTWP) in Kenya

This project is located in the Marsabit region of Kenya. The LWTP is about 300 MW, which is about 20% of the generation capacity of Kenya; this promotes renewable development in Kenya. The cost of the project is about 459 million euro, and nothing was gotten from government coffers meaning it was private investment capital. This is one the challenges facing the development power infrastructure, because private investors are reluctant to invest in businesses with low financial returns [32]. The life span of the project is 20 years and has the capacity reduce carbon emission by 16 million all through the project life, and registering the project with the CDM in a bid to attracting more finances with carbon credits sales. In the quest to boost the financial strength, limits risks and attract private investors, there was a consensus by the developers of the project and national institutions for higher rates and stable price, on the basis of paying the national institutions from the sales of carbon credits. This is different from several projects where the financial strength of the project is based on the sales of carbon credits, which is unreliable because the prices shift due with market forces. The act has greatly increased the sustainability rate as the price from the sale of carbon credits has fallen without affecting the viability of the project.

2.8   Standardized Power Purchase Agreements (SPPAs) (Tanzania)

The power sector in Tanzania is closely related to that of Malawi, plagued with insufficient power generation and excessive demand for electricity when compared to the rest of the LICs in Sub Sahara Africa. Two determinants are involved as support instrument for renewable energy sources and electricity are the priced based instrument and quantity based instrument. These tools create a false market to promote power sources development by price fixing to enable the market determine the quantity of the renewable power sources or fixing the amount of generation of electricity to allow the market determines its prices, these instruments affect the power sectors differently.

Findings from the experiences of projects all over the globe, implies that Feed-in Tariffs (FITs) are the most efficient and viable regulations that can promote sustainable development growth with respect to renewable energy sources, because they offer a stable price for a particular time for the amount of power generated from the sources of renewable energy [33]. These cases are different in the developing countries, due to the limitations of policy reforms, financial constraints and technical abilities [34]. These constraints of clumsy and management delays increase the cost and hinders projects execution, these disrupt the development in the power sectors in the low-income countries of the SSA because it discouragements private investors faces in renewable energy investment [34]. To curb the issue of FITs in Tanzania characterized as a low-income country, the medium of SPPAs was set up to complement the FITs policy to ensure economic development in Tanzania. These SPPAs procurements of network connected volumes and off network connected volumes and the connection between the power sector to the purchases and small energy projects generators. The medium has greatly minimized the rate of different policy requirements in the renewable energy sources, instead improving the use of standardized certification that recognizes the needs of the smaller energy generators. Other countries in the LICs should emulate Tanzania in the deployments of small capacity renewable energy facilities of about 10 MW to be able to promote renewable development in the region and attract more investments [4].

2.9 Countries of SSA

There is need for huge investment in the energy sector of the LICs in SSA countries because traditional means of financing power infrastructure has been constrained with the amount required for financing. For instance, Africa countries requires about US $47.6 billion per year from 2005 to 2015 to finance its power sector. So according to Eberhand et al. [23], the following are the sources used in the financing of power infrastructure in the SSA.

2.9.1 Domestic Finance

Prior to the financial crisis, the fiscal policies and financial base of the SSA were viable with improved economic growth approximately 4% from the year 01–05. Low-income countries allocate about 30% of their annual budgeting to infrastructure development; they are considered to have a fragile tax base. This is low when compared with the rest of developing countries, and suggest a reform in the area of domestic finance to be able to improve domestic generation of resources to boost power infrastructure facilities in the regions.

2.9.2 Development Financial Assistance

The official development assistance (ODA) was framed from development assistance committee (DAC) and is given to the OECD countries in the form of aids to assist in the development of power infrastructure facilities. From 1990 to 2000, the financial assistance for infrastructure development in sub Saharan Africa was stable at US$492 million yearly. This aid was also improved from US$642 million in 2004 to US$810 in 2006, with the sole aim of improving and bridging the financial gap of power infrastructure in the SSA.

2.9.3   Private Finance

Due to the heavy financial capacity needed for power infrastructure development in LICs countries of the SSA, there is a need for private investors participating in the power sector to boost economic growth and development of the region. Private finance has helped boost the financial strength of infrastructure in the region, in the year, 1990 US$40 million was invested in power infrastructure in the SSA, these have since risen to US$1.2 billion in 2008 to restrain the funding gap in financing of power infrastructure in the region. Moreover, this has led to the building of new energy facilities and rehabilitation of old ones to boost the supply of electricity.

2.9.4   Local Capital Market

There is need for local financial market to back power infrastructure investment in the region, by providing cover for the long-term loan to encourage investors in the investing in power infrastructure facilities in the region.

2.9.5   Bank Borrowing

Commercial bank borrowing has greatly with other factors improved the development of energy facilities in the LICs countries; this is evident with the amount borrowed for power facilities by the end of 2006 was above US$2.7 billion dollars. For the countries in the LICs, the capacity of the commercial banks is limited to successfully initiate power infrastructure development. Therefore, there is need for syndicated borrowing to improve infrastructure facilities with the contribution of the commercial banks.

2.9.6   Equity Finance

This involves the sales of energy shares with the aim of getting financial strength in funding power development. The total of about US$55.9 billion was raised through equity finance to boost power facilities in the SSA region, with the aim of realizing a steady power supply and economic development in the region.

2.9.7   Corporate Bonds

This type of power infrastructure financing mechanisms involves the issuing of finance from multinational or foreign corporations with the purpose of meeting the financial needs in the funding power development in the region at large to boost economic growth by alleviating power and improves lives.

3   Research Methodology

This study is carried out with respect to existing literature, from published researchers with the aim of assessing the challenges facing power infrastructure development in the low-income countries (LICs) of Sub-Sahara Africa (SSA). This study mostly focuses on the power/energy infrastructure development in SSA. In addition, this study reviewed power infrastructure literature on challenges, financing and effective legislation in the SSA, quantitative research methodology was used in carrying out this study.

4 Discussions and Findings from Literature

According to the existing literature the following factors were revealed as the challenges in power infrastructure development in the Low-Income Countries (LICs) of Sub-Sahara Africa (SSA). Chirambo [4] revealed that one of the challenges facing the development of power infrastructure in the LICs of SSA is lack of viable power institutions; the lack of skilled and efficient power sector ministries in the LICs of SSA has hindered the development of power infrastructure in the region. No private investor will be willing to invest in a weak and inefficient sector. According to the United Nations report in 2012 on sustainable power development in SSA, the following factors was identified as some of the challenges faced in the financing of power infrastructure in the LICs; lack of adequate finance; due to heavy financial weight involved in the financing of power infrastructure, the LICs lacks adequate private investors participating in the region as a result of these power infrastructure is mainly financed by the government. Unfavorable policy framework; for any development to occur in the power infrastructure sector in any country there must be favorable policies influencing the positive growth of the sector, the LICs lack this factor and has hindered its development. The absence of technological knowledge; the lack of skilled personnel in advancing the technology of the LICs has hindered its growth, thereby restricting its technology growth, which has positive effects on new form of power generation. Tight power generation plan; the nature of the generation plan in the LICs of SSA has restricted its growth factor in power development; this is because no investor is willing to invest much in a platform like that. Low cost of electrification; the low rate of tariffs adopted by the government of the LICs due to the rate of poverty in the region, has however hindered private investor's participation in the region. This is because the aim of every investor is to make profit i.e. value for money.

Furthermore, improving the power infrastructure condition in the LICs in SSA, according to Chirambo [4] there must be efficient policy framework, workable objectives and improved financial atmosphere. Financing through climate finance was used in financing power infrastructure in Tanzania (Lake Turkuna Wind Power Project) and Malawi in the rural electrification programs. According to Eberhand et al. [23] and Chirambo [4], the sources of financing power infrastructure in the LICs in the SSA was observed as the following; domestic finance, development financial assistance, private capital, domestic capital market, bank loans, equity capital, corporate bonds.

5 Conclusion and Recommendations

This article explored literature on power infrastructure development in the SSA and Africa. The explored literature explored showed that there are several challenges facing the development of power infrastructure in LICs in the SSA region, which are; lack of innovative finances, lack of efficient policy regulations. Literature also showed that if power is secured in the region, it will improve the economic growth in the region whereby improving lives, eradicating poverty and employment in the region.

References

1. Yadoo, A., Cruickshank, H.: The role for low carbon electrification technologies in poverty reduction and climate change strategies: a focus on renewable energy mini-grids with case studies in Nepal, Peru and Kenya. Energy Policy **42**, 591–602 (2012)
2. Wicke, B., Sikkema, R., Dornburg, V., Faaij, A.: Exploring land use changes and the role of palm oil production in Indonesia and Malaysia. Land Use policy **28**(1), 193–206 (2011)
3. Glemarec, Y.: Financing off-grid sustainable energy access for the poor. Energy policy **47**, 87–93 (2012)
4. Chirambo, D.: Addressing the renewable energy financing gap in Africa to promote universal energy access: integrated renewable energy financing in Malawi. Renew. Sustain. Energy Rev. **62**, 793–803 (2016)
5. Suberu, M.Y., Mustafa, M.W., Bashir, N., Muhamad, N.A., Mokhtar, A.S.: Power sector renewable energy integration for expanding access to electricity in sub-Saharan Africa. Renew. Sustain. Energy Rev. **25**, 630–642 (2013)
6. Lau, L.C., Lee, K.T., Mohamed, A.R.: Global warming mitigation and renewable energy policy development from the Kyoto protocol to the Copenhagen accord—a comment. Renew. Sustain. Energy Rev. **16**(7), 5280–5284 (2012)
7. Fan, J., Liang, Y.T., Tao, A.J., Sheng, K.R., Ma, H.L., Xu, Y., Wang, C.S., Sun, W.: Energy policies for sustainable livelihoods and sustainable development of poor areas in China. Energy policy **39**(3), 1200–1212 (2011)
8. Gujba, H., Thorne, S., Mulugetta, Y., Rai, K., Sokona, Y.: Financing low carbon energy access in Africa. Energy Policy **47**, 71–78 (2012)
9. Sanoh, A., Kocaman, A.S., Kocal, S., Sherpa, S., Modi, V.: The economics of clean energy resource development and grid interconnection in Africa. Renew. Energy **62**, 598–609 (2014)
10. Odhiambo, N.M.: Energy consumption, prices and economic growth in three SSA countries: a comparative study. Energy Policy **38**(5), 2463–2469 (2010)
11. Duvenage, I., Taplin, R., Stringer, L.C.: Towards implementation and achievement of sustainable biofuel development in Africa. Environ. Dev. Sustain. **14**(6), 993–1012 (2012)
12. Wood, B.T., Sallu, S.M., Paavola, J.: Can CDM finance energy access in least developed countries? Evidence from Tanzania. Clim. Policy **16**(4), 456–473 (2016)
13. Kaygusuz, K.: Energy for sustainable development: a case of developing countries. Renew. Sustain. Energy Rev. **16**(2), 1116–1126 (2012)
14. Jumbe, C.B., Msiska, F.B., Madjera, M.: Biofuels development in Sub-Saharan Africa: are the policies conducive? Energy Policy **37**(11), 4980–4986 (2009)
15. Hailu, Y.G.: Measuring and monitoring energy access: decision-support tools for policymakers in Africa. Energy Policy **47**, 56–63 (2012)
16. Ranjit, L.R., O'Sullivan, K.: A Sourcebook for Poverty Reduction Strategies, vol. 2. The World Bank, Washington (2002)
17. Singh, A., Pant, D., Korres, N.E., Nizami, A.S., Prasad, S., Murphy, J.D.: Key issues in life cycle assessment of ethanol production from lignocellulosic biomass: challenges and perspectives. Bioresour. Technol. **101**(13), 5003–5012 (2010)
18. Roy, J., Ghosh, D., Ghosh, A., Dasgupta, S.: Fiscal instruments: crucial role in financing low carbon transition in energy systems. Curr. Opin. Environ. Sustain. **5**(2), 261–269 (2013)
19. Bazilian, M., Nussbaumer, P., Rogner, H.H., Brew-Hammond, A., Foster, V., Pachauri, S., Williams, E., Howells, M., Niyongabo, P., Musaba, L., Gallachóir, B.Ó.: Energy access scenarios to 2030 for the power sector in Sub-Saharan Africa. Util. Policy **20**(1), 1–16 (2012)

20. Ottinger, R.L., Robinson, N., Tafur, V. (eds.): Compendium of Sustainable Energy Laws. Cambridge University Press, Cambridge (2005)
21. Gamula, G.E., Hui, L., Peng, W.: Development of renewable energy technologies in Malawi (2013)
22. Chaurey, A., Krithika, P.R., Palit, D., Rakesh, S., Sovacool, B.K.: New partnerships and business models for facilitating energy access. Energy Policy **47**, 48–55 (2012)
23. Eberhard, A., Shkaratan, M.: Powering Africa: meeting the financing and reform challenges. Energy Policy **42**, 9–18 (2012)
24. Schut, M., Slingerland, M., Locke, A.: Biofuel developments in Mozambique. Update and analysis of policy, potential and reality. Energy Policy **38**(9), 5151–5165 (2010)
25. Haanyika, C.M.: Rural electrification in Zambia: a policy and institutional analysis. Energy policy **36**(3), 1044–1058 (2008)
26. UNECA: Assessing Progress in Africa toward the Millennium Development Goals. MDG Report (2011)
27. Burian, M., Arens, C.: The clean development mechanism: a tool for financing low carbon development in Africa? Int. J. Clim. Change Strat. Manage. **6**(2), 166–191 (2014)
28. Kaunda, C.S.: Energy situation, potential and application status of small-scale hydropower systems in Malawi. Renew. Sustain. Energy Rev. **26**, 1–19 (2013)
29. Timilsina, G.R., de Gouvello, C., Thioye, M., Dayo, F.B.: Clean development mechanism potential and challenges in Sub-Saharan Africa. Mitigat. Adapt. Strat. Glob. Change **15**(1), 93–111 (2010)
30. Zadek, S.: Beyond climate finance: from accountability to productivity in addressing the climate challenge. Clim. Policy **11**(3), 1058–1068 (2011)
31. Drinkwaard, W., Kirkels, A., Romijn, H.: A learning-based approach to understanding success in rural electrification: insights from micro hydro projects in Bolivia. Energy. Sustain. Dev. **14**(3), 232–237 (2010)
32. Masini, A., Menichetti, E.: Investment decisions in the renewable energy sector: an analysis of non-financial drivers. Technol. Forecast. Soc. Chang. **80**(3), 510–524 (2013)
33. Couture, T., Gagnon, Y.: An analysis of feed-in tariff remuneration models: implications for renewable energy investment. Energy policy **38**(2), 955–965 (2010)
34. Rickerson, W., Hanley, C., Laurent, C., Greacen, C.: Implementing a global fund for feed-in tariffs in developing countries: a case study of Tanzania. Renew. Energy **49**, 29–32 (2013)

Author Index

© Springer International Publishing AG, part of Springer Nature 2019
T. Z. Ahram (Ed.): AHFE 2018, AISC 787, pp. 565–567, 2019.
https://doi.org/10.1007/978-3-319-94229-2

Printed in the United States
By Bookmasters